I0047669

Classical Dynamics of Linear and Nonlinear Systems

Classical Dynamics of Linear and Nonlinear Systems offers a comprehensive exploration of dynamical systems from fundamental principles to advanced applications. This textbook presents a unified treatment of classical dynamics, bridging the gap between linear and nonlinear systems while providing both theoretical foundations and practical applications.

Beginning with a thoughtful classification of dynamical systems, the book systematically builds understanding from particle mechanics to quantum field theory. Following a rigorous analysis of particle dynamics in both configuration and phase spaces (Newtonian, Lagrangian, Hamiltonian and Hamilton-Jacobi formulations), the book provides a detailed examination of molecular and crystalline structures across multiple dimensions. Later chapters conduct an in-depth exploration of nonlinear phenomena and chaos theory with real-world applications and elegant formulations of classical field theories using Lagrangian and Hamiltonian approaches. The final sections of the book provide an accessible introduction to quantum field theory and its relationship to classical systems, in addition to powerful perturbation techniques applicable to both classical and quantum problems. This book transforms abstract theoretical concepts into practical understanding through rigorous mathematical and numerical frameworks and illuminating examples, making it ideally suited for advanced undergraduate and postgraduate students enrolled on physics, applied mathematics, engineering and materials science courses.

Key Features:

- Connects traditional mechanical concepts with modern physics.
- Includes several worked examples, in addition to end-of-chapter problems and further reading to support teaching and learning.
- Features seven appendices covering further topics such as mathematical preliminaries, numerical solutions to first-order and second-order differential equations and the Euler-Lagrange variational principle.

Gyaneshwar P. Srivastava is Emeritus Professor of Theoretical Condensed Matter Physics at Exeter University, UK. In a teaching career of over 45 years he has taught several physics modules, including analytical and chaotic dynamics. His research has concentrated on theoretical and computational studies of the physics of phonons and electrons in crystalline solids, surfaces and nano-structures. He has collaborated with various physicists, both experimentalists and theorists, of international reputation. This has led to over 500 publications, including several review articles and three postgraduate books. He is an Outstanding Referee for APS journals.

Classical Dynamics of Linear and Nonlinear Systems

Gyaneshwar P. Srivastava

CRC Press
Taylor & Francis Group
Boca Raton London New York

CRC Press is an imprint of the
Taylor & Francis Group, an **informa** business

Designed cover image: Gyaneshwar P. Srivastava

First edition published 2026
by CRC Press
2385 NW Executive Center Drive, Suite 320, Boca Raton FL 33431

and by CRC Press
4 Park Square, Milton Park, Abingdon, Oxon, OX14 4RN

CRC Press is an imprint of Taylor & Francis Group, LLC

ISBN: 978-1-032-46793-1 (hbk)
ISBN: 978-1-032-46527-2 (pbk)
ISBN: 978-1-003-38331-4 (ebk)

DOI: 10.1201/9781003383314

Typeset in Nimbus Roman
by KnowledgeWorks Global Ltd.

Publisher's note: This book has been prepared from camera-ready copy provided by the authors.

Jyotiṣkartā yaduśhmasi

(Rigveda I/86/10)

[Show us the light (of knowledge) we long for.]

Dedication

To my family:

Kusum, Jyoti, Deepak, Marina, Elizabeth and Aaron

Contents

Preface

Classical mechanics is one of the oldest topics in science subjects, which is taught to students at various levels from high school onwards and has been covered in perhaps the largest number of textbooks. However, most books discuss only a few aspects of this huge topic, such as the linear oscillatory dynamics of a single or a double pendulum using the vectorial scheme of Newtonian mechanics. In this book I have attempted to describe the classical dynamics of single (*i.e.* non-interacting) and many-body (*i.e.* interacting) systems that are subjected to linear or non-linear forces. Both vectorial (Newtonian) and scalar (Lagrangian and Hamiltonian based analytical) schemes have been adopted. Examples of systems are chosen from physics, materials science, engineering, astrophysics, biophysics, weather science, neuroscience and neural networks and artificial intelligence. The presentation of the subject matter is split in six parts.

Part I (chapter 1) deals with general considerations for discussing classical dynamics. The terms 'mechanics', 'kinematics' and 'dynamics' and their inter-relationships have been clarified. After explaining the concepts of central and non-central forces, conservative and non-conservative systems, and linear, non-linear and chaotic systems, the chapter reminds that there are three scalar alternatives to the vectorial approach based on Newtonian mechanics. The chapter then outlines the scope and coverage of this book.

Part II (chapters 2 and 3) deals with linear dynamics of particles and objects. Chapter 2 presents the Newtonian (vectorial) formulation and the analytical (scalar) formulation developed by Lagrange. Lagrange equations of motion are derived by using the Euler-Lagrange variational principle for *action integral* in configuration space, *i.e.* by using Hamilton's variational principle. Chapter 3 presents the analytical formulations developed by Hamilton and by Hamilton and Jacobi. Hamilton's equations of motion are derived from the application of the Euler-Lagrange variational principle in phase space, *i.e.* by using modified Hamilton's variational principle. Using the concept of generating functions to achieve canonical transformation between two phase spaces, the Hamiltonian formalism has been extended to the Hamilton-Jacobi formalism.

Part III (chapters 4 and 5) deals with the linear dynamics of interacting particles. Chapter 4 discusses the dynamics of small linear molecules and linear atomic chains. Chapter 5 extends the discussion to two-dimensional atomic networks and three-dimensional crystalline structures.

Part IV (chapters 6 and 7) is devoted to non-linear and chaotic dynamics. Chapter 6 presents theoretical and numerical (computational) aspects of non-linear and chaotic dynamics using the concepts of phase orbits and Poincaré map. Discussion of chaotic motion is extended by using the *logistic map* equation, the Feigenbaum diagram and the Lyapunov exponent. Chapter 7 presents applications of the theory presented in chapter 6 from physics, materials science, engineering, astronomy, biophysics, weather science, neuroscience, and neural network and artificial intelligence.

Part V (chapter 8) introduces a many-body version of the classical dynamics of particles. Both the Lagrangian classical field theory and the Hamiltonian classical field theory are described, with applications to topics in electromagnetism and condensed matter physics.

Part VI (chapters 9 and 10) goes beyond classical mechanics. Chapter 9 presents the basic elements of quantum field theory and explains how the quantum-mechanical Hamilton-Jacobi equation reduces to the classical Hamilton-Jacobi equation in the limit when Planck's constant is regarded as a vanishingly small quantity, *i.e.* when $h \to 0$. The second quantisation approach is introduced and applied to quantise bosonic quantum fields of crystal vibrations and of the electromagnetic field in vacuum. Additionally, it explains how a fermionic quantum field can be quantised, with the example of quantising electron waves in solids. Chapter 10 describes some aspects of classical and quantum perturbation theories. These include classical Hamiltonian perturbation theory, classical Hamilton-Jacobi (or, canonical) perturbation theory, classical Poincarè-Lindstedt perturbation theory, Rayleigh-Schrödinger time-independent quantum perturbation theory, Brillouin-Wigner time-independent quantum perturbation theory and Fermi's golden rule in time-dependent quantum perturbation theory.

Worked examples are presented in each of the chapters 2-10. At the end of each of these chapters a number of problems are set for the reader to attempt. Also, a selection of publications for further reading is listed at the end of each chapter

Answers to selected problems have been provided at the end of the book. A solutions manual is available for tutors who adopt this work as a textbook via Routledge's instructors' hub:

https://routledgetextbooks.com/textbooks/instructor_downloads/

which can be searched for using the book's ISBN.

Several colleagues and students have very kindly read different chapters of the manuscript. These include Professor Stephen Jenkins, Professor Hüseyin Tütüncü, Professor Mehmet Çakmak, Professor Jacopo Bertolotti, Dr Iorwerth Thomas, Dr Charles Downing, Dr Ned Taylor and Mr Harry McLean. I am truly grateful to them all for making many useful suggestions for improvement. I would be grateful to readers, students and tutors alike, for their feedback on shortcomings (including errors, omissions and unclear explanations) and usefulness of the book.

I would like to express my thanks to Mrs Rebecca Hodges-Davies, Commissioning Editor for Physics, CRC press, Taylor & Francis Group, Oxford, for inviting me to undertake the writing of this book, and to Dr Danny Kielty and Ms Sanika Shah for their regular encouragement and help in getting the manuscript prepared and submitted well within the agreed timeframe. It is also my great pleasure to thank the three anonymous reviewers of the book proposal whose comments have helped shape the book well, particularly for their suggestion that I should attempt an in-depth coverage of classical dynamics of non-linear systems by considering different topics in science and engineering. I hope they are pleased with the selection and coverage I have made. I would also like to thank my son, Deepak Srivastava (Professor of Molecular Neuroscience, King's College, London), for introducing and teaching me the basics of neuroscience, pointing me towards a couple of books on computational neuroscience and for offering the confocal image of the neural network that is presented in figure 7.38. It would not have been possible for me to bring the writing of the manuscript to fruition so smoothly without the usually unwavering support of my wife Kusum, for which I am truly grateful to her.

Many of the topics covered in this book have been taught by me during 45 years of my teaching career in UK universities (11 years at University of Ulster at Coleraine followed by 34 years at University of Exeter). I sincerely hope that the topics covered in this book prove useful for learning, teaching and applying classical dynamics at undergraduate level, postgraduate level and beyond.

Gyaneshwar P. Srivastava, D Sc
Emeritus Professor of Theoretical Condensed Matter Physics
Department of Physics and Astronomy
University of Exeter, UK
May 2025

Acknowledgements

I am grateful to the following:

Nonlinear Analysis: Modelling and Control for granting permission to use figures 7.6, 7.8. 7.9, 7.10, 7.11, 7.12, 7.13, 7.14, 7.15, 7.16, 7.17, 7.18 and 7.19.

Appl Math Modelling for granting permission to use figure 7.28.

Professor Deepak Srivastava, King's College London for offering figure 7.38.

Part I

General considerations

1 Classification of Dynamics and Scope of This Book

1.1 INTRODUCTION

Classical mechanics, quantum mechanics and statistical mechanics provide the most important theoretical concepts and tools in the study of physical phenomena. Classical mechanics deals with motion of macroscopic bodies. Quantum mechanics is used to study the motion of microscopic bodies. Statistical mechanics deals with temperature-dependent statistical probabilities of events both in classical and quantum regimes.

Historically, classical mechanics is the oldest branch of physical sciences. Two milestones in the formulation of classical physics are the works of Isaac Newton in the seventeenth century for the laws of motion[1] and James Clerk Maxwell in the nineteenth century for the laws of electromagnetism.

This chapter provides a brief introduction to the subject of classical mechanics or dynamics. To do this, we first clarify a few terms that are interchangeably used when discussing the subject of dynamics.

1.2 MECHANICS, KINEMATICS AND DYNAMICS

The term 'mechanics' is used for describing a force-body-motion sequence, and the terms 'kinematics' and 'dynamics' are used for the resolution of mechanics. Kinematics deals with motion in terms of trajectories in space and time. Dynamics deals with changes in motion due to external forces. Usually, no distinction is made between these terms. In this book we will use the phrases 'classical mechanics' and 'classical dynamics' interchangeably.

The main purpose of classical mechanics is to formulate and describe the motion of macroscopic material objects acted upon by external forces. An object may be a single particle (*i.e.* a material point mass), a loose collection of single particles or a rigid body (*i.e.* a condensed collection of single particles). Clearly, for dealing with classical dynamics of an object, we require information about its mass structure and the external force(s) acting upon it. By mass structure we mean specification of the locations of particles making up the object. A complete description of the location of a particle at any instant of time is possible by specifying a frame of reference, or a coordinate system. There are several types of coordinate systems which can be adopted for a given dynamical system. Some commonly used coordinate systems are described in Appendix A, which also provides some other useful mathematical preliminaries.

1.3 CENTRAL AND NON-CENTRAL FORCES

A force acting on a dynamical system can be described either as *central* or as *non-central.*

Central force: Gravitational force, electrostatic force and elastic spring force are some examples of central forces. A central force acts between two particles, directed along the straight line joining the particles and with a magnitude which is a function of the distance between them. Mathematically, a central force between two particles with vector $\boldsymbol{r}$ between them can generally be written

---

[1]There are claims that the essence of Newton's three laws of motion was given much earlier, circa 600 BCE, by Sage Kanada in ancient India [see Mandal (2020) and references therein].

DOI: 10.1201/9781003383314-1

as

$$\boldsymbol{F}(\boldsymbol{r}) = F(r)\frac{\boldsymbol{r}}{r}, \tag{1.1}$$

where $F(r)$ is a function of the distance r. The force is attractive if $F(r) < 0$ and repulsive if $F(r) > 0$.

Non-central force: Nuclear force, magnetic force and frictional force are examples of non-central forces. Unlike a central force, a non-central force is not directed along the straight line joining two particles. It means that a non-central force at a particle involves the action of more than one other particle. For this reason, a non-central force is also called a many-body force. The simplest consideration is a three-body force, in which one particle is acted upon by two particles at the same time. One example is a bond-bending force (see, *e.g.* Stillinger and Weber, 1985), in which the magnitude of the force on particle at $\boldsymbol{R}_i$ due to particles at $\boldsymbol{R}_j$ and $\boldsymbol{R}_k$ is considered as a function of the distances $|\boldsymbol{R}_j - \boldsymbol{R}_i|$ and $|\boldsymbol{R}_k - \boldsymbol{R}_i|$ and the angle θ_{jik} between the vectors $\boldsymbol{R}_j - \boldsymbol{R}_i$ and $\boldsymbol{R}_k - \boldsymbol{R}_i$. Another example of a non-central force is the combination of the central (or 'radial') force and an 'angular force' which depends on the angle that the line joining two particles makes with the equilibrium position of the line (see, *e.g.* de Launay, 1956). We will describe the second type of non-central force with an application in section 5.4.6.

1.4 CONSERVATIVE AND NON-CONSERVATIVE FORCES

An external force acting on a dynamical system can be described either as *conservative* or *non-conservative*. If the work done by a force in moving a system from one position to another remains the same regardless of the path taken, then the force and the system are called conservative. This means that the work done by a conservative force in moving a system around a closed circuit is zero. For a force $\boldsymbol{F}$ to be conservative, it is necessary that its curl vanishes: $\nabla \times \boldsymbol{F} = 0$ (for a detailed proof, see section 2.5 in Barger and Olsson, 1973).

Central forces are conservative. This can be proven by showing that the integral $\int_{r_1}^{r_2} F(r)\hat{\boldsymbol{r}} \cdot \mathrm{d}\boldsymbol{r}$, which is required to obtain the work done in moving a system from $\boldsymbol{r}_1$ to $\boldsymbol{r}_2$, is path independent as it only depends on r and not the direction $\hat{\boldsymbol{r}}$. Appendix B details the proof. A conservative force in the form of a central force is a function of system's displacement vector $\boldsymbol{r}$ alone, *i.e.* $\boldsymbol{F} = \boldsymbol{F}(\boldsymbol{r})$. The velocity and energy of such a system are constant (*i.e.* time independent) and remain so during the history of object's motion. Such a system is also called *autonomous*. If the force depends, in addition to position, on velocity and/or time, *i.e.* $\boldsymbol{F} = \boldsymbol{F}(\boldsymbol{r}, \frac{\mathrm{d}\boldsymbol{r}}{\mathrm{d}t}, t)$, then it is called *non-conservative* or dissipative, and the system is called *non-autonomous* as its energy does not stay constant during motion. This implies that work carried out by a non-conservative force depends on the path taken. Examples of conservative forces are: gravitational force, electrostatic force, elastic spring force, *etc.* Examples of non-conservative forces are frictional force, air resistance, viscous force, *etc.*

1.5 LINEAR, NON-LINEAR AND CHAOTIC DYNAMICS

Dynamics can be classified as *linear* dynamics, *non-linear* dynamics and *chaotic* dynamics. We will make a brief distinction between these.

Consider the magnitude of the force on a one-dimensional dynamical system to depend on the displacement x from its equilibrium in the form

$$|\text{Force}| = f(x). \tag{1.2}$$

If the function $f(x)$ is a linear function of x, *i.e.* $f(x) \propto x$, then the above equation describes linear dynamics. If the function $f(x)$ is a non-linear function of x, such as $f(x) \propto x^j, j > 1$, then the dynamics is non-linear.

Table 1.1
Classification of dynamical systems as linear, non-linear and chaotic.

Linear dynamics	Non-linear dynamics	Chaotic dynamics
Deterministic, with $f(x) \propto x$	Deterministic, with $f(x) \propto x^j, j > 1$	Deterministic as well as non-deterministic, with $f(x)$ subjected to initial conditions
Analytic solutions	Combination of analytic and non-analytic methods (*e.g.* use of approximations and computers)	Non-analytic methods
Regularity	Regularity	Irregularity; Times evolution sensitive to initial conditions

Linear dynamics is characterised with motions that are deterministic and show regularity. Analytic solutions to relevant equations of motion can be obtained. *Non-linear dynamics* is also characterised with motions that are deterministic and show regularity. Relevant equations of motion can be solved by using a combination of analytic and non-analytic methods (*e.g.* using approximations and computers). *Chaotic dynamics* exhibits deterministic as well as non-deterministic features. Chaotic motion is characterised by irregularity, and its time evolution is sensitive to initial conditions. Non-analytic methods are required to solve equations of chaotic motion. The above classifications are summarised in table 1.1.

Linear, non-linear and chaotic dynamical systems are characterised by linear, non-linear and chaotic dynamics, respectively.

1.6 ALTERNATIVE SCHEMES FOR DYNAMICS

Whether linear, non-linear or chaotic in nature, dynamical problems can be expressed and framed for solutions by adopting either a vectorial approach or a scalar approach, as summarised in figure 1.1. The well-known Newtonian mechanics is an example of the vectorial approach, as it uses force and velocity as vector quantities in writing down equations of motion. Scalar approaches to dynamical problems, known as *analytical mechanics*, use scalar quantities (*i.e.* energy) and dynamical relations obtained using differential equations. Examples of scalar approaches include formalisms due to Leibniz, Euler, Lagrange, Hamilton, Jacobi, *etc.* In this book we will discuss the vectorial as well as the analytical schemes of dynamics.

1.7 SCOPE AND COVERAGE OF THE BOOK

This book discusses classical dynamics of non-interacting as well as interacting particles and objects. This has been gently extended to cover some basic concepts in quantum dynamics. It is hoped that the book will prove useful to undergraduate students, postgraduate students and researchers in physics, engineering, materials science and related subjects. In order to help navigate the subject matter in a logically graded manner, the coverage of the material is spread over six parts.

Part I (present chapter) has introduced the topic of classical dynamics. At the outset, it has clarified the terms mechanics, kinematics and dynamics and points out their inter-relationship. It then explained the important concepts of central and non-central forces, conservative and non-conservative forces. A clear distinction between linear, non-linear and chaotic dynamical systems was made. This was followed by a statement that there are several alternative schemes for studying classical

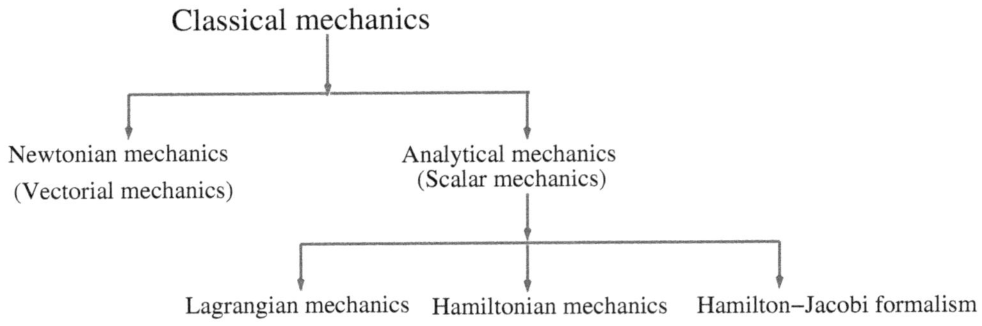

Figure 1.1 Alternative schemes for classical mechanics.

mechanics, which can be grouped as either vectorial or scalar. It pointed out that Newtonian mechanics is a vectorial approach, and Lagrangian, Hamiltonian and Hamilton-Jacobi formalisms are scalar (or, analytical) approaches.

Part II (chapters 2 and 3) describes linear classical dynamics of particles and objects. Chapter 2 begins by presenting the concepts of generalised coordinates, generalised velocities, generalised forces and potential energy in *configuration space*. The usually adopted Newtonian formulation, in configuration space, is discussed with several examples. After defining Lagrangian as scalar (*i.e.* non-vectorial) function, Lagrange equations of motion are derived by using the Euler-Lagrange variational principle for *action integral* in configuration space, *i.e.* by applying Hamilton's variational principle. Chapter 3 starts by expanding the concept of configuration space to that of *phase space*, in which a phase point is located using a phase space coordinate comprised of generalised coordinates and their canonically conjugate momenta. The Hamiltonian of a dynamical system is expressed in phase space as a function of phase space variables. It is stressed that it is inadmissible to have time derivative of coordinates in the expression for Hamiltonian. For example, when writing the Hamiltonian of a free particle of mass m and momentum $p = mv$, the kinetic energy expression $\frac{1}{2}mv^2$ is inadmissible and the appropriate expression is $\frac{1}{2}\frac{p^2}{m}$. Using the phase-space concept, two advanced formalisms of analytical dynamics are presented. Hamiltonian mechanics is described, which is based on the application of the Euler-Lagrange variational principle for action integral in phase space, *i.e.* by using modified Hamilton's variational principle. The Hamiltonian formalism is also expressed using the Poisson bracket notation. Using the concept of generating functions to achieve canonical transformation between two phase spaces, the Hamiltonian formalism has been extended to the Hamilton-Jacobi formalism. The Hamilton-Jacobi equations have been applied and solved for the problem of harmonic oscillations in one dimension, two dimensions and three dimensions. The elliptical nature of Kepler's planetary motion is rigorously derived.

Part III (chapters 4 and 5) deals with linear dynamics of interacting particles forming one-, two- and three-dimensional structures. One-dimensional systems considered are: two coupled harmonic oscillators, linear triatomic molecules, linear chains of large numbers of atoms, infinitely periodic linear atomic chains and semi-infinite atomic linear chains. Consideration of two-dimensional systems includes monatomic and diatomic atomic networks. The discussion of the diatomic atomic network should prove useful for the reader to further develop and study the lattice dynamics of technologically important two-dimensional materials. The theory of lattice dynamics of three-dimensional crystals, including central and non-central inter-atomic forces, has been detailed by considering the diamond structure.

Part IV (chapters 6 and 7) is devoted to non-linear and chaotic dynamics. Chapter 6 presents theoretical and numerical (computational) aspects of non-linear and chaotic dynamics using the concepts of phase orbits and Poincaré map. An important aspect of this part of the book is to emphasise on the importance of computational methods for solving non-linear equations of motion for which generally there are no analytical solutions. Systems considered are: damped one-dimensional harmonic oscillator, periodically driven one-dimensional harmonic oscillator, one-dimensional anharmonic oscillator (Duffing oscillator), damped and driven one-dimensional anharmonic oscillator and van der Pol's oscillator. Discussion of chaotic motion is presented using a double-well Duffing's oscillator, the logistic map equation, the Feigenbaum diagram and the Lyapunov exponent. Using numerical methods, chapter 7 presents a number of interesting and useful applications on topics from engineering, physics, astrophysics, weather science, biophysics, neuroscience and artificial intelligence. These include a chaotic electrical circuit, Euler's three-body problem in astronomy, Lorentz's non-linear weather dynamics, a non-linear dynamical model for normal and abnormal ECG traces, and non-linear neuronal dynamics using linear and quadratic integrate-and-fire models, Izhikevich model, Hindmarsh-Rose model, Hodgkin-Huxley model and Morris-Lecar model. The discussion on computational neuroscience is extended to present a brief discussion of theoretical framework and the general role of neural networks and artificial intelligence in science and beyond.

Part V (chapter 8) deals with a many-body version of the classical dynamics of particles. It presents the basic elements of classical field theory. Both the Lagrangian classical field theory and the Hamiltonian classical field theory are described, with applications in condensed matter and electromagnetism. Examples in condensed matter include periodically arranged atomic linear chains, three-dimensional crystals, elastic linear chains and anharmonic elastic continuum.

Part VI (chapters 9 and 10) goes beyond classical mechanics. Chapter 9 presents some elements of quantum field theory. Using old quantum theory, it shows that Planck's constant h is a quantum of proper action for periodic motion of a conservative system. It also shows that the quantum-mechanical Hamilton-Jacobi equation (Schrödinger equation with its wavefunction expressed in terms of the action integral) reduces to the classical Hamilton-Jacobi equation in the limit when Planck's constant is regarded as a vanishingly small quantity, *i.e.* when $h \to 0$. The chapter then adopts the second quantisation approach to quantise bosonic quantum fields of crystal vibrations and of the electromagnetic field in vacuum. Additionally, it explains how a fermionic quantum field can be quantised, with particular example of quantising electron waves in solids. Finally, chapter 10 describes some aspects of classical and quantum perturbation theories. Topics in classical perturbation theory include Hamiltonian perturbation theory, Hamilton-Jacobi (or, canonical) perturbation theory and Poincarè-Lindstedt perturbation theory. The quantum perturbation theory is explained by considering both time-independent and time-dependent examples. For the time-independent case, the non-degenerate Rayleigh-Schrödinger and Brillouin-Wigner approaches are described. Towards the end, the first-order time-dependent perturbation theory is described, and it is explained how Fermi's transition rule formula for transition probability from an initial state to a continuum of final states can be derived. Two examples of the application of the Fermi golden rule formula are presented by considering the scattering of phonons by mass-defect and by anharmonicity in crystal potential energy.

To help readers, seven appendices have been added at the end of the book. These cover topics such as mathematical preliminaries, central and conservative forces, Euler-Lagrange variational principle, introduction to matrix eigensolutions, reciprocal lattice and Brillouin zone, numerical solutions of first-order and second-order differential equations and analytical solution of second-order linear inhomogeneous differential equations.

Each chapter contains several worked examples and problems and a list of references for further studies. Answers to selected problems have been provided at the end of the book. A solutions manual is available for tutors via Routledge's instructors' hub:

https://routledgetextbooks.com/textbooks/instructor_downloads/

which can be searched for using the book's ISBN.

1.8 ASSUMED FAMILIARITY OF READER

The author assumes that the reader has good grasp of the physics and mathematics usually covered at the first-year university level in UK or its equivalent. Specifically, the reader is expected to have good understanding of vector analysis, differential calculus (partial and full differentiations), differential equations and matrix algebra.

1.9 A NOTE ON THE USE OF VARIABLES AND SYMBOLS

It would have been good to make a thoroughly systematic use of symbols throughout the book. However, this has not been possible. Several variables and symbols have been repeatedly used in different chapters, but carry different meanings. It is important, therefore, for the reader to retain the meaning of symbols in the context of each section in each chapter separately. An attempt has been made to remind the reader when a notation has changed its meaning from one section to another in a chapter.

Part II

Linear dynamics of particles and objects

2 Dynamics of Particles and Objects in Configuration Space

2.1 CONFIGURATION SPACE

For formulating and analysing dynamics of particles and objects it is essential to develop some terminologies.

Configuration: Specification of simultaneous positions of all particles of a system.

Degrees of freedom: Number of independent quantities required to define a configuration uniquely.

Configuration space: A space in which system configuration and degrees of freedom are clearly specified.

2.1.1 GENERALISED COORDINATES

In a n-dimensional space, generalised coordinates are a set of quantities $\boldsymbol{q} \equiv (q_1, q_2, q_3, ..., q_n)$ that completely define configuration of a dynamical system. A generalised coordinate need not have a unit of length.

Consider the motion of a particle located at P on the x-y plane as shown in figure A.1 in Appendix A. The Cartesian coordinates (x, y) are the two obvious generalised coordinates for the particle. The plane polar coordinates (ρ, ϕ) provide an alternative to the cartesian coordinates. In other words, we can specify the configuration of the particle with the help of two generalised coordinates q_1 and q_2, where

$$q_1 = \rho, \qquad q_2 = \phi. \tag{2.1}$$

Of course, we can switch from the Cartesian coordinates to the polar coordinates by expressing

$$\begin{aligned} x &= \rho\cos\phi = x(\rho, \phi), \\ y &= \rho\sin\phi = y(\rho, \phi), \end{aligned} \tag{2.2}$$

and

$$\begin{aligned} q_1 \equiv \rho &= (x^2 + y^2)^{1/2}, \\ q_2 \equiv \phi &= \arctan(y/x). \end{aligned} \tag{2.3}$$

With time considered explicitly, we can express

$$\begin{aligned} x &= \rho\cos(\phi + \omega t) = x(\rho, \phi; t), \\ y &= \rho\sin(\phi + \omega t) = y(\rho, \phi; t), \end{aligned} \tag{2.4}$$

where ω is angular velocity, and

$$\begin{aligned} q_1 \equiv \rho &= (x^2 + y^2)^{1/2}, \\ q_2 \equiv \phi &= \arctan(y/x) - \omega t. \end{aligned} \tag{2.5}$$

Other examples of generalised coordinates include the cylindrical polar coordinates (ρ, ϕ, z) and the spherical polar coordinates $(r\theta, \phi)$, as discussed in Appendix A. Note that being angles the cylindrical polar coordinate ϕ and the spherical polar coordinates θ and ϕ do not have the units of length.

DOI: 10.1201/9781003383314-2

2.1.2 GENERALISED VELOCITIES

Corresponding to each generalised coordinate in configuration space, there is a generalised velocity. The generalised velocity associated with generalised coordinate q_i is $\dot{q}_i = \frac{\partial q_i}{\partial t}$. If there are N particles in a system located at Cartesian coordinates $(x_1, y_1, z_1, ..., x_N, y_N, z_N)$, and there are n generalised coordinates $\boldsymbol{q} = (q_1, q_2, ..., q_n)$, then the Cartesian velocity components can be expressed as

$$\begin{aligned} \dot{x}_1 &= \sum_{i=1}^{n} \frac{\partial x_1}{\partial q_i} + \frac{\partial x_1}{\partial t}, \\ &\vdots \\ \dot{z}_N &= \sum_{i=1}^{n} \frac{\partial z_N}{\partial q_i} + \frac{\partial z_N}{\partial t}. \end{aligned} \tag{2.6}$$

On the right-hand side of these equations, the first n contributions to a Cartesian velocity component come via the time variation of the generalised coordinates (*i.e.* are indirect, or implicit, contributions) and the last term is the direct (or explicit) time variation.

Consider motion of a particle on the x-y plane using the generalised coordinates presented in equation (2.2). It is easy to show that the Cartesian and polar velocity components are related as

$$\begin{aligned} \dot{x} &= \dot{\rho}\cos\phi - \rho\dot{\phi}\sin\phi, \\ \dot{y} &= \dot{\rho}\sin\phi + \rho\dot{\phi}\cos\phi. \end{aligned} \tag{2.7}$$

If, instead, equations (2.4) are used, then

$$\begin{aligned} \dot{x} &= \dot{\rho}\cos(\phi + \omega t) - \rho\dot{\phi}\sin(\phi + \omega t) + \rho\omega\sin(\phi + \omega t), \\ \dot{y} &= \dot{\rho}\sin(\phi + \omega t) + \rho\dot{\phi}\cos(\phi + \omega t) + \rho\omega\cos(\phi + \omega t). \end{aligned} \tag{2.8}$$

Notice that the last terms in equation (2.8) show explicit time variation of the Cartesian velocity components.

2.1.3 GENERALISED FORCES

If a particle's position changes by $\boldsymbol{\delta r} = (\delta x, \delta y, \delta z)$ upon the application of a force $\boldsymbol{F}$, then the work done is

$$\text{work} = \boldsymbol{F} \cdot \boldsymbol{\delta r} = F_x \delta x + F_y \delta y + F_z \delta z. \tag{2.9}$$

Now consider a system of N particles, with the ith particle being displaced by $\boldsymbol{\delta r}_i$. The expression for the work takes the form

$$\text{work} = \sum_{i=1}^{N} (F_x^i \delta x_i + F_y^i \delta y_i + F_z^i \delta z_i), \tag{2.10}$$

where F_x^i is the x component of the force on the ith particle. Let us express δx_i, δy_i and δz_i in terms of generalised coordinates (ignoring explicit time dependence)

$$\begin{aligned} \delta x_i &= \sum_{k=1}^{n} \frac{\partial x_i}{\partial q_k} \delta q_k, \\ \delta y_i &= \sum_{k=1}^{n} \frac{\partial y_i}{\partial q_k} \delta q_k, \\ \delta z_i &= \sum_{k=1}^{n} \frac{\partial z_i}{\partial q_k} \delta q_k, \end{aligned} \tag{2.11}$$

where δq_k is a change in the generalised coordinate q_k. The expression for work can thus be written as

$$\begin{aligned} \text{work} &= \sum_{i=1}^{N}\sum_{k=1}^{n}(F_x^i\frac{\partial x_i}{\partial q_k}+F_y^i\frac{\partial y_i}{\partial q_k}+F_z^i\frac{\partial z_i}{\partial q_k})\delta q_k, \\ &= \sum_{k=1}^{n} Q_k\delta q_k, \end{aligned} \tag{2.12}$$

where

$$Q_k = \sum_{i=1}^{N}(F_x^i\frac{\partial x_i}{\partial q_k}+F_y^i\frac{\partial y_i}{\partial q_k}+F_z^i\frac{\partial z_i}{\partial q_k}) \tag{2.13}$$

is generalised force associated with the generalised coordinate q_k.

2.1.3.1 Generalised force in plane polar coordinates

Consider a particle acted on by a force on the x-y plane. Let us express the Cartesian force acting on the particle as

$$\boldsymbol{F} = F_x\hat{\boldsymbol{x}} + F_y\hat{\boldsymbol{y}}. \tag{2.14}$$

Corresponding to the plane polar coordinates $q_1 = \rho$ and $q_2 = \phi$, as shown in figure A.1 in Appendix A, the generalised force components are

$$Q_{q_1} \equiv Q_\rho = F_x\frac{\partial x}{\partial \rho}+F_y\frac{\partial x}{\partial \rho} = F_x\cos\phi + F_y\sin\phi = F_\rho, \tag{2.15}$$

$$Q_{q_2} \equiv Q_\phi = F_x\frac{\partial x}{\partial \phi}+F_y\frac{\partial y}{\partial \phi} = -F_x\rho\sin\phi + F_y\rho\cos\phi = \rho F_\phi. \tag{2.16}$$

The generalised force $\boldsymbol{Q}$ can thus be written as

$$\boldsymbol{Q} = Q_\rho\hat{\boldsymbol{\rho}} + Q_\phi\hat{\boldsymbol{\phi}}. \tag{2.17}$$

The component Q_ρ, called radial force, acts along the radial direction $\hat{\boldsymbol{\rho}}$ with magnitude $F_\rho = \sqrt{F_x^2+F_y^2}$. The force component Q_ϕ, called angular force, acts in the direction $\hat{\boldsymbol{\phi}}$, which indicates direction of increase in the polar angle ϕ. The magnitude of the angular force Q_ϕ is the torque ρF_ϕ.

2.1.4 POTENTIAL AND POTENTIAL ENERGY

Potential energy $V(\boldsymbol{r})$ of a particle is defined as the work done by a conservative force $\boldsymbol{F}$ when it moves from $\boldsymbol{r}$ to a standard point $\boldsymbol{r}_0$

$$V(\boldsymbol{r}) = \int_{\boldsymbol{r}}^{\boldsymbol{r}_0}\boldsymbol{F}(\boldsymbol{r})\cdot\mathrm{d}\boldsymbol{r} = -\int_{\boldsymbol{r}_0}^{\boldsymbol{r}}\boldsymbol{F}(\boldsymbol{r})\cdot\mathrm{d}\boldsymbol{r}. \tag{2.18}$$

From this definition we immediately see that for a given potential energy $V(\boldsymbol{r})$, there is a corresponding force

$$\boldsymbol{F}(\boldsymbol{r}) = -\boldsymbol{\nabla}V(\boldsymbol{r}), \tag{2.19}$$

where $\boldsymbol{\nabla} = (\frac{\partial}{\partial x}, \frac{\partial}{\partial y}, \frac{\partial}{\partial z})$ is the Del, or Nabla, operator defined in equation (A.19) in Appendix A. For the one-dimensional case, we simply write $F(x) = -\frac{\mathrm{d}V(x)}{\mathrm{d}x}$.

Potential is defined as the potential energy per unit physical quantity under consideration. For example, electrostatic potential $\Phi_{\mathrm{es}}(\boldsymbol{r})$ is the electrostatic potential energy $V_{\mathrm{es}}(\boldsymbol{r})$ per unit test charge and gravitational potential $\Phi_{\mathrm{grav}}(\boldsymbol{r})$ is the gravitational energy $V_{\mathrm{grav}}(\boldsymbol{r})$ per unit mass of a test particle at $\boldsymbol{r}$. The potentials $\Phi_{\mathrm{es}}(\boldsymbol{r})$ and $\Phi_{\mathrm{grav}}(\boldsymbol{r})$ can thus be expressed as

$$\begin{aligned}\text{Electrostatic potential}: \quad \Phi_{\mathrm{es}}(\boldsymbol{r}) &= \frac{V_{\mathrm{es}}(\boldsymbol{r})}{\text{charge}},\\ \text{Gravitational potential}: \quad \Phi_{\mathrm{grav}}(\boldsymbol{r}) &= \frac{V_{\mathrm{grav}}(\boldsymbol{r})}{\text{mass}}.\end{aligned} \tag{2.20}$$

Expressions for the electrostatic potential around a charge $\mathscr{Q}$ and the gravitational potential around a mass M are:

$$\begin{aligned}\Phi_{\mathrm{es}}(\boldsymbol{r}) &= \frac{1}{4\pi\varepsilon_0}\frac{\mathscr{Q}}{r},\\ \Phi_{\mathrm{grav}}(\boldsymbol{r}) &= -G\frac{M}{r},\end{aligned} \tag{2.21}$$

where ε_0 is the permittivity of free space and G is the gravitational constant. Note that while the sign (attractive or repulsive nature) of the electrostatic potential depends on the sign of the charge $\mathscr{Q}$, the gravitational potential is always negative (*i.e.* attractive).

2.2 NEWTONIAN, OR VECTORIAL, MECHANICS

Newtonian mechanics is based on the application of Newton's second law of motion, which can be expressed in the following forms

$$\boldsymbol{F}(\boldsymbol{r}) = m\boldsymbol{a} = m\ddot{\boldsymbol{r}} = m\frac{\mathrm{d}^2\boldsymbol{r}}{\mathrm{d}t^2} \quad \text{rectilinear motion,} \tag{2.22}$$

$$\boldsymbol{\tau}(\boldsymbol{\vartheta}) = I\boldsymbol{\alpha} = I\ddot{\boldsymbol{\vartheta}} = I\frac{\mathrm{d}^2\boldsymbol{\vartheta}}{\mathrm{d}t^2} \quad \text{rotational motion.} \tag{2.23}$$

The above equations can be stated as follows. An object of an inertial mass m acquires linear acceleration $\boldsymbol{a}$ when it is subjected to a force $\boldsymbol{F}$. Also, an object characterised by a moment of inertia I acquires angular acceleration $\boldsymbol{\alpha}$ when it is subjected to a torque $\boldsymbol{\tau}$. Equation (2.22) represents a second-order vector differential equation which, given a functional form of $\boldsymbol{F}(\boldsymbol{r})$, should be solved for $\boldsymbol{r} = \boldsymbol{r}(t)$ as a function of time t. Similarly, equation (2.23) represents a second-order vector differential equation which, given a functional form of $\boldsymbol{\tau} = \boldsymbol{\tau}(\boldsymbol{\vartheta})$, should be solved for $\boldsymbol{\vartheta} = \boldsymbol{\vartheta}(t)$. As pointed out in section 1.5, the functional forms $\boldsymbol{F}(\boldsymbol{r}) \propto \boldsymbol{r}$ and $\boldsymbol{\tau}(\boldsymbol{\vartheta}) \propto \boldsymbol{\vartheta}$ are characteristics of linear dynamical systems. Note that while the rectilinear version of the Newton's equation of motion uses the usual Cartesian coordinate variables x, y and z to define the vector $\boldsymbol{r} = (x,y,z)$, the rotational version uses the angle ϑ as a generalised coordinate.

2.2.1 MOTION UNDER CONSTANT FORCE

Although we are devoting this part of the book to study linear dynamics, let us consider equation (2.22) when force on a particle remains constant during its motion. This means that for a given mass m the magnitude of the acceleration a is a constant. In this case, for motion along the x-axis, equation (2.22) becomes the second-order differential equation

$$\frac{\mathrm{d}^2x}{\mathrm{d}t^2} = a. \tag{2.24}$$

Integration of this equation produces the well-known kinematic result

$$v = u + at, \tag{2.25}$$

where u is the speed at time $t = 0$ and v is the speed at time t. Integrating equation (2.24) twice, we obtain another well-known kinematic result

$$x = ut + \frac{1}{2}at^2, \tag{2.26}$$

where we have considered $x = 0$ at $t = 0$.

2.2.2 THE SIMPLE PENDULUM

One of the simplest examples of linear dynamical systems is the simple pendulum. As illustrated in figure 2.1, it consists of a point mass m suspended from a fixed point by a taut and weightless string of length l, and the only force acting on it is its weight mg, where g is the acceleration due to gravity. We will restrict its motion in the x-z vertical plane. As the motion involves horizontal as well as vertical positions, the pendulum is called *pendulum bob* or simply *bob*. We will discuss its dynamics by applying both the rectilinear and rotational forms of Newton's second law of motion.

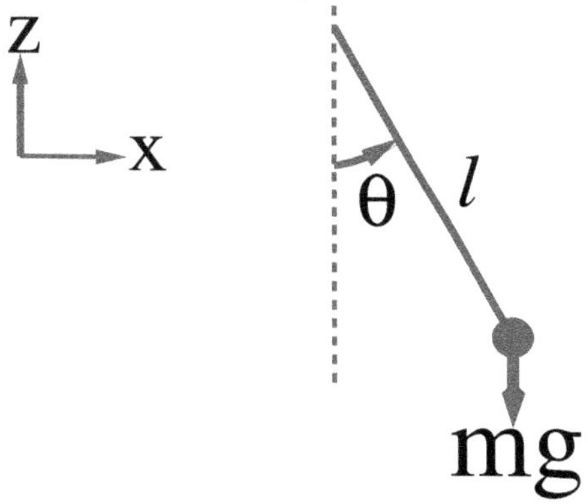

Figure 2.1 The simple pendulum, consisting of a point mass m suspended in the x-z vertical plane from a fixed point by a taut and weightless string of length l. The only force acting on the pendulum is its weight mg, where g is the acceleration due to gravity. When the string is displaced by a small angle θ from the vertical, the linear force on the pendulum is $mg\sin\theta \approx mg\frac{x}{l} = kx$, directed towards its equilibrium.

2.2.2.1 Rectilinear motion

With reference to figure 2.1, consider the bob being pulled out at a small angle θ from its equilibrium and released. As the angle is small, we can relate it to bob's length l and displacement x along the x-axis using $\sin\theta \approx \theta = \frac{x}{l}$. The restoring Hook's force $F = -kx$ acts on the bob towards its equilibrium, where k is a force constant. The equation of motion of the bob then reads

$$m\ddot{x} = -kx. \tag{2.27}$$

As the bob is set in an oscillatory motion about its equilibrium, we can try a solution[1] to the above second-order differential equation either in cosine, or sine, or complex exponential form. In the sine form, we can write

$$x = A\sin(\omega t + \eta), \tag{2.28}$$

[1]See section 3.3.8.1 for a derivation.

where ω is angular frequency, η is a phase factor and A is constant amplitude. Substitution in equation (2.27) gives

$$\omega = \sqrt{\frac{k}{m}}. \tag{2.29}$$

The potential energy corresponding to the force expression $F = -kx$ can easily be expressed as $V = \frac{1}{2}kx^2$. As it is quadratic in x, it is termed *harmonic potential energy*. For this reason, the motion of the bob is simply that of a simple oscillator about its equilibrium and is called *simple harmonic motion*.

2.2.2.2 Rotational motion

From figure 2.1 we note that the motion of the bob can be considered as circular in the x-z plane with the centre at the upper end of the string. The moment of inertia of the bob about its axis of rotation is ml^2. The torque acting at the bottom end of the bob (at distance l), and trying to reduce the swing angle θ, is $-mgl\sin\theta$. Equation (2.23) can then be expressed as

$$I\frac{\mathrm{d}^2\theta}{\mathrm{d}t^2} = -mgl\sin\theta, \tag{2.30}$$

where I is the moment of inertia of the bob for its rotation about the fixed point of the pendulum. Using the trial solution

$$\theta = B\sin(\omega t + \eta) \tag{2.31}$$

with B as constant amplitude, we easily obtain the result, within the small angle approximation ($\sin\theta \approx \theta$),

$$\omega = \sqrt{\frac{g}{l}}. \tag{2.32}$$

2.2.2.3 Time period of oscillations

From equations (2.29) and (2.32) we can establish that the simple harmonic oscillator spring constant k is actually the weight of the bob per unit string length:

$$k = \frac{mg}{l}. \tag{2.33}$$

The period $\mathscr{T}$ of simple harmonic motion is

$$\mathscr{T} = \frac{1}{f} = \frac{2\pi}{\omega} = 2\pi\sqrt{\frac{l}{g}}, \tag{2.34}$$

where f has units of s^{-1} and ω has units of radians per second (rad/s).

2.2.3 EXAMPLES OF APPLICATION OF NEWTONIAN MECHANICS

2.2.3.1 Example 1:

Let us consider a particle of mass m hanging from a vertical spring, as shown in figure 2.2. Let the spring force constant be k.

Due to the weight of the particle, the spring extends by x_0 and occupies its equilibrium at the level shown by the dotted horizontal line in the figure:

$$mg = kx_0. \tag{2.35}$$

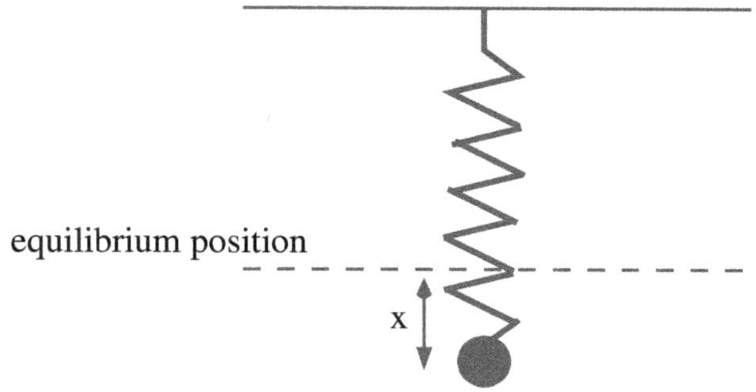

Figure 2.2 A point mass hanging from a vertical spring. The variable x shows the displacement of the mass from its equilibrium position.

If the particle is pulled down by a small distance x below the equilibrium, its motion is described by the equation

$$m\ddot{x} = mg - k(x_0 + x) = -kx \tag{2.36}$$

where we have used equation (2.35). This shows that the particle executes simple harmonic motion around the equilibrium with angular frequency $\omega = \sqrt{k/m}$.

2.2.3.2 Example 2:

Consider a uniform rod of mass m and length l which swings in a vertical plane from a fixed point under gravitational force.

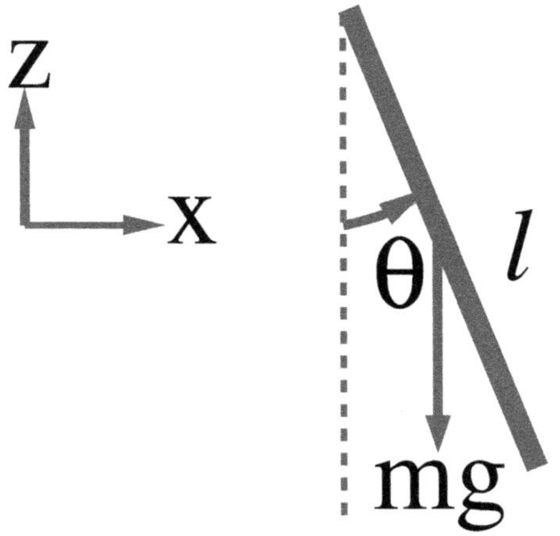

Figure 2.3 A uniform rod of mass m and length l swinging in the x-z vertical plane. The angle θ shows the tilt of the rod from its vertical equilibrium position shown by dotted line. The gravitational force acting through the rod's centre of mass is mg.

The system is shown in figure 2.3. Let us apply the rotational form of Newton's second law of motion. It is easier to use equation (2.30) to solve the motion. We need to work out the torque acting on the rod and remember the expression for the moment of inertia of the rod. From table A.1 in Appendix A we note that the moment of inertia of the rod about an axis through its centre of mass is $ml^2/12$. Using the parallel axis theorem [*cf* equation (A.39) in Appendix A], the moment of inertia about an axis through the fixed point is

$$I = \frac{ml^2}{12} + m\left(\frac{l}{2}\right)^2 = \frac{ml^2}{3}. \tag{2.37}$$

The torque, τ, which acts at the centre of mass of the rod, is

$$\tau = -mg\frac{l}{2}\sin\theta. \tag{2.38}$$

Using equations (2.37) and (2.38) in (2.30), we can obtain the equation of motion

$$\ddot{\theta} = -\frac{3g}{2l}\sin\theta. \tag{2.39}$$

For small angles of oscillation we write $\sin\theta \approx \theta$ and get

$$\ddot{\theta} = -\omega^2\theta, \tag{2.40}$$

where $\omega = \sqrt{\frac{3g}{2l}}$ is the angular frequency.

2.2.3.3 Example 3:

Let us consider a solid ball rolling, without slipping, down an inclined plane, as shown in figure 2.4. Let R be the radius of the ball and α be the incline of the plane. We will apply both the rectilinear and rotational forms of Newton's second law of motion to obtain an expression for the time taken for the ball to roll down a certain distance.

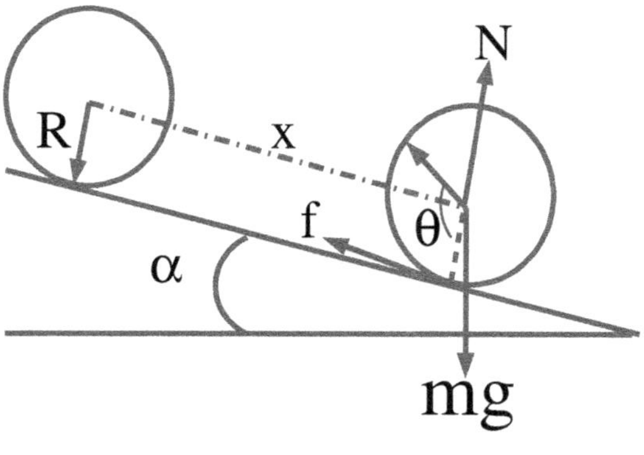

Figure 2.4 A solid ball of mass m and radius R rolling down a plane of inclination α. Starting from rest, the ball travels a distance x so that the radius vector $\boldsymbol{R}$ sweeps an angle θ. The forces acting on the ball are: gravitational force mg, normal reaction N and frictional force f.

Rectilinear version:

Starting from rest at the top of the incline, at time t a radius vector rotates by angle θ and the ball travels the linear distance $x = R\theta$. At that instance, there are three forces acting on the ball: its weight mg, the normal reaction N of the plane and the frictional force f between the ball and the plane, as indicated in figure 2.4. The motion of the ball is governed by the following equations:

$$\text{motion of centre of mass}: \quad m\ddot{x} = mg\sin\alpha - f, \tag{2.41}$$

$$\text{motion about centre of mass}: \quad fR = I_{\text{cm}}\ddot{\theta}, \tag{2.42}$$

$$\text{rolling condition}: \quad x = R\theta, \quad \text{or} \quad \ddot{x} = R\ddot{\theta}, \tag{2.43}$$

where $I_{\text{cm}} = \frac{2}{5}MR^2$ is the moment of inertia of the ball about an axis through its centre of mass. Using these equations, the acceleration of the ball can be expressed as

$$\ddot{x} = \frac{mR^2 g\sin\alpha}{(I_{\text{cm}} + mR^2)} = \frac{5}{7}g\sin\alpha. \tag{2.44}$$

Rotational version:

The torque at the point of contact between the ball and the plane is $mgR\sin\alpha$ and the moment of inertia about an axis through the point of contact is $I = I_{\text{cm}} + mR^2 = \frac{7}{5}mR^2$. The rotational form of the equation of motion in (2.23) thus becomes

$$mgR\sin\alpha = \frac{7}{5}mR^2\ddot{\theta}, \tag{2.45}$$

from which, using the rolling condition in (2.43), we obtain the linear acceleration $\ddot{x} = \frac{5}{7}g\sin\alpha$, the result obtained before.

Energy conservation method:

The result derived above can also be obtained by applying the method of energy conservation. As the ball rolls down the plane by distance x, changes in the energy components are:

$$\text{kinetic energy gain} = \Delta K = \frac{1}{2}m\dot{x}^2 + \frac{1}{2}I_{\text{cm}}\dot{\theta}^2, \tag{2.46}$$

$$\text{potential energy loss} = \Delta V = -mgx\sin\alpha. \tag{2.47}$$

Using the energy conservation condition $\Delta K + \Delta V = 0$, together with the rolling condition $x = R\theta$, we obtain

$$\dot{x}^2 = \frac{2mgR^2\sin\alpha}{(mR^2 + I_{\text{cm}})}x. \tag{2.48}$$

Differentiating this equation with respect to time, we retrieve the expression for acceleration $\ddot{x}$ presented in equation (2.44).

2.2.3.4 Example 4:

Consider a small sphere of radius a at rest in the bottom of a hemispherical bowl of radius R ($R > a$). The sphere is displaced from its rest position by a small distance along the surface of the bowl and released, and no slipping occurs. We will show that the resulting motion is simple harmonic and the length of the equivalent simple pendulum is $\frac{7}{5}(R-a)$.

Figure 2.5 sketches the ball at its rest at the bottom of the hemisphere and at a time t. Starting from equilibrium, the distance rolled by the centre of mass of the ball in time t is $(R-a)\theta$, where θ is the angle swept by the radius R of the hemisphere. During that time the radius a of the ball sweeps angle ϕ. The arclength covered along the hemispherical surface is $x = R\theta = a\phi$. Note that with respect to a fixed axis in space, the ball has turned through the angle $\phi - \theta$. Let f represent the

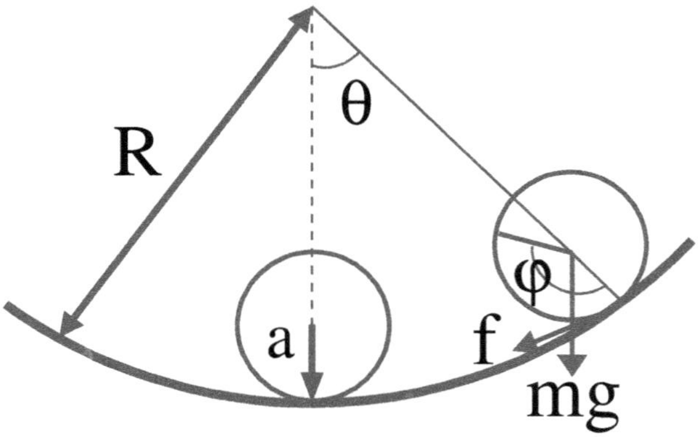

Figure 2.5 Oscillations of a small sphere of mass m and radius a around the bottom of a hemispherical bowl of radius R ($R > a$). The vertical line shows the equilibrium position and the oscillatory position is indicated by the angles θ and ϕ swept by the hemispherical bowl and the solid ball, respectively. The forces acting on the ball are: the gravitational force mg and the frictional force f.

frictional force between the ball and the hemispherical surface. The relevant equations of motion of the ball and its rolling condition are described below:

$$\text{motion of centre of mass}: m(R-a)\ddot{\theta} = -(mg\sin\theta + f), \tag{2.49}$$

$$\text{motion about centre of mass}: fa = I_{\mathrm{cm}}(\ddot{\phi} - \ddot{\theta}), \tag{2.50}$$

$$\text{rolling condition}: R\theta = a\phi, \tag{2.51}$$

where I_{cm} is the moment of inertia of the ball about an axis through its centre of mass.

From equations (2.50) and (2.51), we get

$$fa = (\frac{R}{a} - 1)I_{\mathrm{cm}}\ddot{\theta}. \tag{2.52}$$

Using equations (2.49) and (2.52), we obtain

$$\left(1 + \frac{I_{\mathrm{cm}}}{ma^2}\right)\ddot{\theta} = -\frac{g}{(R-a)}\sin\theta. \tag{2.53}$$

For small angles of rotation $\sin\theta \approx \theta$ so that equation (2.53) reduces to

$$\ddot{\theta} \approx -\frac{g}{(R-a)\left[1 + \frac{I_{\mathrm{cm}}}{ma^2}\right]}\theta = -\frac{5}{7}\frac{1}{(R-a)}g\theta, \tag{2.54}$$

where in the last step $I = \frac{2}{5}ma^2$ has been used. Writing equation (2.54) in the form

$$\ddot{\theta} = -\frac{g}{L_{\mathrm{eff}}}\theta, \tag{2.55}$$

we establish that the effective pendulum length of the ball's simple harmonic motion is $L_{\mathrm{eff}} = \frac{7}{5}(R-a)$.

2.3 LAGRANGIAN, OR SCALAR, MECHANICS

2.3.1 INTRODUCTION

As we noted in the preceding section, the Newtonian mechanics is based on the use of vector quantities, *e.g.* forces. For a system of N particles, the equation of motion in (2.22) expands into $3N$ second-order differential equations, each of which may have to be solved to determine particle positions at a given time. Scalar, or analytical, mechanics, on the other hand, uses scalar quantities, *e.g.* kinetic and potential energies. Analytical mechanics seeks to derive equations of motion using a variational principle, based on calculus of variations. This will be described in this section, with examples of a few applications. We first remind ourselves of two key elements from the calculus of variations described in Appendix C.

2.3.2 THE δ-VARIATION

The δ variation in a function $y(x)$ is constrained by the imposition of no variation at the start point x_1 and the end point x_2 under consideration, *viz* $\delta y(x_1) = \delta y(x_2) = 0$.

2.3.3 EULER-LAGRANGE VARIATIONAL PRINCIPLE AND ITS APPLICATIONS

The Euler-Lagrange variational principle states that the integral

$$J = \int_{x_1}^{x_2} f(y, y', x)\mathrm{d}x, \tag{2.56}$$

of a functional $f(y, y', x)$, with $y = y(x)$ and $y' = \frac{\mathrm{d}y}{\mathrm{d}x}$, is stationary at $y = y_0$, where y_0 is the solution of the partial differential equation

$$\frac{\partial f}{\partial y} - \frac{\mathrm{d}}{\mathrm{d}x}\left(\frac{\partial f}{\partial y'}\right) = 0, \quad x_1 \le x \le x_2, \tag{2.57}$$

with $y(x_1) = \beta_1 = \text{constant}$ and $y(x_2) = \beta_2 = \text{constant}$.

2.3.3.1 Application 1:

Let us apply the Euler-Lagrange variational principle to calculate the *shortest distance between two points on a plane*.

An element of distance $\mathrm{d}s$ between two neighbouring points on the x-y plane is

$$\mathrm{d}s = \sqrt{\mathrm{d}x^2 + \mathrm{d}y^2} = \left[1 + \left(\frac{\mathrm{d}y}{\mathrm{d}x}\right)^2\right]^{1/2}\mathrm{d}x. \tag{2.58}$$

The total length of the curve, between $x = x_1$ and $x = x_2$, is

$$J = \int_{x_1}^{x_2} \mathrm{d}s = \int_{x_1}^{x_2} \left[1 + \left(\frac{\mathrm{d}y}{\mathrm{d}x}\right)^2\right]^{1/2}\mathrm{d}x. \tag{2.59}$$

We can express this in the form of equation (2.57), with

$$f = \sqrt{1 + y'^2}, \quad y' \equiv \frac{\mathrm{d}y}{\mathrm{d}x} \tag{2.60}$$

From this we obtain

$$\frac{\partial f}{\partial y} = 0, \quad \frac{\partial f}{\partial y'} = \frac{y'}{\sqrt{1 + y'^2}} \tag{2.61}$$

The Euler-Lagrange equation in (2.59) [also in (C.11)] then becomes

$$\frac{\mathrm{d}}{\mathrm{d}x}\left(\frac{\partial f}{\partial y'}\right) \equiv \frac{\mathrm{d}}{\mathrm{d}x}\left(\frac{y'}{\sqrt{1+y'^2}}\right) = 0. \tag{2.62}$$

Integration of this gives

$$\frac{y'}{\sqrt{1+y'^2}} = \text{constant}. \tag{2.63}$$

This solution is valid only if

$$y' = a = \text{constant}, \tag{2.64}$$

from which we obtain

$$y = ax + b, \tag{2.65}$$

with b as another constant. Thus, a straight line is an extremum path between two points on a plane. The constants a and b can be determined using the condition that the curve passes through two end points (x_1, y_1) and (x_2, y_2).

As emphasised in Appendix C, the solution of the Euler-Lagrange equation only provides the necessary condition, *viz* the optimum function $y = y(x)$, for the integral J to be an extremum. To prove that the straight line provides the shortest distance between two points, let us first evaluate the distance J_1 between two points x_1 and x_2 along the x axis using equation (2.65):

$$J_1 = \int_{x_1}^{x_2}\left[1+\left(\frac{\mathrm{d}y}{\mathrm{d}x}\right)^2\right]^{1/2}\mathrm{d}x = \int_{x_1}^{x_2}\sqrt{(1+a^2)}\mathrm{d}x = \sqrt{(1+a^2)}(x_2 - x_1). \tag{2.66}$$

Next, consider another form of the path along the x axis, say $y = cx^2 + d$, where c and d are constants. For this choice of path,

$$J_2 = \int_{x_1}^{x_2}\sqrt{(1+4c^2x^2)}\mathrm{d}x = c\left[x\sqrt{\beta^2+x^2} + \beta^2\ln\{\sqrt{\beta^2+x^2}+x\}\right]\Big|_{x_1}^{x_2}. \tag{2.67}$$

where $\beta = \frac{1}{2c}$. It can be verified that $J_1 < J_2$, proving that the distance between two points on a straight line is the shortest. For example, using $y_1 = x$ we can easily show that $J_1 = \int_0^1\left[1+\left(\frac{\mathrm{d}y_1}{\mathrm{d}x}\right)^2\right]^{1/2}\mathrm{d}x = \sqrt{2} = 1.4142$ units, and using $y_2 = x^2$ we can obtain the length $J_2 = \int_0^1\left[1+\left(\frac{\mathrm{d}y_2}{\mathrm{d}x}\right)^2\right]^{1/2}\mathrm{d}x = \frac{\sqrt{5}}{2} + \frac{1}{4}\ln[1+\sqrt{5}/2] + \frac{1}{4}\ln 2 = 1.4789$ units, proving that $J_1 < J_2$.

2.3.3.2 Application 2:

As a second example, we will determine the shape of a freely hanging homogeneous and flexible string suspended from two points. An example of such a system is our washing line, or clothes line.

In order to determine the optimum shape of the chain, we consider its potential energy V as the integral J. Considering y-axis vertically upwards and x-axis as the horizontal line between two poles from which the string is hanging, the potential energy can be expressed as

$$V = \int_0^l \rho g y \mathrm{d}s, \tag{2.68}$$

where ρ is the mass density of the homogeneous string of length l. Obviously, in equilibrium, the potential energy of the string is at its minimum. The extremum condition for this is

$$\delta V = 0 = \delta\int_0^l f\mathrm{d}x, \tag{2.69}$$

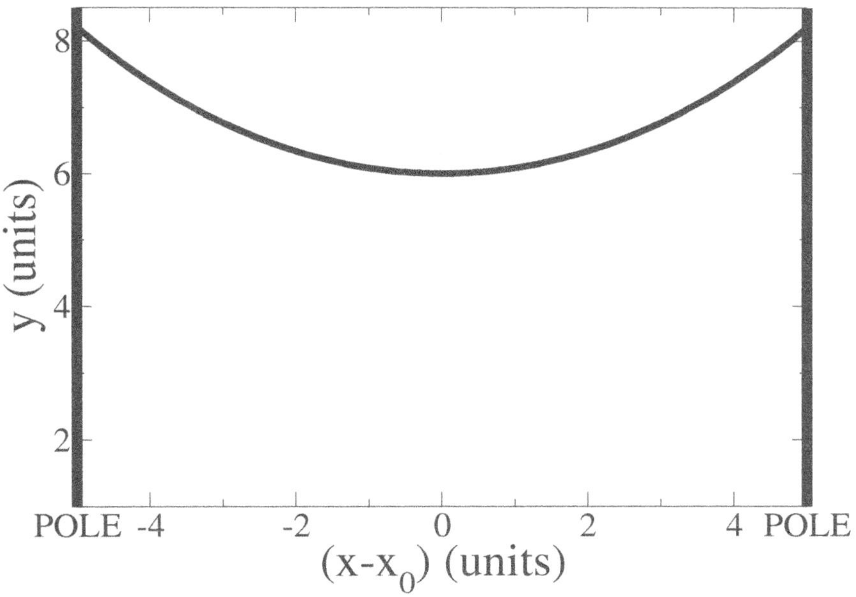

Figure 2.6 A homogeneous and flexible string freely suspended between two poles.

where we have used $\mathrm{d}s = \sqrt{\mathrm{d}x^2 + \mathrm{d}y^2}$, and the functional f is

$$f = y\left(1 + y'^2\right)^{1/2}. \tag{2.70}$$

Using equation (2.70), the Euler-Lagrange equation in (2.57) can be shown to take the form

$$\frac{y'}{y} = \frac{y'y''}{1 + y'^2}. \tag{2.71}$$

Integration of this equation results in

$$y' \equiv \frac{\mathrm{d}y}{\mathrm{d}x} = \frac{1}{a}\sqrt{y^2 - a^2}, \tag{2.72}$$

where a is a constant. Expressing this as $\mathrm{d}x = a\frac{\mathrm{d}y}{\sqrt{y^2 - a^2}}$, we obtain

$$x = a\cosh^{-1}(y/a) + b, \tag{2.73}$$

where b is another constant. From this, we express

$$y = a\cosh\left(\frac{x - b}{a}\right), \tag{2.74}$$

which is the equation of a catenary.

The slope of the curve is zero when $y' = 0 = \frac{y^2 - a^2}{a}$, which gives $y = y_{\mathrm{min}} = a$. The minimum occurs at $x = x_0 = b$, which satisfies $y = y_{\mathrm{min}}\cosh\left(\frac{x - x_0}{y_{\mathrm{min}}}\right) = y_{\mathrm{min}}$. Thus, we can express the equation of the catenary as

$$y = y_{\mathrm{min}}\cosh\left(\frac{x - x_0}{y_{\mathrm{min}}}\right). \tag{2.75}$$

The shape of the catenary with $y_{\mathrm{min}} = 6$ units at $x_0 = 5$ units is shown in figure 2.6. Note that the distance between two poles, $2x_0$ units, must be smaller than the length l of the string.

2.3.4 LAGRANGIAN

Lagrange defined a functional, called *Lagrange's functional* (or, simply *Lagrangian*) L in configuration space, as

$$L = T - V, \tag{2.76}$$

where T and V are, respectively, the kinetic and potential energies of the system under consideration. For conservative systems, V is a function of generalised coordinates and T is a function of generalised velocities. For n generalised coordinates, $V = V(\boldsymbol{q}) \equiv V(q_1, q_2, ..., q_n)$ and for n generalised velocities $T = T(\dot{q}) \equiv T(\dot{q}_1, \dot{q}_2, ..., \dot{q}_n)$, with $q_i = q_i(t)$ and $\dot{q}_i = \frac{\mathrm{d}q_i(t)}{\mathrm{d}t}$.

2.3.5 ACTION INTEGRAL

Having defined the Lagrangian, we now define a functional J, called *action integral* (or, simply *action*), as the time integral of the Lagrangian function:

$$J = \int L\mathrm{d}t. \tag{2.77}$$

2.3.6 HAMILTON'S VARIATIONAL PRINCIPLE: EULER-LAGRANGE VARIATIONAL PRINCIPLE IN CONFIGURATION SPACE

Hamilton applied the Euler-Lagrange variational principle to seek an extremum of the action integral in configuration space.

$$\delta J = \delta \int L(\boldsymbol{q}, \dot{\boldsymbol{q}}, t)\mathrm{d}t = 0. \tag{2.78}$$

This is known as **Hamilton's variational principle**. For the stationarity of the action integral in a n-dimensional configuration space, the following Euler-Lagrange equations must be satisfied

$$\frac{\mathrm{d}}{\mathrm{d}t}\frac{\partial L}{\partial \dot{q}_i} = \frac{\partial L}{\partial q_i}, \quad i = 1, 2, 3, ..., n. \tag{2.79}$$

These are known as the **Lagrange equations of motion**.

Equations of motion for a limited class of non-holonomic systems, for which generalised coordinates are not independent of each other, may also be obtained from Hamilton's principle by using *Lagrange's method of undetermined multipliers*. The reader can find a good discussion of this procedure in the books by Goldstein *et al.* (2002) and Thornton and Marion (2004).

2.3.7 COMPARISON OF LAGRANGIAN MECHANICS WITH NEWTONIAN MECHANICS

Both Lagrangian mechanics and Newtonian mechanics are derived in configuration space, with generalised coordinates as the only set of independent variables. For a system of N particles, the Newtonian mechanics in the three-dimensional Cartesian coordinate system may require solving up to $3N$ second-order differential equations. In contrast, Lagrangian mechanics in configuration space with n generalised coordinates leads to n second-order partial differential equations. Clearly, for a large system with $N >> n$, the Lagrangian mechanics requires solving far fewer number of equations of motion than does the Newtonian mechanics.

It is easy to show that for a single particle, considering Cartesian coordinates, Lagrange's equations of motion are identical to Newton's equations of motion. Consider oscillatory motion along the x-direction, characterised by a spring constant k, of a single particle of mass m. Newton's equation of motion in equation (2.27) can be expressed as

$$m\ddot{x} = -kx,$$

$$
\begin{aligned}
\frac{\mathrm{d}}{\mathrm{d}t}(m\dot{x}) &= -kx,\\
\frac{\mathrm{d}}{\mathrm{d}t}\frac{\partial T}{\partial \dot{x}} &= -kx,\\
\frac{\mathrm{d}}{\mathrm{d}t}\frac{\partial T}{\partial \dot{x}} &= -\frac{\partial V}{\partial x},\\
\frac{\mathrm{d}}{\mathrm{d}t}\frac{\partial (T-V)}{\partial \dot{x}} &= -\frac{\partial V}{\partial x},\\
\frac{\mathrm{d}}{\mathrm{d}t}\frac{\partial (T-V)}{\partial \dot{x}} &= \frac{\partial (T-V)}{\partial x},\\
\frac{\mathrm{d}}{\mathrm{d}t}\frac{\partial L}{\partial \dot{x}} &= \frac{\partial L}{\partial x},
\end{aligned}
\tag{2.80}
$$

which is the Lagrange equation of motion. Note that in writing the above steps we have used the kinetic energy expression $T = \frac{1}{2}m\dot{x}^2$ and the potential energy expression $V = -\int F\mathrm{d}x = \frac{1}{2}kx^2$.

2.3.8 EXAMPLES OF APPLICATION OF LAGRANGIAN MECHANICS

Let us consider the four examples discussed using the Newtonian mechanics in sub-section 2.2.3.

2.3.8.1 Example 1:

Motion of a particle hanging from a vertical spring.

Let us set the zero of the potential energy of the particle at its equilibrium position (shown by the dotted line in figure 2.2). The distance x below the equilibrium is the single independent generalised coordinate for the problem. The expressions for the kinetic and potential energies expressions are

$$
T = \frac{1}{2}m\dot{x}^2, \quad V = \frac{1}{2}kx^2. \tag{2.81}
$$

The Lagrange equation of motion is

$$
\begin{aligned}
\frac{\mathrm{d}}{\mathrm{d}t}\frac{\partial L}{\partial \dot{x}} &= \frac{\partial L}{\partial x},\\
m\ddot{x} &= -kx,
\end{aligned}
\tag{2.82}
$$

which is equation (2.36), obtained by using the Newtonian mechanics.

2.3.8.2 Example 2:

Motion of a uniform rod of mass m and length l swinging in a vertical plane from a fixed point under gravitational force.

The system is shown in figure 2.3. Setting the origin at the top end of the rod, the coordinates of its centre of mass are (x, z), where

$$
x = \frac{l}{2}\sin\theta, \quad z = -\frac{l}{2}\cos\theta \tag{2.83}
$$

and the velocity components are

$$
\dot{x} = \frac{l}{2}\cos\theta\dot{\theta}, \quad \dot{z} = \frac{l}{2}\sin\theta\dot{\theta}. \tag{2.84}
$$

The potential energy of the rod is

$$
V = -mg\frac{l}{2}\cos\theta, \tag{2.85}
$$

and the kinetic energy is

$$\begin{aligned} T &= \text{rectilinear kinetic energy of centre of mass} \\ &\quad + \text{rotational kinetic energy about centre of mass ,} \\ &= \frac{1}{2}m(\dot{x}^2+\dot{z}^2)+\frac{1}{2}I_{\text{cm}}\dot{\theta}^2, \\ &= \frac{1}{2}m\Big(\frac{l}{2}\Big)^2\dot{\theta}^2+\frac{1}{2}\frac{ml^2}{12}\dot{\theta}^2, \\ &= \frac{ml^2}{6}\dot{\theta}^2. \end{aligned} \tag{2.86}$$

Using equations (2.85) and (2.86), the expression for the Lagrangian of the rod reads

$$L=\frac{ml^2}{6}\dot{\theta}^2+mg\frac{l}{2}\cos\theta. \tag{2.87}$$

Lagrange's equation of motion is:

$$\begin{aligned} \frac{\mathrm{d}}{\mathrm{d}t}\frac{\partial L}{\partial\dot{\theta}} &= \frac{\partial L}{\partial\theta}, \\ \frac{\mathrm{d}}{\mathrm{d}t}\Big(\frac{ml^2}{3}\dot{\theta}\Big) &= -mg\frac{l}{2}\sin\theta, \\ \ddot{\theta} &= -\frac{3g}{2l}\sin\theta. \end{aligned} \tag{2.88}$$

For small angles of oscillations, $\sin\theta\approx\theta$, and equation (2.88) reduces to the simple harmonic motion

$$\ddot{\theta}=-\frac{g}{L_{\text{eff}}}\theta, \tag{2.89}$$

where L_{eff} is the effective pendulum length of the rod. The result is identical to that derived in equation (2.40) using the Newtonian mechanics.

2.3.8.3 Example 3:

A solid ball rolling, without slipping, down an incline.

The system is shown in figure 2.4. We set the zero of the potential energy when the ball is at rest at the top of the incline. The potential and kinetic energies of the ball, when it has travelled distance x along the plane, are

$$\begin{aligned} V &= -mgx\sin\alpha, \\ T &= \frac{1}{2}m\dot{x}^2+\frac{1}{2}I_{\text{cm}}\dot{\theta}^2, \\ &= \frac{1}{2}(ma^2+I_{\text{cm}})\dot{\theta}^2. \end{aligned} \tag{2.90}$$

The Lagrange equation of motion can be expressed as

$$\begin{aligned} \frac{\mathrm{d}}{\mathrm{d}t}\frac{\partial L}{\partial\dot{\theta}} &= \frac{\partial L}{\partial\theta}, \\ (ma^2+I_{\text{cm}})\ddot{\theta} &= mga\sin\alpha, \\ \ddot{\theta} &= mga\sin\alpha/(ma^2+I_{\text{cm}}). \\ \ddot{x} &= mga^2\sin\alpha/(ma^2+I_{\text{cm}}), \\ \ddot{x} &= \frac{5}{7}g\sin\alpha, \end{aligned} \tag{2.91}$$

which is what was obtained in equation (2.44) using the Newtonian mechanics.

2.3.8.4 Example 4:

A solid ball of radius a rolling on the inner surface of a hemispherical ball of radius R, where $R > a$.

Figure 2.5 shows the system. Starting at the bottom of the hemisphere, the ball sweeps an ar-clength x in time t. As explained in example 4 in section 2.2.3, the ball turns through $\phi - \theta$ in time t, where $\phi = a/x$ and $\theta = R/x$.

With no loss of generality, we set the zero of potential energy at the vertical height of the centre of the hemisphere. The potential and kinetic energies of the ball can be expressed as

$$\begin{aligned} V &= -mg(R-a)\cos\theta, \\ T &= \text{rectilinear kinetic energy of centre of mass} \\ &\quad + \text{rotational kinetic energy about centre of mass}, \\ &= \frac{1}{2}m(R-a)^2\theta^2 + \frac{1}{2}I_{\text{cm}}(\phi-\theta)^2, \\ &= \frac{1}{2}m(R-a)^2\theta^2 + \frac{2}{10}ma^2\left(\frac{R-a}{a}\right)^2\dot{\theta}^2, \\ &= \frac{7}{10}m(R-a)^2\dot{\theta}^2, \end{aligned} \tag{2.92}$$

where $I_{\text{cm}} = \frac{2}{5}ma^2$ has been substituted.

The Lagrange equation can now be easily written down as

$$\begin{aligned} \frac{\mathrm{d}}{\mathrm{d}t}\frac{\partial L}{\partial\dot{\theta}} &= \frac{\partial L}{\partial\theta}, \\ \frac{7}{5}m(R-a)^2\ddot{\theta} &= -mg(R-a)\sin\theta, \\ \ddot{\theta} &= -\frac{5g}{7(R-a)}\sin\theta. \end{aligned} \tag{2.93}$$

In the small angle approximation this represents the simple harmonic motion

$$\ddot{\theta} = -\frac{g}{L_{\text{eff}}}\theta, \tag{2.94}$$

with the effective pendulum length $L_{\text{eff}} = \frac{7}{5}(R-a)$ as derived using Newton's equation of motion in (2.55).

2.3.9 VELOCITY DEPENDENT POTENTIALS AND LAGRANGE EQUATIONS

In our discussion so far we have considered conservative systems, for which potential is a function of generalised coordinates. What form do Lagrange's equations take when the potential is velocity dependent?

Let us consider, for simplicity, configuration space with a single generalised coordinate q. We note that if $L(q,\dot{q},t) = T(\dot{q}) - V(q)$ is a Lagrangian for a system, then

$$L'(q,\dot{q},t) = L(q,\dot{q},t) + \frac{\mathrm{d}F}{\mathrm{d}t} \tag{2.95}$$

is also an admissible Lagrangian that satisfies Hamilton's principle, where $F(q,t)$ is any continuous and differentiable function of the generalised coordinate q and time t. This is so because if the Hamilton's variational principle $\delta\int_{t_1}^{t_2} L\mathrm{d}t = 0$ is satisfied then $\delta\int_{t_1}^{t_2} L'\mathrm{d}t = 0$ will also be satisfied for an arbitrary F charaterised by $\delta F|_{t_1} = \delta F|_{t_2} = 0$, as required for the δ variation[2].

[2] Also see section 3.3.3

Lagrange's equations for the Lagrangian L' can be expressed as

$$
\begin{aligned}
\frac{\mathrm{d}}{\mathrm{d}t}\frac{\partial L'}{\partial \dot{q}} - \frac{\partial L'}{\partial q} &= 0, \\
\frac{\mathrm{d}}{\mathrm{d}t}\frac{\partial (L+\frac{\mathrm{d}F}{\mathrm{d}t})}{\partial \dot{q}} - \frac{\partial (L+\frac{\mathrm{d}F}{\mathrm{d}t})}{\partial q} &= 0, \\
\frac{\mathrm{d}}{\mathrm{d}t}\frac{\partial L}{\partial \dot{q}} - \frac{\partial L}{\partial q} &= \frac{\partial}{\partial q}\frac{\mathrm{d}F}{\mathrm{d}t} - \frac{\mathrm{d}}{\mathrm{d}t}\frac{\partial}{\partial \dot{q}}\frac{\mathrm{d}F}{\mathrm{d}t}, \\
\frac{\mathrm{d}}{\mathrm{d}t}\frac{\partial L}{\partial \dot{q}} - \frac{\partial L}{\partial q} &= Q,
\end{aligned} \tag{2.96}
$$

where

$$
Q = -\frac{\partial U}{\partial q} + \frac{\mathrm{d}}{\mathrm{d}t}\left(\frac{\partial U}{\partial \dot{q}}\right), \tag{2.97}
$$

and

$$
U(q,\dot{q}) = -\frac{\mathrm{d}F(q,t)}{\mathrm{d}t}. \tag{2.98}
$$

Here L is the Lagrangian containing the potential energy $V(q)$ of the conservative forces, and Q is a *dissipative force* that does not arise from $V(q)$ but from a *velocity-dependent potential energy* (or a *generalised potential energy*) $U = U(q,\dot{q})$. Equation (2.96) should be used when solving problems in the presence of a dissipative force, such as a frictional force.

FURTHER READING

There are many useful textbooks on the subject of Classical Mechanics. The following is a short list of selected textbooks that may be helpful for further reading on the topics covered in this chapter.

1. Calkin M G 1996 *Lagrangian and Hamiltonian Mechanics* (World Scientific: Singapore)
2. Fowles G R and Cassiday G L 2005 *Analytical Mechanics* 7th edn (Thomson: Belmont, CA)
3. Goldstein H, Poole C and Safko J 2002 *Classical Mechanics* 3rd edn (Addison-Wesley: New York)
4. Gregory R D 2006 *Classical Mechanics* (Cambridge University Press: Cambridge)
5. Hand L N and Finch J D 1998 *Analytical Mechanics* (Cambridge University Press: Cambridge)
6. Kibble T W B 1966 *Classical Mechanics* (McGraw-Hill: New York)
7. Landau L D and Lifshitz E M 1986 *Theory of Elasticity* 3rd edn (Butterworth and Heinemann: Oxford)
8. Leech J W 1965 *Classical Mechanics* 2nd edn (Methuen & Co. Ltd.: London)
9. Symon K R 1971 *Mechanics* 3rd edn (Addison-Wesley: Reading MA)
10. Thornton S T and Marion J B 2004 *Classical Dynamics of Particles and Systems* 5th edn (Thomson: Belmont CA)

PROBLEMS

1. Given the information in equation (2.4), prove the results in equation (2.5).
2. Use equation (A.13) to prove the results in equation (A.14).
3. Use equation (2.4) to prove the results in equation (2.8).

4. By considering the n generalised coordinates $\boldsymbol{q}$ as the $3N$ Cartesian coordinates $(x_i, y_i, z_i), i = 1, \ldots N$, use equation (2.13) to show that the generalised force $\boldsymbol{Q}$ reduces to the Cartesian force $\boldsymbol{F}$.
5. Apply the rectilinear form of Newton's second law of motion [*cf* equation (2.22)] to obtain the result in equation (2.37).
6. A cotton real of mass m, radius of gyration k, flange radius b and core radius a, is free to roll on a rough surface.

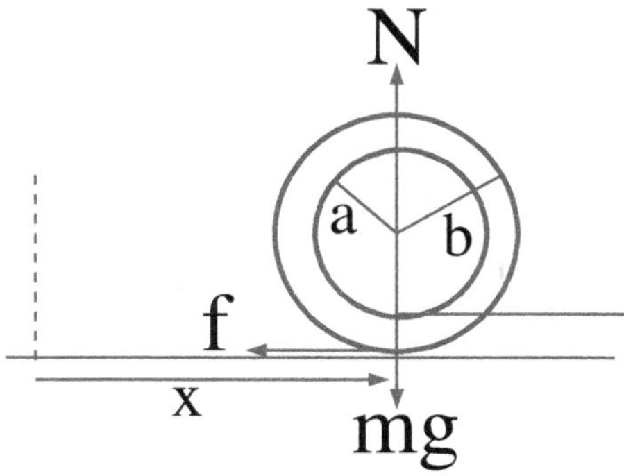

A thin cord is wound round the core, the free end leaving the reel on the underside horizontally. A constant pull P is exerted on the free end in a direction normal to the axis of the reel. The diagram in the figure shows the directions of the weight mg, normal reaction N, the pull P and the frictional force f. If no slipping occurs, which way does the reel roll and what is the linear acceleration?

7. From a simple pendulum of mass m_1 and length l_1, a second mass m_2 is freely suspended by an inextensible string of length l_2, as shown in the figure below.

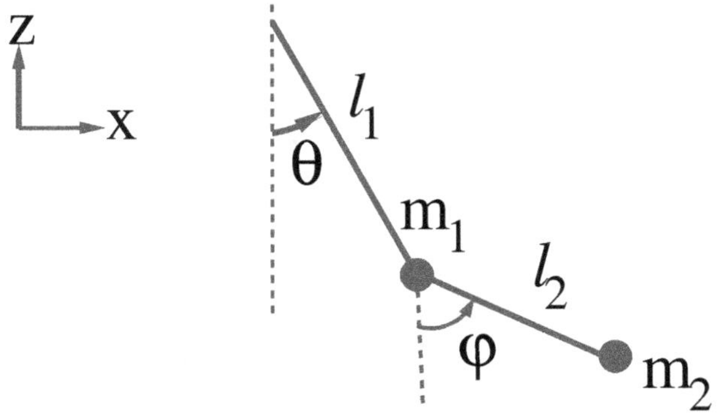

Show that θ and ϕ (the angles the two strings make with the downward vertical) are suitable generalised coordinates for the system. Obtain Lagrange's equations of motion in these variables and show that if both θ and ϕ are small then these equations are:

$$\begin{aligned}
\ddot{\theta} + \frac{g}{l_1}\theta &= -\frac{m_2}{m_1 + m_2}\frac{l_2}{l_1}\ddot{\phi}, \\
\ddot{\phi} + \frac{g}{l_2}\phi &= -\frac{l_1}{l_2}\ddot{\theta}.
\end{aligned}$$

8. The Cartesian coordinates (x, y, z) defining the position of a particle of mass m moving in a force field having potential energy V are given in terms of spherical coordinates (r, θ, ϕ) by the transformation equations

$$x = r\sin\theta\cos\phi, \quad y = r\sin\theta\sin\phi, \quad z = r\cos\theta.$$

Use Lagrange's equations to set up the equations of motion.

9. Suppose it was known experimentally that a particle fell a given distance y_0 in a time $t_0 = \sqrt{2y_0/g}$ but that the time taken to fall distances other than y_0 were not known. Suppose further that the Lagrangian for the problem is known, but that instead of solving the equation of motion for y as a function of time t, it is guessed that the functional form is $y = at + bt^2$. If the constants a and b are adjusted always so that the time taken to fall y_0 is correctly given by t_0, show directly that the integral $\int_0^{t_0} L dt$ is an extremum for real values of the coefficients only when $a = 0$ and $b = g/2$.

10. A particle of mass m is subjected to a force $F = -F_0 \sinh ax$ along the x axis, where $a > 0$. Determine the stable point of the particle's motion. Derive the Lagrange equation of motion and obtain an expression for the frequency of small oscillations around the stable point.

11. The Lagrangian of a non-relativistic electron of mass m and charge e moving with velocity $\boldsymbol{v}$ in an electromagnetic field is given as

$$L = \frac{1}{2}mv^2 + e\Phi - e\boldsymbol{A}\cdot\boldsymbol{v},$$

where the scalar potential Φ and the vector potential $\boldsymbol{A}$ of the electromagnetic field are functions of the electron position $\boldsymbol{r}$ and time t. Show that the momentum of the electron has the generalised form

$$\boldsymbol{p} = m\boldsymbol{v} - e\boldsymbol{A}.$$

12. A small ring of mass M is attached to one end of a uniform rod of length $2a$ and mass $2M$ and is then threaded onto a smooth rigid horizontal wire. The rod is initially held vertical and then is set swinging in the vertical plane containing the wire, as shown in the figure below.

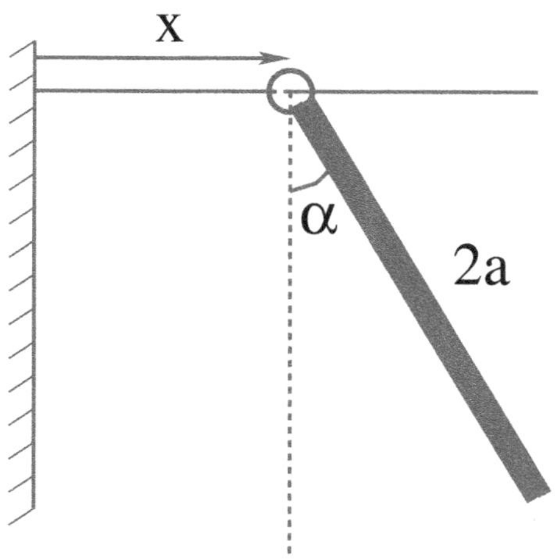

(a) Identify the generalised coordinates used for expressing the configuration of the system.

(b)If in time t the ring moves distance x and the rod acquires a swing angle α, then show that the Lagrangian of the system is

$$L = \frac{1}{2}M\dot{x}^2 + M(\dot{x}^2 + a^2\dot{\alpha}^2 + +2a\dot{x}\dot{\alpha}\cos\alpha) + \frac{M}{3}a^2\dot{\alpha}^2 + 2Mga\cos\alpha,$$

where g is the acceleration due to gravity.

(c)Write down Lagrange's equations of motion for the system.

(d)By solving Lagrange's equations of motion, show that for small values of the angle α the rod executes simple harmonic motion. Obtain an expression for the resulting period of oscillations.

3 Dynamics of Particles and Objects in Phase Space

3.1 PHASE SPACE

3.1.1 CANONICAL, OR CONJUGATE, OR GENERALISED MOMENTUM

For every generalised coordinate q_i we define a *generalised momentum* p_i as

$$p_i = \frac{\partial L}{\partial \dot{q}_i}, \tag{3.1}$$

where $\dot{q}_i$ is the generalised velocity and $L = L(q, \dot{q})$ is the Lagrangian function. As a generalised momentum is directly related to a generalised coordinate, it is also referred to as a *conjugate momentum*, or a *canonical momentum*.

The generalised momentum associated with a Cartesian coordinate is the corresponding linear momentum. The generalised momentum associated with an angular coordinate is the corresponding angular momentum. We explain these statements by considering simple examples. For a particle of mass m at x, characterised with the Lagrangian

$$L = T - V = \frac{1}{2}m\dot{x}^2 - V(x), \tag{3.2}$$

we obtain

$$p_x = \frac{\partial L}{\partial \dot{x}} = m\dot{x} = \text{mass} \times \text{velocity}, \tag{3.3}$$

which is the well-known result for linear momentum, in units of kg·m·s^{-1}. Next, consider the expression $L = \frac{ml^2}{6}\dot{\theta}^2 + mg\frac{l}{2}\cos\theta$ presented in equation (2.87) for the Lagrangian of a rotating rod of mass m and length l. The generalised momentum associated with the angular variable θ is $p_\theta = \frac{ml^2}{3}\dot{\theta}$, in units of kg·m^2·s^{-1}. It should also be noted that for a system with velocity-dependent potential, a generalised momentum will not be identical to linear momentum even if the corresponding generalised coordinate is a Cartesian coordinate.

3.1.2 PHASE SPACE AND CANONICAL VARIABLES

Canonically conjugate generalised coordinate and generalised momentum variables, $\boldsymbol{q} = \{q_1, q_2, q_3, ..., q_n\}$ and $\boldsymbol{p} = \{p_1, p_2, p_3, ..., p_n\}$, are independent variables. This is in contrast to generalised velocity variables $\dot{\boldsymbol{q}} = \{\dot{q}_1, \dot{q}_2, \dot{q}_3, ..., \dot{q}_n\}$, which being derived from generalised coordinates $\boldsymbol{q}$ are not independent variables. As mentioned before, generalised coordinates $\boldsymbol{q}$ span *configuration space*. In the same manner, generalised momenta $\boldsymbol{p}$ span *momentum space*. An extended space comprised of both configuration space and momentum space is known as **phase space**. We may talk about a *phase particle* with these coordinates in the $2n$-dimentional phase space. A particle's dynamics in momentum space can be described by assigning it phase space coordinates $\{\boldsymbol{q}, \boldsymbol{p}\} = \{q_1, q_2, q_3, ..., q_n; p_1, p_2, p_3, ..., p_n\}$. The concept of phase space is very useful in the context of classical mechanics, quantum mechanics as well as statistical mechanics.

Liouville's theorem states that phase particles move as an incompressible fluid. This implies that the phase space occupied by a set of phase particles is constant:

$$\int \mathrm{d}\Gamma = \int \mathrm{d}q_1\mathrm{d}q_2\mathrm{d}q_3...\mathrm{d}p_1\mathrm{d}p_2\mathrm{d}p_3 = \text{constant}. \tag{3.4}$$

DOI: 10.1201/9781003383314-3

In other words, the density of phase space remains constant during the motion of a dynamical system. For the proof of this theorem we refer the reader to the textbook by Gregory (2006).

As it deals with aggregates of particles, Liouville's theorem is particularly useful for applications in classical statistical mechanics.

3.1.3 LEGENDRE TRANSFORMATION

A *Legendre transformation* converts a function of one set of variables to another function of a conjugate set of variables. For example, if $f(x,y)$ is a function of variables x and y so that

$$\begin{aligned} \mathrm{d}f &= \frac{\partial f}{\partial x}\mathrm{d}x + \frac{\partial f}{\partial y}\mathrm{d}y, \\ &= u\mathrm{d}x + v\mathrm{d}y, \end{aligned} \tag{3.5}$$

then we can convert it to another function $g = g(u,y)$ of variables u and y using the Legendre transformation

$$g = f - ux. \tag{3.6}$$

We can show that the transformed function g is a function of the variables u and y. Rewriting equation (3.6) as $g = f - ux = f(x,y) - ux$, we can express

$$\begin{aligned} \mathrm{d}g &= \mathrm{d}f - u\mathrm{d}x - x\mathrm{d}u, \\ &= -x\mathrm{d}u + v\mathrm{d}y, \end{aligned}$$

which confirms that g is function of u and y. Note that the coefficient $u = \frac{\partial f}{\partial x}$ has become a variable at the cost of x.

Considering multi-component canonically conjugate variables $\boldsymbol{q} = \{q_1, q_2, q_3, \ldots, q_n\}$ and $\boldsymbol{p} \equiv \nabla_{\dot{\boldsymbol{q}}} f = \{p_1, p_2, p_3, \ldots, p_n\}$, we can write the Legendre transformation in a generalised form

$$\begin{aligned} g(\boldsymbol{q},\boldsymbol{p}) &= f(\boldsymbol{q},\dot{\boldsymbol{q}}) - \boldsymbol{p}\cdot\dot{\boldsymbol{q}}, \\ &= f(\boldsymbol{q},\dot{\boldsymbol{q}}) - \sum_{i=1}^{n} p_i \dot{q}_i. \end{aligned} \tag{3.7}$$

3.2 HAMILTONIAN MECHANICS

3.2.1 MODIFIED HAMILTON'S VARIATIONAL PRINCIPLE

As described in section 2.3.6, Hamilton's variational principle seeks to obtain the extremum condition for the action functional J by evaluating the time integral of the Lagrangian function over a path in a n-dimensional configuration space $\boldsymbol{q} = \{q_1, q_2, q_3, \ldots, q_n\}$. *Modified Hamilton's variational principle* considers the evaluation of the action functional over a trajectory in phase space $\{\boldsymbol{q},\boldsymbol{p}\} = \{q_1, q_2, q_3, \ldots, q_n; p_1, p_2, p_3, \ldots, p_n\}$. This is done by considering in section 3.1.3 the function $f(y,x,t)$ as the Lagrangian $L(\boldsymbol{q},\dot{\boldsymbol{q}},t)$ and the function $g(y,u,t)$ as $-H(\boldsymbol{q},\boldsymbol{p},t)$ leading to the following Legendre transformation

$$L(\boldsymbol{q},\dot{\boldsymbol{q}},t) = \boldsymbol{p}\dot{\boldsymbol{q}} - H(\boldsymbol{q},\boldsymbol{p},t). \tag{3.8}$$

The function $H(\boldsymbol{q},\boldsymbol{p},t)$ is the Hamilton's function, or simply the Hamiltonian. The modified Hamilton's variational principle thus reads

$$\delta J = \delta\left[\sum_{i=1}^{n} p_i \dot{q}_i - H(\boldsymbol{q},\boldsymbol{p},t)\right]\mathrm{d}t = 0. \tag{3.9}$$

3.2.2 PHYSICAL SIGNIFICANCE OF HAMILTON'S FUNCTION

We first note that we cannot assign any meaningful significance to Lagrange's function, as it is defined as potential energy taken away from kinetic energy: $L = T - V$. Hamilton's function, on the other hand, represents the total energy of the dynamical system under consideration[1]: $H = T + V$. This can be easily proven by considering the simple example of a single particle of mass m at coordinate x possessing linear momentum p_x. Using equation (3.8) we have

$$\begin{aligned} H = p_x\dot{x} - L &= m\dot{x}^2 - \frac{1}{2}m\dot{x}^2 + V(x) \\ &= \frac{1}{2}m\dot{x}^2 + V(x) \\ &= \frac{p_x^2}{2m} + V(x) \\ &= \text{total energy.} \end{aligned}$$

Having established the physical significance of the Hamiltonian of a dynamical system, it is very important to point out that an expression for H must not contain velocity terms such as $\dot{x}$ but only phase space variables such as x and p_x. In other words, the expression $H = \frac{1}{2}m\dot{x}^2 + V(x)$ is inadmissible and should correctly be written as $H = \frac{p_x^2}{2m} + V(x)$.

3.2.3 HAMILTON'S EQUATIONS OF MOTION

Let us write equation (3.9) as

$$\delta J = \delta \int_{t_1}^{t_2} f(\boldsymbol{q}, \dot{\boldsymbol{q}}, \boldsymbol{p}, \dot{\boldsymbol{p}}, t)\mathrm{d}t = 0. \tag{3.10}$$

Then, following the calculus of variation, this leads to the following $2n$ Euler-Lagrange equations:

$$\frac{\mathrm{d}}{\mathrm{d}t}\left(\frac{\partial f}{\partial \dot{q}_i}\right) - \frac{\partial f}{\partial q_i} = 0, \quad i = 1, 2, 3, ..., n, \tag{3.11}$$

$$\frac{\mathrm{d}}{\mathrm{d}t}\left(\frac{\partial f}{\partial \dot{p}_i}\right) - \frac{\partial f}{\partial p_i} = 0, \quad i = 1, 2, 3, ..., n, \tag{3.12}$$

As $f = \boldsymbol{p}\dot{\boldsymbol{q}} - H(\boldsymbol{q}, \boldsymbol{p}, t)$, we have

$$\begin{aligned} \frac{\partial f}{\partial q_i} &= -\frac{\partial H}{\partial q_i}, \quad \frac{\partial f}{\partial \dot{q}_i} = p_i, \\ \frac{\partial f}{\partial p_i} &= \dot{q}_i - \frac{\partial H}{\partial p_i}, \quad \frac{\partial f}{\partial \dot{p}_i} = 0. \end{aligned} \tag{3.13}$$

With these, equations (3.11) and (3.12) read

$$\dot{p}_i + \frac{\partial H}{\partial q_i} = 0, \quad i = 1, 2, 3, ..., n, \tag{3.14}$$

$$\dot{q}_i - \frac{\partial H}{\partial p_i} = 0, \quad i = 1, 2, 3, ..., n. \tag{3.15}$$

These are Hamilton's equations of motion.

[1]This is true for a closed system, which we implicitly assume in our discussion.

3.2.4 SIMILARITY BETWEEN HAMILTON'S EQUATIONS AND NEWTON'S EQUATIONS OF MOTION

Let us write Newton's second law of motion of a particle of mass m at a Cartesian position $\boldsymbol{r} = (r_1, r_2, r_3) = (x, y, z)$ subjected to a force F as

$$m\ddot{r}_i = F_i, \quad i = 1, 2, 3. \tag{3.16}$$

Expressing $m\dot{r}_i$ as the linear momentum component p_i and the force component as the negative of a potential function $V(\boldsymbol{r})$, *viz* $F_i = -\frac{\partial V}{\partial r_i}$, we can rewrite equation (3.16) as

$$\dot{p}_i = -\frac{\partial V}{\partial r_i}. \tag{3.17}$$

Now remembering that the Hamiltonian of the particle in the phase space $(\boldsymbol{r}, \boldsymbol{p})$ is $H = T(\boldsymbol{p}) + V(\boldsymbol{r})$, we express the Hamiton's equation in (3.14) as

$$\dot{p}_i = -\frac{\partial H}{\partial r_i} = -\frac{\partial V}{\partial r_i}. \tag{3.18}$$

Clearly, equations (3.17) and (3.18) are identical. This simple example indicates that Hamilton's equations of motion are essentially Newton's equations of motion, but with a much greater generality.

3.2.5 PROCEDURE FOR CONSTRUCTING HAMILTONIAN

There are five steps for constructing the Hamiltonian of a dynamical system.

Step 1: Choose a set of generalised coordinates $\boldsymbol{q} \equiv \{q_1, q_2, q_3, ..., q_n\}$ and construct the Lagrangian $L(\boldsymbol{q}, \dot{\boldsymbol{q}}, t)$ of the system.

Step 2: Define generalised momenta $\boldsymbol{p} \equiv \{p_1, p_2, p_3, ..., p_n\}$ corresponding to the generalised coordinates.

Step 3: Define the Hamiltonian $H(\boldsymbol{q}, \boldsymbol{p}, t) = \sum_{i=1}^{n} \dot{q}_i p_i - L(\boldsymbol{q}, \dot{\boldsymbol{q}}, t)$. Note that the Hamiltonian expressed at this stage is inadmissible, as it contains the velocity terms $\dot{\boldsymbol{q}}$.

Step 4: Invert the relations $p_i = \frac{\partial L}{\partial \dot{q}_i}$ to obtain $\dot{q}_i = \dot{q}_i(\boldsymbol{q}, \boldsymbol{p}, t)$ for $i = 1, 2, 3, ..., n$.

Step 5: Use the information from step 4 to eliminate $\dot{\boldsymbol{q}}$ from the expression at step 3, so that the Hamiltonian is expressed as $H = H(\boldsymbol{q}, \boldsymbol{p}, t)$.

The resulting form of the Hamiltonian, *viz* . $H = H(\boldsymbol{q}, \boldsymbol{p}, t)$, can be used to obtain Hamilton's equations of motion [*cf* equations (3.14) and (3.15)].

3.2.6 EXAMPLES OF THE APPLICATION OF HAMILTONIAN MECHANICS

3.2.6.1 Example 1:

Motion of a particle hanging from a vertical spring.

This is the same system as shown in figure 2.2 and discussed as the first example using the Newtonian mechanics and the Lagrangian mechanics. Let us set the zero of the potential energy of the particle at its equilibrium (shown by dotted line in figure 2.2). We will go through the five steps listed in the previous section to derive the equation of motion using the Hamiltonian mechanics.

Step 1: Distance x below the equilibrium is a suitable generalised coordinate. Taking the spring constant as k, the kinetic and potential energy expressions are $T=\frac{1}{2}m\dot{x}^2$ and $V=\frac{1}{2}x^2$, as previously presented in equation (2.81). The Lagrangian is thus $L=\frac{1}{2}m\dot{x}^2-\frac{1}{2}kx^2$.

Step 2: The generalised momentum is $p=\frac{\mathrm{d}L}{\mathrm{d}\dot{x}}=m\dot{x}$.

Step 3: Define the Hamiltonian as $H=p\dot{x}-L$.

Step 4: From step 2, we obtain $\dot{x}=\frac{p}{m}$.

Step 5: Using steps 3 and 4, we express the Hamiltonian using the space phase variables x and p as

$$\begin{aligned} H &= p\dot{x}-L, \\ &= \frac{1}{2}m\dot{x}^2+\frac{1}{2}kx^2, \\ &= \frac{p^2}{2m}+\frac{1}{2}kx^2. \end{aligned} \tag{3.19}$$

Hamilton's equations of motion in (3.14) and (3.15) now read

$$\dot{p}=-\frac{\mathrm{d}H}{\mathrm{d}x}=-kx, \tag{3.20}$$

and

$$\dot{x}=\frac{\mathrm{d}H}{\mathrm{d}p}=\frac{p}{m}. \tag{3.21}$$

Equation (3.21) defines the momentum-velocity relationship. And using equations (3.20) and (3.21) we obtain

$$\frac{\mathrm{d}}{\mathrm{d}t}(m\dot{x})=-kx, \tag{3.22}$$

or

$$m\ddot{x}+kx=0, \tag{3.23}$$

which is identical to equations (2.36) and (2.82) and represents the simple harmonic motion of the particle about its equilibrium.

3.2.6.2 Example 2:

Consider the system, called Atwood's machine, shown in figure 3.1. It consists of two masses, m_1 and m_2, attached to the two ends of a light inextensible string which hangs over a frictionless pulley of radius a. The two masses are set in motion by displacing and releasing one of these vertically downwards. The pulley is free to rotate about its axis.

At an instant of time, the masses m_1 and m_2 are, respectively, distances x and $l-x$ below the horizontal line through the centre of the pulley. The total length of the string is $l+\pi a$, with both l and a being constants. Clearly, there is a single coordinate variable x to govern the motion of the system.

The kinetic energy of the system is

$$T=\frac{1}{2}\left(m_1+m_2+\frac{I}{a^2}\right)\dot{x}^2, \tag{3.24}$$

where I is the moment of inertia of the pulley. The potential energy of the dynamical system, which should be considered with reference to the dotted line, is

$$V=-m_1gx-m_2g(l-x), \tag{3.25}$$

where g is acceleration due to gravity.

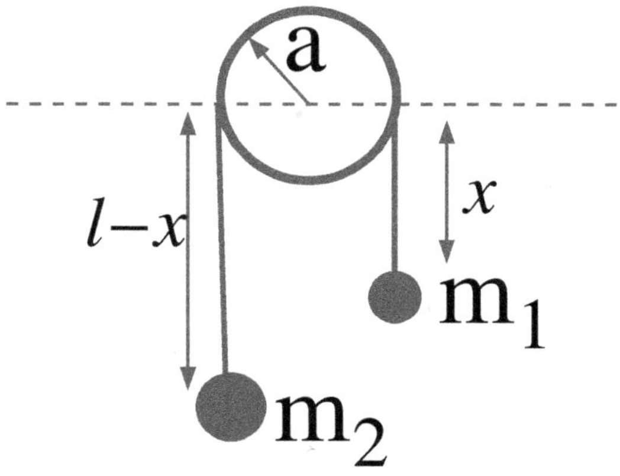

Figure 3.1 Atwood's machine, consisting of two masses m_1 and m_2 attached to the two ends of a light inextensible string which hangs over a frictionless pulley of radius a. The total length of the string is $l + \pi a$.

The conjugate momentum is

$$p = \frac{\partial L}{\partial \dot{x}} = \frac{\partial (T-V)}{\partial \dot{x}} = \left(m_1 + m_2 + \frac{I}{a^2}\right)\dot{x}. \tag{3.26}$$

From this the velocity can be expressed as

$$\dot{x} = \frac{p}{\left(m_1 + m_2 + \frac{I}{a^2}\right)}. \tag{3.27}$$

Using $H = p\dot{x} - L$ and equation (3.27), the admissible form of the Hamiltonian becomes

$$H = \frac{p^2}{2\left(m_1 + m_2 + \frac{I}{a^2}\right)} - m_1 g x - m_2 g (l - x). \tag{3.28}$$

Hamilton's equations of motion are

$$\dot{x} = \frac{\partial H}{\partial p} = \frac{p}{\left(m_1 + m_2 + \frac{I}{a^2}\right)}, \tag{3.29}$$

and

$$\dot{p} = -\frac{\partial H}{\partial x} = (m_1 - m_2) g. \tag{3.30}$$

Using equations (3.29) and (3.30), the acceleration of the system is expressed as

$$\ddot{x} = \frac{(m_1 - m_2) g}{\left(m_1 + m_2 + \frac{I}{a^2}\right)}. \tag{3.31}$$

For a massless pulley, the expression for the acceleration becomes

$$\ddot{x} = \frac{(m_1 - m_2) g}{(m_1 + m_2)}, \tag{3.32}$$

a result that can be obtained using elementary considerations. If $m_1 > m_2$, mass m_1 descends with constant acceleration. If $m_1 < m_2$, mass m_1 ascends with constant acceleration.

3.2.6.3 Example 3:

Consider a particle of mass m performing harmonic oscillation in the x-y plane. In terms of the Cartesian coordinates, the kinetic and potential energies of this two-dimensional harmonic oscillator are

$$T = \frac{1}{2}m(\dot{x}^2 + \dot{y}^2), \tag{3.33}$$

and

$$V = \frac{1}{2}k_1 x^2 + \frac{1}{2}k_2 y^2, \tag{3.34}$$

where k_1 and k_2 are force constants. Let us change the variables x and y to the plane polar coordinates ρ and ϕ, as described in section A.1.2.1 of Appendix A. The Lagrangian $L = T - V$ can then be expressed as

$$L = \frac{1}{2}m(\dot{\rho}^2 + \rho^2\dot{\phi}^2) - V(\rho, \phi). \tag{3.35}$$

The momenta conjugate to ρ and ϕ are

$$\begin{aligned} p_\rho &= \frac{\partial L}{\partial \rho} = m\dot{\rho}, \\ p_\phi &= \frac{\partial L}{\partial \phi} = m\rho^2\dot{\phi}. \end{aligned} \tag{3.36}$$

Obtaining $\dot{\rho}$ and $\dot{\phi}$ from these equations, we express the Hamiltonian as

$$\begin{aligned} H &= p_\rho\dot{\rho} + p_\varphi\dot{\phi} - L, \\ &= \frac{p_\rho^2}{2m} + \frac{p_\phi^2}{2m\rho^2} + V(\rho, \phi). \end{aligned} \tag{3.37}$$

Hamilton's equations of motion are

$$\dot{\rho} = \frac{\partial H}{\partial p_\rho} = \frac{p_\rho}{m}, \tag{3.38}$$

$$\dot{\phi} = \frac{\partial H}{\partial p_\phi} = \frac{p_\varphi}{m\rho^2}, \tag{3.39}$$

$$\dot{p}_\rho = -\frac{\partial H}{\partial \rho} = -\frac{\partial V}{\partial \rho} + \frac{p_\phi^2}{m\rho^3}, \tag{3.40}$$

$$\dot{p}_\phi = -\frac{\partial H}{\partial \phi} = -\frac{\partial V}{\partial \phi}. \tag{3.41}$$

From equations (3.38) and (3.40), the radial acceleration is

$$\ddot{\rho} = \frac{\dot{p}_\rho}{m} = -\frac{1}{m}\frac{\partial V}{\partial \rho} + \frac{p_\phi^2}{m^2\rho^3}, \tag{3.42}$$

and, from (3.39) and (3.41), the angular acceleration is

$$\ddot{\phi} = \frac{\dot{p}_\phi}{m\rho^2} - \frac{2p_\phi\dot{\rho}}{m\rho^3} = -\frac{1}{m\rho^2}\frac{\partial V}{\partial \phi} - \frac{2p_\phi p_\rho}{m\rho^3}. \tag{3.43}$$

For *isotropic harmonic oscillator* $k_1 = k_2 = k$, reducing the potential energy as a function of the single variable ρ, *viz* $V = V(\rho) = \frac{1}{2}k\rho^2$, and the equations of motion are

$$\ddot{\rho} = -\frac{k}{m}\rho + \frac{p_\phi^2}{m^2\rho^3}, \tag{3.44}$$

and

$$\ddot{\phi} = -\frac{2p_\varphi p_\rho}{m\rho^3}. \tag{3.45}$$

3.2.6.4 Example 4:

Let us now consider a three-dimensional harmonic oscillator of mass m. Using Cartesian coordinates, the kinetic and potential energies of this oscillator are

$$T = \frac{1}{2}m(\dot{x}^2 + \dot{y}^2 + \dot{z}^2), \tag{3.46}$$

and

$$V = \frac{1}{2}k_1x^2 + \frac{1}{2}k_2y^2 + \frac{1}{2}k_3z^2, \tag{3.47}$$

where k_1, k_2 and k_3 are force constants. Employing the spherical polar coordinates described in section A.1.2.3 of Appendix A, the Lagrangian of the system can be expressed as

$$L = \frac{1}{2}m\big(\dot{r}^2 + r^2\dot{\theta}^2 + r^2\dot{\phi}^2\sin^2\theta\big) - V(r,\theta,\phi). \tag{3.48}$$

The generalised momenta are

$$\begin{aligned} p_r &= \frac{\partial L}{\partial \dot{r}} = m\dot{r}, \\ p_\theta &= \frac{\partial L}{\partial \dot{\theta}} = mr^2\dot{\theta}, \\ p_\phi &= \frac{\partial L}{\partial \dot{\phi}} = mr^2\dot{\varphi}\sin^2\theta. \end{aligned} \tag{3.49}$$

Using these, we can express the Hamiltonian as

$$\begin{aligned} H &= p_r\dot{r} + p_\theta\dot{\theta} + p_\phi\dot{\phi} - L, \\ &= \frac{p_r^2}{2m} + \frac{p_\theta^2}{2mr^2} + \frac{p_\phi^2}{2mr^2\sin^2\theta} + V(r,\theta,\phi). \end{aligned} \tag{3.50}$$

Hamilton's equations of motion are

$$\begin{aligned} \dot{r} &= \frac{\partial H}{\partial p_r} = \frac{p_r}{m}, \\ \dot{\theta} &= \frac{\partial H}{\partial p_\theta} = \frac{p_\theta}{mr^2}, \\ \dot{\phi} &= \frac{\partial H}{\partial p_\phi} = \frac{p_\phi}{mr^2\sin^2\theta}, \\ \dot{p}_r &= -\frac{\partial H}{\partial r} = \frac{1}{mr^3}\big(p_\theta^2 + \frac{p_\phi^2}{\sin^2\theta}\big) - \frac{\partial V}{\partial r}, \\ \dot{p}_\theta &= -\frac{\partial H}{\partial \theta} = \frac{p_\theta^2\cos\theta}{mr^2\sin^3\theta} - \frac{\partial V}{\partial \theta}, \\ \dot{p}_\phi &= -\frac{\partial H}{\partial \phi} = -\frac{\partial V}{\partial \phi}. \end{aligned} \tag{3.51}$$

With information available for $V(r,\vartheta,\varphi)$, these equations can be used to solve the equation of motion of the oscillator.

In the particular case when $V = V(r,\theta)$, we immediately find that $\dot{p}_\phi = 0$, *i.e.* $p_\phi =$ constant. Further, if V is also independent of θ, and if p_ϕ vanishes, then $\dot{p}_\theta = 0$, *i.e.* $p_\theta =$ constant. Under these restrictions,

$$\dot{p}_r = \frac{\text{constant}}{r^3} - \frac{\partial V}{\partial r}, \tag{3.52}$$

which suggests that the force acting on the oscillator is radial.

3.2.6.5 Example 5:

A particle sliding down a frictionless moveable inclined plane.

Figure 3.2(a) shows a sketch of the system. Consider that the particle starts sliding from the top of the plane. Let α be the incline of the plane, and m and M be masses of the particle and the plane, respectively. There are two generalised coordinates (or, degrees of freedom): distance x moved by the plane and distance x' travelled by the particle, as shown in the figure.

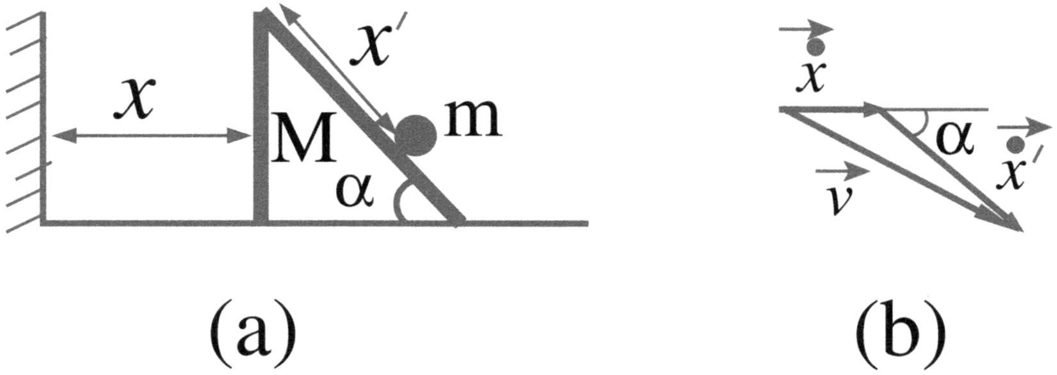

Figure 3.2 (a) A particle of mass m sliding down a moveable inclined plane of a block of mass M. The incline angle of the plane is α. At an instance the block moves by distance x and the particle slides a distance x'. (b) Velocity diagram of the motion of the system.

Using the velocity diagram in figure 3.2(b), we express the square of the speed of the particle as

$$v^2 = \dot{x}^2 + \dot{x}'^2 + 2\dot{x}\dot{x}'\cos\alpha. \tag{3.53}$$

Setting zero of the potential energy of the system at the top of the incline when the particle starts to slide, we write the expressions for the kinetic and potential energies of the system as

$$T = \frac{1}{2}mv^2 + \frac{1}{2}M\dot{x}^2, \tag{3.54}$$

$$V = -mgx'\sin\alpha. \tag{3.55}$$

The Hamiltonian, using equation (3.8), is

$$H = p_x\dot{x} + p_{x'}\dot{x}' - \frac{1}{2}m(\dot{x}^2 + \dot{x}'^2 + 2\dot{x}\dot{x}'\cos\alpha) - \frac{1}{2}M\dot{x}^2 - mgx'\sin\alpha. \tag{3.56}$$

The admissible expression for the Hamiltonian, *viz*. $H = H(x, x', p_x, p_{x'})$, can be obtained by replacing the velocity terms $\dot{x}$ and $\dot{x}'$ in terms of the momenta variables p_x and $p_{x'}$. Using the expressions for the canonical momentum variables, *viz* .

$$p_x = \frac{\partial L}{\partial \dot{x}} = (m+M)\dot{x} + m\dot{x}' \cos\alpha,$$

and

$$p_{x'} = \frac{\partial L}{\partial \dot{x}'} = m(\dot{x}\cos\alpha + \dot{x}'),$$

we obtain

$$\dot{x} = \frac{p_x - p_x' \cos\alpha}{M + m\sin^2\alpha} \tag{3.57}$$

and

$$\dot{x}' = \frac{p_x}{m\cos\alpha} - \frac{(m+M)}{m\cos\alpha}\left(\frac{p_x - p_{x'}\cos\alpha}{M + m\sin^2\alpha}\right). \tag{3.58}$$

The Hamilton's equations of motion are

$$\dot{x} = \frac{\partial H}{\partial p_x}, \qquad \dot{x}' = \frac{\partial H}{\partial p_x'}, \tag{3.59}$$

and

$$\dot{p}_x = -\frac{\partial H}{\partial x}, \quad \dot{p}_{x'} = -\frac{\partial H}{\partial x'}. \tag{3.60}$$

We leave details of the algebra as an exercise and simply write down the expressions describing the accelerations of the particle and the slab as follows

$$\ddot{x} = \frac{-g\cos\alpha\sin\alpha}{\frac{m+M}{m} - \cos^2\alpha}, \tag{3.61}$$

and

$$\ddot{x}' = \frac{g\sin\alpha}{1 - \frac{m}{m+M}\cos^2\alpha}. \tag{3.62}$$

The results in equations (3.61) and (3.62) can be easily verified by using the Lagrange's equations of motion

$$\frac{\mathrm{d}}{\mathrm{d}t}\frac{\partial L}{\partial \dot{x}} = \frac{\partial L}{\partial x} \quad \text{and} \quad \frac{\mathrm{d}}{\mathrm{d}t}\frac{\partial L}{\partial \dot{x}'} = \frac{\partial L}{\partial x'}.$$

3.2.7 CYCLIC, OR IGNORABLE, COORDINATES AND CONSERVATION THEOREMS

A *cyclic coordinate* is one that does not appear explicitly in the expression for the Lagrangian and Hamiltonian of a dynamical system. If q_i is a cyclic coordinate, then

$$\dot{p}_i = \frac{\mathrm{d}}{\mathrm{d}t}\frac{\partial L}{\partial \dot{q}_i} = \frac{\partial L}{\partial q_i} = 0, \tag{3.63}$$

or

$$\dot{p}_i = -\frac{\partial H}{\partial q_i} = 0, \tag{3.64}$$

so that $p_i = c_i =$ constant. We conclude that the momentum conjugate to a cyclic, or ignorable, coordinate is conserved.

3.2.8 CONCEPT OF Δ- VARIATION

So far, when discussing the Hamilton's formulations in configuration space and phase space, we have used the concept of δ-variation. There is another variational principle associated with the Hamiltonian formulation, involving a new type of variation, called the Δ-variation. These two types of variation can be explained both qualitatively as well as mathematically.

The δ-variation: In the δ-variation a varied path in configuration space terminates at the start and end points, at times t_1 and t_2, as the correct path. In other words, for a generalised coordinate q, it is required that $\delta q(t_1) = \delta q(t_2) = 0$.

The Δ-variation: In this type of variation, the constraints of the δ-variation are relaxed, allowing the possibility of varied paths to start and finish at different points and different times.

To derive a mathematical relationship between the δ and Δ variations, let us consider a path passing through a generalised coordinate q which is defined as $q = q(t, \alpha)$, where $\alpha = \alpha_0$ is the correct path. The δ and Δ variations can be expressed as

$$\delta - \text{variation}: \quad \delta q = \left(\frac{\partial q}{\partial \alpha}\right)_{\alpha_0} \mathrm{d}\alpha, \tag{3.65}$$

$$\begin{aligned} \Delta - \text{variation}: \quad \Delta q &= \left(\frac{\mathrm{d}q}{\mathrm{d}\alpha}\right)_{\alpha_0} \mathrm{d}\alpha, \\ &= \left(\frac{\partial q}{\partial \alpha}|_{\alpha_0} + \dot{q}\frac{\mathrm{d}t}{\mathrm{d}\alpha}\right)\mathrm{d}\alpha, \\ &= \delta q + \dot{q}\Delta t. \end{aligned} \tag{3.66}$$

The Δ-variation of a function $f(q,t)$ in configuration space can thus be expressed as

$$\Delta f = \delta f + \frac{\mathrm{d}f}{\mathrm{d}t}\Delta t. \tag{3.67}$$

Stated differently, the Δ and δ operations are related as

$$\Delta = \delta + \Delta t \frac{\mathrm{d}}{\mathrm{d}t}. \tag{3.68}$$

3.2.9 HAMILTON'S CHARACTERISTIC FUNCTION AND PRINCIPLE OF LEAST ACTION

Using equation (3.67), the Δ-variation of the action integral J between times t_1 and t_2 is expressed as

$$\begin{aligned} \Delta J &= \delta J + \frac{\mathrm{d}J}{\mathrm{d}t}\Delta t, \\ &= \delta J + L\Delta t|_{t1_1}^{t_2}, \end{aligned} \tag{3.69}$$

where $\dot{J} = L$ (Lagrangian) has been used. Let us consider a single generalised coordinate q. The first term on the right-hand side can be expressed as

$$\begin{aligned} \delta J &= \delta \int_{t_1}^{t_2} L\mathrm{d}t = \int_{t_1}^{t_2} \delta L \mathrm{d}t, \\ &= \int_{t_1}^{t_2} \left(\frac{\partial L}{\partial q}\delta q + \frac{\partial L}{\partial \dot{q}}\delta \dot{q}\right)\mathrm{d}t, \\ &= \int_{t_1}^{t_2} \left\{\frac{\mathrm{d}}{\mathrm{d}t}\left(\frac{\partial L}{\partial \dot{q}}\right)\delta q + \frac{\partial L}{\partial \dot{q}}\frac{\mathrm{d}}{\mathrm{d}t}(\delta q)\right\}\mathrm{d}t, \\ &= \int_{t_1}^{t_2} \frac{\mathrm{d}}{\mathrm{d}t}\left(\frac{\partial L}{\partial \dot{q}}\delta q\right)\mathrm{d}t, \\ &= \int_{t_1}^{t_2} \frac{\mathrm{d}}{\mathrm{d}t}\left(p\delta q\right)\mathrm{d}t, \end{aligned}$$

$$
\begin{aligned}
&= p\delta q|_{t_1}^{t_2}, \\
&= p\Delta q|_{t_1}^{t_2} - p\dot{q}\Delta t|_{t_1}^{t_2}.
\end{aligned} \tag{3.70}
$$

Using equations (3.69) and (3.70), the *principle of least action* can be expressed as

$$
\Delta J = \Big[p\Delta q - H\Delta t\Big]_{t_1}^{t_2} = 0, \tag{3.71}
$$

where the Hamiltonian H is the energy of the system. From this principle we immediately obtain

$$
[p\Delta q]_{t_1}^{t_2} = [H\Delta t]_{t_1}^{t_2}. \tag{3.72}
$$

As p and q are canonically conjugate variables, this result points out that energy and time are also canonically conjugate variables for conservative systems.

A few simpler forms of the principle of least action can be obtained. Let us first consider the following restrictions. (1) Consider systems for which H is conserved, *i.e.* L and H are not explicit functions of time, (2) the Δ variation is such that H is conserved on the actual path as well as varied paths and (3) $\Delta q = 0$ at the end points of journey (but $\Delta t \neq 0$). Using restriction (3), ΔJ in equation (3.71) reduces to

$$
\Delta J = -H(\Delta t_2 - \Delta t_1). \tag{3.73}
$$

Also, under restrictions (1) and (2),

$$
\begin{aligned}
\Delta J &= \Delta \int_{t_1}^{t_2} [p\dot{q} - H]\mathrm{d}t, \\
&= \Delta \int_{t_1}^{t_2} p\dot{q}\mathrm{d}t - H(\Delta t_2 - \Delta t_1).
\end{aligned} \tag{3.74}
$$

From a comparison of equations (3.73) and (3.74), we can express the principle of least action in the form

$$
\Delta W = 0, \tag{3.75}
$$

where

$$
W = \int_{t_1}^{t_2} p\dot{q}\mathrm{d}t \tag{3.76}
$$

is **Hamilton's characteristic function.**

Let us now add another restriction. If the potential energy V is not velocity dependent and generalised coordinates do not involve time explicitly, then it can be shown that the kinetic energy can be expressed as

$$
T = \frac{1}{2}\sum_{jk} a_{jk}\dot{q}_j\dot{q}_k, \tag{3.77}
$$

where a_{jk} is a symmetric coefficient. The Lagrangian is

$$
L = \frac{1}{2}\sum_{jk} a_{jk}\dot{q}_j\dot{q}_k - V(\boldsymbol{q}). \tag{3.78}
$$

From this a canonical momentum component is

$$
p_l = \frac{\partial L}{\partial \dot{q}_l} = \frac{\partial T}{\partial \dot{q}_l} = \sum_j a_{jl}\dot{q}_j, \tag{3.79}
$$

as a_{ij} is symmetric. From this we obtain

$$
\sum_i p_i\dot{q}_i = \sum_{ij} a_{ij}\dot{q}_i\dot{q}_j = 2T. \tag{3.80}
$$

With all these considerations, the principle of least action becomes

$$\Delta \int_{t_1}^{t_2} T \mathrm{d}t = 0 \quad \text{for conservative systems.} \tag{3.81}$$

If, further, the kinetic energy T is conserved along the paths (restriction 3), then the principle of least action takes the simple form

$$\Delta(t_2 - t_1) = 0. \tag{3.82}$$

This equation states that, if the energy is conserved, then out of all possible paths between two points, the system moves along the particular path for which the transit time is the least (more strictly, an extremum). In geometrical optics this is Fermat's principle: between two points, light ray travels along a path that takes the least amount of time.

3.2.10 LAGRANGE BRACKET, POISSON BRACKET AND JACOBI'S IDENTITY

It is convenient to introduce an abbreviated notation when expressing the total time rate of change of a dynamical function. Two *bracket notations* have been introduced.

Lagrange bracket: Of historical interest is the *Lagrange bracket of two functions* u and v with respect to phase space conjugate variables $\boldsymbol{q}$ and $\boldsymbol{p}$, defined as

$$\{u,v\} = \sum_i \left(\frac{\partial q_i}{\partial u} \frac{\partial p_i}{\partial v} - \frac{\partial p_i}{\partial u} \frac{\partial q_i}{\partial v} \right). \tag{3.83}$$

Poisson bracket: A very useful notation is the *Poisson bracket of two dynamical variables* $X = X(\boldsymbol{q},\boldsymbol{p},t)$ and $Y = Y(\boldsymbol{q},\boldsymbol{p},t)$, defined as

$$[X,Y]_{\boldsymbol{q},\boldsymbol{p}} = \sum_i \left(\frac{\partial X}{\partial q_i} \frac{\partial Y}{\partial p_i} - \frac{\partial X}{\partial p_i} \frac{\partial Y}{\partial q_i} \right). \tag{3.84}$$

In a certain sense, the Lagrange bracket may be considered as the inverse of the Poisson bracket. Consider a dynamical variable $F = F(\boldsymbol{q},\boldsymbol{p},t)$. The total time derivative of F can be written as

$$\begin{aligned} \frac{\mathrm{d}F}{\mathrm{d}t} &= \sum_i \frac{\partial F}{\partial q_i} \dot{q}_i + \sum_i \frac{\partial F}{\partial p_i} \dot{p}_i + \frac{\partial F}{\partial t}, \\ &= \sum_i \left(\frac{\partial F}{\partial q_i} \frac{\partial H}{\partial p_i} - \frac{\partial F}{\partial p_i} \frac{\partial H}{\partial q_i} \right) + \frac{\partial F}{\partial t}, \\ &= [F,H] + \frac{\partial F}{\partial t}. \end{aligned} \tag{3.85}$$

Thus, the equation of motion of the variable F can be expressed using the Poisson bracket notation. The first term on the right-hand side (the Poisson bracket of F and H) is the implicit variation of F with respect to time: it shows the change in F due to the motion of the phase point at which F is evaluated. The second term on the right-hand side expresses the exploit time variation of F.

If F is a constant of motion, then $\frac{\mathrm{d}F}{\mathrm{d}t} = 0$ and $\frac{\partial F}{\partial t} = 0$, and thus $[F,H] = 0$. In other words, the Poisson bracket of a constant of motion with the Hamiltonian of the system is zero.

It is useful to note some identities related to Poisson's bracket.

1. Anti-symmetry:

$$[X,Y] = -[Y,X], \tag{3.86}$$

from which it follows that $[X,X] = 0$ and $[X,c] = 0$, where c is a constant.

2. Linearity: For constants a and b

$$[aX + bY, Z] = a[X,Z] + b[Y,Z]. \tag{3.87}$$

3. Associative nature:

$$[X,YZ] = Y[X,Z]+Z[X,Y], \quad (3.88)$$

$$\frac{\partial}{\partial t}[X,Y] = [\frac{\partial X}{\partial t}]+[X,\frac{\partial Y}{\partial t}]. \quad (3.89)$$

4. *Jacobi's identity*: The sum of the cyclic permutations of the double Poisson bracket of three dynamical variables is zero.

$$[X,[Y,Z]]+[Y,[Z,X]]+[Z,[X,Y]]=0. \quad (3.90)$$

The first three identities are easy to prove. The proof of the Jacobi identity is lengthy and will not be presented here. A consequence of Jacobi's identity can be identified. Let one of the three dynamical variables be taken as the Hamiltonian of a system, say $Z=H$. Then

$$[X,[Y,H]]+[Y,[H,X]]+[H,[X,Y]]=0. \quad (3.91)$$

If the other two variables, *viz* X and Y, are both constants of motion, *i.e.* $[X,H]=0=[Y,H]$, then equation (3.91) becomes

$$[H,[X,Y]]=0, \quad (3.92)$$

i.e. the dynamical variable $[X,Y]$ is also a constant of motion. This is *Poisson's theorem*: the Poisson bracket of any two constants of motion is also a constant of motion. This may be useful in constructing new constants of motion from known ones.

Specially useful Poisson brackets:

(1) *Fundamental, or basic, Poisson brackets*: An important, fundamental, or basic Poisson bracket is that between a generalised coordinate and its conjugate momentum.

$$\begin{aligned}[q_j,p_k] &= \sum_i \frac{\partial q_j}{\partial q_i}\frac{\partial p_k}{\partial p_i} - \sum_i \frac{\partial q_j}{\partial p_i}\frac{\partial p_k}{\partial q_i}, \\ &= \sum_i \delta_{ij}\delta_{ki} - 0, \\ &= \delta_{jk}, \\ &= \begin{cases} 1 & \text{when } j=k, \\ 0 & \text{when } j\neq k. \end{cases}\end{aligned} \quad (3.93)$$

(2) *Hamilton's equations of motion in Poisson bracket form*: When canonical phase space variables $\boldsymbol{q}\equiv\{q_i\}$ and $\boldsymbol{p}\equiv\{p_i\}$ do not explicitly depend on time, *i.e.* $\frac{\partial q_i}{\partial t}=0=\frac{\partial p_i}{\partial t}$, then using equation (3.85) we obtain

$$\dot{q}_i = [q_i,H], \quad (3.94)$$

$$\dot{p}_i = [p_i,H]. \quad (3.95)$$

These are Hamilton's canonical equations of motion for q_i and p_i in Poisson bracket form.

(3) *Equation of motion of Hamiltonian*: If the dynamical variable F in equation (3.85) is taken as the Hamiltonian itself, then

$$\begin{aligned}\frac{\mathrm{d}H}{\mathrm{d}t} &= [H,H]+\frac{\partial H}{\partial t}, \\ &= \frac{\partial H}{\partial t}.\end{aligned} \quad (3.96)$$

In other words, there is no implicit time dependence of the Hamiltonian of a dynamical system.

(4) *Poisson's theorem*: Considering density of phase space, ρ, as a dynamical variable, we write

$$\frac{d\rho}{dt} = [\rho, H] + \frac{\partial \rho}{\partial t}.$$

But, as noted in equation (3.4), ρ is constant for an element of phase space. Thus,

$$\frac{\partial \rho}{\partial t} = -[\rho, H]. \tag{3.97}$$

This is a statement of Liouville's theorem (*cf* section 3.1.2) in the Poisson bracket form.

Poisson brackets associated with angular momentum:

Consider a particle at Cartesian position $\boldsymbol{q} = \boldsymbol{r} = (x, y, z)$ moving with linear momentum $\boldsymbol{p} = (p_x, p_y, p_z)$. Its angular momentum $\boldsymbol{M} = (\mathcal{M}_x, \mathcal{M}_y, \mathcal{M}_z)$ is

$$\boldsymbol{M} = \boldsymbol{r} \times \boldsymbol{p}. \tag{3.98}$$

The following Poisson bracket results can be easily derived

$$\begin{aligned}
[\mathcal{M}_x, \mathcal{M}_y] &= \mathcal{M}_z, \quad [\mathcal{M}_y, \mathcal{M}_z] = \mathcal{M}_x, \quad [\mathcal{M}_z, \mathcal{M}_x] = \mathcal{M}_y, && (3.99)\\
[\mathcal{M}^2, \mathcal{M}_i &= 0, \quad i = x, y, z, && (3.100)\\
[p_i, \mathcal{M}_i] &= 0, \quad i = x, y, z, && (3.101)\\
[p_i, \mathcal{M}^2] &= 0, \quad i = x, y, z, && (3.102)\\
[p_x, \mathcal{M}_y] &= [\mathcal{M}_x, p_y] = p_z, &&\\
[p_y, \mathcal{M}_z] &= [\mathcal{M}_y, p_z] = p_x, &&\\
[p_z, \mathcal{M}_x] &= [\mathcal{M}_z, p_x] = p_y, && (3.103)\\
[p_k, p_l] &= 0. && (3.104)
\end{aligned}$$

Motion of a particle with constant acceleration:

For rectilinear motion with constant acceleration a, we have the equation of motion

$$\frac{du}{dt} = [u, H], \tag{3.105}$$

where u is the speed of a particle of mass m at position x and H is the Hamiltonian

$$H = \frac{p^2}{2m} - max. \tag{3.106}$$

Note that u does not explicitly depend on time t.

A formal solution of equation (3.106) may be obtained by expanding $u(t)$ in a Taylor series around $u = u_0$ at $t = 0$:

$$u(t) = u_0 + t\frac{du}{dt}|_0 + \frac{t^2}{2}\frac{d^2u}{dt^2}|_0 + \frac{t^3}{3!}\frac{d^3u}{dt^3}|_0 + \dots \tag{3.107}$$

From equation (3.105),

$$\frac{du}{dt}|_0 = [u, H]_0 \equiv [u, H]|_{t=t_0},$$

$$\frac{d^2u}{dt^2}|_0 = [[u,[u,H]]_0, \quad etc. \tag{3.108}$$

With these, equation (3.107) can be written as

$$u(t) = u_0 + t[u,H]_0 + \frac{t^2}{2}[[u,H],H]_0 + \frac{t^3}{3!}[[[u,H]H],H]_0 + ... \tag{3.109}$$

Now

$$\begin{aligned} [u,H] &= [\frac{dx}{dt},H] = \frac{d}{dt}[x,H], \\ &= \frac{d}{dt}(\frac{dx}{dx}\frac{dH}{dp} - \frac{dx}{dp}\frac{dH}{dx}), \\ &= a = \text{constant}, \end{aligned} \tag{3.110}$$

and

$$[[u,H],H] = 0, \tag{3.111}$$

etc. From equations (3.109 - 3.111), we obtain

$$u(t) = u_0 + at, \tag{3.112}$$

which the familiar kinematic equation of motion derived in equation (2.25).

3.3 HAMILTON-JACOBI FORMALISM

In section 2.3 we mentioned that the Lagrangian formulation of classical mechanics is developed in configuration space with n generalised coordinates and leads to $2n$ second-order differential equations. In section 3.2 we presented the Hamiltonian formulation which, with $2n$ first-order differential equations using $2n$ phase space variables, provides a more convenient and flexible approach to mechanics. The Hamiltonian formulation can be made even more flexible. This is the essence of the Hamilton-Jacobi formulation. In this section we describe how this can be achieved.

3.3.1 POINT AND CANONICAL TRANSFORMATIONS

Point transformation:

From examples presented in sections 2.3 and 3.2, we note that the formal appearances of Lagrange equations and Hamilton equations are maintained regardless of the choice made for the configuration coordinates. A transformation from one set of n generalised coordinates $\boldsymbol{q} \equiv (q_1, q_2, q_3, ..., q_n)$ to another set of s generalised coordinates $\boldsymbol{Q} \equiv (Q_1, Q_2, Q_3, ..., Q_s)$ can be made

$$\boldsymbol{q} \to \boldsymbol{Q}, \tag{3.113}$$

where

$$Q_i = Q_i(\boldsymbol{q}, t) \tag{3.114}$$

is a function of $\boldsymbol{q}$ and may explicitly depend on time t. Such a transformation is known as a point transformation. Obviously, Lagrange equations as well as Hamilton's equations are invariant, *i.e.* remain unchanged, under point transformations.

Canonical transformation:

As Hamilton's formulation involves $2n$ independent phase space variables $\boldsymbol{q}$ and $\boldsymbol{p}$, a much wider range of transformations is possible

$$(\boldsymbol{q}, \boldsymbol{p}) \to (\boldsymbol{Q}, \boldsymbol{P}), \tag{3.115}$$

where

$$\begin{aligned} Q_i &= Q_i(\boldsymbol{q},\boldsymbol{p},t), \quad i=1,2,3,...,s, \\ P_i &= P_i(\boldsymbol{q},\boldsymbol{p},t), \quad i=1,2,3,...,s. \end{aligned} \tag{3.116}$$

However, not all transformations of the above form allow Hamilton's equations of motion to retain their canonical form. Transformations that leave Hamilton's equations of motion unchanged, *i.e.* such that

$$\dot{Q}_i = \frac{\partial H'}{\partial P_i}, \quad \dot{P}_i = -\frac{\partial H'}{\partial Q_i} \tag{3.117}$$

with some Hamiltonian H', are known as canonical transformations.

3.3.2 CONDITIONS FOR CANONICAL TRANSFORMATIONS

We will present two methods of obtaining conditions for canonical transformations.

Conditions involving partial differentials:

Consider the phase space transformation using a single coordinate and a single momentum for simplicity, *viz* $(q,p) \to (Q,P)$ with $H \to H'$, so that the Hamilton's equations read

$$\dot{q} = \frac{\partial H}{\partial p}, \quad \dot{p} = -\frac{\partial H}{\partial q}, \tag{3.118}$$

$$\dot{Q} = \frac{\partial H'}{\partial P}, \quad \dot{P} = -\frac{\partial H'}{\partial Q}. \tag{3.119}$$

In the above, we have considered $Q = Q(q,p,t)$, $P = P(q,p,t)$ and $H' = H'(Q(q,p),P(q,p))$. With these considerations we can express the first equation in (3.119) as

$$\begin{aligned} \dot{Q} &= \frac{\partial H'}{\partial P}, \\ \frac{\partial Q}{\partial q}\dot{q} + \frac{\partial Q}{\partial p}\dot{p} &= \frac{\partial H'}{\partial q}\frac{\partial q}{\partial P} + \frac{\partial H'}{\partial p}\frac{\partial p}{\partial P}, \\ \frac{\partial Q}{\partial q}\frac{\partial H}{\partial p} - \frac{\partial Q}{\partial p}\frac{\partial H}{\partial q} &= \frac{\partial H'}{\partial q}\frac{\partial q}{\partial P} + \frac{\partial H'}{\partial p}\frac{\partial p}{\partial P}, \\ \frac{\partial Q}{\partial q}\frac{\partial H'}{\partial p} - \frac{\partial Q}{\partial p}\frac{\partial H'}{\partial q} &= \frac{\partial H'}{\partial q}\frac{\partial q}{\partial P} + \frac{\partial H'}{\partial p}\frac{\partial p}{\partial P}, \end{aligned} \tag{3.120}$$

where we have used the point transformations $\frac{\partial H}{\partial q} = \frac{\partial H'}{\partial q}$ and $\frac{\partial H}{\partial p} = \frac{\partial H'}{\partial p}$. From the above equation we obtain the following conditions for canonical transformation

$$\begin{aligned} \left(\frac{\partial Q}{\partial q}\right)\Big|_{q,p} &= \left(\frac{\partial p}{\partial P}\right)\Big|_{q,p}, \\ \left(\frac{\partial Q}{\partial p}\right)\Big|_{q,p} &= -\left(\frac{\partial q}{\partial P}\right)\Big|_{q,p}. \end{aligned} \tag{3.121}$$

There will be two other conditions, which can be obtained by using the equation of motion $\dot{P} = -\frac{\partial H'}{\partial Q}$. These are

$$\begin{aligned} \left(\frac{\partial P}{\partial q}\right)\Big|_{q,p} &= -\left(\frac{\partial p}{\partial Q}\right)\Big|_{q,p}, \\ \left(\frac{\partial P}{\partial p}\right)\Big|_{q,p} &= \left(\frac{\partial q}{\partial Q}\right)\Big|_{q,p}. \end{aligned} \tag{3.122}$$

In above, $\left(\frac{\partial Q}{\partial q}\right)|_{q,p}$ indicates that $\left(\frac{\partial Q}{\partial q}\right)$ is a function of (q,p), *etc.*

Condition involving Poisson brackets:

For a transformation $(q,p) \rightarrow (Q,P)$ with $H \rightarrow H'$ to be canonical, we must be able to write the equations of motion

$$\frac{\mathrm{d}Q}{\mathrm{d}t} = [Q,H]_{q,p} = \frac{\partial Q}{\partial q}\frac{\partial H}{\partial p} - \frac{\partial Q}{\partial p}\frac{\partial H}{\partial q}, \tag{3.123}$$

$$\frac{\mathrm{d}P}{\mathrm{d}t} = [P,H]_{q,p} = \frac{\partial P}{\partial q}\frac{\partial H}{\partial p} - \frac{\partial P}{\partial p}\frac{\partial H}{\partial q}. \tag{3.124}$$

Noting that $\frac{\partial H}{\partial q} = \frac{\partial H'}{\partial q}$ and $\frac{\partial H}{\partial p} = \frac{\partial H'}{\partial p}$, we express

$$\frac{\partial H}{\partial q} = \frac{\partial H'}{\partial q} = \frac{\partial H'}{\partial Q}\frac{\partial Q}{\partial q} + \frac{\partial H'}{\partial P}\frac{\partial P}{\partial q} \tag{3.125}$$

and

$$\frac{\partial H}{\partial p} = \frac{\partial H'}{\partial p} = \frac{\partial H'}{\partial Q}\frac{\partial Q}{\partial p} + \frac{\partial H'}{\partial P}\frac{\partial P}{\partial p}. \tag{3.126}$$

Using equations (3.123)–(3.126), we rewrite the equations of motion as

$$\begin{aligned}\frac{\mathrm{d}Q}{\mathrm{d}t} &= \frac{\partial Q}{\partial q}\left(\frac{\partial H'}{\partial Q}\frac{\partial Q}{\partial p} + \frac{\partial H'}{\partial P}\frac{\partial P}{\partial p}\right) - \frac{\partial Q}{\partial p}\left(\frac{\partial H'}{\partial Q}\frac{\partial Q}{\partial q} + \frac{\partial H'}{\partial P}\frac{\partial P}{\partial q}\right), \\ &= \frac{\partial H'}{\partial P}\left(\frac{\partial Q}{\partial q}\frac{\partial P}{\partial p} - \frac{\partial Q}{\partial p}\frac{\partial P}{\partial q}\right), \\ &= \frac{\partial H'}{\partial P}[Q,P]|_{q,p},\end{aligned} \tag{3.127}$$

and similarly,

$$\begin{aligned}\frac{\mathrm{d}P}{\mathrm{d}t} &= -\frac{\partial H'}{\partial Q}\left(\frac{\partial Q}{\partial q}\frac{\partial P}{\partial p} - \frac{\partial Q}{\partial p}\frac{\partial P}{\partial q}\right), \\ &= -\frac{\partial H'}{\partial Q}[Q,P]|_{q,p}.\end{aligned} \tag{3.128}$$

But, we know that in transformed phase space (Q,P) with the transformed Hamiltonian H', Hamilton's equations of motion are

$$\frac{\mathrm{d}Q}{\mathrm{d}t} = \frac{\partial H'}{\partial P}, \tag{3.129}$$

$$\frac{\mathrm{d}P}{\mathrm{d}t} = -\frac{\partial H'}{\partial Q}. \tag{3.130}$$

For equation (3.127) to become equation (3.129) and equation (3.128) to become equation (3.130), we must have

$$[Q,P]|_{q,p} = 1 \tag{3.131}$$

as the condition for the canonical transformation.

3.3.3 PROCEDURE FOR ACHIEVING CANONICAL TRANSFORMATIONS

It is interesting as well as useful to develop a procedure for achieving canonical transformations. For simplicity, we will discuss this by using a single coordinate and a single canonical momentum in a phase space[2]. Let us consider a canonical transformation from a phase space (q, p) to another phase space (Q,P) such that the Hamiltonian H in the original phase space gets transformed to the Hamiltonian H' in the new phase space. The modified Hamilton's variational principle in the two phase spaces are

$$\delta \int (p\dot{q} - H)\mathrm{d}t = 0, \tag{3.132}$$

$$\delta \int (P\dot{Q} - H')\mathrm{d}t = 0, \tag{3.133}$$

from which we can write

$$\delta \int \{(p\dot{q} - H) - (P\dot{Q} - H')\}\mathrm{d}t = 0. \tag{3.134}$$

As mentioned in section 2.3 and Appendix C, a general feature of the δ-variation is that

$$\delta \int f\mathrm{d}t = 0 \tag{3.135}$$

is satisfied by $f = \frac{\mathrm{d}F}{\mathrm{d}t}$, where F is an arbitrary function subject to fixed values at the start and end of motion:, *viz* $\delta F|_{t_1} = \delta F|_{t_2} = 0.$, Therefore, from equation (3.134) it follows that

$$(p\dot{q} - H) - (P\dot{Q} - H') = \frac{\mathrm{d}F}{\mathrm{d}t}, \tag{3.136}$$

where F is any function of phase space variables with continuous second derivatives. For achieving a canonical transformation from an old phase space to a new phase space, it is meaningful to consider F as a function of equal number of variables from the two phase spaces, bridging the two spaces. In that case, F can be considered as a **generating function** of the canonical transformation.

3.3.4 GENERATING FUNCTIONS FOR FOUR BASIC TYPES OF CANONICAL TRANSFORMATIONS

In general, we can consider four basic types of the generating function for making canonical transformations: $F = F_1(q,Q,t)$, $F = F_2(q,P,t)$, $F = F_3(p,Q,t)$ and $F = F_4(p,P,t)$. Of course, these can be interconnected using appropriate Legendre's transformations.

Type 1 generating function:

Let us consider $F = F_1(q,Q,t)$. Equation (3.136) then can be written as

$$(p\dot{q} - H) - (P\dot{Q} - H') = \frac{\partial F_1}{\partial q}\dot{q} + \frac{\partial F_1}{\partial Q}\dot{Q} + \frac{\partial F_1}{\partial t}. \tag{3.137}$$

As q, Q and t are independent variables, from the above equation we can express

$$p = \frac{\partial F_1}{\partial q}, \tag{3.138}$$

[2]Consideration of canonically conjugate multi coordinates and momenta variables is straightforward by changing terms like qp to $\sum_i q_i p_i$ and $\frac{\partial F}{\partial q}\dot{q}_i$ to $\sum_i \frac{F}{\partial q_i}\dot{q}_i$, *etc.*

$$P = -\frac{\partial F_1}{\partial Q}, \tag{3.139}$$

$$H' = H + \frac{\partial F_1}{\partial t}. \tag{3.140}$$

Equation (3.138) can be solved for Q in terms of q, p and t. Using this information, equation (3.139) can be solved for P in terms of q, p and t. The old Hamiltonian can then be expressed in terms of new phase space variables: $H = H(Q,P,t)$. Finally, using equation (3.140), the transformed Hamiltonian can be obtained as $H' = H'(Q,P,t)$.

The simplest example is $F = F_1 = qQ$. This produces $p = Q$ and $P = -q$, and $H' = H + \frac{\partial F_1}{\partial t}$. If we consider F_1 not to explicitly depend on time, then $H' = H$. Let us, for simplicity, consider the Hamiltonian of a harmonic oscillator $H = \frac{p^2}{2m} + \frac{1}{2}kq^2$. Then the transformed Hamiltonian $H' = H'(Q,P)$ can be obtained, using equations (3.138) and (3.139)

$$\begin{aligned} H' &= H, \\ &= \frac{p^2}{2m} + \frac{1}{2}kq^2, \\ &= \frac{Q^2}{2m} + \frac{1}{2}kP^2. \end{aligned} \tag{3.141}$$

Notice that in the expression for $H' = H'(Q,P)$, the canonical variables Q and P play the role of momentum and coordinate, respectively. This is nothing to be puzzled about, as Q and P are independent variables on equal footing and can be treated either as a coordinate or a momentum variable.

Type 2 generating function:

Some canonical transformations can be achieved by using the $F_2 = F_2(q,P.t)$ type generating function. Such a function can be obtained from the $F_1(q,Q,t)$ function via a Legendre tranformation. It is useful to remember from section 3.1.3 that a function $f(x,y) = \frac{\partial f}{\partial x}\mathrm{d}x + \frac{\partial f}{\partial y}\mathrm{d}y$ can be Legendre transformed to a function $g(u,y) = f - \frac{\partial f}{\partial x}x$. Keeping this in mind, we express, using equation (3.139)

$$\begin{aligned} F_2 &= F_1 - \frac{\partial F_1}{\partial Q}Q, \\ &= F_1 + PQ. \end{aligned} \tag{3.142}$$

Equation (3.136) can then be processed by writing $F = F_1 = F_2(q,P,t) - QP$. We can thus write

$$(p\dot{q} - H) - (P\dot{Q} - H') = \frac{\mathrm{d}(F_2 - PQ)}{\mathrm{d}t}, \tag{3.143}$$

from which the following relations can be obtained

$$p = \frac{\partial F_2}{\partial q}, \tag{3.144}$$

$$Q = \frac{\partial F_2}{\partial P}, \tag{3.145}$$

$$H' = H + \frac{\partial F_2}{\partial t}. \tag{3.146}$$

The simplest example is $F_2 = qP$ with no explicit time dependence, from which $p = \frac{\partial F_2}{\partial q} = P$ and $Q = \frac{\partial F_2}{\partial P} = q$. For the case of harmonic oscillator discussed earlier, the transformed Hamiltonian is

$$H' = H.$$

$$\begin{aligned} &= \frac{p^2}{2m} + \frac{1}{2}kq^2. \\ &= \frac{P^2}{2m} + \frac{1}{2}kQ^2. \end{aligned} \tag{3.147}$$

Notice that in this case Q and P stay as the transformed coordinates and momenta, respectively.

Type 3 generating function:

The third type of generating function is a function of old momentum p, the transformed variable Q and time: $F_3 = F_3(p,Q,t)$. This can be obtained from F_1 through the Legendre transformation: $F_3(p,Q,t) = F_1(q,Q,t) - \frac{\partial F_1}{\partial q}q = F_1 - pq$. Using $F = F_1 = F_3 + pq$ in equation (3.136) we obtain

$$q = -\frac{\partial F_3}{\partial p}, \tag{3.148}$$

$$P = -\frac{\partial F_3}{\partial Q}, \tag{3.149}$$

$$H' = H + \frac{\partial F_3}{\partial t}. \tag{3.150}$$

For the choice $F_3 = pQ$, we get $q = -Q$ and $p = -P$. With these, and assuming no explicit time dependence of F_3, the harmonic oscillator Hamiltonian H in the original phase space discussed above is transformed into

$$\begin{aligned} H' &= H, \\ &= \frac{p^2}{2m} + \frac{1}{2}kq^2. \\ &= \frac{P^2}{2m} + \frac{1}{2}kQ^2. \end{aligned} \tag{3.151}$$

Notice Q and P remain as the transformed coordinates and momenta, respectively.

Type 4 generating function:

The fourth type of generating function is $F_4 = F_4(p,P,t)$. This can be obtained from F_1 using the double Legendre transformation $F_4(p,P,t) = F_1(q,Q,t) - \frac{\partial F_1}{\partial q}q - \frac{\partial F_1}{\partial Q}Q$. Using $F = F_1 = F_4 + \frac{\partial F_1}{\partial q}q + \frac{\partial F_1}{\partial Q}Q$ in equation (3.136) we obtain

$$q = -\frac{\partial F_4}{\partial p}, \tag{3.152}$$

$$Q = \frac{\partial F_4}{\partial P}, \tag{3.153}$$

$$H' = H + \frac{\partial F_4}{\partial t}. \tag{3.154}$$

For the choice $F_4 = pP$, we get $q = -P$ and $Q = p$. With these, and assuming no explicit time dependence of F_4, the harmonic oscillator Hamiltonian H in the original phase space discussed above is transformed into

$$\begin{aligned} H' &= H, \\ &= \frac{p^2}{2m} + \frac{1}{2}kq^2. \end{aligned}$$

$$= \frac{Q^2}{2m} + \frac{1}{2}kP^2. \tag{3.155}$$

Notice that in the new phase space Q and P play the role of momentum and coordinate, respectively.

An inspection of the four types of basic generating functions suggests that if the generating function is a functional of either the coordinates or the momenta in the original and new phase spaces, *viz* either a functional of q and Q (as in F_1), or of p and P (as in F_4), then the new phase space variables Q and P play the roles of momenta and coordinates, respectively.

3.3.5 EXAMPLES OF CANONICAL TRANSFORMATIONS AND GENERATING FUNCTIONS

3.3.5.1 Example 1:

Let us show that the transformation

$$Q = \sqrt{q}\cos(2p), \qquad P = \sqrt{q}\sin(2p)$$

is canonical.

It is sufficient to prove that the Poisson bracket of Q and P with respect to the phase space (q,p) is unity. The Poisson bracket of Q and P is

$$[Q,P]_{q,p} = \frac{\partial Q}{\partial q}\frac{\partial P}{\partial p} - \frac{\partial Q}{\partial p}\frac{\partial P}{\partial q}.$$

From the given information,

$$\begin{aligned} \frac{\partial Q}{\partial q} &= \frac{1}{2}q^{-1/2}\cos(2p), & \frac{\partial Q}{\partial p} &= -2\sqrt{q}\sin(2p), \\ \frac{\partial P}{\partial q} &= \frac{1}{2}q^{-1/2}\sin(2p), & \frac{\partial P}{\partial p} &= 2\sqrt{q}\cos(2p). \end{aligned}$$

The Poisson bracket is then

$$[Q,P]_{q,p} = \cos^2(2p) + \sin^2(2p) = 1,$$

which proves that the transformation is canonical.

3.3.5.2 Example 2:

Let us show that the transformation

$$Q_1 = q_1, \qquad P_1 = p_1,$$

$$Q_2 = p_2, \qquad P_2 = -q_2$$

is canonical.

We need to show that $[Q_i,P_i]_{\mathbf{q},\mathbf{p}} = 1$ for $i = 1,2$.

$$\begin{aligned} [Q_1,P_1]_{\mathbf{q},\mathbf{p}} &= \sum_i \left(\frac{\partial Q_1}{\partial q_i}\frac{\partial P_1}{\partial p_i} - \frac{\partial Q_1}{\partial p_i}\frac{\partial P_1}{\partial q_i}\right), \\ &= \left(\frac{\partial Q_1}{\partial q_1}\frac{\partial P_1}{\partial p_1} + \frac{\partial Q_1}{\partial q_2}\frac{\partial P_1}{\partial p_2}\right) - \left(\frac{\partial Q_1}{\partial p_1}\frac{\partial P_1}{\partial q_1} + \frac{\partial Q_1}{\partial p_2}\frac{\partial P_1}{\partial q_2}\right), \\ &= (1+0) - (0+0), \\ &= 1. \end{aligned}$$

Similarly, it can be shown that $[Q_2,P_2]_{\mathbf{q},\mathbf{p}} = 1$. This proves that the transformation is canonical.

3.3.5.3 Example 3:

Let us show that the transformation

$$Q = \ln\left(\frac{\sin p}{q}\right), \qquad P = q\cot p$$

is canonical and find a generating function for this transformation.

The Poisson bracket $[Q,P]_{q,p}$ is

$$\begin{aligned} [Q,P]_{q,p} &= \frac{\partial Q}{\partial q}\frac{\partial P}{\partial p} - \frac{\partial Q}{\partial p}\frac{\partial P}{\partial q}, \\ &= \left(-\frac{1}{q}\right)\Big(-q(1+\cot^2 p)\Big) - (\cot p)(\cot p), \\ &= (1+\cot^2 p) - \cot^2 p, \\ &= 1, \end{aligned}$$

which proves that the transformation is canonical.

To find a generating function for this canonical transformation, we express

$$\begin{aligned} P &= q\frac{\cos p}{\sin p}, \\ &= \exp(-Q)\cos p \quad \text{from the first given equation,} \\ &= -\frac{\partial F_3}{\partial Q}, \end{aligned}$$

where $F_3 = F_3(p,Q)$ is the type 3 generating function.

Integrating $\frac{\partial F_3}{\partial Q} = -\exp(-Q)\cos p$ we obtain

$$\begin{aligned} F_3(p,Q) &= -\int \mathrm{d}Q\, \exp(-Q)\cos p + \text{constant}, \\ &= \exp(-Q)\cos p + \text{constant}. \end{aligned}$$

Thus $\exp(-Q)\cos p$ is a generating function of the above canonical transformation.

3.3.6 HAMILTON-JACOBI EQUATION

Let us consider the canonical transformation $(\boldsymbol{q},\boldsymbol{p}) \to (\boldsymbol{Q},\boldsymbol{P})$ using the special case of type 2 generating function F_2 such that all transformed phase space variables are constants of motion. Considering a single coordinate and a single momentum for simplicity, this means we consider $(q,p) \to (\alpha,\beta)$ and $F_2(q,P,t) = S(q,\beta,t)$, where α and β are constants of motion. The Hamilton equations of motion then read

$$\begin{aligned} \dot{Q} \equiv \dot{\alpha} &= \frac{\partial H'}{\partial P} \equiv \frac{\partial H'}{\partial \beta} = 0, \\ \dot{P} \equiv \dot{\beta} &= -\frac{\partial H'}{\partial Q} \equiv \frac{\partial H'}{\partial \alpha} = 0. \end{aligned} \tag{3.156}$$

Let us also impose $\frac{\partial H'}{\partial t} = 0$. Then $H' = $ constant, and let us set it to zero. With these considerations, we get from equation (3.146)

$$H + \frac{\partial F_2}{\partial t} = 0,$$

$$H(q, \frac{\partial S}{\partial q}, t) + \frac{\partial S(q, \beta, t)}{\partial t} \quad = \quad 0. \tag{3.157}$$

This is the *Hamilton-Jacobi* equation.

With the consideration of $2n$ conjugate variables in the original phase space $(\boldsymbol{q}, \boldsymbol{p})$ and $2s$ conjugate variables in the transformed phase space $(\boldsymbol{Q}, \boldsymbol{P})$, the special generating function S is a functional of $(n+1)$ variables $(\boldsymbol{q}, t) \equiv (q_1, q_2, q_3, ... q_n, t)$. Thus the Hamilton-Jacobi equation in (3.157) represents first-order differential equations for $(n+1)$ variables. This is clearly a huge improvement over Hamilton's theory, which requires solving $2n$ first-order differential equations of motion for variables $\boldsymbol{q}$ and $\boldsymbol{p}$ [*cf* equations (3.14) and (3.15)].

Since $S = S(\boldsymbol{q}, \boldsymbol{P}, t)$, with $\boldsymbol{P}$ components representing constants of motion, say $\boldsymbol{P} = \boldsymbol{\beta} = (\beta_1, \beta_2, \beta_3, ..., \beta_s)$, a solution for S will be of the type

$$S = f(q_1, q_2, q_3, ..., q_n, \beta_1, \beta_2, \beta_3, ..., \beta_s, t) + A, \tag{3.158}$$

where $\{\beta_i\}$ and A are constants of motion. From equation (3.158) the coordinates $\boldsymbol{Q}$ in the new phase space are obtained as new constants of motion

$$Q_i \equiv \alpha_i = \frac{\partial S}{\partial \beta_i}, \tag{3.159}$$

whose solution gives the coordinates $\{q_i\}$ as functions of time t. The momenta $\{p_i\}$ as functions of time may then be found from $p_i = \frac{\partial S}{\partial q_i}$.

We have now seen that similar to the Lagrange's and Hamilton's equations of motion, the Hamilton-Jacobi equation provides a general method of solving dynamical equations of motion. Let us remind ourselves of the workload involved in applying the three analytical dynamical methods discussed here. Considering n coordinate variables and n momentum variables, the Lagrangian method requires solving n second-order differential equations, the Hamiltonian method requires solving $2n$ first-order differential equations, and the Hamilton-Jacobi method requires solving $(n+1)$ first-order differential equations.

3.3.7 HAMILTON'S PRINCIPAL FUNCTION AND ITS PHYSICAL SIGNIFICANCE

The special generating function $S(\boldsymbol{q}, \boldsymbol{\beta}, t)$ discussed in the previous section is known as **Hamilton's principal function.** It is instructive to gain further insight into the physical significance of this generating function. Considering, for simplicity, a single coordinate and a single momentum, and using $p = \frac{\partial S}{\partial q}$ and $H + \frac{\partial S}{\partial t} = 0$, we express the time rate of S as

$$\begin{aligned} \frac{\mathrm{d}S}{\mathrm{d}t} &= \frac{\partial S}{\partial q}\dot{q} + \frac{\partial S}{\partial t}, \\ &= p\dot{q} + \frac{\partial S}{\partial t}, \\ &= p\dot{q} - H, \\ &= L. \end{aligned} \tag{3.160}$$

From this,

$$S = \int L\mathrm{d}t + \text{constant}, \tag{3.161}$$

making it clear that Hamilton's principal function differs from an indefinite time integral of the Lagrangian at most only by a constant. Thus, an action-like integral (*e.g.* Hamilton's principal function $S = S(q, \beta, t)$) can be used to generate a canonical transformation into a new phase space in which coordinates and momenta that are constants of motion.

We noted in section 3.2.2 that the Hamiltonian of a conservative system does not explicitly depend on time and represents the total energy E. For such a system, equation (3.157) can be written as

$$\frac{\partial S}{\partial t} = -E, \tag{3.162}$$

whose solution can be expressed as

$$S(q,t) = S(q,0) - Et. \tag{3.163}$$

Thus, for a conservative system, the Hamilton-Jacobi equation reads

$$H\left(q, \frac{\partial S(q,0)}{\partial q}\right) = E. \tag{3.164}$$

3.3.8 EXAMPLES OF APPLICATION OF THE HAMILTON-JACOBI METHOD

It should be mentioned that in general it is not easy to solve the Hamilton-Jacobi equation. However, the Hamilton-Jacobi formalism proves very important and useful in further development of classical dynamics. For example, it helps understand the transition from classical to quantum mechanics, as described in section 9.3. It also helps in the development of a classical perturbation method, called the *Hamilton-Jacobi perturbation theory*, or the *canonical perturbation theory*, for solving difficult dynamical problems, which we will describe in section 10.2.2.

In this section we will consider the applications the Hamilton-Jacobi method to the one- and two-dimensional harmonic oscillators, and a particle subjected to a three-dimensional central force. Before proceeding, we note that the $2n$ first-order ordinary differential equations in Hamilton's theory have been reduced to a single first-order partial differential equation for Hamilton's principal function S of $n+1$ variables $(q_1, q_2, q_3, \ldots, q_n, t)$ in the Hamilton-Jacobi theory. Partial differential equations are generally complicated to solve. However, under certain conditions it is possible to employ the method of 'separation of variables' in the Hamilton-Jacobi equation, making it easier to solve. It is useful to note that while the general solution of an ordinary differential equation contains arbitrary constants, the general solution of an ordinary differential equation contains arbitrary functions.

If the Hamiltonian does not contain time explicitly (*i.e.* for conservative systems), a solution for the Hamilton-Jacobi equation, with a single coordinate variable for simplicity, can be tried in the form

$$S(q,\beta,t) = W(q,\beta) - c(\beta)t, \tag{3.165}$$

where $c(\beta)$ is a function of the constant β. Separating S into two functions in this form is known as the method of *additive separation of variables*. The function $W(q,\beta)$ is Hamilton's characteristic function (*cf* section 3.2.9). To see this let us substitute equation (3.165) in the Hamilton-Jacobi equation, which reduces to the form

$$H\left(q, \frac{\partial W}{\partial q}\right) = c(\beta), \tag{3.166}$$

with $p = \frac{\partial W}{\partial q}$, $Q \equiv \alpha = \frac{\partial W}{\partial \beta} - \frac{\partial c(\beta)}{\partial \beta}t$. From here we obtain

$$W = \int p\mathrm{d}q = \int p\dot{q}\mathrm{d}t, \tag{3.167}$$

as defined in equation (3.76).

3.3.8.1 One-dimensional harmonic oscillator

As discussed before, the Hamiltonian in the (q,p) phase space for the one-dimensional harmonic oscillator of mass m is $H(q,p) = \frac{1}{2m}(p^2 + m^2\omega^2 q^2)$, where $\omega = \sqrt{k/m}$ and k is the force constant. For this conservative system, the Hamilton-Jacobi equation is given in (3.166), with the trial solution in the form of equation (3.165). Equation (3.166) suggests that $c(\beta) = E =$ total energy of the system. The Hamilton-Jacobi equation then reads as

$$\frac{1}{2m}\left[\left(\frac{\partial W}{\partial q}\right)^2 + m^2\omega^2 q^2\right] = E. \tag{3.168}$$

From this, we obtain

$$W = \sqrt{2mE}\int \mathrm{d}q\sqrt{1 - \frac{m\omega^2 q^2}{2E}} + g(\beta). \tag{3.169}$$

The choice $g(\beta) = 0$ is good enough, using which we write

$$W = W(q,\beta) = \sqrt{2mE}\int \mathrm{d}q\sqrt{1 - \frac{m\omega^2 q^2}{2E}}, \tag{3.170}$$

and

$$\begin{aligned} S &= W - Et, \\ &= \sqrt{2mE}\int \mathrm{d}q\sqrt{1 - \frac{m\omega^2 q^2}{2E}} - Et. \end{aligned} \tag{3.171}$$

The constant of motion in equation (3.159) is then obtained as

$$\begin{aligned} \alpha &= \frac{\partial S}{\partial \beta} = \frac{\partial S}{\partial E}\eta, \\ &= \left[\frac{1}{\omega}\int \frac{\mathrm{d}q}{\sqrt{\frac{2E}{m\omega^2} - q^2}} - t\right]\eta, \\ &= \left[\frac{1}{\omega}\sin^{-1}\left(q\sqrt{\frac{m\omega^2}{2E}}\right) - t\right]\eta, \end{aligned} \tag{3.172}$$

where $\eta = \frac{\partial c(\beta)}{\partial \beta}$. From this, we obtain

$$\omega\left(\frac{\alpha}{\eta} + t\right) = \sin^{-1}\left(q\sqrt{\frac{m\omega^2}{2E}}\right), \tag{3.173}$$

and thus

$$\begin{aligned} q &= \sqrt{\frac{2E}{m\omega^2}}\sin(\omega t + \alpha\omega/\eta), \\ &= \sqrt{\frac{2E}{m\omega^2}}\sin(\omega t + \psi). \end{aligned} \tag{3.174}$$

Let us express

$$q = \sqrt{\frac{2E}{m\omega^2}}\sin\phi. \tag{3.175}$$

With this, the momentum is

$$p = \frac{\partial S}{\partial q} = \frac{\partial W}{\partial q} = \sqrt{2mE}\cos(\omega t + \psi) = \sqrt{2mE}\cos\phi, \tag{3.176}$$

which obviously is the same as $m\dot{q}$. Equations (3.174) and (3.176) provide the complete solution to the motion of a simple harmonic oscillator. The constant ψ in these equations represents the phase of the motion of the oscillator at the start of motion (*i.e.* at $t = 0$), and ϕ in equations (3.175) and (3.176) is the phase angle at time t.

The phase angle ψ can be determined by using known information about coordinate, momentum and energy at time $t = t_0$, allowing for a complete solution to the problem. Let us write, at $t = t_0$,

$$q = q(t_0) = \sqrt{\frac{2E}{m\omega^2}}\sin(\omega t_0 + \psi), \tag{3.177}$$

$$p = p(t_0) = \sqrt{2mE}\cos(\omega t_0 + \psi), \tag{3.178}$$

$$E = E(t_0) = \frac{p_0^2}{2m} + \frac{1}{2}m\omega^2 q_0^2. \tag{3.179}$$

From equations (3.177) and (3.178)

$$\psi = \tan^{-1}(m\omega q_0/p_0) - \omega t_0. \tag{3.180}$$

It is instructive to examine Hamilton's characteristic function W for the oscillator. We can express it as follows

$$\begin{aligned} W &= \sqrt{2mE}\int \mathrm{d}q\sqrt{1-\frac{m\omega^2 q^2}{2E}}, \\ &= \sqrt{2mE}\int \mathrm{d}q\sqrt{(1-\sin^2\phi)}, \\ &= \frac{E}{\omega}(\phi + \sin\phi\cos\phi). \end{aligned} \tag{3.181}$$

Figure 3.3 shows a plot of W over a cycle. When the position q of the oscillator is taken through one cycle, the phase angle ϕ increases by 2π and W increases by $(\frac{2\pi}{\omega})E$.

Hamilton's principal function can be expressed as

$$\begin{aligned} S &= \sqrt{2mE}\int \mathrm{d}q\sqrt{1-\frac{m\omega^2 q^2}{2E}} - Et, \\ &= 2E\int \mathrm{d}t[\cos^2(\omega t + \psi) - \frac{1}{2}], \\ &= \int \mathrm{d}tL. \end{aligned} \tag{3.182}$$

With this, the Lagrangian can be expressed as

$$L = 2E[\cos^2(\omega t + \psi) - \frac{1}{2}], \tag{3.183}$$

a result that can also be obtained by substituting q and p from equations (3.174) and (3.176) in the expression $L = \frac{1}{2m}(p^2 - m^2\omega^2 q^2)$.

3.3.8.2 Two-dimensional harmonic oscillator

Anisotropic case:

Consider the two-dimensional anisotropic harmonic oscillator of mass m controlled by force constants k_x and k_y in the x- and y- directions, respectively. The Hamiltonian is

$$H = \frac{1}{2m}(p_x^2 + p_y^2 + m^2\omega_x^2 x^2 + m^2\omega_y^2 y^2), \tag{3.184}$$

where $\omega_x^2 = k_x/m$ and $\omega_y^2 = k_y/m$.

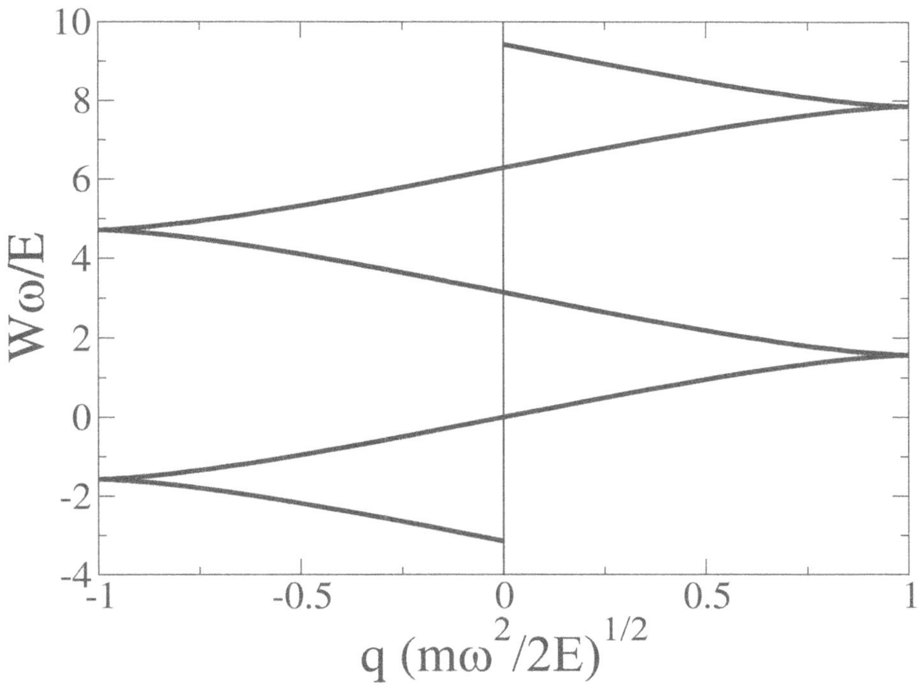

Figure 3.3 Hamilton's characteristic function W for a one-dimensional simple harmonic oscillator.

Substituting $\boldsymbol{q} = (x, y)$ and $\boldsymbol{P} \equiv \boldsymbol{\beta} = (\beta_x, \beta_y)$, and using the concept of additive separation of variables, we express

$$S(x, y, \beta_x, \beta_y, t) = W - Et = W_x(x, \beta_x) + W_y(y, \beta_y) - C_x(\beta)t - C_y(\beta)t, \tag{3.185}$$

where $C_x(\beta) + C_y(\beta) = C(\beta) = E =$ total energy. The Hamilton-Jacobi equation can be expressed as

$$\begin{aligned} H\left(x, y, \frac{\partial S}{\partial x}, \frac{\partial S}{\partial y}\right) + \frac{\partial S}{\partial t} &= 0, \\ \frac{1}{2m}\left[\left(\frac{\partial W}{\partial x}\right)^2 + m^2\omega_x^2 x^2 + \left(\frac{\partial W}{\partial y}\right)^2 + m^2\omega_y^2 y^2\right] &= C_x(\beta) + C_y(\beta). \end{aligned} \tag{3.186}$$

Since the x and y variables are separated, we can express this equation as

$$\begin{aligned} \frac{1}{2m}\left(\frac{\partial W}{\partial x}\right)^2 + \frac{1}{2}m\omega_x^2 x^2 &= C_x, \\ \frac{1}{2m}\left(\frac{\partial W}{\partial y}\right)^2 + \frac{1}{2}m\omega_y^2 y^2 &= C_y. \end{aligned} \tag{3.187}$$

Following the steps for the one-dimensional oscillator, the solutions to these equations are

$$x = \sqrt{\frac{2C_x}{m\omega_x^2}} \sin(\omega_x t + \psi_x),$$

$$\begin{aligned} y &= \sqrt{\frac{2C_y}{m\omega_y^2}}\sin(\omega_y t + \psi_y), \\ p_x &= \sqrt{2mC_x}\cos(\omega_x t + \psi_x), \\ p_y &= \sqrt{2mC_y}\cos(\omega_y t + \psi_y). \end{aligned} \tag{3.188}$$

Isotropic case:

For an isotropic two-dimensional harmonic oscillator we set $k_x = k_y = k$ and $\omega_x = \omega_y = \omega$, and write the Hamiltonian as

$$H = \frac{1}{2m}(p_\rho^2 + \frac{p_\phi^2}{\rho^2} + m\omega^2\rho^2), \tag{3.189}$$

where, following the procedure described in Appendix A, we have converted the Cartesian variables x, y, p_x, p_y to the plane polar coordinates using $x = \rho\cos\phi$, $y = \rho\sin\phi$, $p_x = m\dot{x}$, $p_y = m\dot{y}$, $p_\rho = m\dot{\rho}$, $p_\phi = m\rho^2\dot{\phi}$.

As before, we express $S = W - Et$. Noting that ϕ is an ignorable variable, we can express

$$\begin{aligned} W &= W_1(\rho) + W_2(\phi), \\ &= W_1(\rho) + \int p_\phi \mathrm{d}\phi, \\ &= W_1(\rho) + p_\phi\phi, \end{aligned} \tag{3.190}$$

where p_ϕ, the momentum conjugate to the ignorable variable ϕ, is a constant of motion. The Hamilton-Jacobi equation then becomes

$$\left(\frac{\partial W_1}{\partial\rho}\right)^2 + \frac{p_\phi^2}{\rho^2} + m^2\omega^2\rho^2 = 2mE. \tag{3.191}$$

From the above two equations we obtain

$$W = \int \mathrm{d}\rho\sqrt{(2mE - m^2\omega^2\rho^2 - \frac{p_\phi^2}{\rho^2})} + p_\phi\phi. \tag{3.192}$$

The equations of motion can now be obtained. The results are

$$\rho = \sqrt{\frac{2E}{m\omega^2}}\sqrt{\sin^2\omega t + \sin^2(\omega t + \psi)}, \quad p_\rho = m\dot{\rho}, \tag{3.193}$$

$$\phi = \tan^{-1}\left[\frac{\sin\omega t}{\sin(\omega t + \psi)}\right], \qquad p_\phi = m\rho^2\dot{\phi}. \tag{3.194}$$

The motion in the x-y plane is a **Lissajous figure**, which can be plotted for a chosen value of ψ. For $\psi = \pi/2$, the motion is a circle of radius $\rho_0 = \sqrt{\frac{2E}{m\omega^2}}$.

3.3.8.3 Particle subjected to a central force

Consider a particle of mass m moving in a central potential $V(r)$, a function only of distance r. Using spherical polar coordinates (r, θ, ϕ), we write the Hamiltonian as

$$H = \frac{1}{2m}\left(p_r^2 + \frac{p_\theta^2}{r^2} + \frac{p_\phi^2}{r^2\sin^2\theta}\right) + V(r). \tag{3.195}$$

As before, we express $S = W - Et$, with

$$W = W_r(r) + W_\theta(\theta) + W_\phi(\phi). \tag{3.196}$$

As ϕ is cyclic, we obtain

$$W_\phi = \phi p_\phi, \tag{3.197}$$

where $p_\phi = C(\phi) = \mathscr{M}_z =$ constant of motion. Actually, $\mathscr{M}_z$ is the z-component of the angular momentum. The Hamilton-Jacobi equation then reduces to

$$\left(\frac{\mathrm{d}W_r}{\mathrm{d}r}\right)^2 + \frac{1}{r^2}\left[\left(\frac{\mathrm{d}W_\theta}{\mathrm{d}\theta}\right)^2 + \frac{p_\phi}{\sin^2\theta}\right] + 2mV(r) = 2mE. \tag{3.198}$$

We can express this equation in the form

$$2mr^2\left[\frac{1}{2m}\left(\frac{\mathrm{d}W_r}{\mathrm{d}r}\right)^2 + V - E\right] + \left[\left(\frac{\mathrm{d}W_\theta}{\mathrm{d}\theta}\right)^2 + \frac{\mathscr{M}_z^2}{\sin^2\theta}\right] = 0. \tag{3.199}$$

In order to balance the left-hand side to zero, terms in both square brackets must be recognised as constants of equal and opposite signs. Let us write the term in the second square brackets as

$$\left(\frac{\mathrm{d}W_\theta}{\mathrm{d}\theta}\right)^2 + \frac{\mathscr{M}_z^2}{\sin^2\theta} = \mathscr{M}^2, \tag{3.200}$$

where, for now, $\mathscr{M}^2$ is a constant. With this, equation (3.199) can be expressed as

$$2mr^2\left[\frac{1}{2m}\left(\frac{\mathrm{d}W_r}{\mathrm{d}r}\right)^2 + V + \mathscr{M}^2 - E\right] = 0. \tag{3.201}$$

From equations (3.199), (3.200) and (3.201) we obtain

$$W_r(r) = \int \mathrm{d}r\sqrt{2m(E-V) - \frac{\mathscr{M}^2}{r^2}}, \tag{3.202}$$

$$W_\theta(\theta) = \int \mathrm{d}\theta\sqrt{\mathscr{M}^2 - \frac{\mathscr{M}_z}{\sin^2\theta}}, \tag{3.203}$$

$$W_\phi(\phi) = \mathscr{M}_z\phi. \tag{3.204}$$

With these, the canonical momentum components are obtained as

$$p_r = \frac{\partial W}{\partial r} = \sqrt{2m(E-V) - \frac{\mathscr{M}^2}{r^2}}, \tag{3.205}$$

$$p_\theta = \frac{\partial W}{\partial \theta} = \sqrt{\mathscr{M}^2 - \frac{\mathscr{M}_z^2}{\sin^2\theta}}, \tag{3.206}$$

$$p_\phi = \frac{\partial W}{\partial \phi} = \mathscr{M}_z. \tag{3.207}$$

For a solution of equation (3.202) to be real, it is required that $\mathscr{M}^2 \geq \mathscr{M}_z^2$. Also, from equations (3.203) and (3.206) we can write $\mathscr{M}^2 = p_\theta^2 + \frac{\mathscr{M}_z^2}{\sin^2\theta} = p_\theta^2 + \frac{p_\phi^2}{\sin^2\theta}$, proving that $\mathscr{M}$ is the total angular momentum.

Now the Hamilton-Jacobi equation is reduced to the radial part

$$\frac{\mathrm{d}W_r}{\mathrm{d}r} = \sqrt{2m(E - V(r)) - \frac{\mathscr{M}^2}{r^2}}, \tag{3.208}$$

which, given a form of $V(r)$, can be solved numerically. And, following equation (3.159), the constants of motion can be obtained as follows

$$\alpha_E = \frac{\partial S}{\partial E} = m\int \frac{\mathrm{d}r}{\sqrt{2m(E-V) - \frac{\mathscr{M}^2}{r^2}}} - t, \tag{3.209}$$

$$\alpha_{\mathscr{M}} = \frac{\partial S}{\partial \mathscr{M}} = \int \frac{\mathrm{d}r}{\sqrt{2m(E-V) - \frac{\mathscr{M}^2}{r^2}}}(-\frac{\mathscr{M}}{r^2}) + \int \frac{\mathscr{M}\mathrm{d}\theta}{\sqrt{\mathscr{M}^2 - \frac{\mathscr{M}_z^2}{\sin^2\theta}}}, \tag{3.210}$$

$$\alpha_{\mathscr{M}_z} = \frac{\partial S}{\partial \mathscr{M}_z} = \int \frac{\mathrm{d}\theta}{\sqrt{\mathscr{M}^2 - \frac{\mathscr{M}_z^2}{\sin^2\theta}}}\left(\frac{-\mathscr{M}_z}{\sin^2\theta}\right) + \phi. \tag{3.211}$$

Equation (3.209) describes the radial motion of the particle with time, equation (3.210) describes the relationship between the radial distance r and the polar angle θ, and equation (3.211) describes the relationship between the polar angle θ and the azimuthal angle ϕ.

Kepler problem

Equations (3.209)–(3.211) can be applied to discuss the Kepler problem in Astronomy. We consider the orbital motion of a planet of mass m with angular momentum magnitude $\mathscr{M}$ around a star of mass M. For such a problem, we use the gravitational potential energy $V(r) = -\frac{GMm}{r} = -\frac{k}{r}$, where G is the Gravitational constant.

The integral in equation (3.211) can be evaluated by taking $\mathscr{M}_z$ along the polar (z) axis and making a physically relevant consideration for the direction of the angular momentum with respect to the z axis. Let us denote this as $\phi - \alpha_{\mathscr{M}_z} = \overline{\phi}$. With similar physical consideration, the angular integration in equation (3.210) can also be evaluated. Let us denote this as $\overline{\theta}$. With these considerations, equation (3.210) becomes

$$\begin{aligned} \overline{\theta} - \alpha_{\mathscr{M}} &= \frac{\mathscr{M}}{\sqrt{2m}}\int \frac{\mathrm{d}r}{r^2\sqrt{E + \frac{k}{r} - \frac{\mathscr{M}^2}{2mr^2}}}, \\ &= -\int \frac{\mathrm{d}u}{\sqrt{\frac{2mE}{\mathscr{M}^2} + \frac{2mku}{\mathscr{M}^2} - u^2}}, \end{aligned} \tag{3.212}$$

where $u = \frac{1}{r}$ has been used.

Using the integration formula

$$\int \frac{\mathrm{d}u}{\sqrt{A + Bu + Cu^2}} = \frac{1}{\sqrt{-C}}\cos^{-1}\left(\frac{-(B+2Cu)}{\sqrt{B^2 - 4AC}}\right), \tag{3.213}$$

and setting $A = \frac{2mE}{\mathscr{M}^2}$, $B = \frac{2mk}{\mathscr{M}^2}$, $C = -1$, we obtain

$$\overline{\theta} - \alpha_{\mathscr{M}} = -\cos^{-1}\left(\frac{\frac{\mathscr{M}^2 u}{mk} - 1}{\sqrt{1 + \frac{2E\mathscr{M}^2}{mk^2}}}\right). \tag{3.214}$$

This result can be expressed as

$$\frac{\gamma}{r} = 1 + \varepsilon\cos(\overline{\theta} - \alpha_{\mathscr{M}}), \tag{3.215}$$

where $\gamma = \frac{\mathscr{M}^2}{mk}$ and $\varepsilon = \sqrt{1 + \frac{2E\mathscr{M}^2}{mk^2}}$. For $E < 0$ and $0 < \varepsilon < 1$, this represents the equation of an ellipse, with ε as its eccentricity.

The result derived above from the application of the Hamilton-Jacobi method verifies Kepler's first law of planetary motion, *viz* each planet moves in an elliptic orbit.

FURTHER READING

The following is a short list of selected textbooks which may be helpful for further reading on the topics covered in this chapter.

1. Calkin M G 1996 *Lagrangian and Hamiltonian Mechanics* (World Scientific: Singapore)
2. Fowles G R and Cassiday G L 2005 *Analytical Mechanics* 7th edn (Thomson: Belmont, CA)
3. Goldstein H, Poole C and Safko J 2002 *Classical Mechanics* 3rd edn (Addison-Wesley)
4. Gregory R D 2006 *Classical Mechanics* (Cambridge University Press: Cambridge)
5. Hand L N and Finch J D 1998 *Analytical Mechanics* (Cambridge University Press: Cambridge)
6. Landau L D and Lifshitz E M 1976 *Mechanics* 3rd edn (Pergamon: Oxford)
7. Landau L D and Lifshitz E M 1986 *Theory of Elasticity* 3rd edn (Butterworth and Heinemann: Oxford)
8. Leech J W 1965 *Classical Mechanics* 2nd edn (Methuen & Co. Ltd.: London)
9. Thornton S T and Marion J B 2004 *Classical Dynamics of Particles and Systems* 5th edn (Thomson: Belmont CA)
10. Woodhouse N M J 1987 *Introduction to Analytical Dynamics* (Clarendon: Oxford)

PROBLEMS

1. Consider the thermodynamic relation $\mathrm{dU} = \mathrm{TdS} = \mathrm{PdV}$, where U is the internal energy and S is the entropy of a gas in volume V at temperature T and pressure P. Use a Legendre transformation to convert from $\mathrm{U(S,V)}$ to $\mathrm{H(S,P)}$, where H is the thermodynamic potential.
2. The relativistic Lagrangian for a particle of rest mass m_0 moving along the x axis in a potential $V(x)$ is given by

$$L = m_0c^2\left[1-\left(1-\frac{\dot{x}^2}{c^2}\right)^{1/2}\right] - V(x),$$

where c is the speed of light. Apply the Legendre transformation to show that the Hamiltonian of the particle is given by

$$H = m_0c^2\left[1+\left(\frac{p_x}{m_0c}\right)^2\right]^{1/2} - m_0c^2 + V(x).$$

Derive Hamilton's equations of motion for the particle.

3. The Lagrangian function of a particle moving in an electromagnetic field can be written in the form

$$L = -m(\dot{x}_0^2 - \dot{x}_1^2 - \dot{x}_2^2 - \dot{x}_3^2)^{1/2} - e(\dot{x}_0A_0 - \dot{x}_1A_1 - \dot{x}_2A_2 - \dot{x}_3A_3),$$

where the coordinates x_i $(i = 0,1,2,3)$ are functions of a parameter τ: $x_i = x_i(\tau), \dot{x}_i = dx_i/d\tau$ and $A_i = A_i(x_0,x_1,x_2,x_3)$. Show that the Hamiltonian for the system vanishes.

4. A uniform rod of mass M and length l is fixed to a ceiling and is allowed to oscillate in a vertical plane due to its own weight. Show that the Lagrangian of the system can be expressed as

$$L = \frac{Ml^2}{6}\dot{\theta}^2 + \frac{Mgl}{2}\cos\theta,$$

where θ is the displacement angle from the vertical and g is the acceleration due to gravity. Using this expression for L and using an appropriate Legendre transformation, obtain an expression for the Hamiltonian of the system.
Construct and solve Hamilton's equations of motion for this system. Prove that for small oscillations, the pendulum executes simple harmonic motion. If $l = 0.5$ m, calculate the period of the motion.

5. The Lagrangian of a system can be written as

$$L = \frac{m}{2}(a\dot{x}^2 + 2b\dot{x}\dot{y} + c\dot{y}^2) - \frac{k}{2}(ax^2 + 2bxy + cy^2),$$

where a, b, c, and k are constants but subject to the condition that $b^2 - ac \neq 0$.
(a) Obtain expressions for the generalised-momentum components p_x and p_y.
(b) Derive Hamilton's equations of motion, and show that these represent a simple-harmonic motion in the x-y plane.

6. A particle moves on a plane under the influence of a force whose magnitude is $F = -\frac{b}{r^2}[1 - \frac{\dot{r}^2 - 2\ddot{r}r}{c^2}]$, where r is the distance from the centre of force and the force is directed towards the centre, and b and c are appropriate constants. Determine the generalised potential and hence the Hamiltonian for the system. What are the Hamilton's equations of motion?

7. In cylindrical coordinates (ρ, φ, z) the Lagrangian of a particle constrained to move on the surface of a cylinder of mass M and radius R can be written as

$$L = \frac{1}{2}M(R^2\dot{\varphi}^2 + \dot{z}^2) - \frac{1}{2}k(R^2 + z^2),$$

where k is a constant to describe the potential energy as $V = \frac{1}{2}k(R^2 + z^2)$. Using this expression, and making use of a Legendre transformation, obtain an expression for the Hamiltonian of the particle. Thus find the Hamilton's equations of motion and show that the motion in the z direction is simple harmonic.

8. Derive Hamilton's equations of motion for small amplitude oscillations of a small spherical pebble of radius a around the bottom of a frictionless hemispherical bowl of radius $R >> a$ fixed to a horizontal surface. Obtain an expression for the time period of the oscillation of the pebble.

9. From a pendulum of mass m_1 and length l_1 a second mass m_2 is freely suspended by an inextensible string of length l_2. Both pendula are allowed to oscillate in the vertical plane about their equilibrium positions. The Lagrangian of the system can be expressed as

$$\begin{aligned} L &= \frac{1}{2}m_1 l_1^2\dot{\theta}_1^2 + \frac{1}{2}m_2\left(l_1^2\dot{\theta}_1^2 + l_2^2\dot{\theta}_2^2 + 2l_1 l_2\dot{\theta}_1\dot{\theta}_2\cos(\theta_1 - \theta_2)\right) \\ &\quad + m_1 g l_1\cos\theta_1 + m_2 g(l_1\cos\theta_1 + l_2\cos\theta_2), \end{aligned}$$

where θ_1 and θ_2 are the angles from the vertical made by the first and second pendula, respectively. Outline the steps required for transforming the Lagrangian into the Hamiltonian of the system for the case $m_1 = m_2 = m$ and $l_1 = l_2 = l$. Obtain Hamilton's equations of motion. Using the equations of motion, derive expressions for the angular accelerations of the two bobs.

10. Prove the Poisson bracket identity $[X, YZ] = Y[X, Z] + Z[X, Y]$.

11. Given $H = \frac{p^2}{2m} + V(x)$, use the Hamilton's equation of motion $\dot{x} = [x, H]$ to prove that the linear momentum of a particle of mass m is $p = m\dot{x}$.

12. Given $H = \frac{p^2}{2m} + V(x)$, use the Hamilton's equation of motion $\dot{p} = [p, H]$ to prove that $\dot{p} = F$, where F is the force acting on the particle.

13. Prove $[p_x, \mathscr{M}_y] = p_z$, where p_i and $\mathscr{M}_i$ are, respectively, components of linear and angular momenta of a particle.
14. Prove the result $\sum_k \{F_k, F_l\}[F_k, F_m] = \delta_{lm}$, where $\{,\}$ and $[,]$ indicate Lagrange and Poisson brackets, respectively.
15. The Hamiltonian for a simple problem of one-dimensional motion with a constant acceleration a is

$$H = \frac{p^2}{2m} - max,$$

where x is the position coordinate and p is the linear momentum. If x_0 and p_0 are the values at the start of motion ($t = 0$), prove that the kinematic solution can be expressed as

$$x = x_0 + \frac{p_0 t}{m} + \frac{at^2}{2}.$$

16. Show that the phase space transformation $(q, p) \to (Q, P)$ given by the equations

$$Q = \left(\exp(-2q) - p^2\right)^{1/2}, \quad P = \cos^{-1}\left(p\exp(q)\right)$$

is canonical.
17. The Hamiltonian of a one-dimensional simple harmonic oscillator can be written as

$$H = \frac{1}{2m}(p^2 + m^2\omega^2 q^2),$$

where m is the mass, ω is the circular frequency, q is the position coordinate and p is the momentum. A canonical transformation $(q, p) \to (Q, P)$ of the form

$$p = f(P)\cos Q, \quad q = \frac{f(P)}{m\omega}\sin Q$$

is made. Express the transformed Hamiltonian and show that Q is cyclic. Find the function $f(P)$ and a generating function of the transformation.
18. If the equations

$$Q = q^\alpha \cos(\beta p), \quad P = q^\alpha \sin(\beta p)$$

represent a canonical transformation from a phase space (q, p) to another phase space (Q, P), then find α and β. For these values of α and β obtain an expression for a generating function of the transformation.
19. Obtain an expression for Hamilton's principal function S for the free fall of a particle under gravity.
20. Consider the two-dimensional central force problem in plane polar coordinates (ρ, ϕ). Write down the Hamilton-Jacobi equation for a particle of mass m, and show that the equation is separable in ρ and ϕ. Obtain an expression for Hamilton's principal function S, and using the results prove that the angular momentum is conserved.
21. Derive equation (3.195).
22. Show that, using the plane polar coordinates (r, θ), the equation of an ellipse of semi-major axis a and eccentricity ε and centred at its focus $(-a\varepsilon, 0)$ can be expressed as $\frac{a(1-\varepsilon^2)}{r} = 1 + \varepsilon\cos\theta$.
23. Show that the orbit expressed in equation (3.215) is a circle for $E < 0$ and $\varepsilon = 0$.

Part III

Linear dynamics of interacting particles

4 Dynamics of Linear Molecules and Linear Atomic Chains

4.1 INTRODUCTION

In chapters 2 and 3 we studied dynamics of single and double particles. In this chapter we will study the dynamics of linear molecules and linear chains of atoms. Consideration of linear chains of atoms will be made at two levels: a small number of atoms and infinitely large number of atoms. In order to discuss the dynamics of such systems we will need to develop new concepts. Although in principle any of the dynamical theories described in chapters 2 and 3 can be applied, we will apply the commonly used Newtonian approach.

4.2 TWO COUPLED HARMONIC OSCILLATORS

Let us consider two masses m_1 and m_2 joined together by a spring of force constant Λ and held by springs of force constant λ to fixed positions, all in one line, as shown in figure 4.1. We will assume the spring potential energy in the harmonic form $V(u) = \frac{1}{2}ku^2$, or equivalently the force expression on each atom in the form $F(x) = -ku$, where u is the displacement of the atom from its equilibrium and k is the spring force constant.

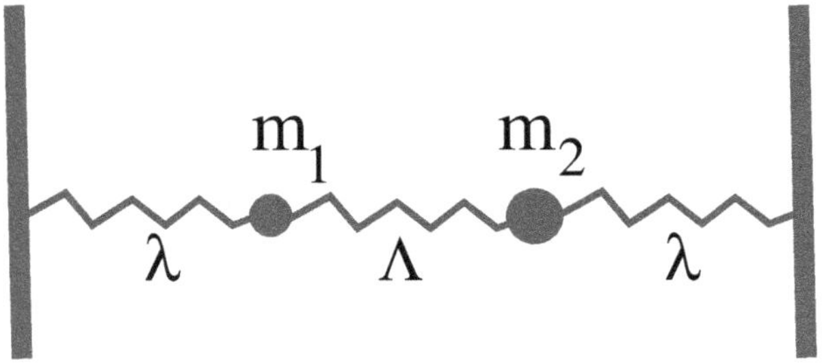

Figure 4.1 Two coupled oscillators. The system consists of two masses m_1 and m_2 joined together by a string of force constant Λ and held by strings of force constant λ to fixed positions, all in one line.

Considering u_1 and u_2 respectively as the displacements of masses m_1 and m_2, the equations of motion of the masses are

$$m_1\ddot{u}_1 = -\lambda u_1 - \Lambda(u_1 - u_2), \tag{4.1}$$

$$m_2\ddot{u}_2 = -\lambda u_2 - \Lambda(u_2 - u_1). \tag{4.2}$$

Following the result derived in equation (3.174) for the one-dimensional harmonic oscillator, we can try solutions of the equations of motion in equations (4.1) and (4.2) as

$$u_1(t) = A_1\exp(i\omega t), \quad u_2(t) = A_2\exp(i\omega t), \tag{4.3}$$

DOI: 10.1201/9781003383314-4

where ω is a double-valued oscillatory frequency and A_1 and A_2 are the vibrational amplitudes of the two atoms. With these trial solutions, equations (4.1) and (4.2) can be expressed in the form of the following matrix eigenvalue equation

$$\begin{bmatrix} \Lambda+\lambda-m_1\omega^2 & -\Lambda \\ -\Lambda & \Lambda+\lambda-m_2\omega^2 \end{bmatrix} \begin{bmatrix} A_1 \\ A_2 \end{bmatrix} = 0. \tag{4.4}$$

Following the discussion in Appendix D, we write this as $\boldsymbol{DA} = \omega^2\boldsymbol{A}$, where $\boldsymbol{A} = (A_1, A_2)$ and for obtaining non-trivial solutions solve the secular equation $|D-\omega^2 I| = 0$. The eigenfrequencies are obtained from the expression

$$\omega^2 = \frac{(m_1+m_2)(\Lambda+\lambda) \pm \sqrt{(m_1+m_2)^2\Lambda^2 + (m_1-m_2)^2(\lambda^2+2\Lambda\lambda)}}{2m_1m_2}. \tag{4.5}$$

We will further discuss the solutions for the case $m_1 = m_2 = m$. The expression in equation (4.5) reduces to (only the solution with the positive root is taken)

$$\omega_{1,2} = \sqrt{\frac{\Lambda+\lambda\pm\Lambda}{m}}. \tag{4.6}$$

Thus, the two vibrational frequencies are

$$\omega_1 = \sqrt{\frac{\lambda}{m}}, \quad \omega_2 = \sqrt{\frac{2\Lambda+\lambda}{m}}. \tag{4.7}$$

As discussed in Appendix D, each of the amplitude components will be given an additional label corresponding to the two eigenvalues. In other words, we will write A_1 as A_{1j} and A_2 as A_{2j}, with $j = 1,2$ corresponding to the frequencies ω_1 and ω_2, respectively. Using the results in equation (4.7), we obtain from equation (4.4),

$$A_{11} = A_{21} \quad \text{for frequency } \omega_1. \tag{4.8}$$

$$A_{12} = -A_{22} \text{ for frequency } \omega_2, \tag{4.9}$$

These results indicate that the two masses oscillate in phase with each other for the lower frequency ω_1 and out of phase with each other for the higher frequency ω_2.

4.3 LINEAR TRIATOMIC MOLECULE

The topic of molecular dynamics is of great interest and has been discussed in most chemistry books. In general, molecular dynamics can take the form of *translational*, *rotational* and *vibrational* degrees of freedom. Here our interest lies in discussing the vibrational (or oscillatory) motion of a small linear molecule. Such a system can have oscillations along the chain containing the atoms (*longitudinal vibrations*), or perpendicular to the chain in a plane containing the atoms (*transverse vibrations*).

We will consider *longitudinal vibrations* of a linear symmetric triatomic molecule. As shown in figure 4.2, we will consider an atom of mass m connected with a harmonic spring of force constant Λ with two equidistant atoms of mass M on both sides.

Following the method used in the previous section, we can write down the equations of motion of the three atoms. Considering the displacements u_1, u_2 and u_3 respectively of the atoms on the left, middle and right, the equations of motion are

$$M\ddot{u}_1 = -\Lambda(u_1-u_2), \tag{4.10}$$

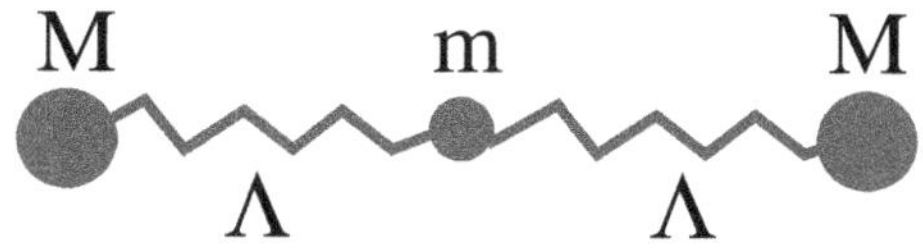

Figure 4.2 A symmetric linear triatomic molecule. The system consists of an atom of mass m connected with a harmonic spring of force constant Λ with two equidistant atoms of mass M on both sides.

$$m\ddot{u}_2 = -\Lambda(u_2 - u_1) - \Lambda(u_2 - u_3), \tag{4.11}$$

$$M\ddot{u}_3 = -\Lambda(u_3 - u_2). \tag{4.12}$$

With the trial solutions

$$u_1 = A_1 \exp(i\omega t), \quad u_2 = A_2 \exp(i\omega t), \quad u_3 = A_3 \exp(i\omega t) \tag{4.13}$$

equations (4.10)–(4.12) can be expressed as

$$\begin{bmatrix} \Lambda - M\omega^2 & -\Lambda & 0 \\ -\Lambda & 2\Lambda - m\omega^2 & -\Lambda \\ 0 & -\Lambda & \Lambda - M\omega^2 \end{bmatrix} \begin{bmatrix} A_1 \\ A_2 \\ A_3 \end{bmatrix} = 0. \tag{4.14}$$

Setting the determinant of the 3×3 matrix to zero provides non-trivial eigenfrequency solutions. These are

$$\omega_1 = 0, \quad \omega_2 = \sqrt{\frac{\Lambda}{M}}, \quad \omega_3 = \sqrt{\frac{\Lambda}{M}\left(1 + \frac{2M}{m}\right)}. \tag{4.15}$$

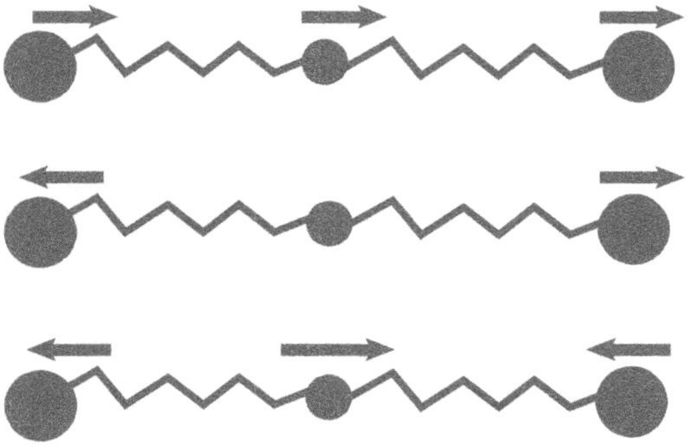

Figure 4.3 Longitudinal characteristic modes of a linear symmetric triatomic molecule consisting of an atom of mass m equidistant from two atoms of mass M each on both sides. The force constant between neighbouring atoms is Λ. Modes (a), (b) and (c) correspond respectively to the characteristic frequencies $\omega = \omega_1 = 0$, $\omega = \omega_2 = \sqrt{\frac{\Lambda}{M}}$ and $\omega = \omega_3 = \sqrt{\frac{\Lambda}{M}\left(1 + \frac{2M}{m}\right)}$. (The results are plotted by considering $2M > m$.) In the chemistry literature, the modes ω_2 and ω_3 are referred to as the symmetric and anti-symmetric stretch modes, respectively.

The phase relationships between atomic oscillations can be obtained by substituting each the three eigenvalues ω_1, ω_2 and ω_3 in equation (4.14).

For $\omega = \omega_1 = 0$, we obtain

$$A_{11} = A_{21} = A_{31}, \tag{4.16}$$

which means that all three atoms in the molecule oscillate in phase.

For $\omega = \omega_2 = \sqrt{\frac{\Lambda}{M}}$, we obtain

$$A_{12} = -A_{32}, \quad A_{22} = 0. \tag{4.17}$$

For $\omega = \omega_3 = \sqrt{\frac{\Lambda}{M}\left(1 + \frac{2M}{m}\right)}$, we obtain

$$A_{13} = -\frac{m}{2M}A_{23} = A_{33}. \tag{4.18}$$

Figure 4.3 illustrates these characteristic modes of the linear symmetric triatomic molecule. The dynamical eigenmode corresponding to the zero frequency ($\omega_1 = 0$) and equal amplitude for all three atoms clearly represents translational motion of the molecule and is of no interest from oscillatory viewpoint. Thus, the triatomic linear molecule possesses two vibrational frequencies ω_2 and ω_3. The mode with frequency ω_2 results from the antiphase vibrations of equal amplitude for the masses at the two ends, while the atom in the middle remains stationary. For the mode with frequency ω_3 the atoms at the two ends vibrate in phase with equal amplitude but in antiphase with the atom in the middle. The magnitude of the amplitude of the atom in the middle is $2M/m$ times that of an atom at the end. In the plot we have used a slightly larger arrow for the vibration of the middle atom, which indicates the choice $2M > m$. In the chemistry literature, the modes ω_2 and ω_3 are referred to as the symmetric and anti-symmetric stretch modes, respectively. Clearly, the symmetric stretch requires less energy than the antisymmetric stretch.

We mentioned at the start of this section that we have considered pure longitudinal vibration of a linear triatomic molecule. Considerations of both longitudinal and transverse vibrations are required for explaining stretch as well as bending modes of triatomic molecules in general. For example, CO_2 (for which $m < M$ is true) is a linear triatomic molecule in its free state. However, it becomes a bent molecule in its excited state or when it is chemisorbed on solid surfaces. Inclusion of transverse vibration is required for explaining the bending mode of CO_2, which is of great practical importance in the greenhouse gas context.

4.4 LINEAR CHAIN OF A LARGE NUMBER OF ATOMS

In the previous two sections we considered the longitudinal oscillatory dynamics of a small number of atoms along a line. In the present and further sections in this chapter we will consider the oscillatory dynamics of linear chains of a very large number of atoms. We are faced with an immediate question. Do we deal with the motion of all atoms, including the atoms at the two ends of the chain? In other words, we have to think whether the end (or surface) effect is important or not. If it is, then we should expect dynamical results for both bulk (*i.e.* interior) as well as end (*i.e.* surface) atoms. If the surface effect is unimportant, then we should mathematically model the system in such a manner that eliminates it. Both scenarios are cherished topics of *lattice dynamics* in condensed matter physics. Detailed discussion can be found in several textbooks, including Born and Huang (1954) and Srivastava (2022). Here, we will provide a simple coverage of both scenarios.

Let us for now consider a very large number of oscillators of mass m joined together along a straight line by a harmonic spring of force constant Λ, as shown in figure 4.4. Let u_{n-1}, u_n and u_{n+1} represent the displacements of the $(n-1)$th, nth and $(n+1)$th atoms, respectively. Following the discussion in the previous two sections, the equation of motion of the nth atom is

$$m\ddot{u}_n = -\Lambda(u_n - u_{n-1}) - \Lambda(u_n - u_{n+1}). \tag{4.19}$$

For a small number of coupled oscillators, considered in the previous sections, we sought solutions as a function of time in the form $u_n = u_n(t)$. For a very large number of oscillators, we should

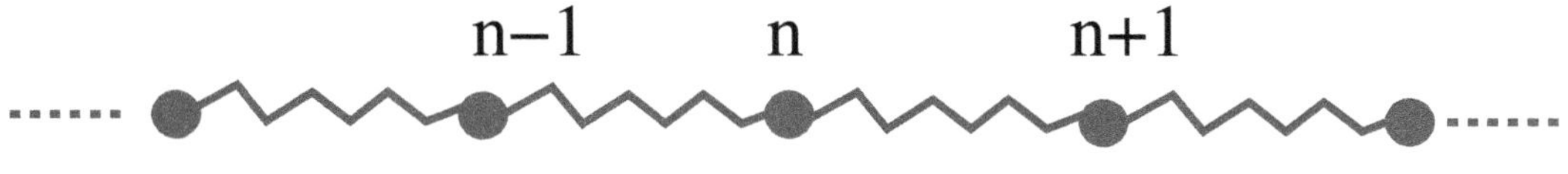

Figure 4.4 A very large number of equidistant harmonic oscillators coupled together along a straight line. Each oscillator has mass m and the indices $n-1$, n and $n+1$ indicate locations of neighbouring oscillators. The spring constant is Λ.

look for solutions of the form $u_n = u_n(x,t)$, where x is distance of the nth oscillator along the chain. Using the method of *multiplicative separation of variables*, we will express

$$u(x,t) = u_x(x)u_t(t). \tag{4.20}$$

The solution for the time part, similar to those in the previous sections [*cf* equation (3.174)], is of the form $u_t(t) \propto \exp(i\omega t)$. After the substitution of the time part of the solution, equation (4.19) takes the form

$$-m\omega^2 u_n(x) = \Lambda\left[u_{n-1}(x) + u_{n+1}(x) - 2u_n(x)\right]. \tag{4.21}$$

This represents a set of linear difference equations which could be solved for $u_n(x)$ in terms of $u_{n-1}(x)$ and $u_{n+1}(x)$. We note that this equation is the finite difference version of a linear differential equation of the type $\frac{\mathrm{d}^2 f(x)}{\mathrm{d}x^2} + q^2 f(x) = 0$, which has a solution of the form $f(x) \propto \exp(iqx)$ [*cf* equation (3.174)]. Putting things together, solutions to equation (4.19) can hence be considered in the form

$$u_n(x,t) = A\exp[i(qx - \omega t)], \tag{4.22}$$

where q represents a momentum variable[1] conjugate to the coordinate x. Such a sinusoidal wave travels in the (x-p) phase space as well as the (ω-t) phase space, with q pointing in the direction of increasing x. In one-dimensional case q is a wavenumber, and in higher-dimensional cases it turns into a wavevector.

It is useful to mention here that a rigorous analysis of the solution of the oscillatory motion of a very large number of coupled harmonic oscillators, either in one dimension or in a higher dimension, as a superposition of simple harmonic motions in the form of equation (4.21) can be performed elegantly by introducing *new generalised coordinates* and their *conjugate momenta*, and making a Fourier decomposition. The new coordinates thus generated are known as *normal coordinates*. Use of normal coordinates helps present expressions for the Lagrangian and Hamiltonian of a dynamical system in diagonal form. In other words, the Lagrangian and Hamiltonian are expressed as a simple sum of contributions from simple oscillators, without any cross-terms. We refrain from explaining the procedure here, but refer the interested reader to advanced textbooks [such as Ziman (1960), Srivastava (2022)] and section 9.4.2 of this book.

4.5 INFINITELY PERIODIC LINEAR ATOMIC CHAIN

We will now consider linear chains containing an infinite number of atoms as harmonic oscillators arranged periodically. We will also assume that only nearest-neighbour forces are significant. In order to develop the lattice dynamical theory for such a system, we will invoke a few new concepts,

---

[1]Note that in contrast to previous chapters, in this chapter and the next, q (or $\boldsymbol{q}$) will be used for momentum, not a generalised coordinate. Similarly, x (or $\boldsymbol{r}$) will be used for coordinate. The change in notation is unfortunate, but is unavoidable.

such as *lattice, periodic, cyclic or Born–von Kármán boundary condition, translational symmetry, unit cell, Brillouin zone, dispersion of vibrational energy of normal modes.*

Lattice and crystal structure: A *Bravais lattice* is a mathematical realisation in coordinate space, consisting of a regular arrangement of 'points' with identical surroundings. A crystal structure is easily visualised by assigning a *basis* of atom(s) to each lattice point. Mathematically, we can write

$$crystal = lattice + basis.$$

The significance of the basis will become clear when we discuss the dynamics of different types of atomic chain in this chapter and crystal structure in the next chapter.

Translational symmetry: Starting from one lattice point we can reach out to any other lattice point using the concept of *translational symmetry.*

Unit cell: A region containing one full lattice point is a primitive *unit cell.* Each primitive unit cell in a crystal contains a basis (or, collection of atoms).

Reciprocal lattice: Corresponding to a chosen lattice, there is a canonically defined *reciprocal lattice*, realised in momentum space.

Brillouin zone: A region spread symmetrically and containing a single reciprocal lattice point is called the *Brillouin zone.* It is more accurately called the first, or central, Brillouin zone. Clearly, this region is a primitive unit cell of the reciprocal lattice.

We have provided a brief description of the concepts related to the reciprocal lattice and the Brillouin zone in Appendix E.

Dispersion relation

Equation (4.22) reveals that oscillatory motions of a large number of particles comprising a system can be viewed as travelling waves of the form $\exp[i(\boldsymbol{q}\cdot\boldsymbol{r}-\omega t)]$, where $\boldsymbol{r}$ is a position vector in real space, $\boldsymbol{q}$ is the conjugate momentum space variable, ω is the vibrational frequency and t is the time. For small molecules we examine the relationship between $\boldsymbol{r}$ and t. It becomes very convenient, however, to discuss the dynamical behaviour of an infinitely large number of particles arranged in a periodic structure by examining the relationship between the variables ω and $\boldsymbol{q}$. The dependence of ω on $\boldsymbol{q}$ in the form $\omega = \omega(\boldsymbol{q})$ is called *dispersion relation*, and a plot of ω as a function of $\boldsymbol{q}$ is called a *dispersion curve.*

Example of a one-dimensional system

Figure 4.5(a) shows a sketch of the real space lattice which is used to guide arrangement of atoms for a one-dimensional monatomic chain. Also indicated is one choice of the unit cell. Clearly, there is just one atom per unit cell. Figure 4.5(b) shows the corresponding reciprocal lattice and the (central, or first) Brillouin zone. If the length of the real space primitive unit cell is a then the size of the Brillouin zone is $2\pi/a$. The real space primitive translation vector can be expressed as $\boldsymbol{a} = a\hat{\boldsymbol{x}}$, where $\hat{\boldsymbol{x}}$ is the direction along the atomic chain. The primitive translation vector of the reciprocal lattice is $\boldsymbol{b} = \frac{2\pi}{a}\hat{\boldsymbol{x}}$. The Brillouin zone is the region between $-\frac{1}{2}\boldsymbol{b}$ and $\frac{1}{2}\boldsymbol{b}$. Independent values of the momentum variable $\boldsymbol{q}$ span the Brillouin zone. The vector $\boldsymbol{T} = j\boldsymbol{a}$, where j is any integer, is a real space translation vector. Similarly, the vector $\boldsymbol{G} = l\boldsymbol{b}$, where l is an integer, is a reciprocal space translation vector. It is easy to verify that real space and reciprocal space translation vectors are linked through the relationship $\exp(i\boldsymbol{G}\cdot\boldsymbol{T}) = 1$.

Born–von Kármán boundary condition

A mathematical scheme to avoid dealing with discussion of vibrations of the ends of a finite chain is the so-called *periodic, cyclic, or Born–von Kármán boundary condition.* This amounts to considering the $(N+1)$st atom as the 1st atom. It is assumed that the Nth and $(N+1)$st atoms are joined by an additional spring which is identical to the spring in the chain. Graphically, this

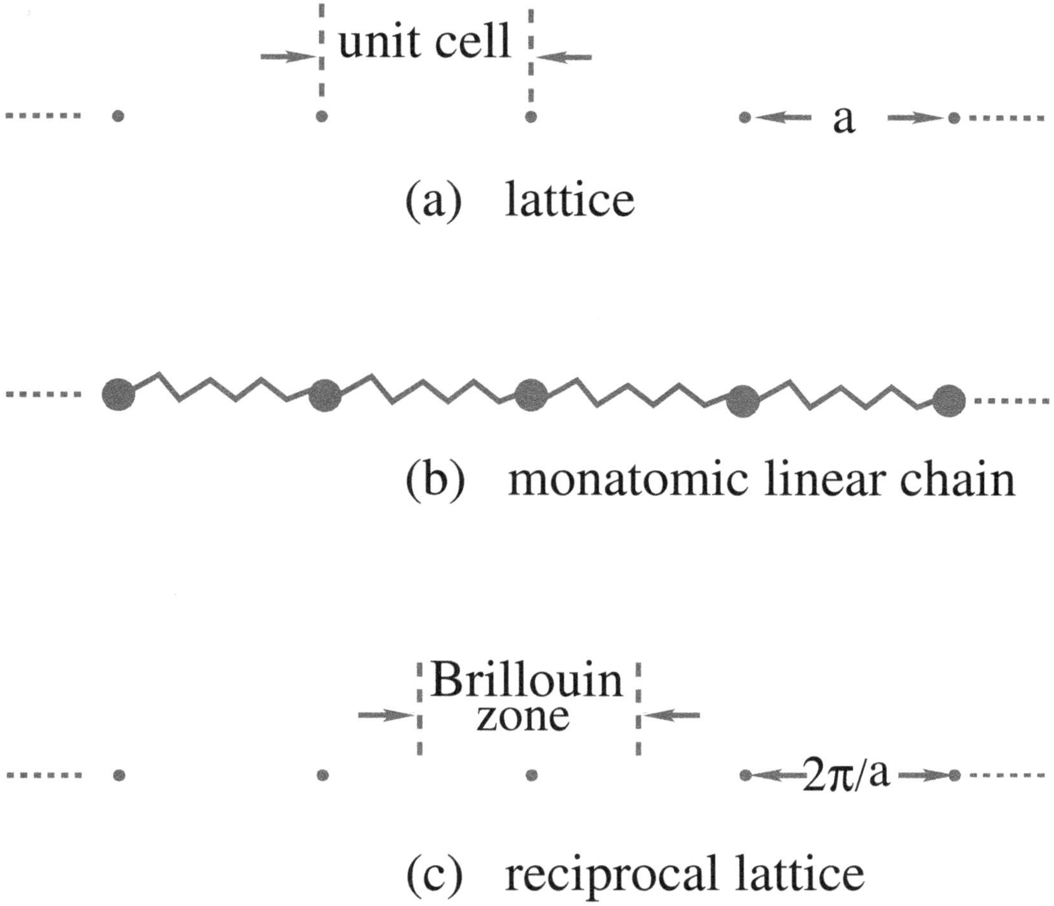

Figure 4.5 (a) A one-dimensional coordinate space lattice, with one choice of unit cell marked. (b) An infinitely large linear chain of identical atoms, with one atom per unit cell. The atoms are connected by a harmonic spring. (c) Reciprocal lattice of the one-dimensional direct space lattice in (a), with the Brillouin zone marked.

means that the chain has been folded into a loop, as shown in figure 4.6. The atoms in the chain can vibrate either along the chain (longitudinal polarisation) or along one of two directions perpendicular to the chain (transverse polarisations), subject to the periodic boundary condition for atomic displacements $u_1 = u_{N+1}$. Using the solution $u_n(x,t) = A\exp[i(qx - \omega t)]$, we then obtain $\exp(iNqa) = 1 \equiv \exp(2\pi ni)$, where n is any integer. Noting that the Brillouin zone boundaries are at $\pm\frac{\pi}{a}$, the allowed N independent values of q are $q = 0, \pm\frac{2\pi}{Na}, \pm\frac{4\pi}{Na}, \cdots, \frac{\pi}{a}$. The interval between successive values of q is $\Delta q = \frac{2\pi}{Na}$. The number of modes per unit range of q is $\frac{Na}{2\pi}$. For an infinitely large lattice $N \to \infty$ and q turns into a continuous variable inside the Brillouin zone.

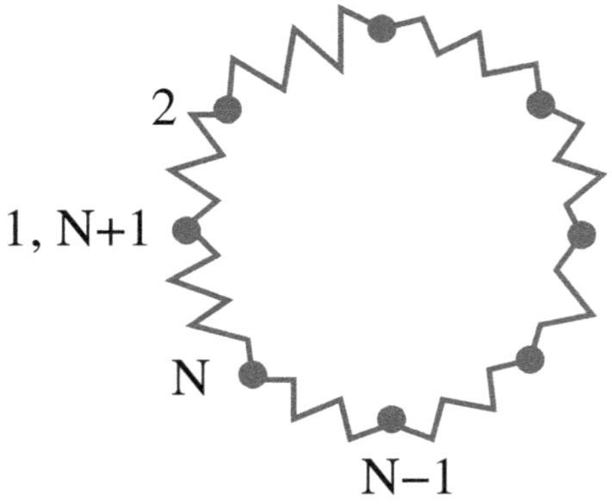

Figure 4.6 The Born–von Kármán boundary condition applied to an infinitely long linear chain of atoms to avoid dealing with 'end' or 'surface' effects. In this mathematical scheme an infinitely long chain of N atoms is folded onto itself, so that the $(N+1)$st atom is the same as the 1st atom. This requires the assumption that there is an additional spring, which is identical to that in the chain, joining the Nth and $(N+1)$st atoms.

4.5.1 MONATOMIC LINEAR CHAIN

As explained earlier, a monatomic chain implies a basis comprising a single atom. Let us consider a monatomic linear chain with atomic mass m, nearest-neighbour separation a and nearest-neighbour harmonic force constant Λ. Let the jth atom be at distance ja from a reference lattice site. As shown in figure 4.7, let u_{n-1}, u_n and u_{n+1} represent longitudinal displacements from equilibrium of the $(n-1)$th, nth and $(n+1)$th atoms, respectively. The equation of motion of the nth atom, presented in equation (4.19), is

$$m\ddot{u}_n = -\Lambda\left(2u_n - u_{(n-1)} - u_{(n+1)}\right). \tag{4.23}$$

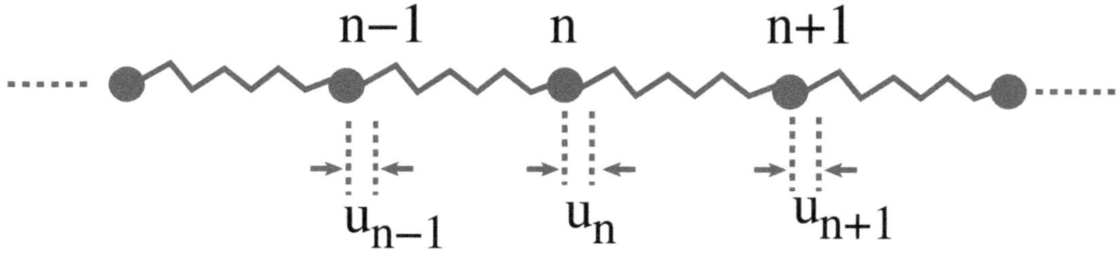

Figure 4.7 A monatomic linear chain with atoms connected by a harmonic spring. u_{n-1}, u_n and u_{n+1} indicate longitudinal displacements from equilibrium of the $(n-1)$st, nth and $(n+1)$st atoms in the chain.

Following equation (4.22) we seek solution for the displacement of the jth atom in the form

$$u_j = A\exp[i(ja - \omega t)], \tag{4.24}$$

where A is the amplitude of the motion. Substitution of equation (4.24) in equation (4.23) produces the eigenvalue result

$$\omega^2 = \frac{2\Lambda}{m}(1 - \cos qa). \tag{4.25}$$

From this we obtain the following relationship between frequency ω and wavenumber q (known as the dispersion relation)

$$\omega = 2\sqrt{\frac{\Lambda}{m}}|\sin\frac{qa}{2}|. \tag{4.26}$$

Figure 4.8 shows the dispersion curve for the longitudinal normal modes of the chain. In accordance with the translation symmetry, we only need to examine the part of the curve within the Brillouin zone. We can also notice that the curve shows the inversion symmetry, *viz* $\omega(-q) = \omega(q)$. This means that the essential feature of the dispersion curve lies in the irreducible part of the Brillouin zone, *viz* for q values in the range $[0, \pi/a]$. The frequency is zero for the infinitely long wave (*i.e.* when $q = 0$). The maximum frequency is $\omega_{\max} = 2\sqrt{\frac{\Lambda}{m}}$ and occurs when the wavenumber reaches the Brillouin zone boundary (*i.e.* for the shortest wavelength corresponding to $q = \pm\pi/a$). A few interesting features of the results can be noted.

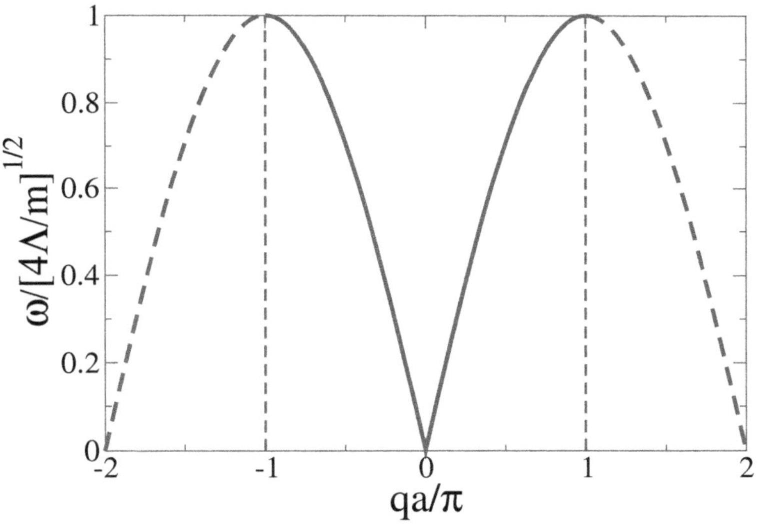

Figure 4.8 Dispersion relation (ω *vs* q) for normal modes of a monatomic linear chain. m and Λ represent, respectively, atomic mass and interatomic harmonic force constant. The portion within the first Brillouin zone is shown with the solid curve.

(i) While a simple harmonic oscillator oscillates with frequency $\omega_{\text{SHO}} = \sqrt{\frac{\Lambda}{m}}$, a system comprising an infinitely large number of simple harmonic oscillators coupled to their neighbours along a line vibrates with frequencies in the range $0 \leq \omega \leq 2\sqrt{\frac{\Lambda}{m}}$.

(ii) We can define two types of velocity associated with the vibrations of the chain. **Phase velocity** v_p is defined as the ratio of the frequency and the wavenumber. **Group velocity** v_g is defined as the gradient of the frequency along the q-axis Using the dispersion relation in equation (4.26), the expressions for the two types of velocity are

$$v_p = \omega/q = a\sqrt{\frac{\Lambda}{m}}\frac{\sin(\frac{qa}{2})}{(\frac{qa}{2})}, \tag{4.27}$$

$$v_g = \frac{d\omega}{dq} = a\sqrt{\frac{\Lambda}{m}}\cos(\frac{qa}{2}). \tag{4.28}$$

The group velocity vanishes at the Brillouin zone boundaries, for which $q = \pm\frac{\pi}{2}$, *i.e.* when the vibrational wavelength becomes twice the interatomic distance: $\lambda = 2a$. Using equation (4.24) it can be noted that at the Brillouin zone boundaries the vibrational phases of two neighbouring atoms differ by π. In other words, the lattice wave becomes a standing wave. This can be interpreted in terms of the Bragg diffraction condition $2d\sin\theta = n\lambda$, with $n = 1$, $d = a$ and $\theta = 90^o$.

(iii) In the long wavelength limit $qa << 1$, there is no distinction between the two types of velocity and the dispersion relation simplifies to $\omega = vq$, where $v = v_p = v_g$. In this limit we can write $u_n = u(x)$ and $u_{n\pm1} = u(x\pm a)$, with x considered as a continuous variable, and express $u(x\pm a)$ in a Taylor series

$$u(x\pm a) \approx u(x) \pm a\frac{\mathrm{d}u(x)}{\mathrm{d}x} + \frac{1}{2}a^2\frac{\mathrm{d}^2u(x)}{\mathrm{d}x^2}. \tag{4.29}$$

With these substitutions, equation (4.23) can be expressed in the form of a sound wave (or acoustic wave) equation

$$\frac{1}{v^2}\frac{\mathrm{d}^2u}{\mathrm{d}t^2} = \frac{\mathrm{d}^2u(x)}{\mathrm{d}x^2}. \tag{4.30}$$

Clearly, the vibrations of the linear chain can be described as those of an elastic continuum with sound velocity $v = a\sqrt{\frac{\Lambda}{m}}$. Moving from an atomic model to that of an elastic continuum, for long wavelengths the inter-atomic force constant Λ should be replaced in terms of a second-order elastic constant. We will deal with the dynamics of the elastic continuum in later chapters.

4.5.2 DIATOMIC LINEAR CHAIN

In the previous sub-section we studied the longitudinal vibrations of a monatomic linear chain of N atoms, subjected to the periodic boundary condition so that the $(N+1)$th atom is the same as the 1st atom. Each atom, of mass m, was separated from its neighbour by a distance a. We now consider a linear chain of $2N$ atoms subjected to the periodic boundary condition so the $(2N+1)$th atoms is the same as the 1st atom. Let alternate atoms have different masses (m and M), and each atom be separated from its neghbouring atoms by a distance a. This structure of the chain can be constructed by considering a linear lattice with primitive unit cell size $2a$ and a basis of two atoms (of masses m and M), as shown in figure 4.9. As before, we will consider harmonic forces operating between nearest neighbours only, with force constant Λ. Following the explanation for linear the monatomic chain in the previous sub-section and the discussion in Appendix E, it is easy to verify that the Brillouin zone for this lattice (which has a unit cell of size $2a$) is the region $[-\frac{\pi}{2a}, \frac{\pi}{2a}]$, with $[0, \frac{\pi}{2a}]$ being its irreducible portion. Note that this region is half the size of the Brillouin zone for the monatomic chain, for which the unit cell size is a.

Considering longitudinal displacements from equilibrium of atomic mass m as u and of atomic mass M as u', we can write down the equations of motion for the two basis atoms as

$$m\ddot{u}_{2n} = -\Lambda(2u_{2n} - u'_{2n+1} - u'_{2n-1}), \tag{4.31}$$

$$M\ddot{u}'_{2n+1} = -\Lambda(2u'_{2n+1} - u_{2n} - u_{2n+2}), \tag{4.32}$$

where Λ is the spring force constant.

We seek solutions of the forms

$$u_{2n} = A_1\exp[i(2nqa - \omega t)], \tag{4.33}$$

$$u'_{2n+1} = A_2\exp[i((2n+1)qa - \omega t)], \tag{4.34}$$

with A_1 and A_2 being the vibrational amplitudes for the atoms of masses m and M, respectively. With these, equations (4.31) and (4.32) turn into the following coupled eigenvalue equations

$$(2 - \omega^2 m)A_1 - \Lambda[\exp(iqa) + \exp(-iqa)]A_2 = 0, \tag{4.35}$$

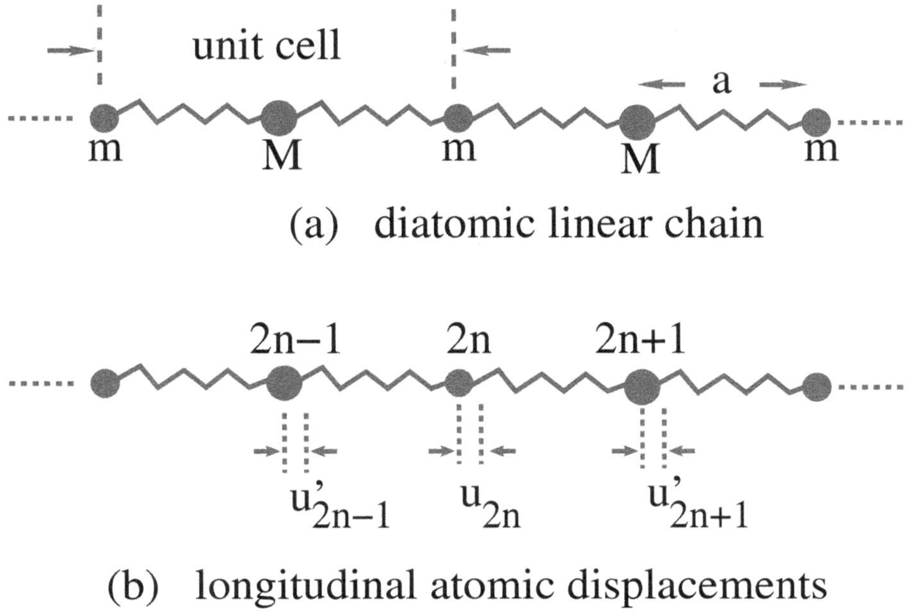

Figure 4.9 An infinitely long diatomic linear chain of masses m and M. Panel (a) shows the nearest neighbour distance a and the unit cell size $2a$. Panel (b) shows the longitudinal displacements u'_{2n-1}, u_{2n} and u'_{2n+1} of masses M, m and M for atoms labelled $2n-1$, $2n$ and $2n+1$, respectively.

$$(2-\omega^2 M)A_2 - \Lambda[\exp(iqa)+\exp(-iqa)]A_1 = 0. \tag{4.36}$$

These equations can be expressed in the following matrix form

$$\begin{bmatrix} (2\Lambda/m)-\omega^2 & -(2\Lambda/m)\cos(qa) \\ -(2\Lambda/M)\cos(qa) & (2\Lambda/M)-\omega^2 \end{bmatrix} \begin{bmatrix} A1 \\ A2 \end{bmatrix} = 0. \tag{4.37}$$

Non-trivial eigenvalue results, obtained by setting the determinant of the 2×2 matrix in equation (4.37) to zero, are

$$\omega^2 = \Lambda\left(\frac{1}{m}+\frac{1}{M}\right) \pm \Lambda\left[\left(\frac{1}{m}+\frac{1}{M}\right)^2 - \frac{2}{mM}(1-\cos(2qa))\right]^{1/2}. \tag{4.38}$$

Also, from equations (4.35) and (4.36), the ratio of the vibrational amplitudes of the two basis atoms can be expressed as

$$\frac{A_1}{A_2} = \frac{2\Lambda\cos(qa)}{2\Lambda-\omega^2 m} = \frac{2\Lambda - M\omega^2}{2\Lambda\cos(qa)}. \tag{4.39}$$

A simple computation effort is required to obtain the dispersion relation $\omega = \omega(q)$ from equation (4.38). It may also be interesting to examine the dispersion relation as a function of the mass ratio M/m. With this in mind, the expression in equation (4.38) can be written as

$$\omega^2 = \frac{\omega_0^2}{2}\left\{1 \pm \left[1 - \frac{4R}{(1+R)^2}\sin^2(qa)\right]^{1/2}\right\}, \tag{4.40}$$

where $R = M/m$ and $\omega_0^2 = 2\Lambda(\frac{1}{m} + \frac{1}{M})$. Corresponding to the two signs in this equation, there are two branches of the normal mode dispersion curve of the diatomic linear chain. The dispersion relation for $R = 2$ has been plotted in figure 4.10(a). Of course, for $R = 1$ (*i.e.* $m = M$) the chain turns into the monatomic chain discussed in the previous sub-section and the upper branch of the dispersion curve in 4.10(b) can be 'unfolded' to recover the dispersion curve of the monatomic linear chain over the full range of the Brillouin zone as shown in figure 4.8.

It is interesting to discuss the two vibrational frequency branches for the diatomic chain and the corresponding amplitudes of atomic displacements. This can be easily done at the Brillouin zone centre, close to the zone centre and at the boundary of the zone.

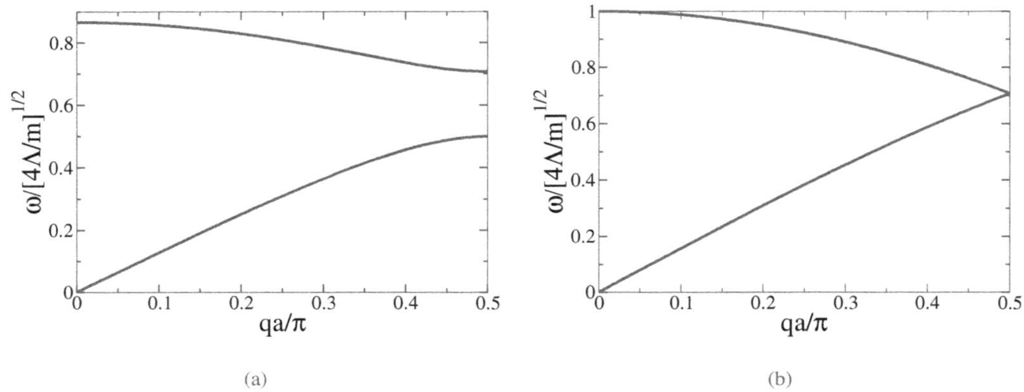

Figure 4.10 Dispersion relation for normal modes of a diatomic linear chain, calculated with the mass ratios (a) $M/m = 2$ and (b) $M/m = 1$. The nearest neighbour force constant is Λ. Two changes in the spectrum can be noticed when $M/m > 1$: the maximum frequency (at the zone centre) drops and a gap in the spectrum opens at the zone boundary.

Zone centre: At the zone centre $q = 0$ and the two frequencies and the corresponding amplitude ratios are:

$$\omega_1 = 0, \quad \tfrac{A_1}{A_2} = 1, \tag{4.41}$$

$$\omega_2 = \sqrt{2\Lambda(\frac{1}{m} + \frac{1}{M})}, \quad \tfrac{A_1}{A_2} = -1. \tag{4.42}$$

These results indicate that the two basis atoms vibrate in phase with each other for the lower frequency and out of phase with each other for the higher frequency.

Zone boundary: At the zone boundary $q = \frac{\pi}{2a}$ and the two frequencies and the corresponding amplitude ratios are:

$$\omega_1 = \sqrt{2\Lambda/M}, \quad \tfrac{A_1}{A_2} = 0, \tag{4.43}$$

$$\omega_2 = \sqrt{2\Lambda/m}, \quad \tfrac{A_1}{A_2} = \infty. \tag{4.44}$$

Consequently, the lower frequency is produced by only the vibrations of heavier masses, while the lighter masses are stationary. For the higher frequency only the lighter masses vibrate, while the heavier masses are stationary. The results of the dynamics of this system also mean that there are no travelling waves with real values of q inside the gap between these frequencies.

Close to the zone centre: Close to the zone centre we set $2qa << 1$ and expand $\cos(2qa) \approx 1 - \frac{(2qa)^2}{2}$. With this, equation (4.38) can be approximated as

$$\omega^2 = \Lambda\left(\frac{1}{m}+\frac{1}{M}\right) \pm \Lambda\left(\frac{1}{m}+\frac{1}{M}\right)\left[1-\frac{2mM}{(m+M)^2}(qa)^2\right] \tag{4.45}$$

With this, the dispersion relations for the two vibrational branches can be expressed as

$$\omega_1 = \sqrt{\frac{2\Lambda}{(m+M)}}qa, \tag{4.46}$$

$$\omega_2 = \sqrt{2\Lambda(\frac{1}{m}+\frac{1}{M})}\left[1-\frac{mM}{2(m+M)^2}(qa)^2\right]. \tag{4.47}$$

The lower branch, with dispersion relation $\omega = vq$ and in-phase vibrations of the two basis atoms, is the *acoustic branch*, The higher branch, with dispersion relation of the form $\omega_{\max} - \omega \propto q^2$ and with out of phase vibrations of the two basis atoms, is called the *optical branch*. It is so called because if the two basis atoms were charged with opposite polarities (*i.e.* for an ionic linear diatomic chain), this motion would set up an oscillating dipole moment, which can interact with the infrared region of the optical light spectrum.

4.5.3 TRIATOMIC LINEAR CHAIN

As a further example, we consider a triatomic linear chain. Let us say that there are $3N$ atoms subjected to the periodic boundary condition so the $(3N+1)$st atom is the same as the 1st atom. We consider a linear lattice with primitive unit cell size $3a$ and a basis of three atoms, of masses m_1, m_2 and m_3, as shown in figure 4.11. We further consider a as the inter-atomic separation and Λ as the harmonic force constant between nearest neighbours[2]. The irreducible part of the first Brillouin zone for this lattice is the region $[0, \frac{\pi}{3a}]$.

Considering longitudinal displacements from equilibrium of the basis atoms, of mass m_1 at location numbered $3n$, of mass m_2 at location numbered $3n+1$ and of mass m_3 at location numbered $3n+2$, the equations of motion are

$$m_1\ddot{u}_{3n} = -\Lambda(2u_{3n} - u'_{3n+1} - u''_{3n-1}), \tag{4.48}$$

$$m_2\ddot{u}'_{3n+1} = -\Lambda(2u'_{3n+1} - u''_{3n+2} - u_{3n}), \tag{4.49}$$

$$m_3\ddot{u}''_{3n+2} = -\Lambda(2u''_{3n+2} - u_{3n+3} - u'_{3n+1}). \tag{4.50}$$

To solve these equations, we try solutions of the forms

$$u_{3n} = \frac{B_1}{\sqrt{m_1}}\exp[i(3nqa - \omega t)], \tag{4.51}$$

$$u'_{3n+1} = \frac{B_2}{\sqrt{m_2}}\exp[i((3n+1)qa - \omega t)], \tag{4.52}$$

$$u''_{3n+2} = \frac{B_3}{\sqrt{m_3}}\exp[i((3n+2)qa - \omega t)], \tag{4.53}$$

---

[2] Kesavaswamy and Krishnamurthy (1978) have studied the general case with alternating force constants.

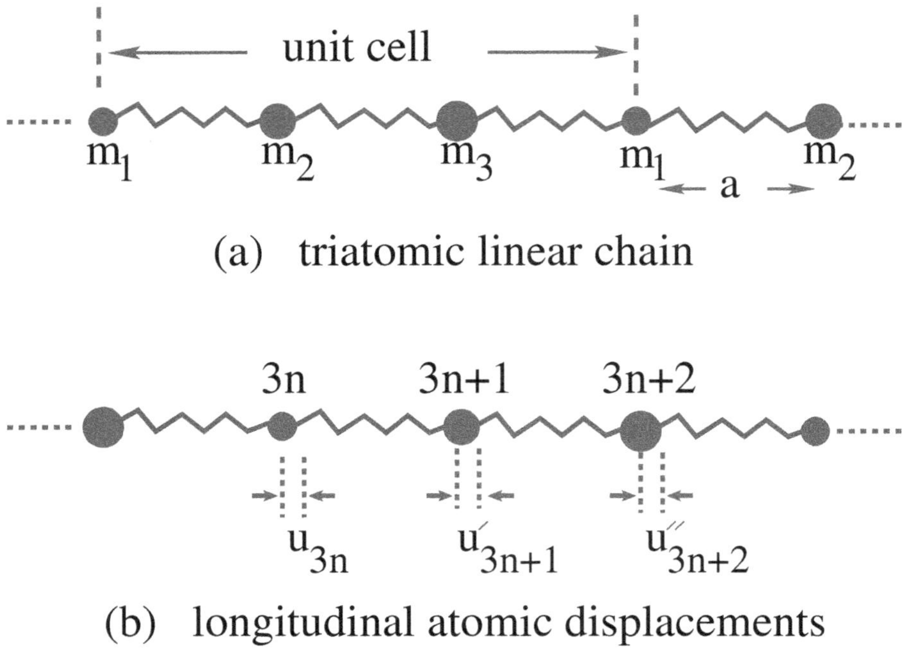

Figure 4.11 Triatomic linear chain. Panel (a) shows the nearest neighbour distance a and the unit cell of size $3a$ containing three masses m_1, m_2 and m_3. Panel (b) shows the longitudinal atomic displacements u_{3n}, u'_{3n+1} and u''_{3n+2} for atoms numbered $3n$, $3n+1$ and $3n+2$ with masses m_1, m_2 and m_3, respectively..

where B_1, B_2 and B_3 represent vibrational amplitudes. With these, the equations of motion in equations (4.48)-(4.50) can be expressed in the following matrix equation form

$$\begin{bmatrix} \frac{2\Lambda}{m_1} - \omega^2 & -\frac{\Lambda}{\sqrt{m_1 m_2}}\exp(iqa) & -\frac{\Lambda}{\sqrt{m_1 m_3}}\exp(-iqa) \\ -\frac{\Lambda}{\sqrt{m_2 m_1}}\exp(-iqa) & \frac{2\Lambda}{m_2} - \omega^2 & -\frac{\Lambda}{\sqrt{m_2 m_3}}\exp(iqa) \\ -\frac{\Lambda}{\sqrt{m_3 m_1}}\exp(iqa) & -\frac{\Lambda}{\sqrt{m_3 m_2}}\exp(-iqa) & \frac{2\Lambda}{m_3} - \omega^2 \end{bmatrix} \begin{bmatrix} B_1 \\ B_2 \\ B_3 \end{bmatrix} = 0. \tag{4.54}$$

Notice that the choice of solution in equations (4.48)-(4.50) has resulted in the generation of a Hermitian matrix in equation (4.54)[3]. The eigensolutions for an arbitrary q value in the Brillouin zone can be obtained by solving equation (4.54) numerically (see, *e.g.* Press *et al.* 1992).

Essentially, the eigenvalues are obtained by equating the determinant of the square matrix in equation (4.54) to zero. The resulting expression is

$$\omega^6 - 2\Lambda\left(\frac{1}{m_1} + \frac{1}{m_2} + \frac{1}{m_3}\right)\omega^4 + 3\Lambda^2\left(\frac{1}{m_1 m_2} + \frac{1}{m_1 m_3} + \frac{1}{m_2 m_3}\right)\omega^2 - \frac{2\Lambda^3}{m_1 m_2 m_3}\left[1 - \cos(3qa)\right] = 0. \tag{4.55}$$

Compact analytical solutions can only be obtained in special cases. In general, equation (4.55) is a bi-cubic equation (*i.e.* it becomes cubic in the variable $y = \omega^2$) and so the analytical solutions

[3]While a Hermitian matrix has complex numbers, its eigenvalues are always real.

can be found using Cardano's general cubic formula. We will consider $m_1 = m_2 = m_3 = m$ and examine the eigenvalues at the Brillouin zone centre ($q = 0$) and the zone edge ($q = \pi/3a$). For $m_1 = m_2 = m_3 = m$ equation (4.55) reads

$$\omega^6 - 6\left(\frac{\Lambda}{m}\right)\omega^4 + 9\left(\frac{\Lambda}{m}\right)^2\omega^2 - 2\left(\frac{\Lambda}{m}\right)^3\left[1 - \cos(3qa)\right] = 0. \tag{4.56}$$

At the zone centre ($q = 0$) the solutions are easily obtained as:

$$\omega_1 = 0, \quad \omega_2 = \omega_3 = \sqrt{\frac{3\Lambda}{m}}. \tag{4.57}$$

It is not so straight forward to obtain the solutions at the zone edge. Substituting $q = \pi/3a$, equation (4.56) reads

$$\omega^6 - 6\left(\frac{\Lambda}{m}\right)\omega^4 + 9\left(\frac{\Lambda}{m}\right)^2\omega^2 - 4\left(\frac{\Lambda}{m}\right)^3 = 0. \tag{4.58}$$

To solve this equation, we make the substitution (Kesavaswamy and Krishnamurthy, 1978)

$$\omega^2 = \gamma\eta + \frac{2\Lambda}{m}, \tag{4.59}$$

and express equation (4.58) in the form

$$\eta^3 - 3\left(\frac{\Lambda}{\gamma m}\right)^2\eta - 2\left(\frac{\Lambda}{\gamma m}\right)^3 = 0. \tag{4.60}$$

This is a depressed cubic equation of the form

$$\eta^3 - \frac{3}{4}\eta - \frac{1}{4}\cos(3\theta) = 0, \tag{4.61}$$

from which, remembering $\cos(3\theta) = 4\cos^3\theta - 3\cos\theta$, we deduce $\eta = \cos\theta$. Comparing equation (4.60) with equation (4.61), we thus obtain

$$\gamma = \frac{2\Lambda}{m}, \quad \text{and} \quad \left(\frac{2\Lambda}{\gamma m}\right)^3 = \cos(3\theta). \tag{4.62}$$

From these,

$$\cos(3\theta) = 1, \tag{4.63}$$

so that $\theta = 0, \frac{2\pi}{3}$ and $\frac{4\pi}{3}$. Thus, the solutions $\omega^2 = \gamma\eta + \frac{2\Lambda}{m} = \frac{2\Lambda}{m} + \gamma\cos\theta$ of equation (4.58) produce the results

$$\omega_1 = \sqrt{\frac{\Lambda}{m}}, \quad \omega_2 = \sqrt{\frac{\Lambda}{m}}, \quad \omega_3 = \sqrt{\frac{4\Lambda}{m}}. \tag{4.64}$$

Figure 4.12 shows computational results for the dispersion curves, obtained with two different mass considerations for the basis: panel (a) for $m_1 = m_3 = m$, $m_2 = 2m$ and panel (b) for $m_1 = m_2 = m_3 = m$.

For the case $m_1 = m_2 = m_3 = m$ and referring to the Brillouin zone for the triatomic chain, as analytically derived in equations (4.57) and (4.64), the frequencies are zero and $\sqrt{3\Lambda/m}$ (doubly-degenerate) at the zone centre, and $\sqrt{\Lambda/m}$ (doubly-degenerate) and $\sqrt{4\Lambda/m}$ at the zone edge. The upper two branches in figure 4.12(b) can be 'unfolded' to recover the dispersion curve in figure 4.8 for the monatomic linear chain.

For unequal basis masses, the degeneracies at the zone centre and the zone edge are lifted. As seen in figure 4.12(a), for $m_1 = m_3 = m$ and $m_2 = 2m$ there are two non-zero frequencies $0.707\sqrt{4\Lambda/m}$ and $0.866\sqrt{4\Lambda/m}$ at the zone centre and three non-zero frequencies $0.383\sqrt{4\Lambda/m}$,

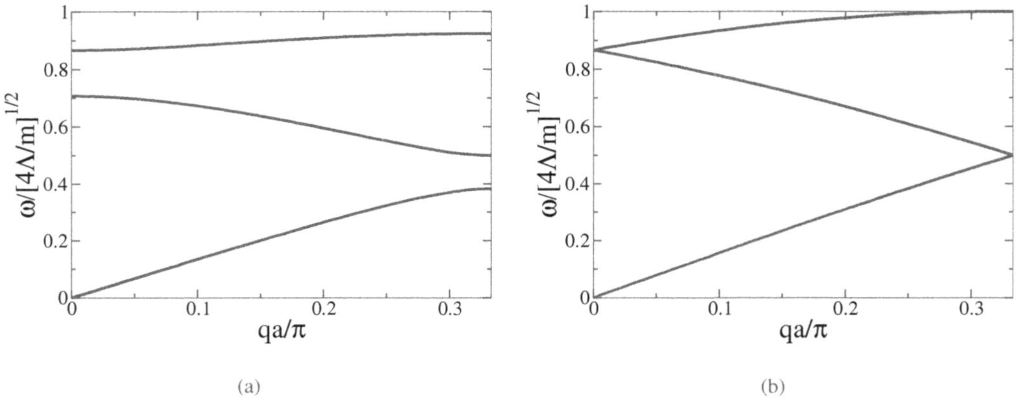

Figure 4.12 Computed dispersion relation for normal modes of a triatomic linear chain, obtained with the mass considerations for the basis (a) $m_1 = m_3 = m, m_2 = 2m$ and (b) $m_1 = m_2 = m_3 = m$. The nearest-neighbour force constant is Λ. It can be noticed that when the consideration $m_1 = m_2 = m_3 = m$ changes to $m_1 = m_3 = m$ and $m_2 = 2m$, the highest frequency (at the zone boundary) decreases and two band gaps appear (the upper one at the zone centre and the lower one at the zone boundary).

$0.5\sqrt{4\Lambda/m}$ and $0.923\sqrt{4\Lambda/m}$ at the zone edge. Thus, in general, there are two 'gaps' in the dispersion curve for a linear triatomic chain with a basis formula AB_2 with unequal masses for the basis atoms A and B.

It will be left as a computational exercise to examine the effect of alternating force constants, and the combined effect of alternating force constants and different masses of basis atoms, on the dispersion curves of a linear triatomic chain.

From the results obtained for the monatomic, diatomic and triatomic linear chains, we are in a position to make a couple of general points regarding longitudinal vibrational dynamics. (1) While isolated molecules vibrate with a fixed number of frequency(ies), formation of an atomic chain generates an spectrum of frequencies. (2) For a linear chain with n atoms/cell, there will be one longitudinal acoustic branch (the lowest lying branch) and (n-1) longitudinal optical branches. If all basis atoms in the unit cell have different masses, then neighbouring branches can potentially have a frequency gap.

4.6 SEMI-INFINITE LINEAR ATOMIC CHAIN

In the previous section we studied the normal modes of vibrations in infinitely periodic linear atomic chains. The cyclic, or periodic, boundary condition was invoked to ensure that all atoms had identical surroundings, and there was no 'end', or 'surface' atom to deal with. If we are interested in studying the vibrational mode of the end atom in a semi-infinite chain, then rather than the periodic boundary condition an open boundary condition would have to be envoked. We will consider two cases, using a monatomic linear chain and a diatomic linear chain.

4.6.1 SEMI-INFINITE MONATOMIC LINEAR CHAIN

In this sub-section we will study atomic vibrations in a semi-infinite monatomic linear chain, as shown in figure 4.13. In this chain there are N atoms, numbered 1 through to N, each of mass m and separated from each other by a spring of length a and harmonic force constant Λ. Let us attach, or adsorb, an atom (*i.e.* 0th atom) of mass m_0 at one end of the chain, and let the spring constant between this atom and the 1st atom be Λ_0. The adsorbate will become 'surface' atom when $m_0 = m$. The theory of this system was developed the 1950s (see, Wallis, 1957). A detailed account can be found in Wallis (1964, 1974). Here we present a simplified and brief discussion.

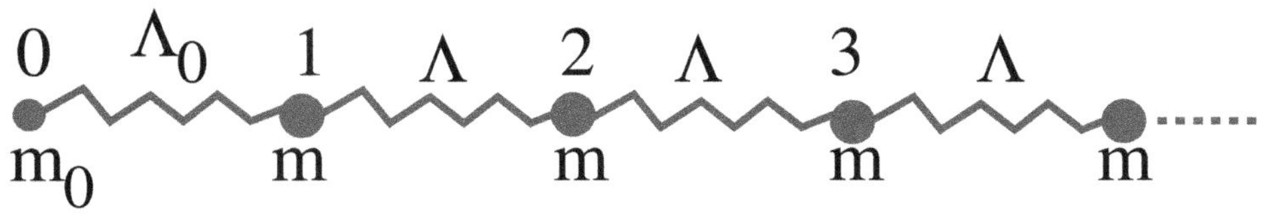

Figure 4.13 A semi-infinite monatomic linear chain, with an adsorbate attached at one end. The inter-atomic distance is a, all atoms in the interior of the chain have mass m and connected with a spring constant Λ, and the adsorbate atom of mass m_0 is attached to the chain with force constant Λ_0.

The equations of motion for the atoms are

$$m_0 \ddot{u}_0 = -\Lambda_0(u_0 - u_1), \tag{4.65}$$

$$m\ddot{u}_1 = -\Lambda_0(u_1 - u_0) - \Lambda(u_1 - u_2), \tag{4.66}$$

$$m\ddot{u}_n = -\Lambda(2u_n - u_{n-1} - u_{n+1}), \quad n \geq 2, \tag{4.67}$$

where u_n is the displacement of the nth atom from its equilibrium position.

We noted in section 4.5.1 that the vibrational frequency spectrum for an infinitely long monatomic linear chain continuously covers the range $(0, \sqrt{4\Lambda/m})$. The interesting question is whether there would be any solutions above this frequency range with or without an adatom at one end of a semi-infinite linear chain. To examine this, let us write the displacements in the form

$$u_0 = A' \exp(i\omega t), \tag{4.68}$$

$$u_n = A(-1)^n \exp(-\xi n + i\omega t), \quad n \geq 1, \tag{4.69}$$

where ξ is an attenuation constant, and A' and A are the vibrational amplitudes for the adsorbate and bulk atoms, respectively. Equation (4.69) indicates that the displacements decay exponentially from the surface atom (adatom) towards the bulk atoms (*i.e.* interior of the chain). Substituting equation (4.69) into equation (4.67), we obtain the following necessary condition for the frequency to be related to the attenuation constant ξ

$$m\omega^2 = 2\Lambda(1 - \cosh\xi). \tag{4.70}$$

The attenuation constant ξ can be obtained by substituting equations (4.68) and (4.69) into equations (4.65) and (4.66), using equation (4.70), and setting the determinant of the coefficients of A and A' equal to zero. The resulting equation is

$$[\Lambda_0 - \Lambda\{1 + \exp(\xi)\}]\left[\Lambda_0 - 2\frac{m_0}{m}\Lambda(1 + \cosh\xi)\right] - \Lambda_0{}^2 = 0. \tag{4.71}$$

For ξ to be real and positive, the following inequality must be satisfied

$$\frac{\Lambda_0}{\Lambda} > \frac{4(\frac{m_0}{m})}{\left[2(\frac{m_0}{m}) + 1\right]}. \tag{4.72}$$

With the real and positive value of the attenuation constant taken as ξ_0, the surface mode frequency is determined from

$$\omega_S = \sqrt{\frac{2\Lambda}{m}(1 + \cosh\xi_0)}. \tag{4.73}$$

Noting that $\cosh \xi_0 > 1$ for real and positive ξ_0, this frequency is clearly larger than the maximum bulk frequency $\sqrt{4\Lambda/m}$.

From equation (4.72) it is easily noted that if the mass of the adsorbate atom is the same as that the chain atom ($m_0 = m$) then the existence of a localised surface mode (a mode outside the bulk frequency spectrum) requires the force constant Λ_0 to be larger than the inter-atomic force constant Λ in the chain such that $\Lambda_0/\Lambda > 4/3$. If, on the other hand, the force constant Λ_0 remains the same as Λ then the existence of a localised surface mode requires the adsorbate atom to be at least half lighter than the chain atom, *viz* $m_0/m < 1/2$.

The location of a localised surface mode above the bulk vibrational spectrum of a monatomic linear chain is not contrary to Rayleigh's theorem (Rayleigh, 1885), according to which reduction in coordination of a surface atom depresses the normal-mode frequency to a non-negative value. As the monatomic linear chain is characterised by a gap-less frequency spectrum from a maximum down to zero, there is no possibility of the production of a Rayleigh surface mode for a linear atomic chain.

4.6.2 SEMI-INFINITE DIATOMIC LINEAR CHAIN

Compared to the semi-infinite monatomic linear chain, the semi-infinite diatomic linear chain offers more possibilities for the development of surface vibrational modes. These are discussed by Wallis (1964) and Hori and Asahi (1964). Here we will study a semi-infinite diatomic linear chain with the lighter of the two basis masses at the free end. Figure 4.14 shows the free end of such a chain, with $m < M$. We assume inter-atomic distance a and inter-atomic force constant Λ, without any change in the surface force constant.

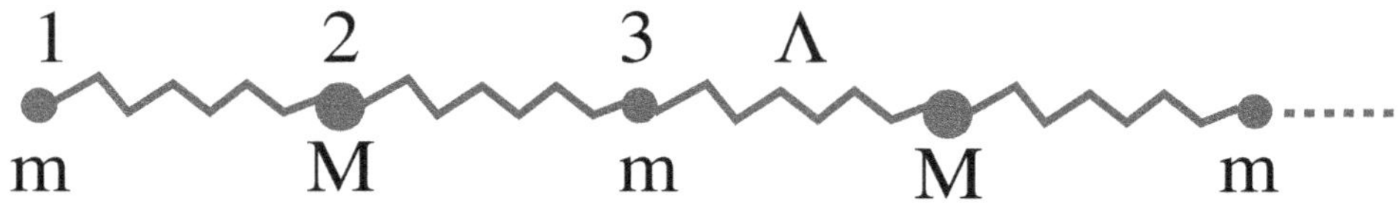

Figure 4.14 A semi-infinite diatomic linear chain with alternating masses m and $M(> m)$ connected by springs of inter-atomic length a and force constant Λ.

The equations of motion are

$$m\ddot{u}_1 = -\Lambda(u_1 - u_2), \tag{4.74}$$

$$M\ddot{u}_{2n} = -\Lambda(2u_{2n} - u_{2n-1} - u_{2n+1}) \quad n \geq 1, \tag{4.75}$$

$$m\ddot{u}_{2n+1} = -\Lambda(2u_{2n+1} - u_{2n} - u_{2n+2}) \quad n \geq 2, \tag{4.76}$$

where u_n is the displacement from equilibrium of the atom numbered n, *etc*. The displacements can be expressed as attenuated or damped oscillations

$$u_{2n} = A(-1)^n (m/M)^n \exp(i\omega t), \tag{4.77}$$

$$u_{2n+1} = A(-1)^{n+1} (m/M)^{n+1} \exp(i\omega t), \tag{4.78}$$

where A is time independent amplitude. We will not go into detail of derivation, but simply mention that Wallis (1957, 1964, 1974) showed that there exists a surface solution to equations (4.74)-(4.76), with the surface mode frequency

$$\omega_S = \sqrt{\Lambda\left(\frac{1}{m} + \frac{1}{M}\right)}. \tag{4.79}$$

Clearly, this surface mode lies in the gap between the acoustic range $(0, \sqrt{2\Lambda/M})$ and the optical range $\left(\sqrt{2\Lambda/m}, \sqrt{2\Lambda(1/m+1/M)}\right)$ for an infinite and periodic diatomic linear chain (*cf* section 4.5.2). In terms of squared frequencies, this mode lies exactly in the middle of the acoustic-optical band gap. As the surface (or end) atom has reduced coordination, according to Rayleigh's theorem its frequency should be lower than that a bulk (or chain) frequency. This suggests that the surface mode arises from an optical mode, which has dropped into the acoustic-optical gap.

FURTHER READING

1. Ashcroft N W and Mermin N D 1976 *Solid State Physics* (Sounders College: Philadelphia)
2. Born M and Huang K 1954 *Dynamical Theory of Crystal Lattices* (Oxford: Clarendon)
3. Goldstein H, Poole C and Safko J 2002 *Classical Mechanics* 3rd edn (Addison-Wesley: New York)
4. Hori J and Asahi T 1964 *Prog Theo Phys* **31** 49
5. Kesavaswamy K and Krishnamurthy N 1978 *Am J Phys* **46** 815
6. Kittel C 1996 *Introduction to Solid State Physics* 7th edn (Wiley: New York)
7. Srivastava G P 2022 *The Physics of Phonons* 2nd edn (CRC: Oxon)
8. Wallis R F 1957 *Phys Rev* **105** 540
9. Ziman J M 1060 *Electrons and Phonons* (Oxford: Clarendon)

PROBLEMS

1. Derive equations (4.1) and (4.2) using the Lagrangian method of mechanics.
2. The lower frequency, $\omega_1 = \sqrt{\frac{\lambda}{m}}$, for the system of two-coupled oscillators in figure 4.1 is independent of the interatomic force constant Λ. Explain why?
3. Using the results in equation (4.7), derive the result $\omega_2 = \sqrt{\frac{\lambda}{m}(1+2\eta)}$, in the limit $\eta = \frac{\Lambda}{2\lambda} << 1$.
4. Assuming that interactions with the same coupling strength are allowed beyond the nearest neighbours, show that equation (4.26) for the dispersion relation for a monatomic linear chain can be generalised to the form

$$\omega = 2\sqrt{\frac{\Lambda}{m}\sum_{n>0}\sin^2(\frac{qan}{2})}.$$

5. Consider a monatomic linear chain with alternate force constants Λ_1 and Λ_2. Show that equation (4.25) should be modified to

$$\omega^2 = \frac{\Lambda_1+\Lambda_2}{m} \pm \frac{1}{m}\sqrt{\Lambda_1^2+\Lambda_2^2+2\Lambda_1\Lambda_2\cos(qa)}.$$

6. Derive equations (4.10), (4.11) and (4.12) using the Lagrangian approach.
7. Consider a triatomic linear chain with basis atomic masses m_1, $m_2 > m_1$, $m_3 = m_1$ and the chain force constant Λ. Show that the non-zero vibrational frequencies at the zone centre are: $\sqrt{\Lambda(1/m_1+2/m_2)}$ and $\sqrt{3\Lambda/m_1}$.
8. Derive the inequality in equation (4.72).

5 Dynamics of Two- and Three-Dimensional Crystals

5.1 INTRODUCTION

In the previous chapter we discussed the dynamics of infinite and semi-infinite one-dimensional atomic chains. The equations of motion and their solutions, for vibrational waves to propagate along the chain, were obtained by considering the longitudinal displacement of atoms, *i.e.* expansion and compression of atomic chains. It was mentioned that wave propagation could also be examined by considering a transverse displacement of atoms. That would correspond to plucking the system in a direction normal to the atomic chain. A wave propagating along the chain arising from the longitudinal displacement of atoms is *longitudinally polarised.* A wave propagating along the chain arising from a transverse displacement of atoms is *transversally polarised.* For a chain of N atoms, there would be N normal modes for any polarisation (be it the longitudinal or a transverse). For a system comprised of higher dimensions, such as two- and three-dimensional atomic chains, it would be possible to have waves of both longitudinal and transverse polarisations. The theory discussed in the previous chapter would need to be extended to incorporate these possibilities. The concept of the wavenumber q used for a linear chain would have to change to that of $\boldsymbol{q}$-vector and, similarly, the displacement u would need to be considered as vector $\boldsymbol{u}$. For both two- and three-dimensional structures with N_0 number of unit cells and p number of atoms per unit cell, there would be $3p$ branches of the dispersion curve, each with N_0 distinct $\boldsymbol{q}$-vectors (or normal modes). The lowest three of these will be acoustic branches such that $\omega(\boldsymbol{q}) \to 0$ as $\boldsymbol{q} \to \boldsymbol{0}$. Remaining branches will be optical such that $\omega(\boldsymbol{q}) \to$ constant as $\boldsymbol{q} \to \boldsymbol{0}$. In this chapter we will discuss the linear dynamics of two- and three-dimensional crystals, without worrying about surface effects.

As mentioned in the previous chapter, dispersion relations of normal modes are usually presented along symmetry direction in the Brillouin zone of the crystalline system under consideration. Appendix E provides a discussion of the Brillouin zone for two- and three-dimensional lattices.

In order to proceed, we need to generalise the simple non-vectorial force expression $F = -\Lambda u$, used in the previous chapter for one-dimensional structures, to a suitable vectorial form. Let us consider harmonic interaction between 0th atom with displacement $\boldsymbol{u}_0$ and nth atom with displacement $\boldsymbol{u}_n$, as shown in figure 5.1.

Let $\boldsymbol{R}_n$ be the vector separating these atoms and $\hat{\boldsymbol{\varepsilon}}_n = \boldsymbol{R}_n/R_n$ be a unit vector parallel to $\boldsymbol{R}_n$. If Λ_n is the force constant between the two atoms, then the central force experienced by the 0th atom due to the nth atom is

$$\boldsymbol{F}_n = -\Lambda_n \left[\hat{\boldsymbol{\varepsilon}}_n \cdot (\boldsymbol{u}_0 - \boldsymbol{u}_n)\right] \hat{\boldsymbol{\varepsilon}}_n. \tag{5.1}$$

With these considerations, the equation of motion for the 0th atom can be written as

$$m\ddot{\boldsymbol{u}}_0 = \sum_n \boldsymbol{F}_n, \tag{5.2}$$

where the sum over n includes neighbours that need to be considered (*e.g.* nearest and/or next-nearest). Note that $\hat{\boldsymbol{\varepsilon}}_n \cdot (\boldsymbol{u}_0 - \boldsymbol{u}_n)$ is the component of the displacement of atom 0 with respect to that of atom n along the central direction $\boldsymbol{R}_n$.

DOI: 10.1201/9781003383314-5

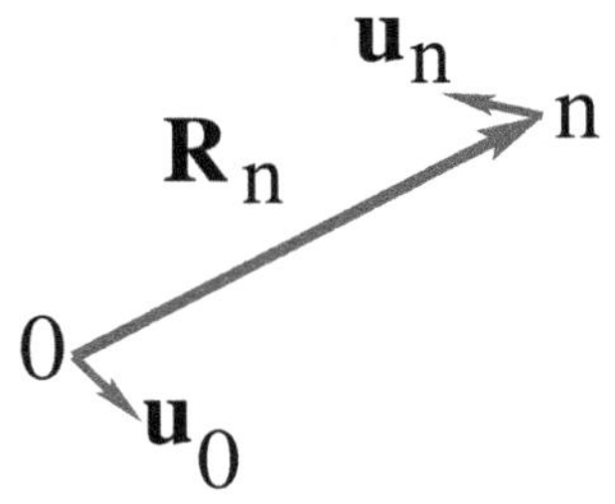

Figure 5.1 Distance and displacement vectors between two atoms. Here $\boldsymbol{u}_0$ and $\boldsymbol{u}_n$ show, respectively, displacement vectors of the 0th and nth atoms and $\boldsymbol{R}_n$ is the vector separating the two atoms.

5.2 MONATOMIC TWO-DIMENSIONAL NETWORK

We will discuss the linear dynamics of three types of periodic two-dimensional networks. We mentioned earlier that, in principle, atomic displacements for two-dimensional crystals could take place in three mutually orthogonal directions (corresponding to longitudinal and two transverse polarisations). However, in this section we restrict discussion to only two orthogonal directions in the plane of the two-dimensional crystal (corresponding to longitudinal and one transverse polarisations)[1].

5.2.1 SQUARE NETWORK

We first consider a periodic monatomic two-dimensional square network. The lattice for such a system is shown in figure 5.2(a). We take the primitive unit cell as a square with sides of length a and the primitive translation vectors as $\boldsymbol{a}_1 = a\hat{\boldsymbol{x}}$ and $\boldsymbol{a}_2 = a\hat{\boldsymbol{y}}$. The corresponding Brillouin zone is a square with sides of length $2\pi/a$. The irreducible part of this zone is the triangle ΓXM, as shown in figure 5.2(b). The symmetry direction Γ-X is along the x-axis, *viz* $[1,0]$, and the symmetry direction Γ-M is $[1,1]$. Figure 5.2(c) shows the atomic positions (basis of one atom per lattice point). As we are considering a monatomic network, there is a single atom per unit cell, which we take as the atom marked 0. Shown in figure 5.2(c) are the four nearest neighbours (1-4) and four next-nearest neighbours (5-8) to atom 0. We consider atomic mass m and harmonic force constants Λ_1 between nearest neighbours and Λ_2 between next-nearest neighbours.

The equation of motion of the single atom in the unit cell is

$$m\ddot{\boldsymbol{u}}_0 = -\Lambda_1 \sum_{n=1}^{4} [\hat{\boldsymbol{\varepsilon}}_n \cdot (\boldsymbol{u}_0 - \boldsymbol{u}_n)]\hat{\boldsymbol{\varepsilon}}_n - \Lambda_2 \sum_{n=5}^{8} [\hat{\boldsymbol{\varepsilon}}_n \cdot (\boldsymbol{u}_0 - \boldsymbol{u}_n)]\hat{\boldsymbol{\varepsilon}}_n. \tag{5.3}$$

The atomic displacement vectors can be expressed as

$$\boldsymbol{u}_n = \boldsymbol{A}\exp\left[i(\boldsymbol{q}\cdot\boldsymbol{R}_n - \omega t)\right], \tag{5.4}$$

where $\boldsymbol{A}$ is the amplitude vector.

Using the Cartesian coordinate system, we express $\boldsymbol{q} = (q_x, q_y)$, $\boldsymbol{A} = (A_x, A_y)$ and set the origin at atom 0. As shown in figure 5.2(c), the coordinates of the nearest and next-nearest atoms to atom 0 at $\boldsymbol{R}_0 = (0,0)$ are: $\boldsymbol{R}_1 = a(1,0)$, $\boldsymbol{R}_2 = a(-1,0)$, $\boldsymbol{R}_3 = a(0,1)$, $\boldsymbol{R}_4 = a(0,-1)$, $\boldsymbol{R}_5 = a(1,1)$, $\boldsymbol{R}_6 = a(-1,-1)$, $\boldsymbol{R}_7 = a(-1,1)$ and $\boldsymbol{R}_8 = a(1,-1)$. The corresponding unit vectors are: $\hat{\boldsymbol{\varepsilon}}_1 = (1,0)$, $\hat{\boldsymbol{\varepsilon}}_2 = (-1,0)$, $\hat{\boldsymbol{\varepsilon}}_3 = (0,1)$, $\hat{\boldsymbol{\varepsilon}}_4 = (0,-1)$, $\hat{\boldsymbol{\varepsilon}}_5 = (\frac{1}{\sqrt{2}}, \frac{1}{\sqrt{2}})$, $\hat{\boldsymbol{\varepsilon}}_6 = (-\frac{1}{\sqrt{2}}, -\frac{1}{\sqrt{2}})$, $\hat{\boldsymbol{\varepsilon}}_7 = (-\frac{1}{\sqrt{2}}, \frac{1}{\sqrt{2}})$ and $\hat{\boldsymbol{\varepsilon}}_8 =$

[1] All three mutually orthogonal polarisation directions will be considered in the next section.

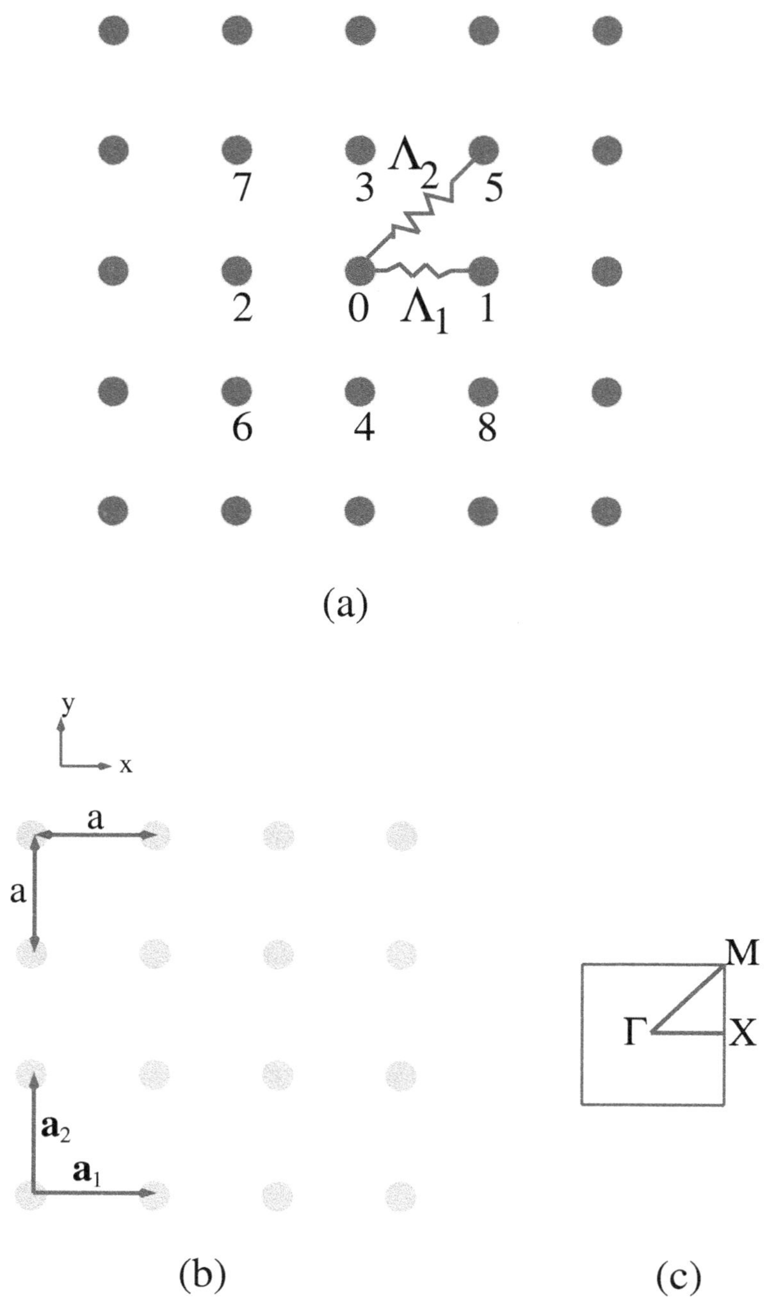

Figure 5.2 Schematic illustration of: (a) a monatomic two-dimensional crystal with nearest inter-atomic distance a, (b) two-dimensional square lattice and its primitive translation vectors $\boldsymbol{a}_1$ and $\boldsymbol{a}_2$, and (c) the corresponding Brillouin zone. Γ-X and Γ-M are the principal symmetry directions $[1,0]$ and $[1,1]$, respectively. The area ΓXM represents the irreducible part of the zone.

$(\frac{1}{\sqrt{2}}, -\frac{1}{\sqrt{2}})$. Linear forces acting on atom 0 from displacements of neighbouring atoms can easily be evaluated using equation (5.1). For example, forces $\boldsymbol{F}_1$ and $\boldsymbol{F}_8$ are,

$$\begin{aligned} \boldsymbol{F}_1 &= -\Lambda_1 \left[\hat{\boldsymbol{\varepsilon}}_1 \cdot (\boldsymbol{u}_0 - \boldsymbol{u}_1) \right] \hat{\boldsymbol{\varepsilon}}_1, \\ &= -\Lambda_1 \left[1 - \exp\left(i\boldsymbol{q} \cdot \boldsymbol{R}_1 \right) \right] (\hat{\boldsymbol{\varepsilon}}_1 \cdot \boldsymbol{u}_0) \hat{\boldsymbol{\varepsilon}}_1, \\ &= -\Lambda_1 \left[1 - \exp(i\boldsymbol{q} \cdot \boldsymbol{R}_1) \right] (A_x, 0) \exp(-i\omega t). \end{aligned} \tag{5.5}$$

and

$$\begin{aligned} \boldsymbol{F}_8 &= -\Lambda_2 \left[\hat{\boldsymbol{\varepsilon}}_8 \cdot (\boldsymbol{u}_0 - \boldsymbol{u}_8) \right] \hat{\boldsymbol{\varepsilon}}_8, \\ &= -\Lambda_2 \left[1 - \exp\left(i\boldsymbol{q} \cdot \boldsymbol{R}_8 \right) \right] (\hat{\boldsymbol{\varepsilon}}_8 \cdot \boldsymbol{u}_0) \hat{\boldsymbol{\varepsilon}}_8, \\ &= -\Lambda_2 \left[1 - \exp(\mathrm{i}\boldsymbol{q} \cdot \boldsymbol{R}_8) \right] \left(\frac{A_x}{\sqrt{2}} - \frac{A_y}{\sqrt{2}} \right) \left(\frac{1}{\sqrt{2}}, -\frac{1}{\sqrt{2}} \right) \exp(-i\omega t). \end{aligned} \tag{5.6}$$

Equation (5.3) then can be expressed as the following matrix eigenvalue equation

$$\omega^2 A_\alpha = \sum_\beta D_{\alpha\beta} A_\beta, \quad \alpha, \beta = x, y, \tag{5.7}$$

where the elements of the 2×2 dynamical matrix D are

$$\begin{aligned} D_{xx} &= \frac{2}{m} [\Lambda_1 (1 - \cos q_x a) + \Lambda_2 (1 - \cos q_x a \cos q_y a)], \\ D_{xy} &= \frac{2\Lambda_2}{m} \sin q_x a \sin q_y a, \\ D_{yx} &= D_{xy}, \\ D_{yy} &= \frac{2}{m} [\Lambda_1 (1 - \cos q_y a) + \Lambda_2 (1 - \cos q_x a \cos q_y a)]. \end{aligned} \tag{5.8}$$

For a chosen wavevector $\boldsymbol{q}$, normal mode frequencies can be obtained by solving the secular equation

$$\left| D_{\alpha\beta} - \omega^2 \delta_{\alpha\beta} \right| = 0. \tag{5.9}$$

As D is a 2×2 matrix, there will be two non-negative values of ω^2 (and hence two real values of ω) for each $\boldsymbol{q}$. In other words, there will be two branches of the ω *vs* $\boldsymbol{q}$ curve, which we can identify by assigning indices $s = 1$ and $s = 2$. The corresponding eigen-displacement vector components $A_x(\boldsymbol{q}, s)$ and $A_y(\boldsymbol{q}, s)$ can then be obtained from equation (5.7).

By setting $q_x = q_y = 0$, we immediately notice that both solutions correspond to $\omega = 0$, and that the directions of the amplitude vectors $\boldsymbol{A}(\boldsymbol{q} = 0, s = 1)$ and $\boldsymbol{A}(\boldsymbol{q} = 0, s = 2)$ become indeterminate.

Figure 5.3 shows the dispersion curves along the principal symmetry directions Γ-X and Γ-M for the choice $\Lambda_2/\Lambda_1 = 0.5$.

Eigensolutions along Γ-X:

To examine solutions along the symmetry direction Γ-X, we set $q_x = q$ and $q_y = 0$. With this $D_{xy} = D_{yx} = 0$ and the eigensolutions (eigenvalues ω^2 and eigenvectors $\boldsymbol{e}$) are

$$\text{for } s = 1: \quad \omega^2(\mathsf{TA}) = \frac{2}{m} \Lambda_2 (1 - \cos qa), \quad \hat{\boldsymbol{e}} = \boldsymbol{A}/A = (0, 1), \tag{5.10}$$

$$\text{for } s = 2: \quad \omega^2(\mathsf{LA}) = \frac{2}{m} (\Lambda_1 + \Lambda_2)(1 - \cos qa), \quad \hat{\boldsymbol{e}} = \boldsymbol{A}/A = (1, 0). \tag{5.11}$$

Close to the zone centre, $q \to 0$, both solutions are of the form $\omega \propto q$, indicating the acoustic nature of both branches. Further, as $\boldsymbol{q}$ is along $[1, 0]$, the branches described by equations (5.10) and (5.11) are transverse acoustic (TA) and longitudinal acoustic (LA), respectively.

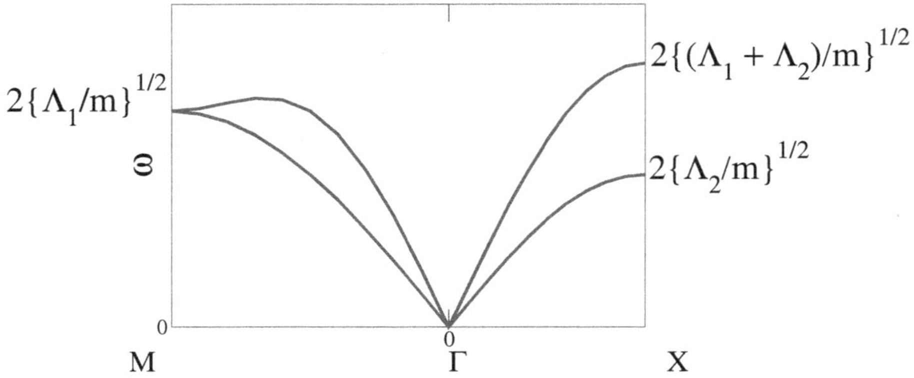

Figure 5.3 Dispersion curves (ω *vs* $\boldsymbol{q}$) for the normal modes of a monatomic square crystal along two principal symmetry directions. The atomic mass is m and the next nearest-neighbour force constant is taken as half of the nearest neighbour force constant ($\Lambda_2/\Lambda_1 = 0.5$).

Along this symmetry direction, the maximum frequencies for the TA and LA branches are $2(\Lambda_2/m)^{1/2}$ and $2[(\Lambda_1+\Lambda_2)/m]^{1/2}$, respectively. Clearly, along this direction the TA branch exists only because of the next-nearest coupling. Also, with consideration only of the nearest neighbour interaction (*i.e.* when $\Lambda_2 = 0$), the dispersion relation for the LA branch becomes identical to the expression in equation (4.25) and the dispersion curve becomes identical to that plotted in figure 4.8 for the monatomic linear chain.

Eigensolutions along Γ-M:

Along the symmetry direction Γ-M, we set $q_x = q_y = q/\sqrt{2}$. The eigensolutions (eigenvalues ω^2 and eigenvectors $\boldsymbol{e}$) can be worked out as follows

$$\omega^2(\text{TA}) = \frac{2}{m}\Lambda_1\big(1-\cos\frac{qa}{\sqrt{2}}\big), \quad \hat{\boldsymbol{e}}(\text{TA}) = \big(-\frac{1}{\sqrt{2}}, \frac{1}{\sqrt{2}}\big), \tag{5.12}$$

and

$$\omega^2(\text{LA}) = \frac{2}{m}\big[\Lambda_1\big(1-\cos\frac{qa}{\sqrt{2}}\big) + \Lambda_2\big(1-\cos\sqrt{2}qa\big)\big], \quad \hat{\boldsymbol{e}}(\text{LA}) = \big(\frac{1}{\sqrt{2}}, \frac{1}{\sqrt{2}}\big). \tag{5.13}$$

Clearly, similar to the Γ point, at the M point the two solutions are the same, indicating that the directions of the eigenvectors become indeterminate and any pair of orthogonal unit vectors can be acceptable.

Along this symmetry direction, only the nearest neighbour coupling (*i.e.* Λ_1) contributes to the dispersion relation of the TA branch. In contrast, both the nearest and next-nearest couplings (*i.e.* Λ_1 as well as Λ_2) contribute to the dispersion relation of the LA branch. Both branches become degenerate at the M point with the frequency $2(\Lambda_1/m)^{1/2}$. As long as the nearest-neighbour force constant is weaker than four times the second nearest-neighbour force constant (*i.e.* for $\Lambda_1 < 4\Lambda_2$) the maximum frequency of the LA branch along Γ-M is higher than the frequency at the M point.

5.2.2 MONATOMIC HEXAGONAL NETWORK

Let us now consider a monatomic hexagonal system, with atomic mass m, interatomic distance a and nearest neighbour harmonic force constant Λ. Figure 5.4 illustrates the primitive unit cell, the Brilluion zone, and the nearest neighbour of the single atom (labelled 0) in the unit cell.

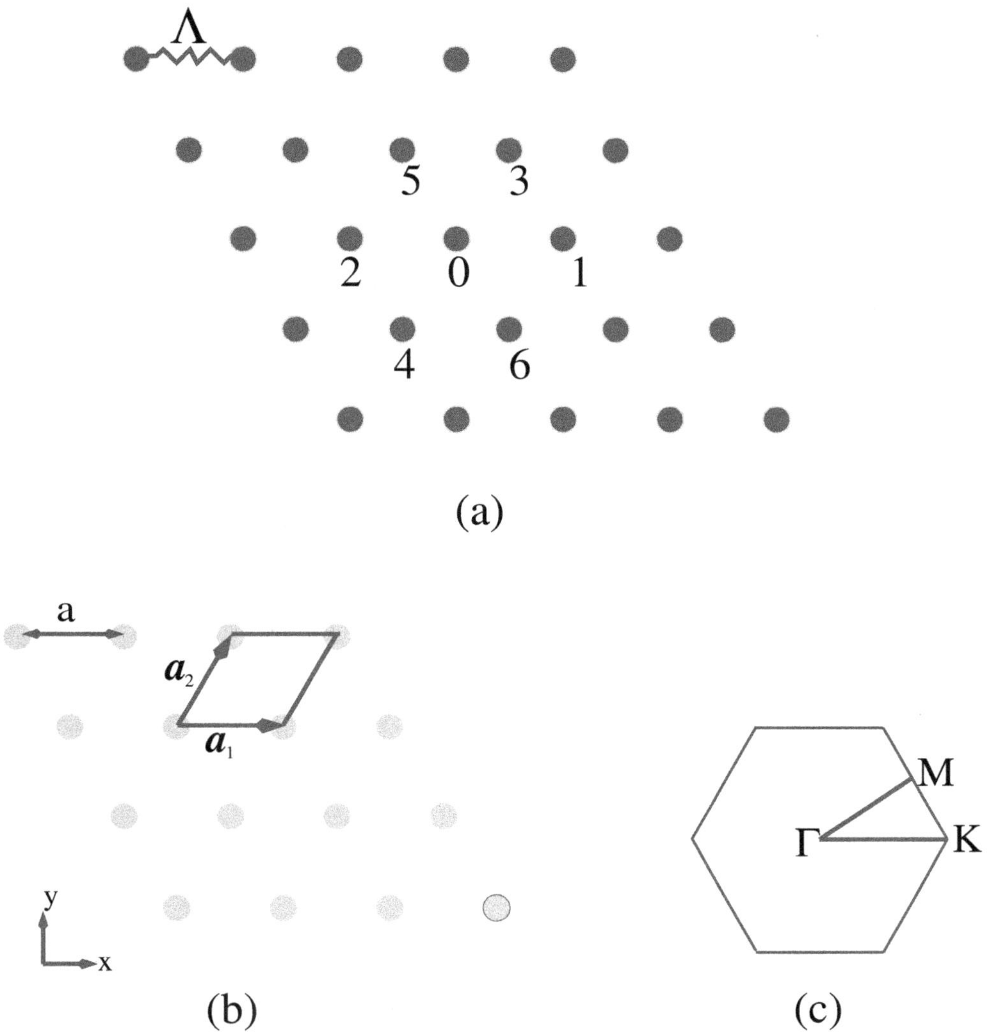

Figure 5.4 Schematic illustration of: (a) a monatomic two-dimensional hexagonal crystal with nearest interatomic distance a and nearest-neighbour force constant Λ, (b) two-dimensional hexagonal lattice and its primitive translation vectors $\boldsymbol{a}_1$ and $\boldsymbol{a}_2$, and (c) the corresponding Brillouin zone. Γ-M and Γ-K are the principal symmetry directions and the area ΓKM represents the irreducible part of the zone.

Using the Cartesian x-y coordinate system, the primitive translation vectors can be taken as $\boldsymbol{a}_1 = a(1,0)$ and $\boldsymbol{a}_2 = a(\frac{1}{2}, \frac{\sqrt{3}}{2})$. For this choice, the primitive translation vectors of the corresponding reciprocal hexagonal lattice are $\boldsymbol{b}_1 = \frac{2\pi}{a}(1, -\frac{1}{\sqrt{3}})$ and $\boldsymbol{b}_2 = \frac{2\pi}{a}(0, \frac{2}{\sqrt{3}})$. The irreducible part of the Brillouin zone is the triangle ΓKM, where $\Gamma = (0,0)$, $M = \frac{2\pi}{a}(0, \frac{1}{\sqrt{3}})$ and $K = \frac{2\pi}{a}(\frac{1}{3}, \frac{1}{\sqrt{3}})$.

The vectors connecting atom 0 to its nearest neighbours are $\boldsymbol{R}_1 = a(1,0)$, $\boldsymbol{R}_2 = a(-1,0)$, $\boldsymbol{R}_3 = a(\frac{1}{2}, \frac{\sqrt{3}}{2})$, $\boldsymbol{R}_4 = a(-\frac{1}{2}, -\frac{\sqrt{3}}{2})$, $\boldsymbol{R}_5 = a(-\frac{1}{2}, \frac{\sqrt{3}}{2})$ and $\boldsymbol{R}_6 = a(\frac{1}{2}, -\frac{\sqrt{3}}{2})$. The corresponding unit vectors $\hat{\boldsymbol{\varepsilon}}_n = \boldsymbol{R}_n/R_n$ can be easily worked out. The force exerted by the nth neighbour on atom 0 can be worked out from the expression $\boldsymbol{F}_n = -\Lambda[\hat{\boldsymbol{\varepsilon}}_n \cdot (\boldsymbol{u}_0 - \boldsymbol{u}_n)]\hat{\boldsymbol{\varepsilon}}_n$, where u_j is the displacement of the jth atom.

Clearly, we need to obtain an expression for $\eta_n = \hat{\boldsymbol{\varepsilon}}_n \cdot (\boldsymbol{u}_0 - \boldsymbol{u}_n)$, the component of the displacement of atom 0 with respect to that of atom n along the central direction $\boldsymbol{R}_n$. Let us obtain expressions for η_1 and η_3. With atomic displacements expressed as

$$\boldsymbol{u}_n = \boldsymbol{A}\exp[i(\boldsymbol{q}\cdot\boldsymbol{R}_n - \omega t)] = (A_x, A_y)\exp[i(\boldsymbol{q}\cdot\boldsymbol{R}_n - \omega t)], \tag{5.14}$$

and following the steps in the previous sub-section, we obtain

$$\begin{aligned}
\eta_1 &= \hat{\boldsymbol{\varepsilon}}_1 \cdot (\boldsymbol{u}_0 - \boldsymbol{u}_1), \\
&= (1,0)\cdot(A_x, A_y)\left[1 - \exp\left(i\boldsymbol{q}\cdot\boldsymbol{R}_1\right)\right]\exp(-i\omega t), \\
&= A_x\left[1 - \exp\left(i\boldsymbol{q}\cdot\boldsymbol{R}_1\right)\right]\exp(-i\omega t), \\
&= A_x\left[1 - \exp(iq_x a)\right]\exp(-i\omega t),
\end{aligned} \tag{5.15}$$

and

$$\begin{aligned}
\eta_3 &= \hat{\boldsymbol{\varepsilon}}_3 \cdot (\boldsymbol{u}_0 - \boldsymbol{u}_3), \\
&= (\frac{1}{2}, \frac{\sqrt{3}}{2})\cdot(A_x, A_y)\left[1 - \exp\left(i\boldsymbol{q}\cdot\boldsymbol{R}_3\right)\right]\exp(-i\omega t), \\
&= (\frac{A_x}{2} + \frac{\sqrt{3}A_y}{2})\left[1 - \exp\left(i\boldsymbol{q}\cdot\boldsymbol{R}_3\right)\right]\exp(-i\omega t), \\
&= (\frac{A_x}{2} + \frac{\sqrt{3}A_y}{2})\left[1 - \exp\left(i\frac{a}{2}(q_x + \sqrt{3}q_y)\right)\right]\exp(-i\omega t).
\end{aligned} \tag{5.16}$$

Expressions for the remaining displacement components, *viz* η_2, η_4-η_6, can be obtained in a similar manner.

With the expression for the total force acting on atom 0, its equation of motion

$$m\ddot{\boldsymbol{u}}_0 = -\Lambda\sum_{n=1}^{6}[\hat{\boldsymbol{\varepsilon}}_n\cdot(\boldsymbol{u}_0 - \boldsymbol{u}_n)]\hat{\boldsymbol{\varepsilon}}_n = -\Lambda\sum_{n=1}^{6}\eta_n\hat{\boldsymbol{\varepsilon}}_n, \tag{5.17}$$

can be expressed as the eigenvalue equation (5.7), with the following elements of 2×2 dynamical matrix D:

$$\begin{aligned}
D_{xx} &= \frac{\Lambda}{m}\left(3 - 2\cos q_x a - \cos\frac{q_x a}{2}\cos\frac{\sqrt{3}q_y a}{2}\right), \\
D_{xy} &= -\frac{\sqrt{3}\Lambda}{m}\sin\frac{q_x a}{2}\sin\frac{\sqrt{3}q_y a}{2}, \\
D_{yx} &= D_{xy}, \\
D_{yy} &= \frac{3\Lambda}{m}\left(1 - \cos\frac{q_x a}{2}\cos\frac{\sqrt{3}q_y a}{2}\right).
\end{aligned} \tag{5.18}$$

The dispersion curves (ω *vs* $\boldsymbol{q}$) along the principal symmetry directions Γ-M and Γ-K are plotted in figure 5.5. The lower branch (transverse acoustic) spans the frequency range $(0, \sqrt{\frac{2\Lambda}{m}})$ along Γ-M and $(0, 3\sqrt{\frac{\Lambda}{2m}})$ along Γ-K. The upper branch (longitudinal acoustic) rises up to $\sqrt{\frac{6\Lambda}{m}}$ along Γ-M. The upper branch become degenerate with the lower branch at the K point, with frequency $3\sqrt{\frac{\Lambda}{2m}}$.

5.3 DIATOMIC TWO-DIMENSIONAL NETWORK: GRAPHENE-LIKE STRUCTURE

In the previous two sections we noted that the dynamical matrix for monatomic two-dimensional systems is of size 2×2 when only in-plane atomic displacements are considered. When out-of-plane atomic displacements are also included in the consideration, then the size of the dynamical

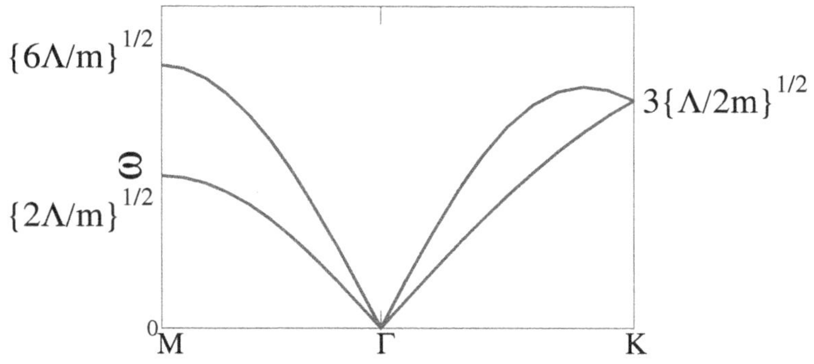

Figure 5.5 Dispersion curves (ω *vs* q) for the normal modes of a monatomic hexagonal crystal along two principal symmetry directions. The atomic mass is m and the nearest-neighbour force constant is Λ.

matrix will become 3×3. Obviously, the size of the dynamical matrix will be 6×6 for diatomic two-dimensional systems, when both in-plane and out-of-plane atomic displacements are taken into account. This can be explained mathematically by writing the force expression in equation (5.1) in the following general form

$$F_\alpha(n) = -\sum_\beta \Phi_{\alpha\beta}(n;n')u_\beta(n'), \tag{5.19}$$

where $\alpha, \beta = x, y, z$. Here $F_\alpha(n)$ is αth component of force on the nth atom due βth displacement component $u_\beta(n')$ of the n'th atom, and Φ is a 3×3 harmonic force constant matrix. The size of the dynamical matrix for a system containing p atoms per unit cell is $3p \times 3p$.

Let us consider graphene, the most celebrated two-dimensional material with two carbon atoms per hexagonal unit cell. This is shown in figure 5.6. The hexagonal primitive translation vectors may be taken as $\boldsymbol{a}_1 = a(1,0)$ and $\boldsymbol{a}_2 = a(\frac{1}{2}, \frac{\sqrt{3}}{2})$, with a as the hexagonal lattice constant. The corresponding primitive translation vectors of the reciprocal lattice are $\boldsymbol{b}_1 = \frac{2\pi}{a}(1, -\frac{1}{3})$ and $\boldsymbol{b}_2 = \frac{2\pi}{a}(0, \frac{2}{\sqrt{3}})$. The two basis atoms are positioned at $\boldsymbol{R}_1 = (0,0)$ and $\boldsymbol{R}_2 = \frac{1}{3}(\boldsymbol{a}_1 + \boldsymbol{a}_2) = a(\frac{1}{2}, \frac{1}{2\sqrt{3}})$. The vector separating the two basis atoms is $\boldsymbol{d} = \frac{1}{3}(\boldsymbol{a}_1 + \boldsymbol{a}_2)$. The nearest neighbours to atom 1 are located at $\boldsymbol{R}_2 = \boldsymbol{d}$, $\boldsymbol{R}_{2'} = \boldsymbol{d} - \boldsymbol{a}_1$ and $\boldsymbol{R}_{2''} = \boldsymbol{d} - \boldsymbol{a}_2$. And the nearest neighbours to atom 2 are at $\boldsymbol{R}_1 = \boldsymbol{0}$, $\boldsymbol{R}_{1'} = \boldsymbol{a}_1$ and $\boldsymbol{R}_{1''} = \boldsymbol{a}_2$.

The equations of motion for the two basis atoms can be set up by following the procedure described in the previous two sections. It is useful to note that neighbouring carbon atoms are strongly bonded with sp^2 electronic orbitals in the plane of the graphene sheet and are characterised with out-of-plane weak π bonding. We will not go through the details here, but simply state that the size of the dynamical matrix will be 4×4 if only in-plane atomic displacements are considered, and 6×6 if two in-plane (longitudinal and one transverse) and one out-of-plane (another transverse) atomic displacements are considered. Interested readers may follow the discussion presented in Sahoo and Mishra (2012), who constructed the force constant matrix by considering interactions up to the 4th nearest neighbours. There are six branches of the dispersion curve of the normal modes. These include the in-plane acoustic branches TA and LA, an out-of-plane acoustic branch called ZA, longitudinal optical (LO), in-plane transverse optical (TO), and the out-of-plane transverse branch called ZO.

Here we will discuss the dispersion of ZA and ZO modes, which arise from flexural vibrations of atoms normal to the graphene sheet. Writing $u_{n,z}$ as the out-of-plane displacement of nth atom, and for simplicity taking Λ as the out-of-plane component of force constant, we write the equations of

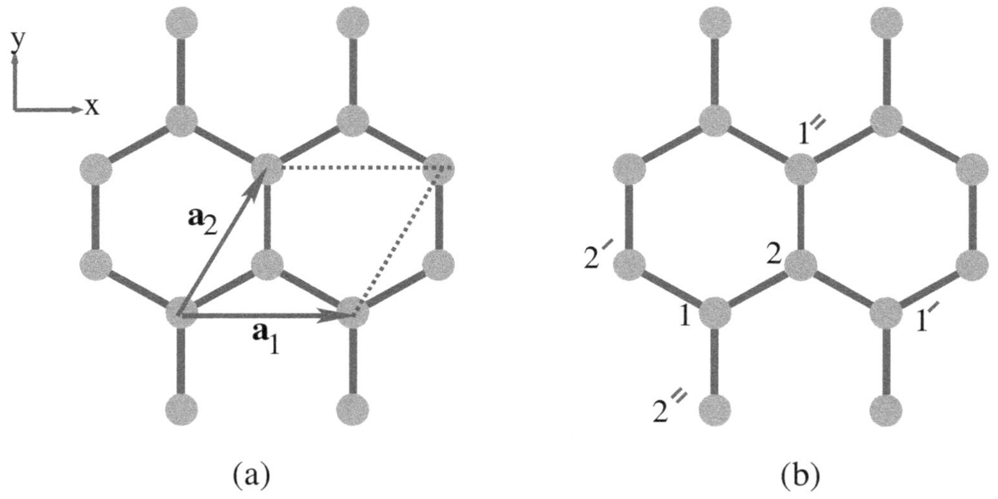

Figure 5.6 Atomic structure of graphene. As shown in panel (a), a possible choice of the hexagonal unit cell is the parallelogram formed by the primitive translation vectors $\boldsymbol{a}_1$ and $\boldsymbol{a}_2$. As shown in panel (b), the nearest neighbours of the basis atom 1 are 2, 2′ and 2″, and the nearest neighbours of the basis atom 2 are 1, 1′ and 1″.

motion of the two basis atoms as

$$m\ddot{u}_{1,z} = -\Lambda(3u_{1,z} - u_{2,z} - u_{2',z} - u_{2'',z}), \tag{5.20}$$

and

$$m\ddot{u}_{2,z} = -\Lambda(3u_{2,z} - u_{1,z} - u_{1',z} - u_{1'',z}). \tag{5.21}$$

Considering trial displacements, for an in-plane wave vector $\boldsymbol{q}$, as

$$\begin{aligned} u_{1,z} &= A_{1,z}\exp[i\boldsymbol{q}\cdot\boldsymbol{R}_1 - \omega t] = A_{1,z}\exp[-i\omega t], \\ u_{2,z} &= A_{2,z}\exp[i\boldsymbol{q}\cdot\boldsymbol{R}_2 - \omega t] = A_{2,z}\exp[i\boldsymbol{q}\cdot\boldsymbol{d} - \omega t], \end{aligned} \tag{5.22}$$

we can express equations (5.20) and (5.21) as

$$-m\omega^2 A_{1,z} = -\Lambda[3A_{1,z} - A_{2,z}e^{i\boldsymbol{q}\cdot\boldsymbol{d}}(1 + e^{-i\boldsymbol{q}\cdot\boldsymbol{a}_1} + e^{-i\boldsymbol{q}\cdot\boldsymbol{a}_2})], \tag{5.23}$$

$$-m\omega^2 A_{2,z} = -\Lambda[3A_{2,z} - A_{1,z}e^{-i\boldsymbol{q}\cdot\boldsymbol{d}}(1 + e^{i\boldsymbol{q}\cdot\boldsymbol{a}_1} + e^{i\boldsymbol{q}\cdot\boldsymbol{a}_2})]. \tag{5.24}$$

These equations can be written in matrix form as

$$\begin{bmatrix} 3\Lambda - m\omega^2 & -\Lambda e^{i\boldsymbol{q}\cdot\boldsymbol{d}}(1 + e^{-i\boldsymbol{q}\cdot\boldsymbol{a}_1} + e^{-i\boldsymbol{q}\cdot\boldsymbol{a}_2}) \\ -\Lambda e^{-i\boldsymbol{q}\cdot\boldsymbol{d}}(1 + e^{i\boldsymbol{q}\cdot\boldsymbol{a}_1} + e^{i\boldsymbol{q}\cdot\boldsymbol{a}_2}) & 3\Lambda - m\omega^2 \end{bmatrix} \begin{bmatrix} A_{1,z} \\ A_{2,z} \end{bmatrix} = 0. \tag{5.25}$$

Setting the determinant of the 2×2 matrix to zero, the eigenvalues can be obtained by solving the equation

$$(3\Lambda - m\omega^2)^2 - \Lambda^2\left[3 + 2\cos\boldsymbol{q}\cdot\boldsymbol{a}_1 + 2\cos\boldsymbol{q}\cdot\boldsymbol{a}_2 + 2\cos\boldsymbol{q}\cdot(\boldsymbol{a}_1 - \boldsymbol{a}_2)\right] = 0. \tag{5.26}$$

In figure 5.7(a) we have plotted the dispersion curves for the flexural acoustic (ZA) and flexural optical (ZO) branches along the paths Γ-M-K-Γ. There are two important features around the symmetry point K. Firstly, both ZA and ZO modes have the same frequency at the K point in the

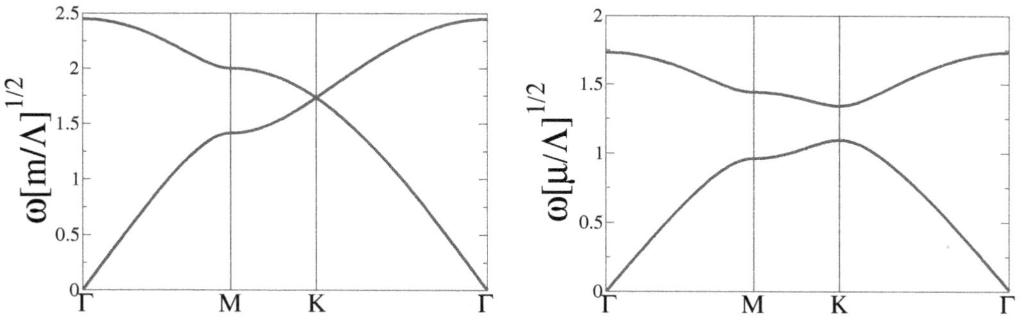

(a) (b)

Figure 5.7 (a) Dispersion curves for the ZA and ZO branches of normal modes in graphene, with atomic mass m and the out-of-plane force constant Λ. Plotted is the reduced-dimensional frequency $\omega\sqrt{(m/\Lambda)}$. (b) Dispersion curves for the ZA and ZO branches of normal modes in a graphene-like two-dimensional structure with different basis atomic masses, *e.g.* in monolayer materials such as BN, BP, *etc.* Plotted is the reduced-dimensional frequency $\omega\sqrt{(\mu/\Lambda)}$, where μ is the reduced mass, and the mass ratio is arbitrarily taken as $m_2/m_1 = 1.5$.

hexagonal Brillouin zone. Secondly, both branches show *linear* dispersion around K. Both these features arise due to the *inversion symmetry* present in graphene. The K point is known as the Dirac point. A view of the dispersion plot (ω *vs* q, with $\boldsymbol{q} = (q_x, q_y)$), of the flexural branches close to the K point is that of a cone, called the Dirac cone[2]. The concepts of Dirac point and Dirac cone are very important with regards to the topological property of a material.

There are several two-dimensional materials which have graphene-like structure but with non-identical basis masses (*i.e.* when the inversion symmetry is broken). Examples include, monolayers of boron nitride (BN), boron phosphide (BP), *etc.* In such structures, there is no inversion symmetry. The question is what happens to the frequencies of the ZA and ZO modes at K and to their linear dispersion around K. For basis atomic masses m_1 and m_2, equation (5.26) becomes

$$\begin{aligned}&(3\Lambda - m_1\omega^2)(3\Lambda - m_2\omega^2)\\ &-\Lambda^2\left[3 + 2\cos\boldsymbol{q}\cdot\boldsymbol{a}_1 + 2\cos\boldsymbol{q}\cdot\boldsymbol{a}_2 + 2\cos\boldsymbol{q}\cdot(\boldsymbol{a}_1 - \boldsymbol{a}_2)\right] = 0.\end{aligned} \tag{5.27}$$

This can be expressed in the form

$$\begin{aligned}&\frac{m_1 m_2}{\Lambda^2}\omega^4 - 3\frac{m_1 + m_2}{\Lambda}\omega^2 + 9\\ &-\left[3 + 2\cos\boldsymbol{q}\cdot\boldsymbol{a}_1 + 2\cos\boldsymbol{q}\cdot\boldsymbol{a}_2 + 2\cos\boldsymbol{q}\cdot(\boldsymbol{a}_1 - \boldsymbol{a}_2)\right] = 0.\end{aligned} \tag{5.28}$$

Figure 5.7(b) shows the dispersion curves when the mass of one of the basis atoms is 1.5 times that of the other basis atom. It is found that the degeneracy at K is split and the dispersion around K is no longer linear. Clearly, under this consideration we cannot speak of Dirac point or Dirac cone.

5.4 THREE-DIMENSIONAL CRYSTALS

The dynamical matrix for a three-dimensional crystal can be set up by following the procedure described for one-dimensional and two-dimensional systems in sections 5.2 and 5.3. This is a cele-

[2]Dirac point and Dirac cone are also found in the electronic structure of graphene, and much for the same reason.

brated topic of lattice dynamics, which has been discussed in many books and review articles. A detailed discussion of this topic is far beyond the scope of the present book but interested readers may wish to follow discussions presented in several previous publications, including Born and Huang (1954), Maradudin *et al.* (1971) and Srivastava (2022). By making use of the lattice translational symmetry and the Born–von Kármán periodic boundary condition, in this section we will show how the *central force* procedure can be extended to set up the dynamical matrix for three-dimensional cubic systems. We will then extend the discussion by describing and applying de Launey's *non-central force* concept as a combination of 'radial' and 'angular' forces (de Launay, 1956) to set up the dynamical matrix for the cubic *diamond structure*.

5.4.1 MONATOMIC SIMPLE CUBIC (SC) STRUCTURE

The simple cubic structure can be described by using the simple cubic (SC) lattice and a basis of one atom. In other words, we can 'decorate' each lattice site with an atom. The primitive unit cell is a cube, with lattice constant a and translation vectors $\boldsymbol{a}_1 = a(1,0,0)$, $\boldsymbol{a}_2 = a(0,1,0)$ and $\boldsymbol{a}_3 = a(0,0,1)$. For the single atom inside the unit cell considered at $(0,0,0)$, there are six nearest neighbours at $\pm a(1,0,0)$, $\pm a(0,1,0)$ and $\pm a(0,0,1)$. The harmonic force on this atom can be worked out by extending the procedure described in section 5.2.1. Using Λ as the force constant and equation (5.4) for atomic displacements, the components of the total force on the atom at $(0,0,0)$ from the nearest neighbours are

$$\begin{aligned} F_x &= -2\Lambda A_x(1-\cos q_x a)\exp(-i\omega t), \\ F_y &= -2\Lambda A_y(1-\cos q_y a)\exp(-i\omega t), \\ F_z &= -2\Lambda A_z(1-\cos q_z a)\exp(-i\omega t). \end{aligned} \tag{5.29}$$

The 3×3 dynamical matrix can very easily be constructed by using these expressions. As discussed in section 5.2.1, it would be important to include force contributions from next nearest neighbours for a realistic description of normal mode dispersion relation.

5.4.2 MONATOMIC BODY CENTRED CUBIC (BCC) STRUCTURE

Similar to the SC structure, there is a single atom in the primitive unit cell for the BCC structure. The primitive unit cell, with cubic lattice constant a, can be defined using the translation vectors $\boldsymbol{a}_1 = \frac{a}{2}(-1,1,1)$, $\boldsymbol{a}_2 = \frac{a}{2}(1,-1,1)$ and $\boldsymbol{a}_3 = \frac{a}{2}(1,1,-1)$. For the single atom inside the unit cell considered at $\boldsymbol{R}_0 = (0,0,0)$, there are eight nearest neighbours at $\boldsymbol{R}_1 = \frac{a}{2}(1,1,1)$, $\boldsymbol{R}_2 = -\boldsymbol{R}_1$, $\boldsymbol{R}_3 = \frac{a}{2}(-1,1,1)$, $\boldsymbol{R}_4 = -\boldsymbol{R}_3$, $\boldsymbol{R}_5 = \frac{a}{2}(1,-1,1)$, $\boldsymbol{R}_6 = -\boldsymbol{R}_5$, $\boldsymbol{R}_7 = \frac{a}{2}(1,1,-1)$ and $\boldsymbol{R}_8 = -\boldsymbol{R}_7$. Expressing atomic displacements as $\boldsymbol{u}_n = \boldsymbol{A}\exp(i\boldsymbol{q}\cdot\boldsymbol{R}_n)\exp(-i\omega t)$, considering the nearest neighbour central force constant as Λ and using equation (5.1), the following expressions can be obtained for forces on atom 0 due to atoms $n = 1,2,\cdots,8$:

$$\begin{aligned} \boldsymbol{F}_1 &= -\frac{\Lambda}{3}[(A_x+A_y+A_z)(1-\mathrm{e}^{i\boldsymbol{q}\cdot\boldsymbol{R}_1})\exp(-i\omega t)](1,1,1), \\ \boldsymbol{F}_2 &= -\frac{\Lambda}{3}[(-A_x-A_y-A_z)(1-\mathrm{e}^{i\boldsymbol{q}\cdot\boldsymbol{R}_2})\exp(-i\omega t)](-1,-1,-1), \\ \boldsymbol{F}_3 &= -\frac{\Lambda}{3}[(-A_x+A_y+A_z)(1-\mathrm{e}^{i\boldsymbol{q}\cdot\boldsymbol{R}_3})\exp(-i\omega t)](-1,1,1), \\ \boldsymbol{F}_4 &= -\frac{\Lambda}{3}[(A_x-A_y-A_z)(1-\mathrm{e}^{i\boldsymbol{q}\cdot\boldsymbol{R}_4})\exp(-i\omega t)](1,-1,-1), \\ \boldsymbol{F}_5 &= -\frac{\Lambda}{3}[(A_x-A_y+A_z)(1-\mathrm{e}^{i\boldsymbol{q}\cdot\boldsymbol{R}_5})\exp(-i\omega t)](1,-1,1), \\ \boldsymbol{F}_6 &= -\frac{\Lambda}{3}[(-A_x+A_y-A_z)(1-\mathrm{e}^{i\boldsymbol{q}\cdot\boldsymbol{R}_6})\exp(-i\omega t)](-1,1,-1), \\ \boldsymbol{F}_7 &= -\frac{\Lambda}{3}[(A_x+A_y-A_z)(1-\mathrm{e}^{i\boldsymbol{q}\cdot\boldsymbol{R}_7})\exp(-i\omega t)](1,1,-1), \end{aligned}$$

$$F_8 = -\frac{\Lambda}{3}[(-A_x - A_y + A_z)(1-e^{i\mathbf{q}\cdot\mathbf{R}_8})\exp(-i\omega t)](-1,-1,1). \tag{5.30}$$

From these, the x-component of the total force on atom 0 is

$$\begin{aligned}
F_x &= -\frac{\Lambda}{3}\exp(-i\omega t) \\
&\quad \times\big[2(A_x+A_y+A_z) - 2(A_x+A_y+A_z)\cos\mathbf{q}\cdot\mathbf{R}_1 \\
&\quad +2(A_x-A_y-A_z) - 2(A_x-A_y-A_z)\cos\mathbf{q}\cdot\mathbf{R}_3 \\
&\quad +2(A_x-A_y+A_z) - 2(A_x-A_y+A_z)\cos\mathbf{q}\cdot\mathbf{R}_5 \\
&\quad +2(A_x+A_y-A_z) - 2(A_x+A_y-A_z)\cos\mathbf{q}\cdot\mathbf{R}_7\big], \\
&= -\frac{\Lambda}{3}\exp(-i\omega t) \\
&\quad \times\big[8A_x - 2A_x(\cos\mathbf{q}\cdot\mathbf{R}_1 + \cos\mathbf{q}\cdot\mathbf{R}_3 + \cos\mathbf{q}\cdot\mathbf{R}_5 + \cos\mathbf{q}\cdot\mathbf{R}_7) \\
&\quad -2A_y(\cos\mathbf{q}\cdot\mathbf{R}_1 - \cos\mathbf{q}\cdot\mathbf{R}_3 - \cos\mathbf{q}\cdot\mathbf{R}_5 + \cos\mathbf{q}\cdot\mathbf{R}_7) \\
&\quad -2A_z(\cos\mathbf{q}\cdot\mathbf{R}_1 - \cos\mathbf{q}\cdot\mathbf{R}_3 + \cos\mathbf{q}\cdot\mathbf{R}_5 - \cos\mathbf{q}\cdot\mathbf{R}_7)\big].
\end{aligned} \tag{5.31}$$

Writing $\mathbf{q} = (q_x, q_y, q_z)$, we have

$$\begin{aligned}
\mathbf{q}\cdot\mathbf{R}_1 &= \frac{a}{2}(q_x+q_y+q_z), \quad \mathbf{q}\cdot\mathbf{R}_3 = \frac{a}{2}(-q_x+q_y+q_z), \\
\mathbf{q}\cdot\mathbf{R}_5 &= \frac{a}{2}(q_x-q_y+q_z), \quad \mathbf{q}\cdot\mathbf{R}_7 = \frac{a}{2}(q_x+q_y-q_z).
\end{aligned} \tag{5.32}$$

Expressions for F_y and F_z can similarly be obtained. These can be used in the equation of motion $m\ddot{\mathbf{u}}_0 = \sum_{n=1}^{8}\mathbf{F}_n$, where m is the mass of the atom, to obtain the expression for the dynamical matrix.

5.4.3 MONATOMIC FACE CENTRED CUBIC (FCC) STRUCTURE

Similar to the SC and BCC structures, there is a single atom in the primitive unit cell for the FCC structure. The primitive unit cell, with cubic lattice constant a, can be defined using the translation vectors $\mathbf{a}_1 = \frac{a}{2}(0,1,1)$, $\mathbf{a}_2 = \frac{a}{2}(1,0,1)$ and $\mathbf{a}_3 = \frac{a}{2}(1,1,0)$. For the single atom inside the unit cell considered at $\mathbf{R}_0 = (0,0,0)$, there are twelve nearest neighbours at $\mathbf{R}_1 = \frac{a}{2}(0,1,1)$, $\mathbf{R}_2 = -\mathbf{R}_1$, $\mathbf{R}_3 = \frac{a}{2}(1,0,1)$, $\mathbf{R}_4 = -\mathbf{R}_3$, $\mathbf{R}_5 = \frac{a}{2}(1,1,0)$, $\mathbf{R}_6 = -\mathbf{R}_5$, $\mathbf{R}_7 = \frac{a}{2}(0,-1,1)$, $\mathbf{R}_8 = -\mathbf{R}_7$, $\mathbf{R}_9 = \frac{a}{2}(-1,0,1)$, $\mathbf{R}_{10} = -\mathbf{R}_9$, $\mathbf{R}_{11} = \frac{a}{2}(-1,1,0)$ and $\mathbf{R}_{12} = -\mathbf{R}_{11}$. Considering force constant Λ, forces on atom 0 due to its nearest neighbours are

$$\begin{aligned}
\mathbf{F}_1 &= -\frac{\Lambda}{2}[(A_y+A_z)(1-e^{i\mathbf{q}\cdot\mathbf{R}_1})\exp(-i\omega t)](0,1,1), \\
\mathbf{F}_2 &= -\frac{\Lambda}{2}[(-A_y-A_z)(1-e^{i\mathbf{q}\cdot\mathbf{R}_2})\exp(-i\omega t)](0,-1,-1), \\
\mathbf{F}_3 &= -\frac{\Lambda}{2}[(A_x+A_z)(1-e^{i\mathbf{q}\cdot\mathbf{R}_3})\exp(-i\omega t)](1,0,1), \\
\mathbf{F}_4 &= -\frac{\Lambda}{2}[(-A_x-A_z)(1-e^{i\mathbf{q}\cdot\mathbf{R}_4})\exp(-i\omega t)](-1,0,-1), \\
\mathbf{F}_5 &= -\frac{\Lambda}{2}[(A_x+A_y)(1-e^{i\mathbf{q}\cdot\mathbf{R}_5})\exp(-i\omega t)](1,1,0), \\
\mathbf{F}_6 &= -\frac{\Lambda}{2}[(-A_x-A_y)(1-e^{i\mathbf{q}\cdot\mathbf{R}_6})\exp(-i\omega t)](-1,-1,0), \\
\mathbf{F}_7 &= -\frac{\Lambda}{2}[(-A_y+A_z)(1-e^{i\mathbf{q}\cdot\mathbf{R}_7})\exp(-i\omega t)](0,-1,1), \\
\mathbf{F}_8 &= -\frac{\Lambda}{2}[(A_y-A_z)(1-e^{i\mathbf{q}\cdot\mathbf{R}_8})\exp(-i\omega t)](0,1,-1), \\
\mathbf{F}_9 &= -\frac{\Lambda}{2}[(-A_x+A_z)(1-e^{i\mathbf{q}\cdot\mathbf{R}_9})\exp(-i\omega t)](-1,0,1), \\
\mathbf{F}_{10} &= -\frac{\Lambda}{2}[(A_x-A_z)(1-e^{i\mathbf{q}\cdot\mathbf{R}_{10}})\exp(-i\omega t)](1,0,-1),
\end{aligned}$$

$$\begin{aligned} \boldsymbol{F}_{11} &= -\frac{\Lambda}{2}[(-A_x+A_y)(1-e^{i\boldsymbol{q}\cdot\boldsymbol{R}_{11}})\exp(-i\omega t)](-1,1,0), \\ \boldsymbol{F}_{12} &= -\frac{\Lambda}{2}[(A_x-A_y)(1-e^{i\boldsymbol{q}\cdot\boldsymbol{R}_{12}})\exp(-i\omega t)](1,-1,0). \end{aligned} \tag{5.33}$$

These expressions can be used in the equation of motion $m\ddot{\boldsymbol{u}}_0 = \sum_{n=1}^{12} \boldsymbol{F}_n$, where m is the mass of the atom, to obtain the expression for the dynamical matrix.

5.4.4 DIATOMIC THREE-DIMENSIONAL STRUCTURE: CAESIUM CHLORIDE

The caesium chloride (CsCl) structure adopts the simple cubic (SC) lattice with a basis of two atoms, one at $(0,0,0)$ and the other at $\frac{a}{2}(1,1,1)$, where a is the cubic lattice constant. Each basis atom has eight nearest neighbours. The coordinates of the nearest neighbours of the basis atom at $(0,0,0)$ are $\frac{a}{2}(1,1,1)$, $\frac{a}{2}(-1,1,1)$, $\frac{a}{2}(1,-1,1)$, $\frac{a}{2}(1,1,-1)$, $\frac{a}{2}(1,-1,-1)$, $\frac{a}{2}(-1,1,-1)$, $\frac{a}{2}(-1,-1,1)$ and $\frac{a}{2}(-1,-1,-1)$. The coordinates of the nearest neighbours for the other basis atom can similarly be established. The procedure described in the previous sub-sections can be adopted to obtain expressions of forces on the basis atoms, and eventually set up the dynamical equations of motion.

5.4.5 DIATOMIC THREE-DIMENSIONAL STRUCTURE: ROCKSALT

The rocksalt, or sodium chloride (NaCl), structure adopts the face centred cubic (FCC) lattice with a basis of two atoms, one at $(0,0,0)$ and the other at $\frac{a}{2}(1,1,1)$, where a is the cubic lattice constant. Each basis atoms has six nearest neighbours. The coordinates of the nearest neighbours of the basis atom at $(0,0,0)$ are $\pm\frac{a}{2}(1,0,0)$, $\pm\frac{a}{2}(0,1,0)$ and $\pm\frac{a}{2}(0,0,1)$. The coordinates of the nearest neighbours of the other basis atom can similarly be established. The procedure described in the previous sub-sections can be adopted to obtain expressions of forces on the basis atoms, and eventually set up the dynamical equations of motion.

5.4.6 DIATOMIC THREE-DIMENSIONAL STRUCTURES: ZINCBLENDE AND DIAMOND

Many technologically important semiconductors grow in the form of three-dimensional cubic crystals with their atoms arranged in the *diamond* and the *zincblende* structures. Figure 5.8 shows atomic arrangement in the zincblend structure. This three-dimensional structure can be described by using the FCC lattice and a basis of two non-identical atoms, as described further. The zincblend structure turns into the diamond structure when the two basis atoms are identical chemical entities. In this sub-section we will discuss the linear dynamics of the diamond structure.

The FCC lattice can be generated by placing points at the tips of vectors

$$\boldsymbol{R} = n_1\boldsymbol{a}_1 + n_2\boldsymbol{a}_2 + n_3\boldsymbol{a}_3, \quad n_i = 0, \pm1, \pm2, \pm3, \cdots, \tag{5.34}$$

where $\boldsymbol{a}_1$, $\boldsymbol{a}_2$ and $\boldsymbol{a}_3$ are the primitive unit vectors

$$\boldsymbol{a}_1 = \frac{a}{2}(0,1,1), \quad \boldsymbol{a}_2 = \frac{a}{2}(1,0,1), \quad \boldsymbol{a}_3 = \frac{a}{2}(1,1,0) \tag{5.35}$$

and a is the cubic *lattice constant*. The two basis atoms per FCC unit cell can be taken at the positions $\boldsymbol{\tau}_1 = (0,0,0)$ and $\boldsymbol{\tau}_2 = \frac{a}{4}(1,1,1)$. Figure 5.9 shows the conventional cubic unit cell and the primitive translation vectors $\boldsymbol{a}_1$, $\boldsymbol{a}_2$ and $\boldsymbol{a}_3$ of the FCC lattice. The FCC unit cell is the parallelopiped formed by these translation vectors. The two basis atoms inside the FCC unit cell are labelled 1 and 2. It should be mentioned that the two basis atoms can be positioned in a flexible manner in relation to an FCC lattice point. Two choices are commonly made. In one choice one of the basis atoms is placed at the lattice site $(0,0,0)$ and the other basis atom is placed at $\frac{a}{4}(1,1,1)$, *viz* at a quarter distance along the cubic diagonal direction. In the second choice a lattice point lies halfway between the

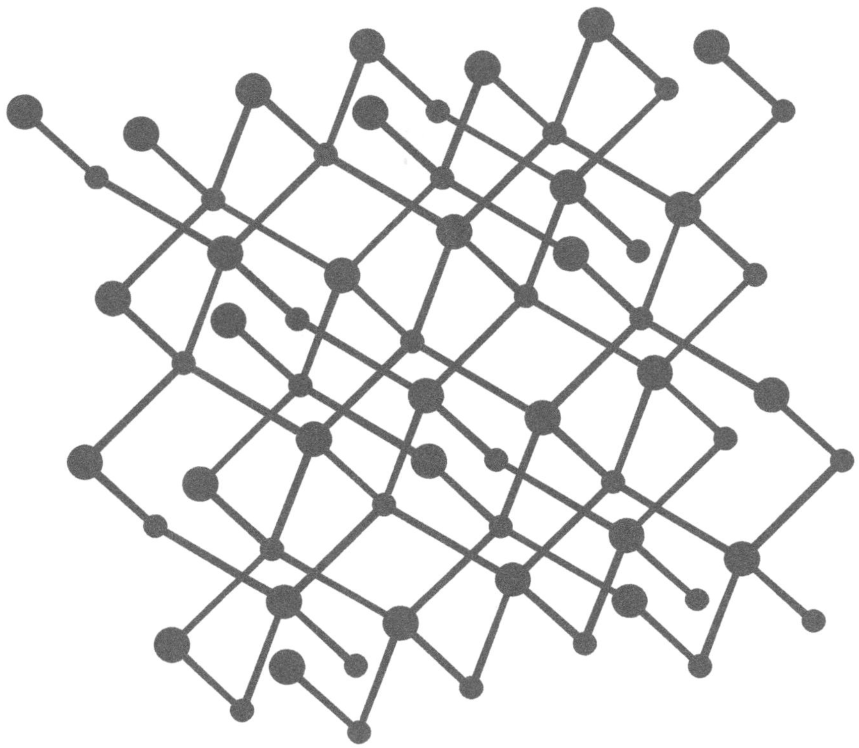

Figure 5.8 The **zincblende crystal structure**, containing two types of atoms (shown as small and big filled circles). The zincblende structure turns into the **diamond structure** when all atoms are identical chemical entities.

two basis atoms, *viz* a lattice site at $(0,0,0)$ is decorated with the basis atoms at $\pm\frac{a}{8}(1,1,1)$. The second choice is beneficial when performing numerical calculations of properties of materials with the diamond structure, as then the atomic geometry is characterised with the inversion symmetry with respect to the lattice point at the origin.

Consider atomic mass m, nearest neighbour central (also called radial) force constant Λ, and the basis atoms at $\boldsymbol{\tau}_1 = (0,0,0)$ and $\boldsymbol{\tau}_2 = \frac{a}{4}(1,1,1)$. The nearest neighbours of the basis atom 1 are at

$$\boldsymbol{\tau}_2 = \frac{a}{4}(1,1,1), \boldsymbol{\tau}_{2'} = \frac{a}{4}(1,-1,-1), \boldsymbol{\tau}_{2''} = \frac{a}{4}(-1,1,-1), \boldsymbol{\tau}_{2'''} = \frac{a}{4}(-1,-1,1). \tag{5.36}$$

Let us express displacements of the basis atoms as

$$\boldsymbol{u}_1 = \boldsymbol{U}\mathrm{e}^{-i\omega t}, \quad \boldsymbol{u}_2 = \boldsymbol{V}\mathrm{e}^{i(\boldsymbol{q}\cdot\boldsymbol{\tau}_2 - \omega t)}, \tag{5.37}$$

where $\boldsymbol{U} = (U_x, U_y, U_z)$ and $\boldsymbol{V} = (V_x, V_y, V_z)$ are displacement amplitude vectors, and $\boldsymbol{q}$ is the wave propagation vector.

Following equation (5.1), we can write down the central (radial) force on atom 1 due to its four nearest neighbours. The force due to atom 2 is

$$\boldsymbol{F}(1,2) \quad = \quad -\Lambda\boldsymbol{\varepsilon}(1,2)\cdot(\boldsymbol{u}_1 - \boldsymbol{u}_2)]\boldsymbol{\varepsilon}(1,2),$$

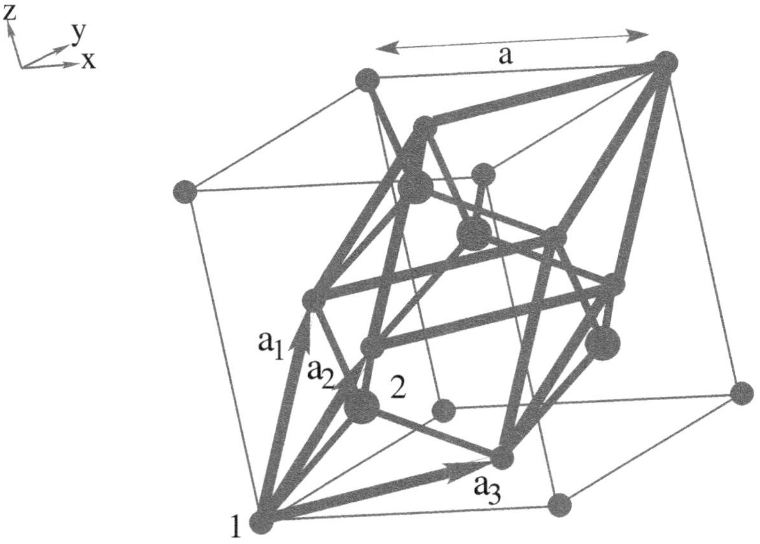

Figure 5.9 The cubic unit cell (marked with thin lines) and the FCC unit cell (marked with thick lines) of the zincblende structure. The FCC primitive unit vectors $\boldsymbol{a}_1$, $\boldsymbol{a}_2$ and $\boldsymbol{a}_3$ are indicated with arrows. The symbol a represent the cubic lattice constant. There are two non-identical basis atoms inside the primitive unit cell, labelled 1 and 2 and shown as the small and big filled circles. The cubic unit cell contains eight atoms (four of each of the two basis types).

$$\begin{aligned} &= -\Lambda\frac{1}{\sqrt{3}}(1,1,1)\cdot(\boldsymbol{U}-\boldsymbol{V}e^{i\boldsymbol{q}\cdot\boldsymbol{\tau}_2})e^{-i\omega t}\frac{1}{\sqrt{3}}(1,1,1), \\ &= -\frac{\Lambda}{3}[U_x+U_y+U_z-(V_x+V_y+V_z)e^{i\boldsymbol{q}\cdot\boldsymbol{\tau}_2}]e^{-i\omega t}(1,1,1). \end{aligned} \tag{5.38}$$

Similarly, the forces due to the other three nearest neighbours can be expressed as

$$\boldsymbol{F}(1,2') = -\frac{\Lambda}{3}[U_x-U_y-U_z-(V_x-V_y-V_z)e^{i\boldsymbol{q}\cdot\boldsymbol{\tau}_{2'}}]e^{-i\omega t}(1,-1,-1), \tag{5.39}$$

$$\boldsymbol{F}(1,2'') = -\frac{\Lambda}{3}[-U_x+U_y-U_z-(-V_x+V_y-V_z)e^{i\boldsymbol{q}\cdot\boldsymbol{\tau}_{2''}}]e^{-i\omega t}(-1,1,-1), \tag{5.40}$$

$$\boldsymbol{F}(1,2''') = -\frac{\Lambda}{3}[-U_x-U_y+U_z-(-V_x-V_y+V_z)e^{i\boldsymbol{q}\cdot\boldsymbol{\tau}_{2'''}}]e^{-i\omega t}(-1,-1,1). \tag{5.41}$$

The nearest neighbours of the basis atom 2 are at

$$\boldsymbol{\tau}_1 = (0,0,0), \quad \boldsymbol{\tau}_{1'} = \frac{a}{4}(0,1,1), \quad \boldsymbol{\tau}_{1''} = \frac{a}{4}(1,0,1), \quad \boldsymbol{\tau}_{1'''} = \frac{a}{4}(1,1,0). \tag{5.42}$$

The central (radial) force on atom 2 due to atom 1 is

$$\begin{aligned} \boldsymbol{F}(2,1) &= -\Lambda[\hat{\boldsymbol{\varepsilon}}(2,1)\cdot(\boldsymbol{u}_2-\boldsymbol{u}_1)]\hat{\boldsymbol{\varepsilon}}(2,1), \\ &= -\Lambda\frac{1}{\sqrt{3}}(-1,-1,-1)\cdot(\boldsymbol{V}e^{i\boldsymbol{q}\cdot\boldsymbol{\tau}_2}-\boldsymbol{U}e^{i\boldsymbol{q}\cdot\boldsymbol{\tau}_1})e^{-i\omega t}\frac{1}{\sqrt{3}}(-1,-1,-1), \\ &= -\frac{\Lambda}{3}[(-V_x-V_y-V_z)e^{i\boldsymbol{q}\cdot\boldsymbol{\tau}_2}-(-U_x-U_y-U_z)e^{i\boldsymbol{q}\cdot\boldsymbol{\tau}_1}] \end{aligned}$$

$$e^{-i\omega t}(-1,-1,-1). \quad (5.43)$$

Similarly, the forces due to the other nearest neighbours can be expressed as

$$\begin{aligned} \boldsymbol{F}(2,1') &= -\frac{\Lambda}{3}[(-V_x+V_y+V_z)e^{i\boldsymbol{q}\cdot\boldsymbol{\tau}_2}-(-U_x+U_y+U_z)e^{i\boldsymbol{q}\cdot\boldsymbol{\tau}_{1'}}] \\ &\quad e^{-i\omega t}(-1,1,1), & (5.44) \\ \boldsymbol{F}(2,1'') &= -\frac{\Lambda}{3}[(V_x-V_y+V_z)e^{i\boldsymbol{q}\cdot\boldsymbol{\tau}_2}-(U_x-U_y+U_z)e^{i\boldsymbol{q}\cdot\boldsymbol{\tau}_{1''}}] \\ &\quad e^{-i\omega t}(1,-1,1), & (5.45) \\ \boldsymbol{F}(2,1''') &= -\frac{\Lambda}{3}[(V_x+V_y-V_z)e^{i\boldsymbol{q}\cdot\boldsymbol{\tau}_2}-(U_x+U_y-U_z)e^{i\boldsymbol{q}\cdot\boldsymbol{\tau}_{1'''}}] \\ &\quad e^{-i\omega t}(1,1,-1). & (5.46) \end{aligned}$$

As the diamond and zincblende are open crystal structures, with the packing fraction of approximately 0.34, it is important to consider angular deviation of the springs from the lines connecting the nearest neighbours. In other words, an *angular force* should be included in addition to the radial force that we have considered so far. The combination of the radial and angular forces is a **non-central force** of the type described by de Launay (1956). Let us consider figure 5.1. When considering an angular (or non-radial) force, we require components of $\boldsymbol{u}_0$ and $\boldsymbol{u}_n$ which are perpendicular to $\boldsymbol{R}_n$ (*i.e.* normal to the line joining atoms 0 and n). The cross product $(\boldsymbol{\varepsilon}_n \times \boldsymbol{u}_0) \times \boldsymbol{\varepsilon}_n$ represents a vector with component of $\boldsymbol{u}_0$ perpendicular to $\boldsymbol{\varepsilon}_n$ with magnitude $|\boldsymbol{\varepsilon}_n \times \boldsymbol{u}_0|$. We will represent the **angular harmonic force constant** as Λ'. The total non-central force on atom 0 due to atom n is,

$$\begin{aligned} \boldsymbol{F}_n &= F_n^{\text{radial}} + F_n^{\text{angular}}, \\ &= -\Lambda\left[\hat{\boldsymbol{\varepsilon}}_n\cdot(\boldsymbol{u}_0-\boldsymbol{u}_n)\right]\hat{\boldsymbol{\varepsilon}}_n - \Lambda'\left[\hat{\boldsymbol{\varepsilon}}_n\times(\boldsymbol{u}_0-\boldsymbol{u}_n)\right]\times\hat{\boldsymbol{\varepsilon}}_n, \\ &= -\Lambda\left[\hat{\boldsymbol{\varepsilon}}_n\cdot(\boldsymbol{u}_0-\boldsymbol{u}_n)\right]\hat{\boldsymbol{\varepsilon}}_n - \Lambda'(\boldsymbol{u}_0-\boldsymbol{u}_n) + \Lambda'\left[\hat{\boldsymbol{\varepsilon}}_n\cdot(\boldsymbol{u}_0-\boldsymbol{u}_n)\right]\hat{\boldsymbol{\varepsilon}}_n, \\ &= -\Lambda'(\boldsymbol{u}_0-\boldsymbol{u}_n) - (\Lambda-\Lambda')\left[\hat{\boldsymbol{\varepsilon}}_n\cdot(\boldsymbol{u}_0-\boldsymbol{u}_n)\right]\hat{\boldsymbol{\varepsilon}}_n. & (5.47) \end{aligned}$$

The non-central force on atom 1 due to its nearest neighbours follows the following equations

$$\begin{aligned} e^{i\omega t}\boldsymbol{F}(1,2) &= -\frac{\Lambda-\Lambda'}{3}\left[U_x+U_y+U_z-(V_x+V_y+V_z)h_1\right](1,1,1) \\ &\quad -\Lambda'(\boldsymbol{U}-\boldsymbol{V}h_1), & (5.48) \\ e^{i\omega t}\boldsymbol{F}(1,2') &= -\frac{\Lambda-\Lambda'}{3}\left[U_x-U_y-U_z-(V_x-V_y-V_z)h_2\right](1,-1,-1) \\ &\quad -\Lambda'(\boldsymbol{U}-\boldsymbol{V}h_2), & (5.49) \\ e^{i\omega t}\boldsymbol{F}(1,2'') &= -\frac{\Lambda-\Lambda'}{3}\left[-U_x+U_y-U_z-(-V_x+V_y-V_z)h_3\right](-1,1,-1) \\ &\quad -\Lambda'(\boldsymbol{U}-\boldsymbol{V}h_3), & (5.50) \\ e^{i\omega t}\boldsymbol{F}(1,2''') &= -\frac{\Lambda-\Lambda'}{3}\left[-U_x-U_y+U_z-(-V_x-V_y+V_z)h_4\right](-1,-1,1) \\ &\quad -\Lambda'(\boldsymbol{U}-\boldsymbol{V}h_4), & (5.51) \end{aligned}$$

where

$$h_1 = e^{i\boldsymbol{q}\cdot\boldsymbol{\tau}_2}, \quad h_2 = e^{i\boldsymbol{q}\cdot\boldsymbol{\tau}_{2'}}, \quad h_3 = e^{i\boldsymbol{q}\cdot\boldsymbol{\tau}_{2''}}, \quad h_4 = e^{i\boldsymbol{q}\cdot\boldsymbol{\tau}_{2'''}}. \quad (5.52)$$

From above, the Cartesian components of the total force on atom 1 are

$$\begin{aligned} F_x &= [-4AU_x + AV_xc_1 + BV_yc_2 + BV_zc_3]e^{-i\omega t}, & (5.53) \\ F_y &= [-4AU_y + BV_xc_2 + AV_yc_1 + BV_zc_4]e^{-i\omega t}, & (5.54) \\ F_z &= [-4AU_z + BV_xc_3 + BV_yc_4 + AV_zc_1]e^{-i\omega t}, & (5.55) \end{aligned}$$

where

$$\begin{aligned} c_1 &= h_1 + h2 + h_3 + h_4, \quad c_2 = h_1 - h2 - h_3 + h_4, \\ c_3 &= h_1 - h2 + h_3 - h_4, \quad c_4 = h_1 + h2 - h_3 - h_4, \end{aligned} \tag{5.56}$$

and

$$A = \frac{\Lambda + 2\Lambda'}{3}, \quad B = \frac{\Lambda - \Lambda'}{3}. \tag{5.57}$$

The parameters A and B are referred to as the two **general force constants** of the diamond structure. When the angular force is ignored, *i.e.* when $\Lambda' = 0$, then $A = B = \frac{\Lambda}{3} = A_1$ is the single central force constant. Further discussion on these force constants can be found in the publications by Smith (1948) and Herman (1959).

The components of the total non-central force on the second basis atom can similarly be obtained. Having done that, the equations of motion of the two basis atoms can be expressed as a 6×6 secular equation $|D - \omega^2 I| = 0$, with the dynamical matrix expressed as

$$D = \frac{4}{m} \begin{bmatrix} A & 0 & 0 & -Af_1 & -Bf_2 & -Bf_3 \\ 0 & A & 0 & -Bf_2 & -Af_1 & -Bf_4 \\ 0 & 0 & A & -Bf_3 & -Bf_4 & -Af_1 \\ -Af_1^* & -Bf_2^* & -Bf_3^* & A & 0 & 0 \\ -Bf_2^* & -Af_1^* & -Bf_4^* & 0 & A & 0 \\ -Bf_3^* & -Bf_4^* & -Af_1^* & 0 & 0 & A \end{bmatrix}, \tag{5.58}$$

where

$$\begin{aligned} f_1 &= c_x c_y c_z - i s_x s_y s_z, \quad f_2 = -s_x s_y c_z + i c_x c_y s_z, \\ f_3 &= -s_x c_y s_z + i c_x s_y c_z, \quad f_4 = -c_x s_y s_z + i s_x c_y c_z, \end{aligned} \tag{5.59}$$

with

$$c_x = \cos(q_x a/4), \quad s_x = \sin(q_x a/4), \quad etc. \tag{5.60}$$

Note that the matrix expressed above is the transpose of the dynamical matrix presented in the works of de Launay (1956) and Parrott (1969).

Figure 5.10 shows the Brillouin zone for the FCC lattice. It is a truncated octahedron, with six square faces and eight hexagonal faces. One of its 48 irreducible parts is shown and the symmetry points indicated. One of the three principal symmetry directions is the line Γ-X, where $\Gamma = (0,0,0)$ is the centre of the zone and $X = \frac{2\pi}{a}(1,0,0)$ is the centre of one of the square faces. The characteristic equation $|D - \omega^2 I| = 0$ can be solved numerically for any choice of the wavevector $\boldsymbol{q}$ inside the Brillouin zone. However, we will show that it is possible to solve it analytically for wave propagation along the Γ-X symmetry direction. Along this direction $q_y = q_z = 0$, so that $c_y = c_z = 1$ and $s_y = s_z = 0$ and the dynamical matrix becomes

$$D = \frac{4A}{m} \begin{bmatrix} 1 & 0 & 0 & -c_x & 0 & 0 \\ 0 & 1 & 0 & 0 & -c_x & -i\eta s_x \\ 0 & 0 & 1 & 0 & -i\eta s_x & -c_x \\ -c_x & 0 & 0 & 1 & 0 & 0 \\ 0 & -c_x & i\eta s_x & 0 & 1 & 0 \\ 0 & i\eta s_x & -c_x & 0 & 0 & 1 \end{bmatrix}, \tag{5.61}$$

where $\eta = B/A$. The characteristic equation $|D - \omega^2 I| = 0$ can be solved for ω^2 by expanding the 6×6 determinant. This can be done either by straightforwardly employing the Laplace expansion

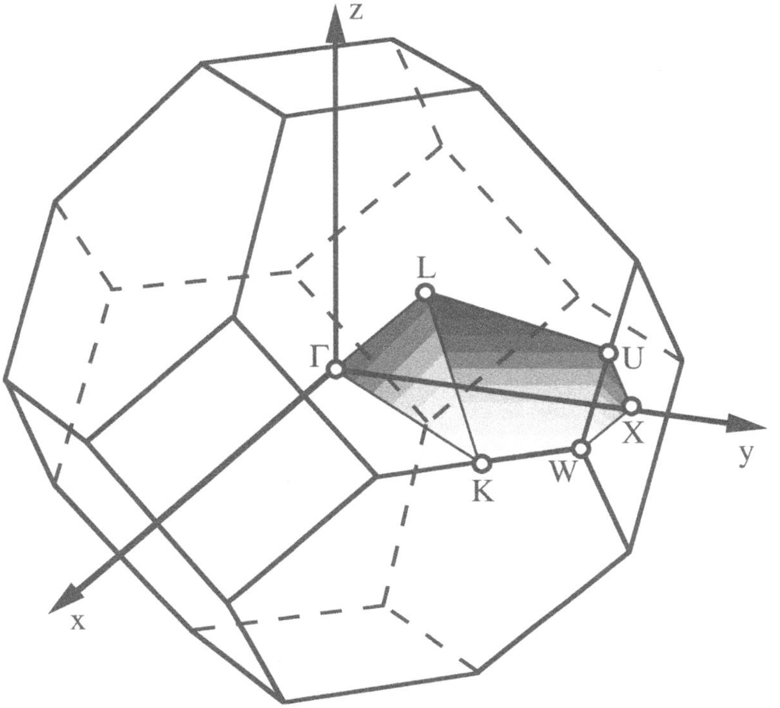

Figure 5.10 The Brillouin zone for the FCC lattice. It is a truncated octahedron, with six square faces and eight hexagonal faces. The symmetry points of the zone are labelled. The shaded pentahedral region is an irreducible part of the zone. The volume of the irreducible part, based on the point group symmetry of the FCC lattice, is $(\frac{1}{48})$th that of the full zone.

method, or by employing a reduction method discussed in de Launay (1956) and Parrott (1969). The following results can be obtained

$$\omega^2 = \frac{4A}{m}(1 \pm c_x), \tag{5.62}$$

$$\omega^2 = \frac{4A}{m}\left[1 \pm \sqrt{c_x^2 + \eta^2 s_x^2}\right], \quad \text{doubly degenerate.} \tag{5.63}$$

It can be shown that the zone-centre eigenvalues are $\omega^2_{\text{TA}}(\Gamma) = \omega^2_{\text{LA}}(\Gamma) = 0$ and $\omega^2_{\text{TO}}(\Gamma) = \omega^2_{\text{LO}}(\Gamma) = \frac{8A}{m}$. The eigenvalues at the X zone edge are $\omega^2_{\text{TA}}(X) = \frac{4A}{m}(1-\eta)$, $\omega^2_{\text{LA}}(X) = \omega^2_{\text{LO}}(X) = \frac{4A}{m}$ and $\omega^2_{\text{TO}}(X) = \frac{4A}{m}(1+\eta)$. The TA and TO branches are doubly degenerate.

Figure 5.11 shows the dispersion of normal modes along Γ-X. Numerical calculations were made with the choice $\eta = B/A = 0.5$. The LA branch lies higher than the two degenerate TA branches. The LO branch becomes degenerate with the LA branch at the X symmetry point. The two degenerate TO branches lie above the LO branch.

It should be mentioned that, while the simple formulation described here gets the essence of the dispersive curves, it is important to consider inter-atomic interaction involving several nearest neighbours for obtaining dispersion curves that can match experimental measurements. We have merely shown the basic steps of applying the Newtonian mechanics to the harmonic ball-and-spring model for setting up the dynamical matrix of a three-dimensional crystal structure, such as the cubic diamond structure, and will not attempt to take the discussion any further.

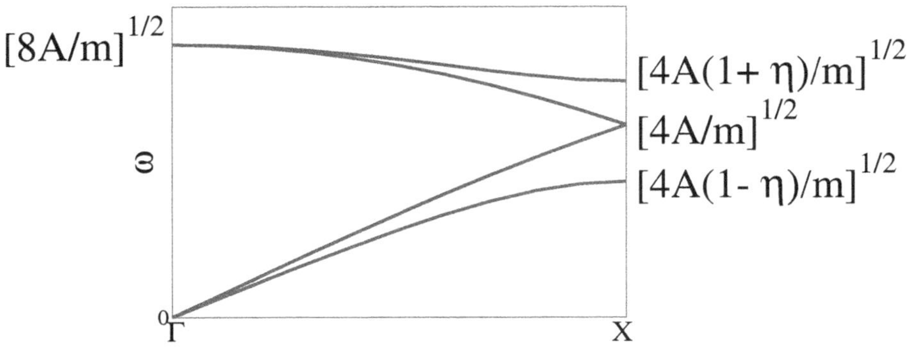

Figure 5.11 Dispersion curves along Γ-*X* (*i.e.* the [001] symmetry direction) for normal modes in the cubic diamond structure, using the nearest-neighbour general force model. Analytic solutions of the zone-centre and zone-edge frequencies are expressed in terms of the atomic mass m and the general force constants A and B, with $\eta = B/A$.

FURTHER READING

1. de Launay J 1956 *Solid State Physics* **2** 220
2. Kittel C 1996 *Introduction to Solid State Physics* 7th edn (Wiley: New York)
3. Srivastava G P 2022 *The Physics of Phonons* 2nd edn (CRC: Oxon)
4. Sahoo R and Mishra R (2012) *J Expt and Theor Physics* **114** 805.
5. Herman F J 1959 *J Phys Chem Solids* **8** 405
6. Parrott J E 1969 *Solid State Theory - Methods and Applications* (Ed. P T Landsberg) (Wiley - Interscience: London)
7. Smith H M J 1948 *Phil Trans Roy Soc Lond* **241** 105

PROBLEMS

1. Obtain expressions for the group velocity of the TA and LA normal modes in a monatomic two-dimensional square crystal along the Γ-*X* symmetry direction. Find the ratio of these two speeds for the choices $\Lambda_2/\Lambda_1 = 0.5$ and 1.0 and comment on the results.
2. Derive the results presented in equations (5.10) and (5.11).
3. Show that along the direction *X*-*M*, where $q_x = \pi/a$ and $q_y = q'$, the eigensolutions of the secular equation in equation (5.9) are:

$$\text{for } s = 1: \quad \omega^2 = \frac{2}{m}[\Lambda_1(1-\cos q'a)+\Lambda_2(1+\cos \boldsymbol{q}'a)], \quad \boldsymbol{e} = \boldsymbol{A}/A = (0,1).$$

$$\text{for } s = 2: \quad \omega^2 = \frac{2}{m}[2\Lambda_1+\Lambda_2(1+\cos q'a)], \quad \boldsymbol{e} = \boldsymbol{A}/A = (1,0).$$

   Based on the directions of the eigenvectors, comment on the use of words ‘transverse’ and ‘longitudinal’ polarisations along this direction.
4. Using the dynamical matrix in equation (5.18), prove that the eigenvalue results for the monatomic hexagonal crystal are $\omega_1^2 = \frac{2\Lambda}{m}$ and $\omega_2^2 = \frac{6\Lambda}{m}$ at the *M* point and $\omega_1^2 = \omega_2^2 = \frac{9\Lambda}{2m}$ at the *K* point.
5. Provide a step by step derivation of equation (5.27).

6. Use equation (5.28) to estimate the flexural spring constant Λ for monolayer boron nitride (BN) which is characterised by splitting between the ZA and ZO modes at the K point of approximately 3 THz.
7. Using equation (5.61), derive the following eigenvalue expressions at the zone centre (Γ) and the zone edge point X: $\omega^2_{\mathrm{TA}}(\Gamma) = \omega^2_{\mathrm{LA}}(\Gamma) = 0$, $\omega^2_{\mathrm{TO}}(\Gamma) = \omega^2_{\mathrm{LO}}(\Gamma) = \frac{8A}{m}$, $\omega^2_{\mathrm{TA}}(X) = \frac{4A}{m}(1-\eta)$, $\omega^2_{\mathrm{LA}}(X) = \omega^2_{\mathrm{LO}}(X) = \frac{4A}{m}$ and $\omega^2_{\mathrm{TO}}(X) = \frac{4A}{m}(1+\eta)$, where $\eta = B/A$ and A and B are the two force constant parameters of the general force model for the cubic diamond structure.

Part IV

Non-linear dynamics

6 Non-linear Motion and Chaos – Theory

6.1 INTRODUCTION

In Parts I and II of this book we studied the classical dynamics of single and interacting particles subjected to linear and time-independent forces. In general, dynamics of real dynamical systems is governed by non-linear and/or time-dependent forces. In this Part we will consider effects of non-linearity and time dependence of forces when discussing the dynamics of particles and systems. As mentioned at the beginning of the book, systems subjected to time-independent forces are called autonomous systems. In order to analyse the dynamics of autonomous and non-autonomous systems we will develop a few terminologies as we go along. We will try to present examples from different branches of science and engineering.

6.2 CHARACTERISING LINEAR, NON-LINEAR, CHAOTIC AND FRACTAL SYSTEMS

In section 1.5, and particularly in table 1.1, we made a brief distinction between linear, non-linear and chaotic dynamics. Here we will expand upon the explanations.

The magnitude of force acting upon a linear dynamical system is of the form $F(u) \propto u$, where u is the amount of displacement from system's equilibrium. The dynamics of such a system is regular and deterministic. As we saw in Parts I and II, the dynamics of such a system can either be studied analytically or using a computational model.

The magnitude of the force acting upon a non-linear dynamical system can depend on displacement u, momentum p (or $\dot{u}$) and time t in the general form $F(u^j, p^l, t)$, with $j, l \geq 1$. Unlike linear systems, motions of non-linear systems cannot in general be solved analytically. Regular and deterministic non-linear dynamics can be studied by using a combination of analytical methods and non-analytical methods (*e.g.* using approximations and computational models).

Chaotic dynamics is governed by motion exhibiting deterministic as well as non-deterministic characteristics. A chaotic motion shows irregularity and its time evolution is sensitive to initial conditions. Examples of chaos can be found in all areas of science and engineering. Some examples are: electrical circuits, motion of planets, vibrations of atoms in condensed matter, motion of electrons in atoms and solids, irregular heart beats, *etc.* Approximate methods and computational tools are required to solve chaotic equations of motion.

A chaotic system can become a fractal system, which is characterised by self-similarity. Chaos and fractals can be connected by **strange attractors**. We will briefly discuss the term 'strange attractors' later in this chapter.

Details of motion of linear, non-linear and chaotic systems can be examined by presenting solutions of the dynamical problem in **space-time** (plots of displacement vector $\boldsymbol{r}$ against time t) and **phase-space** (plots of momentum $\boldsymbol{p}$ against displacement $\boldsymbol{r}$), and analysing these using **Poincaré sections** and **Poincaré maps**. As described in section 3.1.2, we remind ourselves that phase space is a $2n$-dimensional space spanned by generalised coordinates $\boldsymbol{q} \equiv \{q_i\}$ and generalised momenta $\boldsymbol{p} \equiv \{p_i\}$, with $i = 1, 2, 3, \cdots, n$. We also note that as time varies, a phase point (q_i, p_i) moves along a certain phase path in phase space. For different initial conditions, the motion is described by different phase paths. A phase path represents the complete time history of the dynamics of the system under consideration. Construction of all possible phase paths constitutes the **phase diagram** of the

DOI: 10.1201/9781003383314-6

system. Thus, we require a $(2n+1)$-dimensional space with variables $\{q_i, p_i, t\}$ to fully construct the phase diagram.

6.3 PHASE DIAGRAM FOR ONE-DIMENSIONAL HARMONIC OSCILLATOR

6.3.1 PHASE ORBITS

Let us discuss the concepts underlying **phase diagram**, **Poincarè section** and **Poincarè map** by considering the simplest of linear dynamical systems, the one-dimensional harmonic oscillator.

Writing $q = x$ for the displacement coordinate and $p = m\dot{x}$ for the linear momentum, the total energy of the pendulum bob of mass m is $\frac{p^2}{2m} + \frac{1}{2}kx^2$, where k is the spring constant. As the system is autonomous, *i.e.* the expression for the force acting on the bob does not explicitly contain the time variable t, its total energy is conserved to a constant value E. In other words, we can write

$$\begin{aligned} \text{const.} = E &= \frac{p^2}{2m} + \frac{1}{2}kx^2, \\ 1 &= \frac{p^2}{2mE} + \frac{x^2}{2E/k}. \end{aligned} \tag{6.1}$$

This is the equation of an ellipse in the (p-x) phase space

$$\frac{x^2}{a^2} + \frac{p^2}{b^2} = 1, \tag{6.2}$$

with $a = \sqrt{2E/k}$ and $b = \sqrt{2mE}$ as semi x-axis and semi p-axis, respectively. With the choice $E = 1$, $m = 1$ and $k = 1$ in appropriate units, this becomes the equation of a circle of unit radius: $x^2 + p^2 = 1$.

Following discussions in section 2.2.2, the equation of motion of the bob is

$$\frac{\mathrm{d}^2x}{\mathrm{d}t^2} + \omega^2 x = 0. \tag{6.3}$$

The solutions of this dynamical system can be expressed as [*cf* equation (3.174)]

$$x = A\cos(\omega t + \psi), \tag{6.4}$$

and

$$p = -\omega A\sin(\omega t + \psi), \tag{6.5}$$

where ψ is a phase angle.

Let us, for simplicity, consider $\omega = 1$ and $\psi = 0$ and write $x = \cos(t)$ and $p = -\sin(t)$. The x-*vs*-t and p-*vs*-t plots are presented in figures 6.1(a) and 6.1(b), respectively. As expected, both $x(t)$ and $p(t)$ exhibit periodic repetition when t changes by a multiple of 2π.

6.3.2 POINCARE SECTIONS AND MAP

As it is a one-dimensional system, *i.e.* $n = 1$, a complete description of the dynamics of the system would require sketching a $(2n+1)$-dimensional (*i.e.* 3-dimensional) phase diagram, using x, p and t as axes. Poincaré invented a simple scheme for representing phase space diagrams of dynamical systems by considering a series of two-dimensional *slices*, or *sections*, for phase space orbits at different times t. Collecting 'snapshots' of phase orbits from Poincaré sections taken at different times, and projecting them onto one of the x-p planes, generates what is called a **Poincaré map**. For the one-dimensional simple harmonic oscillator, each point on a phase orbit repeats itself when time t changes by a multiple of the time period T, *i.e.* when t changes by $nT = 2\pi n/\omega$, where n is

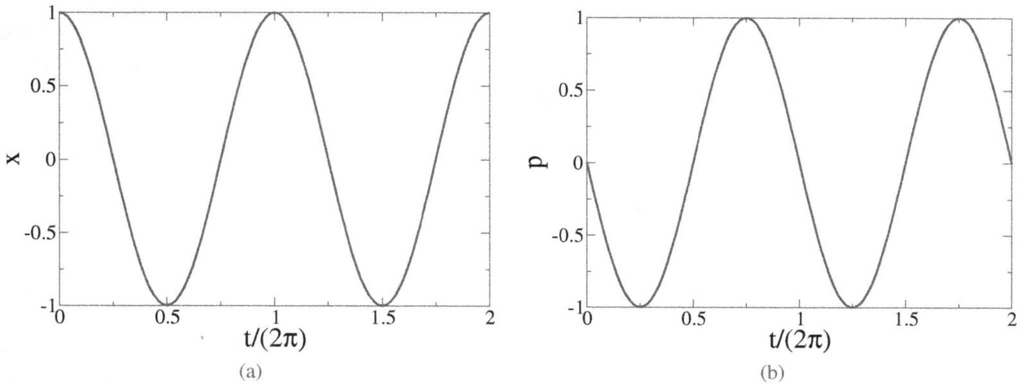

Figure 6.1 Plots for the one-dimensional simple harmonic oscillator using the simplified expressions $x = cos(t)$ and $p = -sin(t)$ for equations (6.4) and (6.5(, respectively: (a) coordinate x against time t and (b) linear momentum p against time t.

an integer and ω is the circular frequency of the motion. Consequently, there is a single point on the Poincaré map for the one-dimensional simple harmonic oscillator. Using the equations $x = \cos(t)$ and $p = -\sin(t)$ discussed in section 6.3, the periodic phase orbit is shown in figure 6.2(a) and the Poincaré map is shown in figure 6.2(b). The single point representing the Poincaré map is consistent with the strictly periodic motion of the simple harmonic oscillator.

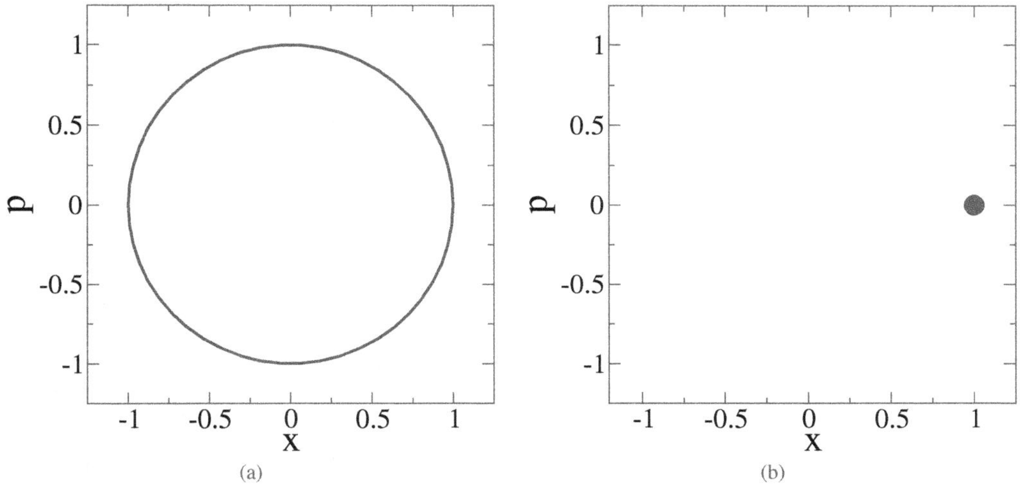

Figure 6.2 (a) The periodic phase orbit for the one-dimensional simple harmonic oscillator. (b) The Poincaré map of the orbit in (a). For plotting these, the simplified expressions $x = \cos(t)$ and $p = -\sin(t)$ for equations 6.4 and 6.5, respectively, were used.

6.4 NON-LINEAR FORCES AND ANHARMONIC POTENTIALS

It is helpful to discuss the concepts of non-linear force and non-harmonic potential (usually called **anharmonic potential**). The potential energy of a one-dimensional system can be expressed as a function of its position x in the form of a Taylor series

$$\begin{aligned} V(x) &= V(x_0) + \frac{\mathrm{d}V}{\mathrm{d}x}|_{x_0} u + \frac{1}{2}\frac{\mathrm{d}^2V}{\mathrm{d}x^2}|_{x_0} u^2 + \frac{1}{3!}\frac{\mathrm{d}^3V}{\mathrm{d}x^3}|_{x_0} u^3 + \cdots, \qquad (6.6) \\ &= \frac{1}{2}\frac{\mathrm{d}^2V}{\mathrm{d}x^2}|_{x_0} u^2 + \frac{1}{3!}\frac{\mathrm{d}^3V}{\mathrm{d}x^3}|_{x_0} u^3 + \cdots, \\ &= \frac{1}{2}ku^2 + \frac{1}{3!}k'u^3, \qquad (6.7) \end{aligned}$$

where $x - x_0 = u$ is a small displacement around x_0. In writing the second line above, we have considered $x = x_0$ as system's equilibrium position and thus set $V(x = x_0)$ as a constant reference level. Also, as there should be zero force on the system in its equilibrium, we set $\frac{\mathrm{d}V}{\mathrm{d}x}|_{x_0} = 0$. In equation (6.7) the first term is the harmonic term and the second term is the cubic anharmonic term. The coefficient k is the harmonic force constant and the coefficient k' is the cubic force constant. The corresponding force expression, following equation (2.19), is

$$\begin{aligned} F(x) &= -\frac{\mathrm{d}V}{\mathrm{d}x}|_{x_0} u - \frac{1}{2}\frac{\mathrm{d}^3V}{\mathrm{d}x^3}|_{x_0} u^2 + \cdots, \\ &= -ku - \frac{1}{2}k'u^2. \qquad (6.8) \end{aligned}$$

The first term is the linear force and the second term represents the lowest-order non-linear force.

For a three-dimensional system, the potential energy can be written as

$$V(\boldsymbol{r}) = \frac{1}{2}\sum_{\alpha\beta} \Phi_{\alpha,\beta} u_\alpha u_\beta + \frac{1}{3!}\sum_{\alpha\beta\gamma} \Psi_{\alpha,\beta,\gamma} u_\alpha u_\beta u_\gamma + \text{higher order terms}, \qquad (6.9)$$

and the corresponding force expression can be obtained from

$$F_\alpha(\boldsymbol{r}) = -\nabla_\alpha V(\boldsymbol{r}), \qquad (6.10)$$

where $\boldsymbol{u} = \boldsymbol{r} - \boldsymbol{r}_0$ is the displacement vector from the equilibrium position $\boldsymbol{r}_0$ and α, β, γ represent the x, y and z components, $\boldsymbol{\Phi}$ is a second-order matrix representing harmonic force constants and $\boldsymbol{\Psi}$ is a third-order matrix representing cubic anharmonic force constants, and so on.

6.5 PLANE PENDULUM

The plane pendulum consists of a point mass m suspended from a fixed point by a taut and weightless string of length l, which is allowed to oscillate in a vertical plane, as shown in figure 6.3. Let us remind ourselves that we considered this system in figure 2.1 and allowed it to oscillate in the x-z plane with small values of the angle θ. In the present discussion, we do not restrict θ to be small.

The only force acting on the bob is gravitational, and the rotational version of its equation of motion presented in equation (2.30) can be expressed as

$$\ddot{\theta} + \omega_0^2 \sin\theta = 0, \quad \text{with} \quad \omega_0^2 = \frac{g}{l}. \qquad (6.11)$$

Let us set the zero of the potential energy of the system at $\theta = 0$, when the pendulum is pointing straight downwards. When the bob is at angle θ from its vertical equilibrium position, expressions for the force acting on it and the corresponding potential energy are

$$F = -mg\sin\theta \quad \text{and} \quad V = mgl(1 - \cos\theta). \qquad (6.12)$$

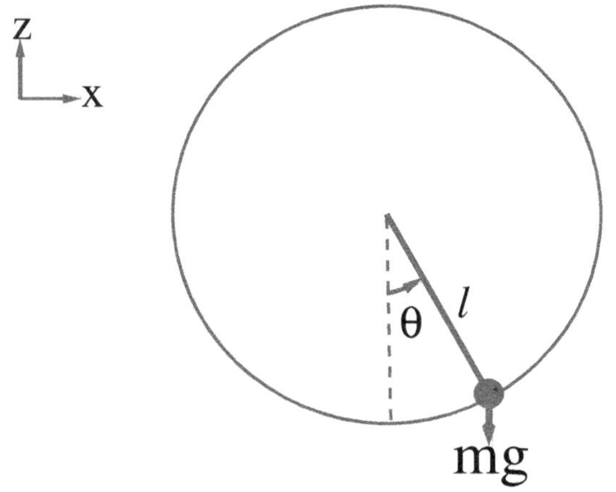

Figure 6.3 A plane pendulum. It consists of a point mass m suspended from a fixed point by a taut and weightless string of length l, which is allowed to oscillate in a vertical plane,

From these, expressions for linear force and harmonic potential energy, corresponding to small angles, are $F = -mg\theta$ and $V = \frac{1}{2}mgl\theta^2$. Figure 6.4 shows a plot of these expressions. It is clear that the small angle approximation, used for simple harmonic motion, is not valid for $\theta > \pi/4$.

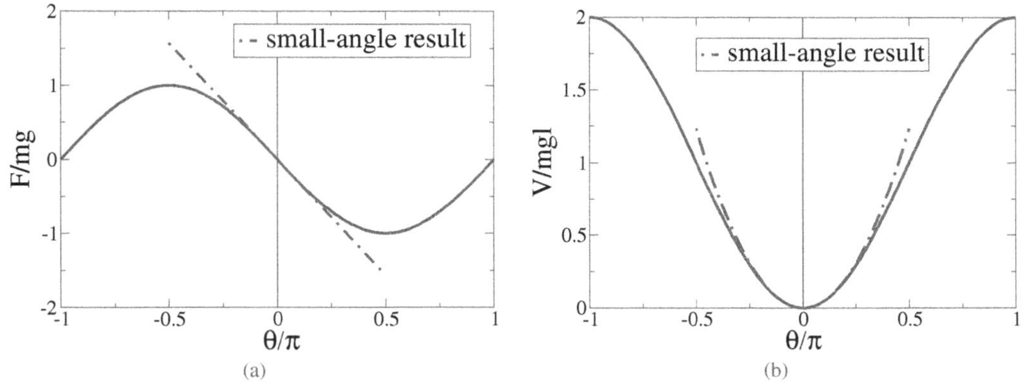

Figure 6.4 Plots of (a) the force $F = -mg\sin\theta$ and (b) the corresponding potential energy $V = mgl(1-\cos\theta)$ for the plane pendulum of mass m and length l oscillating with the angle θ in a vertical plane. It is clear that the small-angle results, shown as the dot-dash curves, are not valid for $\theta > \pi/4$.

The total energy of this conservative system is

$$\begin{aligned} \text{constant} = E = T + V &= \frac{1}{2}I\dot{\theta}^2 + mgl(1-\cos\theta), \\ &= \frac{1}{2}ml^2\dot{\theta}^2 + 2mgl\sin^2\left(\frac{\theta}{2}\right), && (6.13) \\ &= \frac{p_\theta^2}{2ml^2} + mgl(1-\cos\theta), && (6.14) \end{aligned}$$

where p_θ is the generalised momentum corresponding to θ.

Figure 6.5 shows a plot of the phase orbits in the range $-\pi \leq \theta \leq \pi$ for different values of the total energy E. The orbits are periodic in θ (as the potential is). Thus, the plots can be expanded by using periodic repetition outside the range $(-\pi, \pi)$. Identical orbits will result in the regions $-3\pi \leq \theta \leq -\pi$, $\pi \leq \theta \leq 3\pi$, and so on.

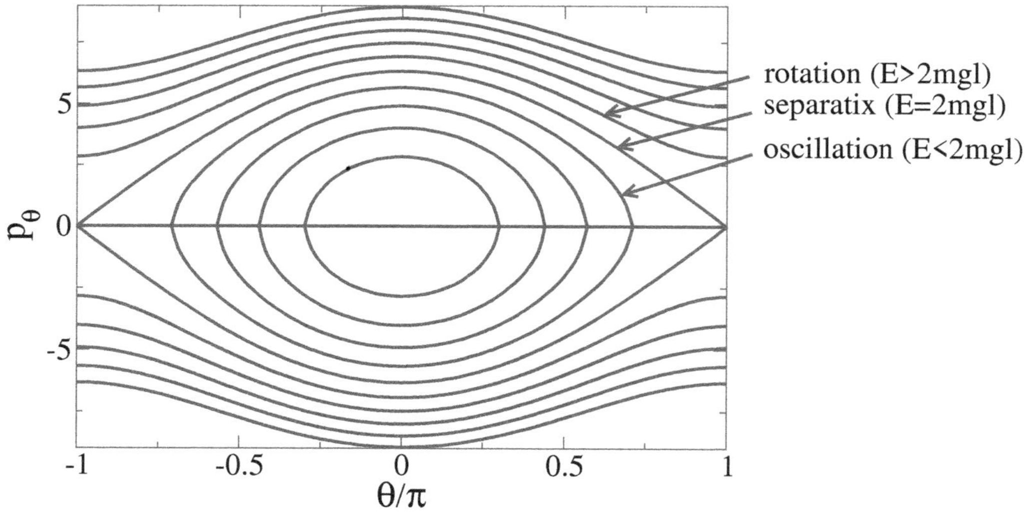

Figure 6.5 Phase orbits for a plane pendulum of mass m and length l. The energy values for the orbits from the innermost towards the outermost are $0.4mgl$, $0.8mgl$, $1.2mgl$, $1.6mgl$, $2mgl$, $2.4mgl$, $2.8mgl$, $3.2mgl$, $3.6mgl$ and $4mgl$, respectively.

We will analyse the phase orbits for three ranges of total energy: (a) $E < 2mgl$, (b) $E > 2mgl$ and (c) $E = 2mgl$.

(a) $E < 2mgl$: If $\theta = \theta_0$ at the highest point of the motion in the range $E < 2mgl$, then $T(\theta = \theta_0) = \frac{1}{2}ml^2\dot{\theta}_0^2 = 0$ and $V(\theta = \theta_0) = E_\theta = 2mgl\sin^2(\frac{\theta_0}{2})$. Using these, we can express the kinetic energy $T = E_\theta - V$ as

$$\frac{1}{2}ml^2\dot{\theta}_0^2 = 2mgl\sin^2\left(\frac{\theta_0}{2}\right) - 2mgl\sin^2\left(\frac{\theta}{2}\right), \tag{6.15}$$

from which we express

$$\frac{p_\theta^2}{4m^2l^3g} = \sin^2\left(\frac{\theta_0}{2}\right) - \sin^2\left(\frac{\theta}{2}\right). \tag{6.16}$$

In general, closed elliptical orbits result for $-\pi < \theta < \pi$ and $E < 2mgl$. For small values of θ and θ_0, we reduce equation (6.16) to the form

$$\frac{p_\theta^2}{m^2l^3g} + \theta^2 = \theta_0^2. \tag{6.17}$$

This indicates that the phase orbits in the p_θ-θ phase space will be elliptical, similar to what we have already seen in section 6.3.

The coordinates $\theta = \cdots, -2\pi, 0, 2\pi, \cdots$ along the θ-axis are positions of *stable equilibrium*. These are known as **attractors**, to which the bob is "attracted" when it is damped. The coordinates $\cdots, -\pi, \pi, \cdots$ are positions of *unstable equilibrium.*

(b) $E > 2mgl$: Phase orbits for $E > 2mgl$ are no longer closed. Though still periodic, these correspond to the pendulum executing complete revolutions about its support axis. In other words, for such higher energies, the bob's motion is rotational.

(c) $E = 2mgl$: This value of energy provides the largest closed orbit, with zero momentum at $\theta = \pm n\pi$. This orbit separates locally bounded motion from locally unbounded motion. For this reason, it is called a **separatrix**. Orbits in the vicinity of a separatrix are extremely sensitive to initial conditions of motion of the bob.

6.6 ONE-DIMENSIONAL ANHARMONIC, OR DUFFING, OSCILLATOR

An example of a one-dimensional anharmonic oscillator can easily be modelled by truncating the force and potential energy expressions for the plane pendulum in equation (6.12) at the lowest order anharmonic term. This means that we expand $\sin\theta = \theta - \theta^3/3! + \cdots$ and $\cos\theta = 1 - \theta^2/2! + \theta^4/4! + \cdots$, and consider

$$F = -mg\theta + \frac{1}{3!}mg\theta^3 \quad \text{and} \quad V = \frac{1}{2}mgl\theta^2 - \frac{1}{4!}mgl\theta^4. \tag{6.18}$$

Note that the above expressions can easily be re-written using the linear coordinate (*i.e.* displacement from equilibrium) $x = l\theta$ as

$$F = -k_{\text{har}}x + \frac{1}{3!}k_{\text{anh}}x^3 \quad \text{and} \quad V = \frac{1}{2}k_{\text{har}}x^2 - \frac{1}{4!}k_{\text{anh}}x^4, \tag{6.19}$$

with the harmonic and anharmonic force constants

$$k_{\text{har}} = \frac{mg}{l} \quad \text{and} \quad k_{\text{anh}} = \frac{mg}{l^3}. \tag{6.20}$$

Plots of the expressions in equation (6.18) are presented in figure 6.6.

The one-dimensional system described here is also known as the Duffing oscillator, and the equation

$$m\frac{\mathrm{d}^2x}{\mathrm{d}t^2} + k_{\text{har}}x - \frac{1}{3!}k_{\text{anh}}x^3 = 0 \tag{6.21}$$

satisfies Duffing's equation (Duffing, 1918).

Phase orbits for the anharmonic oscillator described by equation (6.18) can be plotted by following the procedure described for the plane pendulum. The results in the (p_θ-θ) plane are shown in figure 6.7. Closed orbits, for energy values $0.4mgl$, $0.8mgl$ and $1.2mgl$, indicate periodic oscillations of the oscillator about $\theta = 0$. Other orbits shown by solid curves correspond to unbounded motion of the oscillator. The orbit for the energy $1.5mgl$ is the separatrix between closed and open orbits. Note that this energy is different from the separatrix energy of $2mgl$ for the plane pendulum. This is due to the use of the truncated trigonometric function of the plane pendulum to model the anharmonic oscillator. There are three equilibrium points, with the (θ, p_θ) coordinates $(-\sqrt{3!}, 0)$, $(0,0)$ and $(\sqrt{3!}, 0)$, which translate as $(-0.780\pi, 0)$, $(0,0)$ and $(0.780\pi, 0)$, respectively. The separatrix 'terminates' at the equilibrium points $\pm(0.780\pi, 0)$.

As we have considered θ linearly related to the linear coordinate x and p_θ linearly related to the linear momentum $p_x = m\dot{x}$, a similar plot will appear in the (p_x-x) plane. Using either the equation of motion in (6.19) or the equation $E = \text{constant} = \frac{p_x^2}{2m} + V(x)$, it can be shown that there are three equilibrium points for the oscillator. Using the (x, p_x) coordinates, these are at $(0,0)$ and $(\pm\sqrt{3!\frac{k_{\text{har}}}{k_{\text{anh}}}}, 0)$. This will be set as an exercise to the reader.

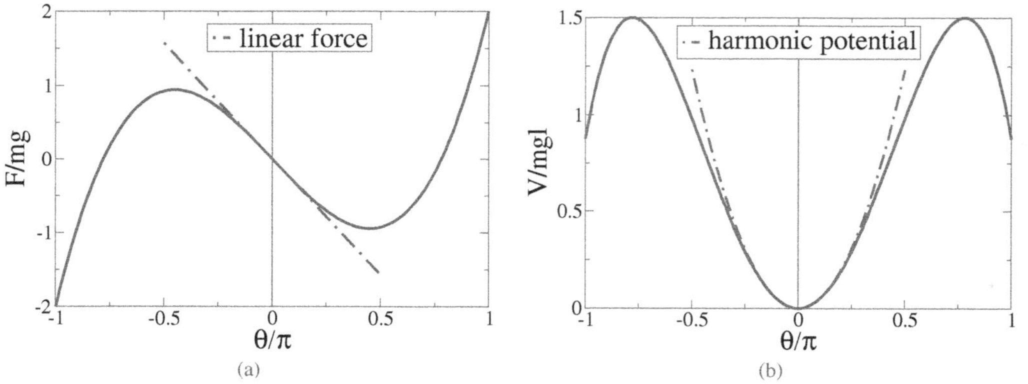

Figure 6.6 Plot of (a) force $F = -mg\theta + \frac{1}{3!}mg\theta^3$ and (b) the corresponding potential energy $V = \frac{1}{2}mgl\theta^2 - \frac{1}{4!}mgl\theta^4$ for a one-dimensional anharmonic oscillator. For comparison, the linear force and the harmonic potential energy are shown by the dot-dash curves.

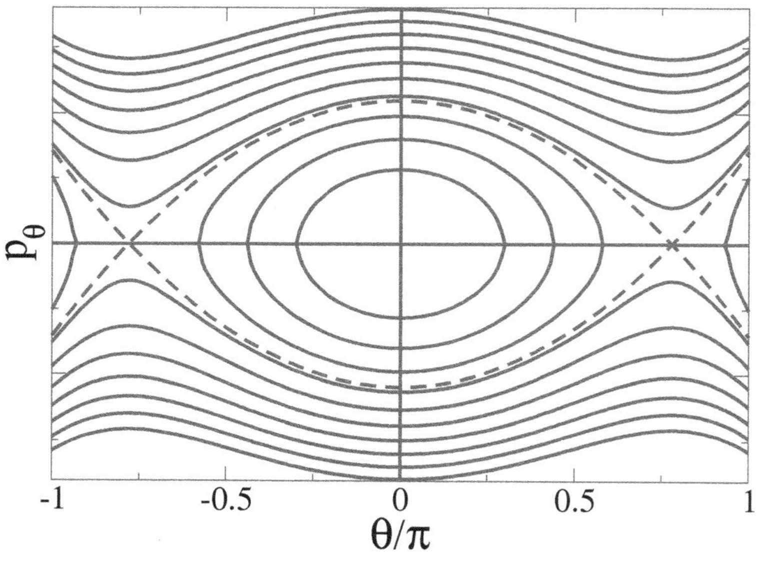

Figure 6.7 Phase orbits for the one-dimensional anharmonic oscillator under the influence of the force $F = -mg\theta + \frac{1}{3!}mg\theta^3$, where m is the oscillator mass, g is acceleration due to gravity and θ is the angle of oscillation. The orbits with energies $0.4mgl$, $0.8mgl$ and $1.2mgl$ are bound and periodic, while the orbits with energies larger than $1.5mgl$ are unbounded. The separatrix between the bound and unbound orbits, for energy $1.5mgl$, is shown by the dotted curves. The equilibrium points for the separatrix are $\pm(0.780\pi, 0)$.

6.7 DAMPED ONE-DIMENSIONAL HARMONIC OSCILLATOR

As discussed previously, the equation of motion $m\frac{d^2x}{dt^2} + kx = 0$ of the one-dimensional simple harmonic oscillator of characteristic angular frequency $\omega = \sqrt{k/m}$ has solution $x = A\cos(\omega t + \psi)$, containing two parameters A (amplitude) and ψ (phase angle). This is an idealised system, as motion of the oscillator is likely to be subjected to a damping term which is proportional to $\frac{dx}{dt}$. The equation of motion of a one-dimensional damped harmonic oscillator is, thus

$$m\frac{d^2x}{dt^2}+b\frac{dx}{dt}+kx=0, \tag{6.22}$$

where b is the *damping constant*. This is an example of a linear homogeneous second-order differential equation, for which a two-parameter trial solution of the type

$$x=Ce^{\alpha x} \tag{6.23}$$

can always be attempted, with C either a constant or a function of x and α a real or complex constant. Substitution of this trial function in equation (6.22) gives

$$Ce^{\alpha}\left(m\alpha^2+b\alpha+k\right)=0. \tag{6.24}$$

For non-trivial solution of this equation, the values of the constant α are

$$\alpha=-\frac{b}{2m}\pm\sqrt{\left(\frac{b}{2m}\right)^2-\frac{k}{m}}. \tag{6.25}$$

With this, the solution now reads

$$x=C\,e^{-\frac{b}{2m}t}\,e^{\left(\pm\sqrt{(\frac{b}{2m})^2-\frac{k}{m}}\right)t}, \tag{6.26}$$

with C remaining as a single parameter.

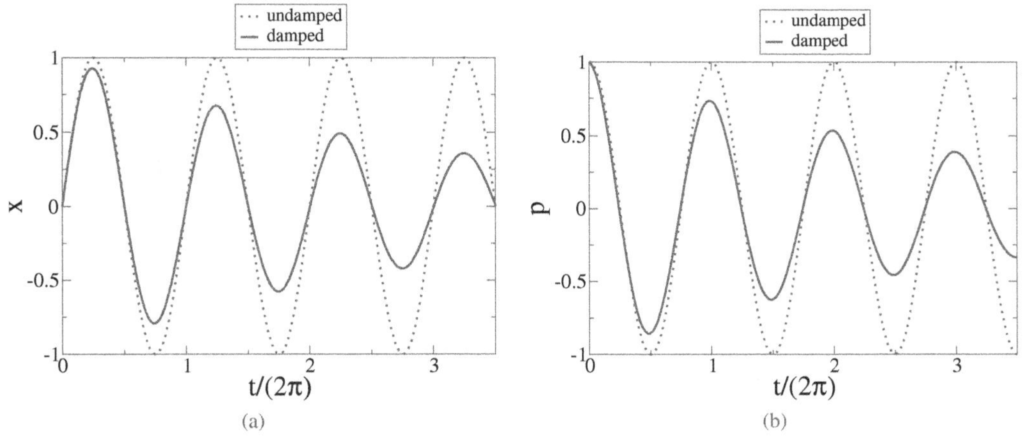

Figure 6.8 Plots for the one-dimensional damped harmonic oscillator of mass m: (a) coordinate x against time t and (b) linear momentum p against time t, using the analytic solutions $x=Ae^{-\frac{b}{2m}t}\cos(\omega' t-\phi)$, and $p=-mA\,e^{-\frac{b}{2m}t}\left[\frac{b}{2m}\cos(\omega' t-\phi)+\omega'\sin(\omega' t-\phi)\right]$. The parameters are chosen as: $A=1, m=1, k=1, b=0.1, \omega=1, \omega'=(1-b^2)^{1/2}$ and $\phi=\pi/2$. The corresponding plots for the undamped oscillator (*i.e.* with $b=0$) are shown for comparison.

Three physical situations may arise

$$\text{light damping}: \quad \left(\frac{b}{2m}\right)^2-\frac{k}{m}<0,$$

$$\text{critical damping:} \quad \left(\frac{b}{2m}\right)^2 - \frac{k}{m} = 0,$$
$$\text{heavy damping:} \quad \left(\frac{b}{2m}\right)^2 - \frac{k}{m} > 0.$$

Here we will discuss the light damping case, and express

$$x = C\,\mathrm{e}^{-\frac{b}{2m}t}\,\mathrm{e}^{\pm j\omega' t}, \tag{6.27}$$

where

$$j = \sqrt{-1} \quad \text{and} \quad \omega' = \sqrt{\frac{k}{m} - \left(\frac{b}{2m}\right)^2}. \tag{6.28}$$

As there are two values of x in equation (6.27), it is possible to rewrite the solution as

$$x = A\,\mathrm{e}^{-\frac{b}{2m}t}\cos(\omega' t - \phi), \tag{6.29}$$

where A is the maximum amplitude of the oscillation and ϕ is a phase factor. The expression for the linear momentum of the oscillator is, then

$$p = -mA\,\mathrm{e}^{-\frac{b}{2m}t}\left[\frac{b}{2m}\cos(\omega' t - \phi) + \omega' \sin(\omega' t - \phi)\right]. \tag{6.30}$$

Figure 6.8 shows the variation of the displacement x and momentum p with time t with the parameters chosen as $A = 1, m = 1, k = 1, b = 0.1, \omega = 1, \omega' = (1-b^2)^{1/2}$ and $\phi = \pi/2$. For comparison, the corresponding plots for the undamped oscillator are also shown. Clearly, due to damping, the amplitudes of the oscillatory behaviour of both phase space variables (*i.e.* x and p) decrease exponentially with time as $\mathrm{e}^{-\frac{b}{2m}t}$. Figure 6.9(a) shows the phase orbit in the x-p plane. The phase orbit clearly is quite different from that of the undamped harmonic oscillator. For the undamped oscillator, the phase orbit repeats itself along the closed circular path with its characteristic time period of 2π. When damping is introduced, the closed circular orbit is turned into an inward spiralling phase orbit. It can be verified that for $t \to \infty$, the orbit will terminate at the centre of the phase space (*i.e.* at $(0,0)$). Figure 6.9(b) presents the Poincaré map of the orbits shown in figure 6.9(a). Whereas the map is a single point for the undamped oscillator, it is spread as three points, in decreasing momentum direction, for the damped oscillator.

6.8 POINCARE-BENDIXSON THEOREM AND LIMIT CYCLE

In the previous section we noticed that the phase orbit is closed for a simple harmonic oscillator but is unclosed and spirals inwards for a damped simple harmonic oscillator. The famous Poincaré–Bendixson theorem states that, for an autonomous system a phase path must either (1) close itself in a cycle, (2) be open and terminate at an equilibrium point, or (3) tend to a *limit cycle*. A limit cycle separates closed orbits from unclosed orbits. A limit cycle is *stable* or *attracting* if it is approached by neighbouring orbits.

In section 6.6 we noted the non-linear nature of the Duffing oscillator. There is another famous example of non-linear oscillator: the van der Pol's oscillator (van der Pol, 1926) which satisfies the second-order differential equation

$$\frac{\mathrm{d}^2x}{\mathrm{d}t^2} + \varepsilon(x^2 - a^2)\frac{\mathrm{d}x}{\mathrm{d}t} + x = 0, \tag{6.31}$$

where a is a parameter and ε may be positive or negative with magnitude smaller or larger than 1. It is possible to solve this equation approximately when $\varepsilon << 1$. This is described in the original paper by van der Pol, but we will not present it here. In contrast, it is not possible to obtain an

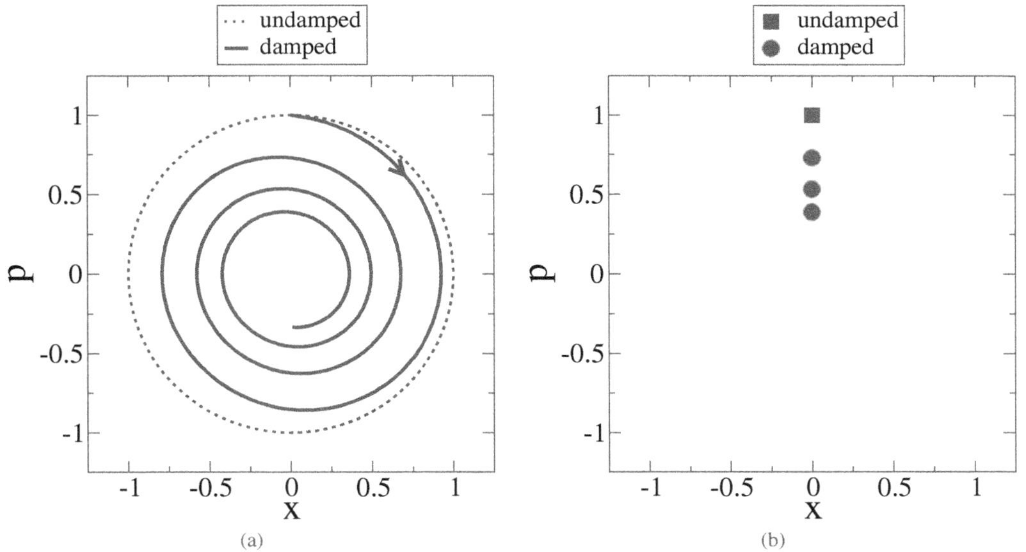

Figure 6.9 (a) Plot of phase orbits for the one-dimensional damped harmonic oscillator using the analytic solutions $x = Ae^{-\frac{b}{2m}t}\cos(\omega' t - \phi)$, and $p = -mA\, e^{-\frac{b}{2m}t}\left[\frac{b}{2m}\cos(\omega' t - \phi) + \omega' \sin(\omega' t - \phi)\right]$, with the parameters chosen as $A = 1, m = 1, k = 1, b = 0.1, \omega = 1, \omega' = (1 - b^2)^{1/2}$ and $\phi = \pi/2$. The corresponding plot for the undamped oscillator (*i.e.* with $b = 0$) is shown by the dotted curve for comparison. Panel (b) shows the Poincaré map for the damped and undamped oscillators.

approximate analytical solution for this equation when $\varepsilon >> 1$. It is, however, possible to obtain a graphical solution for any value of ε.

Let us consider obtaining a numerical solution to the van der Pol equation, where ε is a small positive parameter. We can express equation (6.31) as the following two coupled first-order differential equations

$$\frac{dx}{dt} = y, \tag{6.32}$$

$$\frac{dy}{dt} = -x - \varepsilon(x^2 - a^2)y. \tag{6.33}$$

Both of these equations can be solved by using one of many numerical analysis techniques. A simple scheme is to use the Euler-Cromer method (Cromer, 1981), which we have described in Appendix F.

Phase orbits of different shapes can be expected for different choices of ε and a. Figure 6.10 shows two spiralling phase orbits, both plotted with the choices $\varepsilon = 0.1$ and $a^2 = 1.2$, and the initial momentum condition $\frac{dx}{dt}(0) = 0$. The orbits in dotted and solid curves are plotted with the coordinate initial conditions $x(0) = 1.0$ and $x(0) = 3.0$, respectively. The orbit in the dotted curve spirals outwards and the orbit in the solid curve spirals inwards, both equilibrating at the limit cycle.

6.9 PERIODICALLY DRIVEN ONE-DIMENSIONAL HARMONIC OSCILLATOR

Let us examine the dynamics of a one-dimensional oscillator subjected to a periodically driving force. For this purpose, we set mass $m = 1$ and the harmonic force constant $k_{har} = 1$ and write the

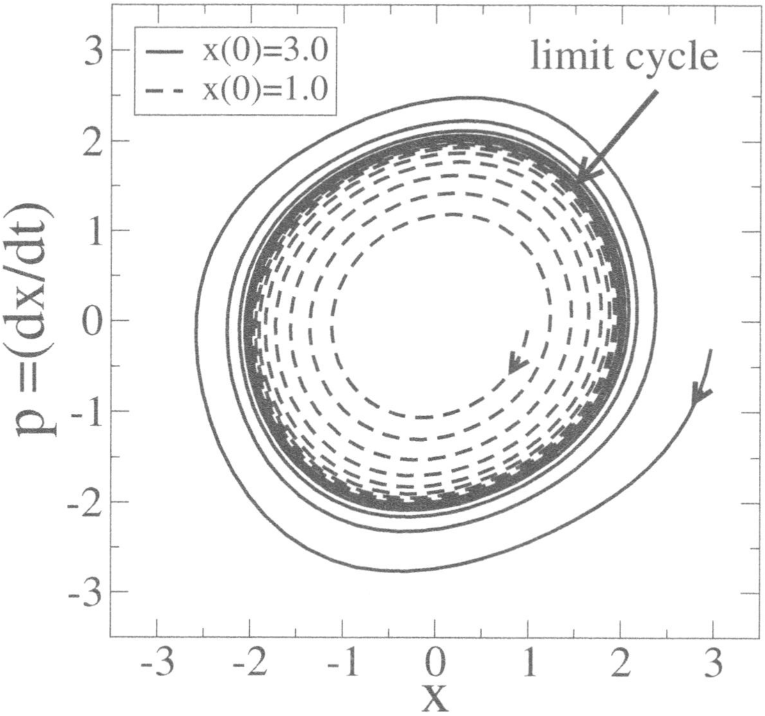

Figure 6.10 Phase orbits for the van der Pol oscillator for the choice of parameters $\varepsilon = 0.1$, $a^2 = 1.2$ and $\frac{dx}{dt}(0) = 0$ in equations $\frac{dx}{dt} = y$ and $\frac{dy}{dt} = -x - \varepsilon(x^2 - a^2)y$. The orbits shown in dotted and solid curves are plotted with the initial coordinate choices $x(0) = 1.0$ and $x(0) = 3.0$, respectively. The orbit in dotted curve spirals outwards and the orbit in solid curve spirals inwards. Both orbits terminate at the limit cycle.

equation of motion of the oscillator in the form

$$\frac{d^2x}{dt^2} = -x + F\cos(\Omega t), \tag{6.34}$$

where the periodically applied force has amplitude F and frequency Ω. Obviously, both F and Ω will affect the dynamics of the oscillator.

A general solution of equation (6.34) is[1]

$$\begin{aligned} x(t) &= A\cos(\omega t + \psi) + B\cos(\Omega t + \phi), \\ &= x_c + x_p \end{aligned} \tag{6.35}$$

where A, B, ψ and ϕ are constants to be determined from boundary (*initial*) conditions. In the above, the first and second terms are known, respectively, as the complementary solution (x_c) and a particular solution (x_p). Recall that the first term is the solution of the simple harmonic dynamical system (*i.e.* when $F = 0$), as presented in equations (6.4) and (6.5). The constant B can be expressed

[1] Also see Appendix G.

in terms of F, ω and Ω. Substituting $x_p = B\cos(\Omega t + \phi)$ into equation (6.34) we obtain

$$F = B(\omega^2 - \Omega^2). \tag{6.36}$$

The particular solution in equation (6.35) can therefore be written as

$$x_p = \frac{F}{\omega^2 - \Omega^2}\cos(\Omega t - \phi). \tag{6.37}$$

Clearly, both the amplitude and time period of the driven oscillator will be different from that of the undriven oscillator. It is obvious from equation (6.37) that as the driving frequency Ω approaches the natural frequency ω of the undriven oscillator, the amplitude of the driven oscillator will increase hugely, leading to the phenomenon of **resonance**.

We will employ the Euler-Cromer numerical method to solve equation (6.34) and study the dynamics of the periodically forced oscillator, with the initial conditions $x(t = 0) = 1.0$ and $\frac{\mathrm{d}x}{\mathrm{d}t}(t = 0) = 0$. It would be interesting to examine the motion as the force strength F varies for a chosen angular frequency Ω. In the same spirit, it would be interesting to examine the motion for different values of Ω for a chosen value of F.

Let us first set a small value of the force amplitude, say $F = 0.05$. Figures 6.11 and 6.12 show the time variation of displacement x, time variation of momentum $p = \frac{\mathrm{d}x}{\mathrm{d}t}$ (as $m = 1$), orbits in the (x-p) phase space and the Poincaré map for $\Omega = 0.5$ and $\Omega = 0.25$, respectively. As expected, the periodicity of oscillations has doubled for $\Omega = 0.5$ and quadrupled for $\Omega = 0.25$.

Figure 6.13 shows the results for $F = 0.5$ and $\Omega = 0.5$. The periodicities with $\omega = 1$ and $\Omega = 0.5$ are more clearly noticed. The small and big phase orbits correspond, respectively, to the small and big amplitudes in the x *vs* t and p *vs* t plots.

Let us now examine the results presented in figure 6.14 for $\Omega = 1.0 = \omega$ and two different values of the force amplitude, *viz* $F = 0.05$ and $F = 0.5$. We can clearly note the changes as F increases. Firstly, as mentioned before, application of a periodic force modulates the time period of the oscillator. Secondly, the displacement and momentum amplitudes increase as the force amplitude F increases. Consequently, the coverage of the phase-space area also increases. In the resonance condition ($\Omega = \omega$) the Poincaré map covers a much larger phase space region when a larger value of F is considered. In particular, for the numerical simulation considered here, starting from the initial point $(x, p) = (1, 0)$ the Poincaré map extends up to $(1.4, 6.1)$ when $F = 0.05$ and up to $(5.2, 61.2)$ when $F = 0.5$.

6.10 DYNAMICS OF ONE-DIMENSIONAL ANHARMONIC OSCILLATOR

The equation

$$\frac{\mathrm{d}^2x}{\mathrm{d}t^2} = -x + c_3x^3 \tag{6.38}$$

describes the motion of an oscillator subjected to the anharmonic force which varies as $-x + c_3x^3$. This force is less (greater) than the linear force of the form $-x$ if $c_3 > 0$ (< 0) and the system is said to be soft (hard). The potential corresponding to this force varies as $\frac{1}{2}x^2 - \frac{1}{4}c_3x^4$. Figure 6.15 shows this anharmonic potential for the soft and hard systems.

Numerical solutions, using the Euler-Cromer algorithm, of equation (6.38) for $c_3 = \frac{1}{6}$ and $c_3 = -\frac{1}{6}$ are presented in figures 6.16 and 6.17. The presence of the cubic anharmonic force term generates waves of frequency 3ω (and its multiples)[2]. This makes the time period 2π of the oscillator to increase by approximately $1.07 \times 2\pi$ when it becomes the soft system ($c_3 = \frac{1}{6}$) and to decease by approximately $0.94 \times 2\pi$ when becomes the hard system ($c_3 = -\frac{1}{6}$). Whereas the phase

[2]Also see section 10.2.3

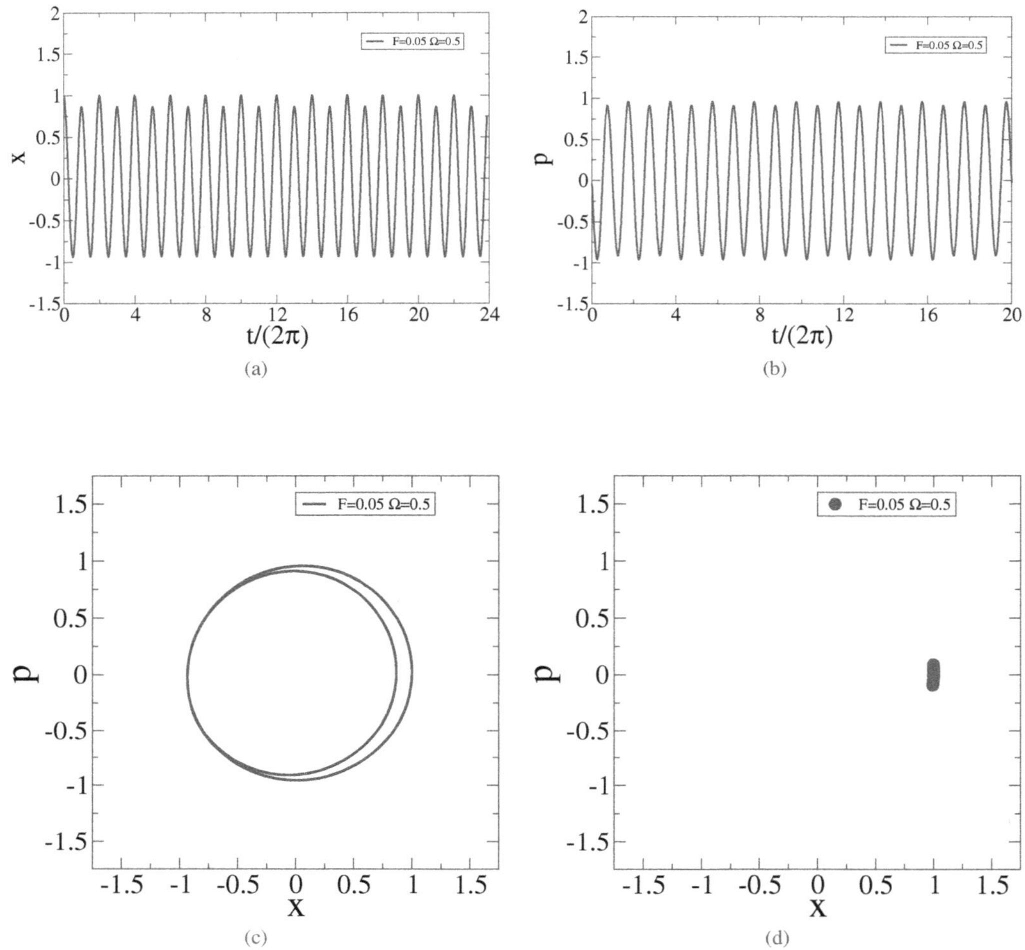

Figure 6.11 Dynamics of a harmonic oscillator driven by a periodic force of amplitude $F = 0.05$ and frequency $\Omega = 0.5$ in the equation $\frac{d^2x}{dt^2} = -x + F\cos(\Omega t)$. The initial phase space choice is $x(0) = 1.0$ and $p(0) = \frac{dx}{dt}(0) = 0.0$. The left and right panels in the upper row show the time variation of the displacement x and momentum $p = \frac{dx}{dt}$, respectively. The left and right panels in the lower row show the phase-space orbits and the Poincaré map, respectively.

orbit of the harmonic oscillator is circular, the phase orbit of the anharmonic oscillator is elliptical. In comparison to the radius of the phase orbit of the harmonic oscillator, the minor radius of the elliptical orbit (*i.e.* along the momentum direction) is smaller (larger) for the soft (hard) anharmonic oscillator.

6.11 DAMPED AND DRIVEN ONE-DIMENSIONAL ANHARMONIC OSCILLATOR

In the previous sections we have studied the effects of each of damping, anharmonicity and periodic driving on the motion of a one-dimensional harmonic oscillator. Here, we study the motion of the oscillator when it is subjected to all of these factors. For this purpose, we consider a periodically

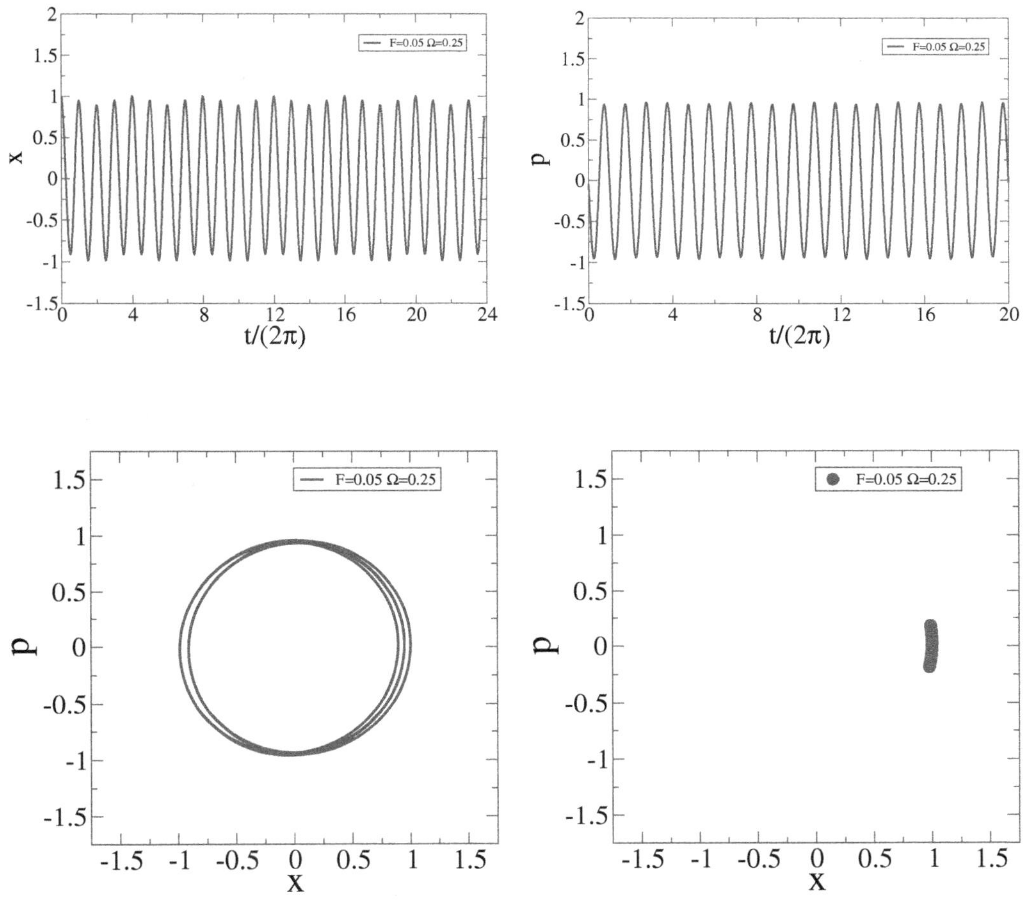

Figure 6.12 Dynamics of a harmonic oscillator driven by a periodic force of amplitude $F = 0.05$ and frequency $\Omega = 0.25$ in the equation $\frac{d^2x}{dt^2} = -x + F\cos(\Omega t)$. The initial phase space choice is $x(0) = 1.0$ and $p(0) = \frac{dx}{dt}(0) = 0.0$. The left and right panels in the upper row show the time variation of the displacement x and momentum $p = \frac{dx}{dt}$, respectively. The left and right panels in the lower row show the phase-space orbits and the Poincaré map, respectively.

forced Duffing oscillator, the equation of motion for which is in the form

$$\frac{d^2x}{dt^2} = -h_1\dot{x} - h_2x - h_3x^3 + F\cos(\Omega t), \tag{6.39}$$

with h_1 as the damping coefficient, h_2 as the harmonic stiffness constant, h_3 as the anharmonic stiffness constant and $F\cos(\Omega t)$ as the force applied to the system. As mentioned in the previous sub-section, the system with $h_2 > 0, h_3 < 0$ is known as the soft system and the system with $h_2 > 0, h_3 > 0$ is known as the hard system. The system with $h_2 < 0, h_3 > 0$ is known as a double-well potential system for the reason explained later.

We note from figure 6.17 in section 6.10 that the dynamics of undamped and undriven soft as well as hard anharmonic oscillators is characterised by bound orbits in the phase space. Here, we will study the dynamics of the damped and driven hard system (with $h_2 = 1.0$ and $h_3 = 0.3$) and

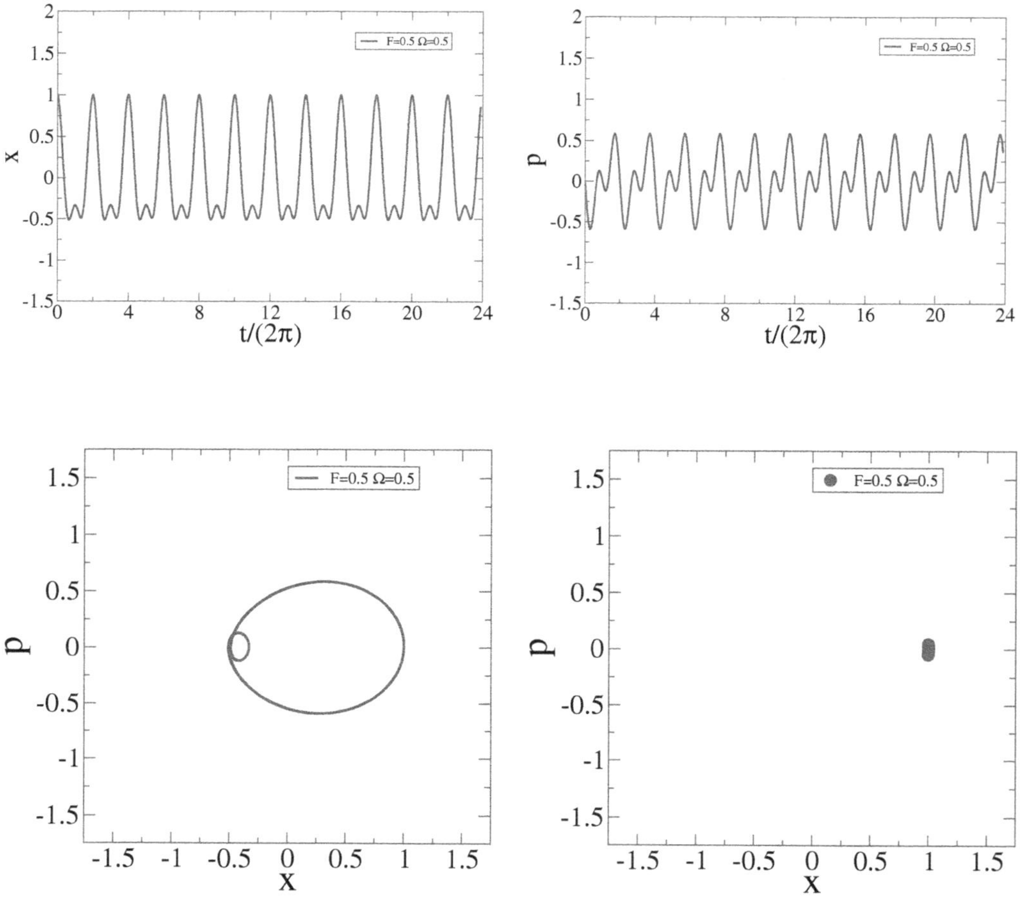

Figure 6.13 Dynamics of a harmonic oscillator driven by a periodic force of amplitude $F = 0.5$ and frequency $\Omega = 0.5$ in the equation $\frac{d^2x}{dt^2} = -x + F\cos(\Omega t)$. The initial phase space choice is $x(0) = 1.0$ and $p(0) = \frac{dx}{dt}(0) = 0.0$. The left and right panels in the upper row show the time variation of the displacement x and momentum $p = \frac{dx}{dt}$, respectively. The left and right panels in the lower row show the phase-space orbits and the Poincaré map, respectively.

the damped and driven double-well system (with $h_2 = -1.0$ and $h_3 = 0.3$). For both cases, we set the damping and driving parameters as $h_1 = 0.02, F = 0.5, \Omega = 0.5$, and the initial condition as $x(0) = -0.05, \frac{dx}{dt}(0) = 0.0$. The results from the application of the Euler-Cromer algorithm are shown in figures 6.18-6.20. The simulations involved 5000 time (t) steps with interval $\Delta t = 0.05$.

The results presented in figures 6.18 and 6.19 clearly suggest that the dynamics of the driven Duffing oscillator with hard springs shows a behaviour that is a mixture of periodic and chaotic motions. This is consistent with a detailed numerical study by Fang and Dowell (1987) who found that both periodic and chaotic behaviours exist in a stable hard-spring Duffing system. This behaviour is different from the results in figure 6.13 for which we did not include the damping and anharmonic terms (*i.e.* with $h_1 = 0$ and $h_3 = 0$). Whereas the plot of the Poincaré map showed a simple structure in figure 6.13, the hard-spring Duffing oscillator is characterised with a much richer structure. It can be shown that for small values of the driving amplitude F, there are three equilibria: two stable at

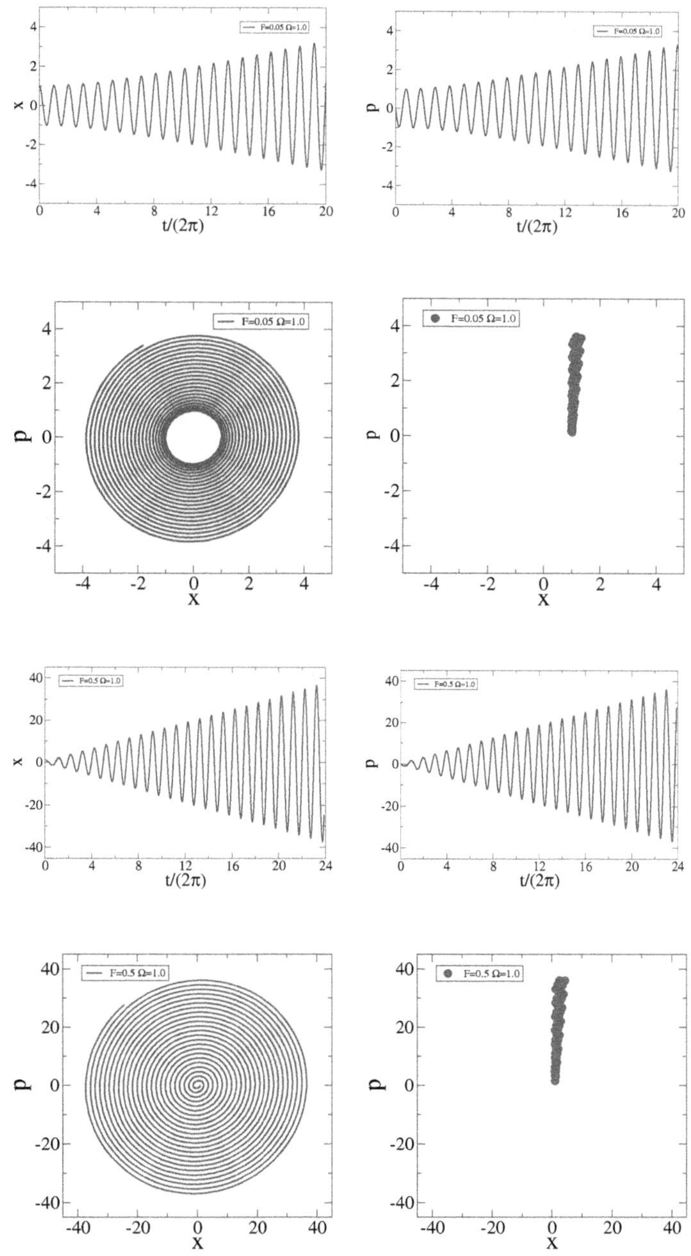

Figure 6.14 Comparison of the dynamics of a harmonic oscillator driven by a periodic force of frequency $\Omega = 1.0$ and force amplitudes $F = 0.05$ and $F = 0.5$ in the equation $\frac{d^2x}{dt^2} = -x + F\cos(\Omega t)$. The initial phase space choice is $x(0) = 1.0$ and $p(0) = \frac{dx}{dt}(0) = 0.0$. The left and right panels in the first and third rows show the time variation of the displacement and momentum, respectively. The left and right panels in the second and fourth rows show the phase-space orbits and the Poincaré map, respectively.

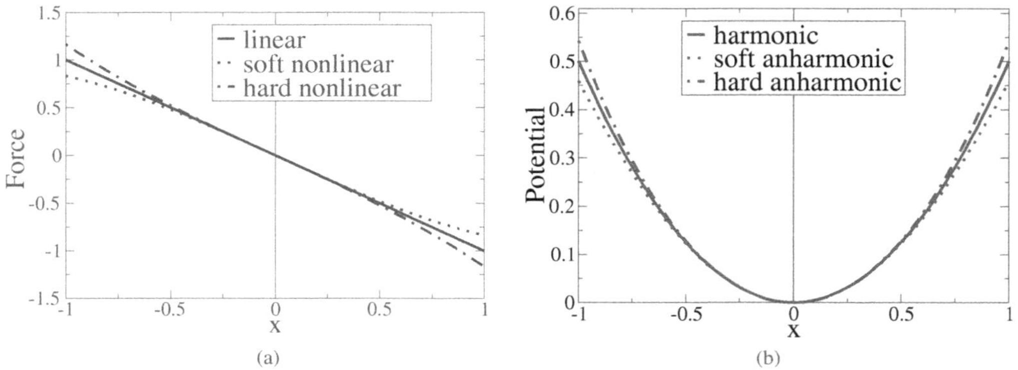

Figure 6.15 Anharmonic force and potential for the soft and hard systems. In panel (a) the solid, dotted and dashed curves show plots of the linear force $-x$, soft non-linear force $-x+\frac{1}{6}x^3$ and hard non-linear force $-x-\frac{1}{6}x^3$, respectively. Panel (b) shows the potentials corresponding to the forces in (a).

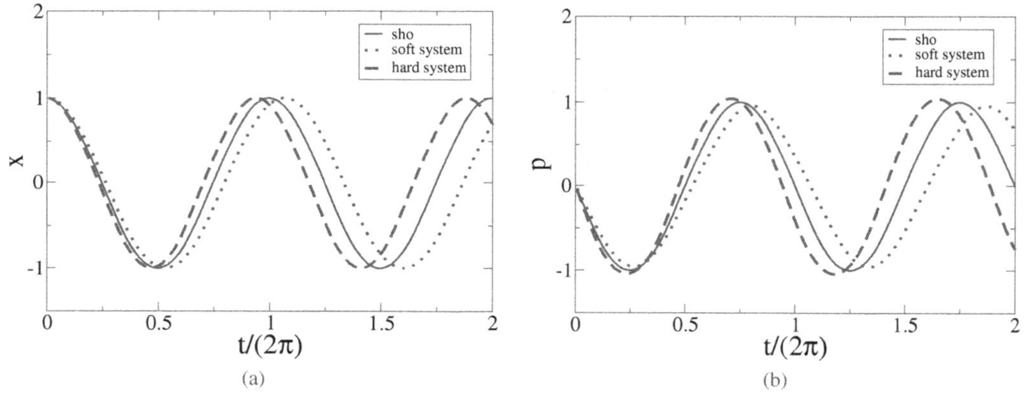

Figure 6.16 Dynamics of an anharmonic oscillator for which force varies as $-x+c_3x^3$. The numerical simulation is carried out with the initial phase space choice as $x(0) = 1.0$ and $p(0) = \frac{dx}{dt}(0) = 0.0$. The panels (a) and (b) show the time variation of the displacement x and momentum $p = \frac{dx}{dt}$ map, respectively. The solid, dotted and dot-dashed curves are for the simple harmonic oscillator, the soft system ($c_3 = 1/6$) and the hard system ($c_3 = -1/6$), respectively.

approximately $(x,p) = \left(\pm(h_2/h_3)^{1/2}, 0\right)$ and one unstable (a saddle) at $(x,p) = (0,0)$. Proof of this will be left as an exercise for the reader (see Problems section). A detailed study of the motion of this system has been carried out by Homes (1979).

Figures 6.20 and 6.21 show the dynamics of the double-well Duffing's oscillator. The time evolutions of the distance x and the momentum p show a mixture of periodic and non-periodic patterns. The oscillator performs motion in each of the two wells as well as around the two wells. For this reason such an oscillator is known as a double-well system. Consistent with the orbital behaviour, the Poincaré map shows a complex structure.

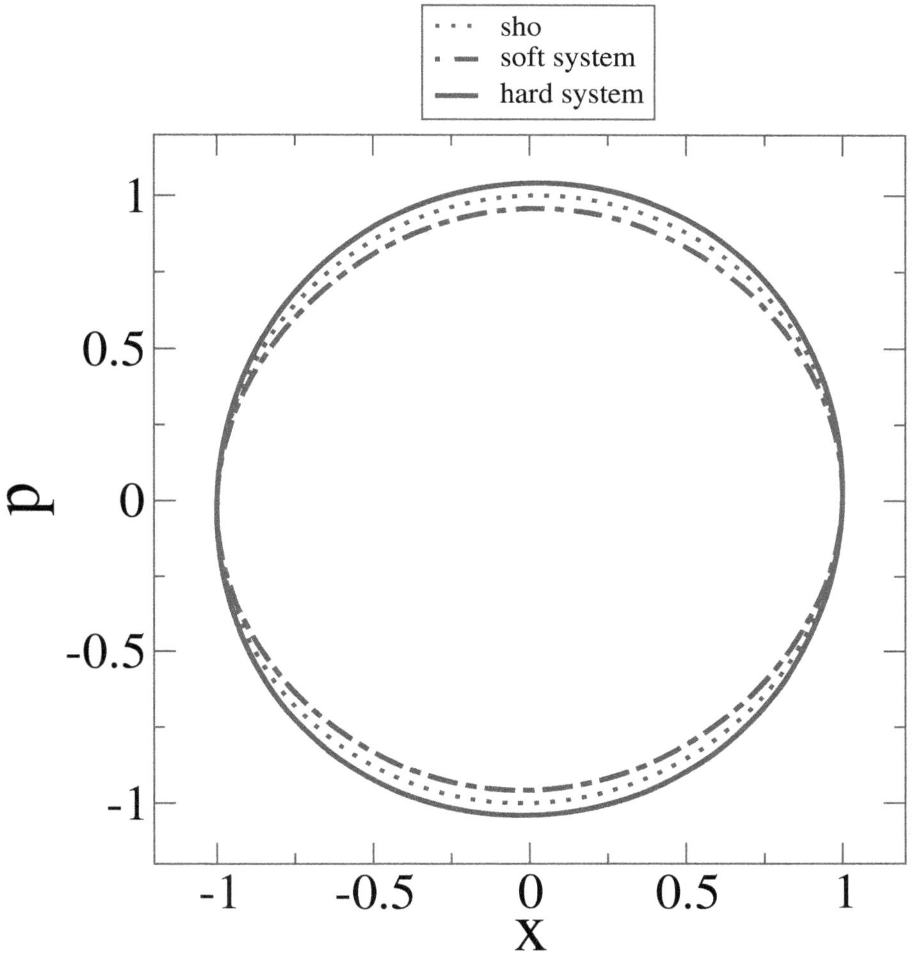

Figure 6.17 Phase-space orbits of an anharmonic oscillator for which force varies as $-x + c_3x^3$. The numerical simulation is carried out with the initial phase space choice as $x(0) = 1.0$ and $p(0) = \frac{dx}{dt}(0) = 0.0$. The middle (thin solid) curve, the innermost (dotted) curve and the thick solid (outermost) curve are for the simple harmonic oscillator, the soft system ($c_3 = 1/6$) and the hard system ($c_3 = -1/6$), respectively.

Simulations with different initial conditions do not change the long-term behaviour of the hard-spring Duffing oscillator. However, different initial conditions result in very different dynamical behaviours of the double-well Duffing oscillator. This we show in figure 6.22 by plotting the time evolution of the displacement variable x for the two systems (the hard-spring and the double-well systems) with the initial conditions $x(0) = 1.0$ and $x(0) = 0.99$ (keeping the other parameters of figures 6.18-6.21 unaltered). The overall behaviour of the hard-spring oscillator is maintained with only minor changes appearing in the results. In contrast, there are drastic changes in the behaviour of the double-well oscillator with only a 1% change in the initial condition. This is a clear indication that the double-well Duffing oscillator can exhibit chaotic behaviour. We will discuss this aspect in some detail in the next chapter (see, section 7.2.4).

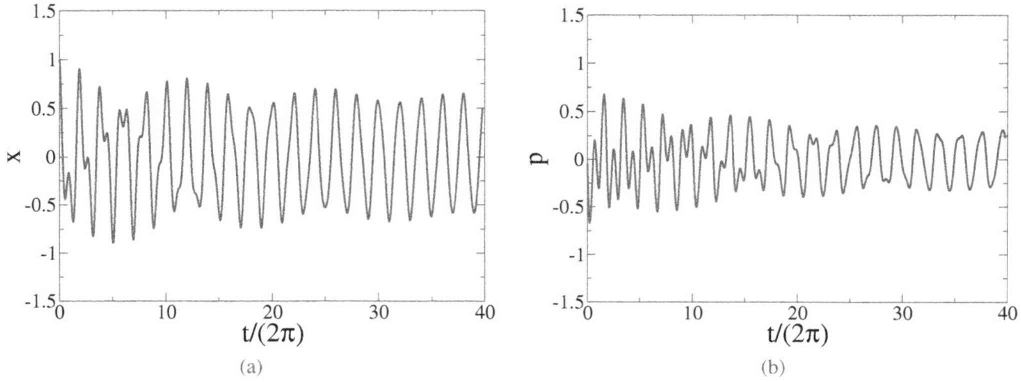

Figure 6.18 Dynamics of hard-spring Duffing's oscillator. Parameters used in the numerical evaluation of equation $\frac{d^2x}{dt^2} = -h_1\dot{x} - h_2x - h_3x^3 + F\cos(\Omega t)$ are: $h_1 = 0.02, h_2 = 1.0, h_3 = 0.3, F = 0.5, \Omega = 0.5, x(0) = 1.0, \frac{dx}{dt}(0) = 0.0$. The panels (a) and (b) show the time variation of the displacement x and momentum $p = \frac{dx}{dt}$, respectively.

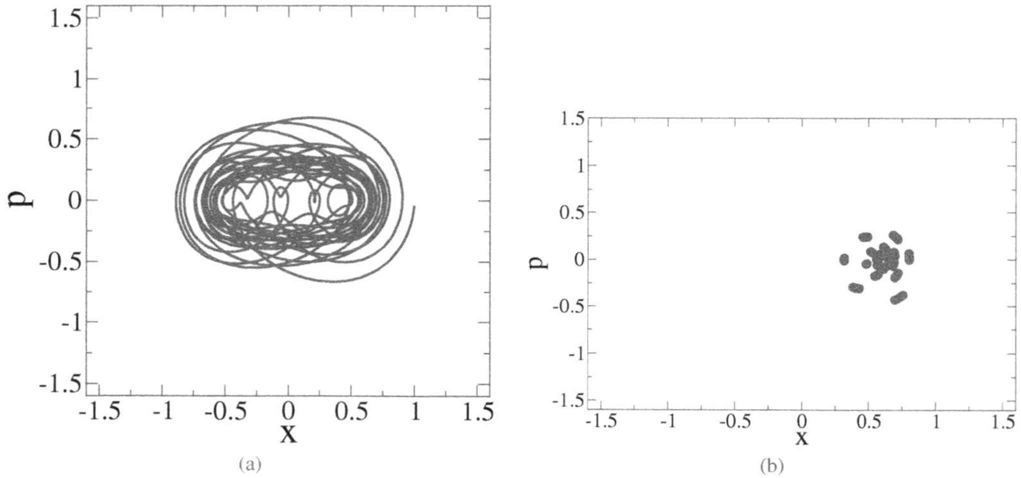

Figure 6.19 Dynamics of hard-spring Duffing's oscillator. Parameters used in the numerical evaluation of equation $\frac{d^2x}{dt^2} = -h_1\dot{x} - h_2x - h_3x^3 + F\cos(\Omega t)$ are: $h_1 = 0.02, h_2 = 1.0, h_3 = 0.3, F = 0.5, \Omega = 0.5, x(0) = 1.0, \frac{dx}{dt}(0) = 0.0$. The panels (a) and (b) show the phase-space orbits and the Poincaré map, respectively.

6.12 THE LOGISTIC MAP

The essence of solving a dynamical equation of motion is to investigate the temporal behaviour of the phase-space variables q (coordinates) and p (momenta), given initial conditions. In the previous few sections we noted that the dynamics of a non-linear system can exhibit both periodic and non-periodic features. For a one-dimensional dynamical system, the Poincaré map provides a two-dimensional map as a periodic snap shot of a phase-space point (x, p). However, as we can

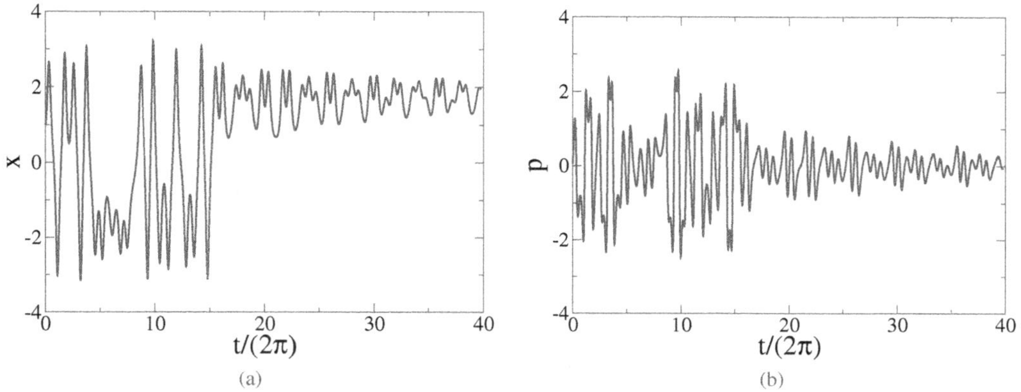

Figure 6.20 Dynamics of a double-well Duffing's oscillator. Parameters used in the numerical evaluation of equation $\frac{d^2x}{dt^2} = -h_1\dot{x} - h_2x - h_3x^3 + F\cos(\Omega t)$ are: $h_1 = 0.02, h_2 = -1.0, h_3 = 0.3, F = 0.5, \Omega = 0.5, x(0) = 1.0, \frac{dx}{dt}(0) = 0.0$. The panels (a) and (b) show the time variation of the displacement x and momentum $p = \frac{dx}{dt}$, respectively.

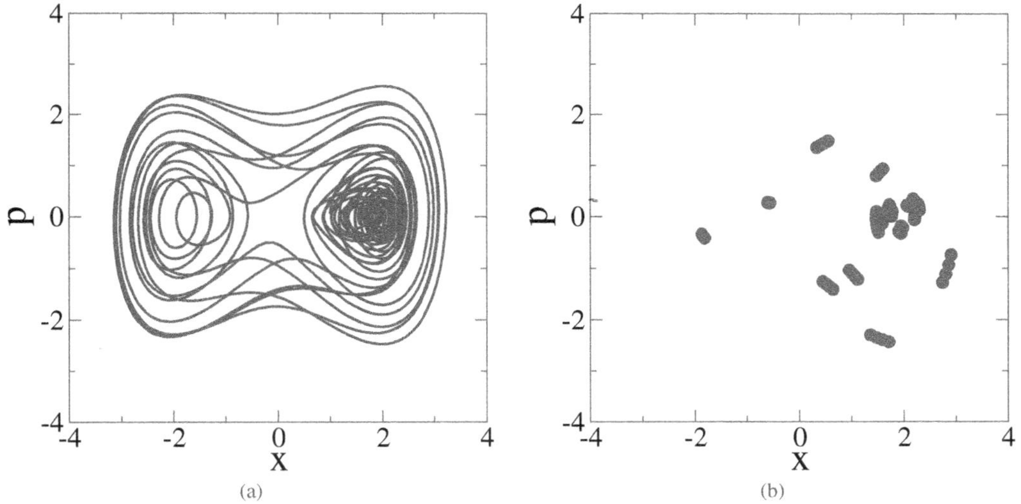

Figure 6.21 Dynamics of a double-well Duffing's oscillator. Parameters used in the numerical evaluation of equation $\frac{d^2x}{dt^2} = -h_1\dot{x} - h_2x - h_3x^3 + F\cos(\Omega t)$ are: $h_1 = 0.02, h_2 = -1.0, h_3 = 0.3, F = 0.5, \Omega = 0.5, x(0) = 1.0, \frac{dx}{dt}(0) = 0.0$. The panels (a) and (b) show the phase-space orbits and the Poincaré map, respectively.

appreciate by looking at the results presented for the Duffing's oscillators in figures 6.18 to 6.21, it would be a very time consuming numerical exercise to decipher and detail periodic and non-periodic parts of the dynamics. Simple mathematical and computational techniques have been employed to study the progression of non-linear systems, and in particular examine their chaotic behaviour. A

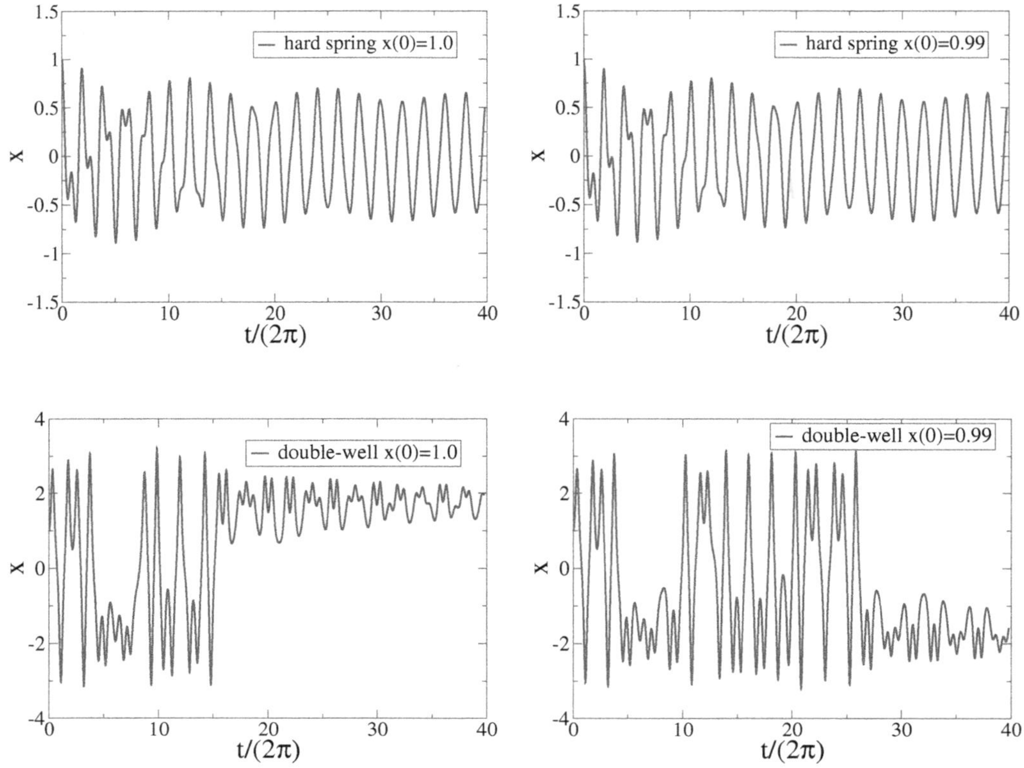

Figure 6.22 Time variation of the displacement variable x with two closely different initial choices $x(0) = 0.99$ and $x(0) = 1.0$ (keeping other parameters identically the same) for the hard-spring and double-well Duffing oscillators. While there are only mild changes for the hard system, there are drastic changes for the double-well system.

discrete map can be considered in an iterative form (or, a difference equation form)

$$s_{n+1} = f(s_n, r), \tag{6.40}$$

where f is a function of the phase point s_n at the nth stage and r represents a set of *tuning parameters* (*e.g.* the damping factor, the anharmonic factor, the frequency and strength of the driving force, *etc.*). For studying time evolution, s_n represents a phase point at $t = t_n$, *viz* $s_n = s(t_n)$.

Hénon (1976) developed a mathematical method for obtaining a two-dimensional discrete map of a dynamical system. Writing the coordinates of the phase point s_n as $s_n = (x_n, y_y)$ and the tuning parameter r as $r = (r_1, r_2)$, the Hénon map is defined by the equations

$$\begin{aligned} x_{n+1} &= 1 - r_1 x_n^2 + y_n, \\ y_{n+1} &= r_2 x_n. \end{aligned} \tag{6.41}$$

A much simpler scheme is to employ the *logistic expression* for the function f

$$f(s_n, r) = r s_n (1 - s_n). \tag{6.42}$$

With this choice, the one-dimensional equation (6.40) can be expressed as the *logistic equation*, or *logistic map*, or *quadratic iterator* (Feigenbaum, 1978)

$$s_{n+1} = rs_n(1-s_n). \tag{6.43}$$

In this one-dimensional scheme, the variable s can be taken to represent the physical quantity of interest, such as the coordinate x or the momentum p of the one-dimensional oscillator[3]. As $f=0$ for $s=0$ and $s=1$, the dynamical variable s is generally restricted to lie in the range $0 \leq s \leq 1$. The converged solution, *viz* $s_{n+1}=s_n$, is called the limiting value s_∞, or the fixed point, and this happens for $s_\infty = 0$ and $s_\infty = 1-1/r$. The solution $s_\infty = 0$ is trivial and the solution $s_\infty = 1-1/r$ is valid only if $r>1$. Although there is no upper limit on the value of r, interesting and useful dynamical features can be extracted by considering the range $1<r<4$. In this range of r-values, the maximum value of the function $f(s,r)$ lies in the interval $[0,1]$.

The stability of the non-trivial fixed point s_∞ can be determined by examining the function $f(s,r)$ around it (Li and Yorke, 1975; José and Saletan, 1998). If in some interval $I = [s_\infty - \varepsilon, s_\infty + \varepsilon]$ we have $|f(s)-y| < |s-y|$ for all $s \in I$ then the function $f(s)$ is said to be asymptotically stable. Assuming f to be differentiable at y, this behaviour is met by the simple condition $|\frac{\mathrm{d}f(s,r)}{\mathrm{d}s}|_{s_\infty} < 1$. This gives, at the non-trivial fixed point $s_\infty = 1-1/r$

$$\left|\frac{\mathrm{d}f(s,r)}{\mathrm{d}s}\right|_{s_\infty} = 2-r, \tag{6.44}$$

which implies that the solution is stable for $1<r<3$ and unstable for $r>3$. Goldstein, Poole and Safko (2002) have reached this conclusion by using a different derivation.

6.12.1 BIFURCATION AND FEIGENBAUM DIAGRAM

A plot of the non-trivial fixed point s_∞ against the tuning parameter r is called the Feigenbaum diagram. We present numerical results for the logistic map described by equation (6.43). The number of iterations was taken up to 100, starting with the initial value $s_0 = 0.3$. In agreement with the discussion following equation (6.44), there is a single fixed point for $1<r<3$. Figure 6.23 shows the Feigenbaum diagram for $2.8 \leq r \leq 3.566$. The numerical results show that there are 2, 4, 8 and 16 stable fixed points when the value of r is set to 3.0, 3.5, 3.55 and 3.566, respectively. The appearance of the double solution, *i.e.* development of *bistability*, is known as **bifurcation**. Similarly, the appearance of the four-fold solution is known as *double bifurcation*. In general, development of a 2^k-fold stable fixed point is called 2^k*-fold bifurcation.* No clear identification of stable fixed points was made for $r>3.566$. For this reason, the region $3.570... < r < 4$ is often associated with *chaos*.

For further discussion of the bifurcation diagram for the logistic, or quadratic, map we refer the reader to the review article by Olsen and Deng (1985).

Feigenbaum (1978) found that a large class of recursion relations $s_{n+1} = rf(s_n)$ exhibits an infinite number of bifurcations. Feigenbaum also showed that there are two universal constants associated with such relations

$$\delta_r = \lim_{n\to\infty} \frac{r_{n+1}-r_n}{r_{n+2}-r_{n+1}}, \tag{6.45}$$

$$\delta_s = \lim_{n\to\infty} \frac{s_{n+1}-s_n}{s_{n+2}-s_{n+1}}. \tag{6.46}$$

From numerical studies of the logistic equation, Feigenbaum established

$$\delta_r = 4.6692016..., \tag{6.47}$$

$$\delta_s = 2.50290787.... \tag{6.48}$$

[3]Equation (6.43) can be obtained from the Hénon map equation in (6.41) by setting $y_n = rs_n - 1$ and $r_2 = 0$ and writing $r_1 = r$.

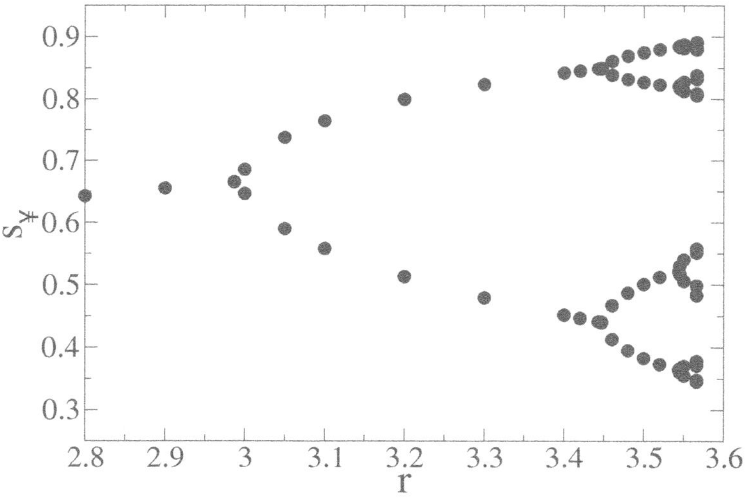

Figure 6.23 Feigenbaum diagram for the iterative solution of the logistic equation $s_{n+1} = rs_n(1 - s_n)$, with the initial value $s_0 = 0.3$.

May and Oster (1980) derived the analytic approximation $\delta_r \sim 2(1+\sqrt{2}) = 4.8284$, which is approximately 3% larger than the value obtained by Feigenbaum. It should be mentioned that different values of the Feigenbaum constants δ_r and δ_s should be expected for different forms of the general iterative equation $s_{n+1} = rf(s_n)$.

Figure 6.24 shows plots of the numerical solution of the logistic map equation. A few observations can be made. First, larger the value of the tuning parameter r, larger is the number of iterations required to achieve converged solution. Second, a periodic solution is not achieved for values of r close to 4. Third, after a few initial iterations, a fixed stable periodic solution is obtained regardless of the starting choice s_0. Fourth, very different aperiodic solutions are obtained for different choices of the initial choice s_0. The last point can be appreciated from the two plots for $r = 3.90$, in that evolution of the solutions for both $s_0 = 0.4$ and $s_0 = 0.30001$ are very different from that for $s_0 = 0.3$. Remarkably, the solutions for $r = 0.30000$ and $r = 0.30001$ are similar for several iterations, but start to diverge after that. The bottom right panel in the figure shows that the solutions start to diverge at iteration number $n = 16$. The clear message from the last observation is, then, that the solution of the logistic equation for the tuning parameter r close to 4, and hence of aperiodic motion, is very sensitive to the choice of the initial condition set for s_0.

Let us consider $r = 2$. As the variable s lies in the interval $[0, 1]$, it is not meaningful to consider the initial choices $s_0 = 0.0$ and $s_0 = 1.0$. The choice $s_0 = 0.5$, is uninteresting, as it will produce the final result $s_\infty = 0.5$ even at zeroth iteration. With this in mind, let us further examine how the iteration cycle progresses when we consider a wide range of choice for the initial condition s_0. For this purpose, we consider the logical equation and choose s_0 to take values 0.2, 0.4, 0.6 and 0.8. The results are presented in figure 6.25. As noted earlier, regardless of the initial choice, the converged result is $s_\infty = 0.5$. Two further points are interesting to note. First, with each starting point, a well converged result is obtained at the fourth iteration. Second, the results for $s_0 = 0.2$ and $s_0 = 0.8$ coincide at the first iteration. Similarly, the results for $s_0 = 0.4$ and $s_0 = 0.6$ are the same at the first iteration. These observations are consistent with the symmetry of the logistic equation.

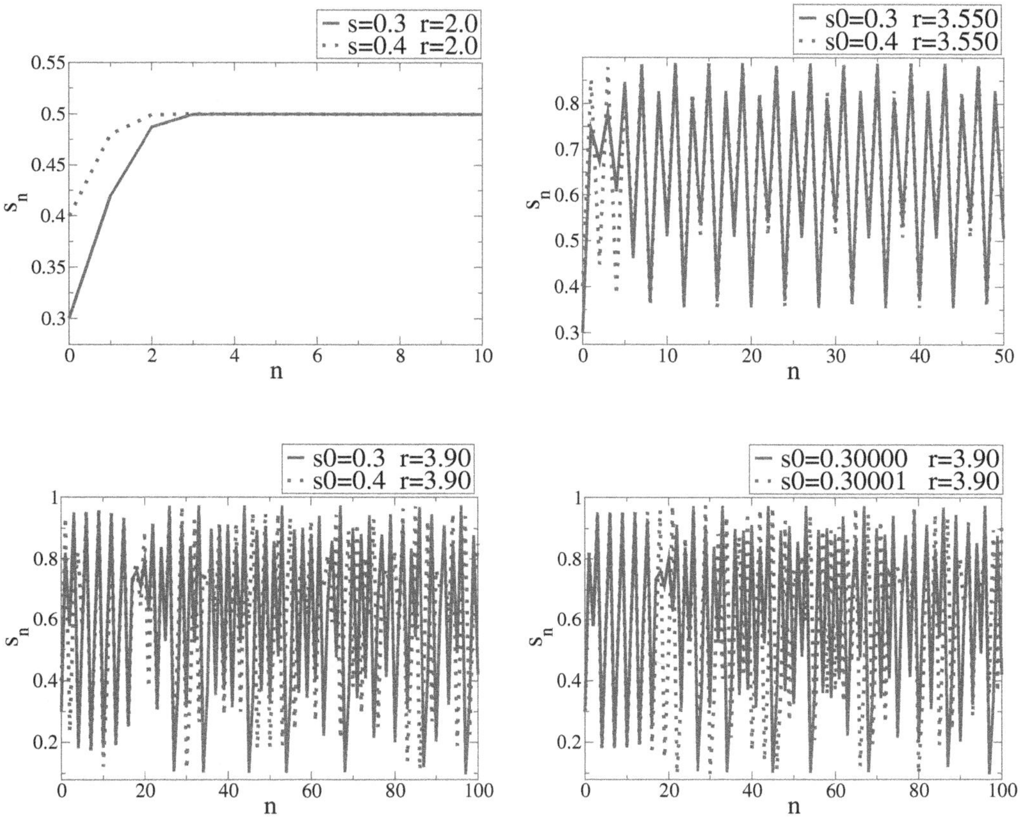

Figure 6.24 Plots of solutions of the logistic equation $s_{n+1} = rs_n(1 - s_n)$ for different choices of the tuning parameter r and initial condition s_0. Sensitivity of the solution with respect to the choice of the starting parameter s_0 for large r values leading to aperiodic motion is evident.

6.12.2 LYAPUNOV EXPONENT

The discussion in the previous sub-section clarifies that the solutions of the logistic equation $s_{n+1} = rs_n(1 - s_n)$ in the region $3.570... < r < 4$ can be considered as aperiodic. Also, we remind ourselves that if a dynamical behaviour exhibits sensitive dependence on initial conditions, then it is called chaotic. A quantitative measure of chaotic behaviour is given by the Lyapunov (or, Liapunov, or Liapounoff) exponent λ (Liapounoff, 1949), for which we derive an expression below.

Let us consider s_0 and $s_0 + \varepsilon$ as two extremely nearby initial points (*i.e.* with $\varepsilon \to 0$), and let $f(s_0)$ and $f(s_0 + \varepsilon)$ be their maps. The difference in the maps of the two points becomes, after n iterations,

$$d_n = f(s_n + \varepsilon) - f(s_n). \tag{6.49}$$

For a given r, the Lyapunov exponent is defined by the equation

$$|d_n| = |\varepsilon| e^{n\lambda}, \quad n \to \infty. \tag{6.50}$$

Using equations (6.49) and (6.50), we can express

$$\lambda = \lim_{n\to\infty} \frac{1}{n} \ln \left| \frac{d_n}{\varepsilon} \right|,$$

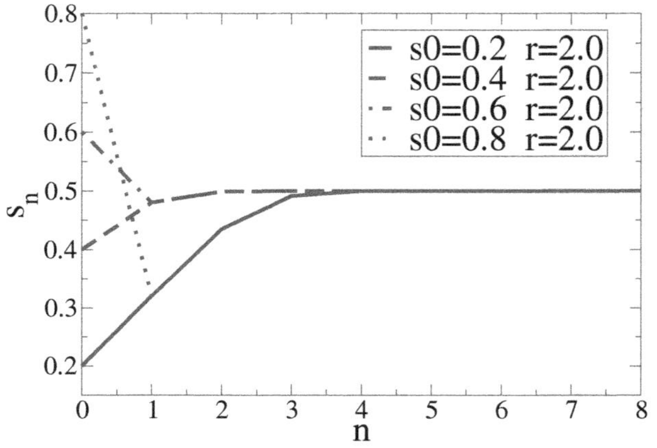

Figure 6.25 Plots of solutions of the logistic equation $s_{n+1} = rs_n(1 - s_n)$ for a fixed r value and different choices of the initial condition s_0.

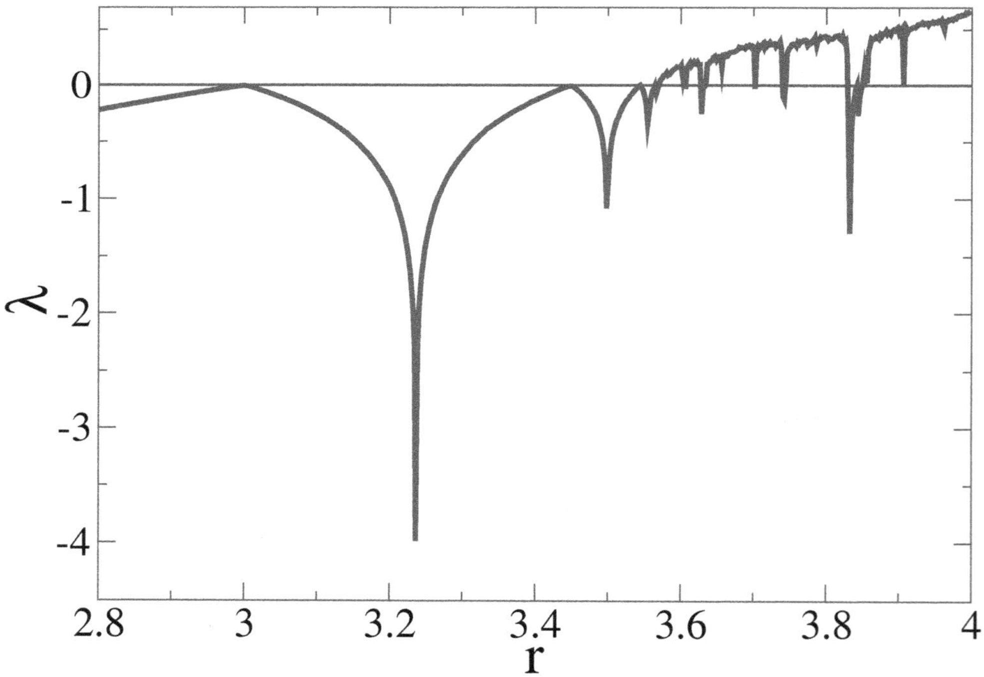

Figure 6.26 Plot of the Lyapunov exponent λ for the logistic map $s_{n+1} = rs_n(1 - s_n)$ against the tuning parameter r. The Lyapunov exponent λ is the average value of the slope of the map for a given value of r.

$$
\begin{aligned}
&= \lim_{n\to\infty} \frac{1}{n} \ln \left| \frac{f(s_n+\varepsilon) - f(s_n)}{\varepsilon} \right|, \\
&\equiv \lim_{n\to\infty} \frac{1}{n} \ln \left| \lim_{\varepsilon\to 0} \frac{f(s_n+\varepsilon) - f(s_n)}{\varepsilon} \right|, \\
&= \lim_{n\to\infty} \frac{1}{n} \ln \left| \frac{\mathrm{d}f(s_n)}{\mathrm{d}s} \right|. \qquad (6.51)
\end{aligned}
$$

Noting that $f(s_n)$ is obtained by iterating the map $f(s_0)$ n times, *i.e.* $f(s_n) = f(f(...(f(s_0))...)$, and using the derivative chain rule, equation (6.51) can be expressed as the following computationally useful formula

$$
\begin{aligned}
\lambda &= \lim_{n\to\infty} \frac{1}{n} \ln \left| \prod_{i=0}^{n-1} \frac{\mathrm{d}f(s_i)}{\mathrm{d}s} \right|, \\
&= \lim_{n\to\infty} \frac{1}{n} \ln \left| \frac{\mathrm{d}f(s_{n-1})}{\mathrm{d}s} \right| \left| \frac{\mathrm{d}f(s_{n-2})}{\mathrm{d}s} \right| \cdots \left| \frac{\mathrm{d}f(s_1)}{\mathrm{d}s} \right| \left| \frac{\mathrm{d}f(s_0)}{\mathrm{d}s} \right|, \\
&= \lim_{n\to\infty} \frac{1}{n} \sum_{i=0}^{n-1} \ln \left| \frac{\mathrm{d}f(s_i)}{\mathrm{d}s} \right|. \qquad (6.52)
\end{aligned}
$$

In other words, the Lyapunov exponent is the average absolute value of the slope of the map $s_{n+1} = f(s_n)$ for a given tuning parameter r. A negative value of λ means that two adjacent points come closer together and the system is stable. A positive value of λ indicates that the system is unstable.

Figure 6.26 shows a plot of the Lyapunov exponent against the tuning parameter r for the logistic map $s_{n+1} = rs_n(1-s_n)$. Note that $\frac{\mathrm{d}f}{\mathrm{d}s} = r(1-2s)$ for all values of s in the logistic map. Numerical results were obtained by using 1000 iterations to solve equation (6.43) for values of r in steps of 0.002 in the range 2.8–4.0. Consistent with the discussion in the previous sub-section, negative values of λ, indicating stable logical maps, exist for $3.57 < r$. The r-values for which λ goes to zero from below correspond to the pitchfork bifurcation points seen in figure 6.20. Generally, larger positive values of λ are found as r becomes larger than 3.57, except for a few 'dips' below zero. In the interval $3.57 < r < 3.844$, there are a few narrow intervals of r for which λ is negative.

FURTHER READING

1. Thornton S T and Marion J B 2004 *Classical Dynamics of Particles and Systems* 5th edn (Thomson: Belmont CA)
2. Calkin M G 1996 *Lagrangian and Hamiltonian Mechanics* (World Scientific: Singapore)
3. van der Pol B 1926 *Phil Mag* **2** 978
4. Cromer A 1981 *Am J Phys* **49** 455
5. Duffing G 1918 *Erzwungene Schwingungen bei Veränderlicher Eigenfrequenz und ihre Technische Bedeutung* (Vieweg: Braunschweig)
6. Lea S M 2004 *Mathematics for Physicists* (Thompson Learning: London)
7. Fang T and Dowell E H 1987 *Int J Non-Linear Mechanics* **22** 401
8. Homes P 1979 *Phil Trans Roy Soc London: A. Math & Phys Sciences* **292** 419
9. May R M and Oster G F 1980 *Phys Lett* **78A** 1
10. Hénon M 1976 *Commun Math Phys* **50** 69
11. José J V and Saletan E J 1998 *Classical Dynamics – A Contemporary Approach* (Cambridge University Press: Cambridge)
12. Olsen L F and Deng H 1985 *Quart Rev Biophys* **18** 165
13. Liapounoff M A 1949 *Problême Général du Mouvement* (Princeton University Press: Princeton)
14. Shaw R 1981 *Z. Naturforsch.* **36**A 80

15. Li T-Y and Yorke J A 1975 *Am Math Mon* **82** 985
16. Strogatz S H 1994 *Nonlinear Dynamics and Chaos* (Westview: Cambridge MA)

PROBLEMS

1. Sketch an elliptical phase orbit of a simple harmonic oscillator using the phases-space coordinates $x = a\cos\phi$, $p = b\sin\phi$ and $a/b = \sqrt{2}$. Sketch Poincaré maps when the oscillator is at the phase-orbit coordinates $a(1,0)$, $a(1/\sqrt{2},1/2)$, $a(0,1/\sqrt{2})$, $a(-1/2,1/2)$ and $a(-1,0)$.
2. The innermost phase orbit of the plane pendulum in figure 6.5 has been plotted for the total energy $E = 0.4mgl$. Determine the corresponding maximum angle of oscillation.
3. Obtain an expression for the phase orbit of the plane pendulum in the (p_θ-θ) plane if the total energy is $E > 2mgl$.
4. Obtain an expression for the separatix orbit of the plane pendulum in the (p_θ-θ) plane.
5. Prove that the equilibrium points for the one-dimensional anharmonic discussed in section 6.6 lie along the x-axis of the x-p_x plane at $x = 0$ and $x = \pm\sqrt{3!\frac{k_{\text{har}}}{k_{\text{anh}}}}$, where k_{har} and k_{anh} are the harmonic and lowest-order anharmonic force constants, respectively.
6. Consider a one-dimensional harmonic oscillator of mass 0.1 kg and force constant $k = 100$ N/m. Calculate its frequency when its motion is subjected to a damping constant $b = 5$ kg/s.
7. The amplitude of a damped, one-dimensional harmonic oscillator decreases to $1/e$ of its initial value after 10 periods. Show that its frequency is approximately $[1 - 1/(800\pi^2)]$ times the frequency of the corresponding undamped oscillator.
8. A damped one-dimensional harmonic oscillator satisfies the equation

$$m\ddot{x} + b\dot{x} + kx = 0,$$

with $m = 1$, $b = 1$ and $k = 1$ in appropriate units. Using equations (6.29) and (6.30), and considering $\phi = \pi/2$, show that, at time $t = \pi/\sqrt{3}$, the coordinates (x, p) in the phase space are $Ae^{-\pi/2\sqrt{3}}(1, \frac{1}{2})$.
9. Using equation (6.29), show that in the weak damping limit ($\omega\tau \to \infty$, where $\tau = 2m/b$) the energy at time t of the damped one-dimensional harmonic oscillator is given by $E(t) = \frac{1}{2}kA^2e^{-2t/\tau}$.
10. Using the displacement and momentum expressions in equations (6.29) and (6.30), derive an expression for the total energy of a damped harmonic oscillator. Considering the weak damping condition $\frac{b}{2m} << \frac{k}{m}$ so that $\omega' = \omega_0 = \sqrt{\frac{k}{m}}$ and setting $\phi = 0$, prove that the expression for the total energy at time t reduces to $E(t) = \frac{1}{2}mA^2e^{-2\beta t}(\omega_0^2 + \beta^2\cos^2\omega_0 t + 2\beta\omega_0\cos\omega_0 t\sin\omega_0 t)$, where $\beta = \frac{b}{2m}$.
11. Prove that the equation

$$\frac{d^2x}{dt^2} = -a\frac{dx}{dt} + bx - cx^3,$$

where $a, b, c > 0$, possesses three stable phase-space points: $(0,0)$ and $(\pm\sqrt{b/c}, 0)$.
12. Plot the quadratic map $y = x(1-x)$ in the x-interval $[0,1]$ and indicate its crossing with the curve $y = x$.
13. Plot the *sine map* $x_{n+1} = \sin(\pi x_n)$ for $0 \le x \le 1$ and discuss its similarities and differences with the logistic map $x_{n+1} = x_n(1 - x_n)$.

14. The following iterative equations

$$\begin{aligned} s_{n+1} &= 2rs_n \quad \text{for } 0 < s < 1/2, \\ &= 2r(1-s_n) \quad \text{for } 1/2 < s < 1 \end{aligned}$$

represent the *tent map*. Show analytically that the Lyapunov exponent for this map is $\lambda = \ln(2r)$ and that the system behaves chaotically for $r > \frac{1}{2}$.

15. Plot the logistic equation $s_{n+1} = 2.8s_n(1-s_n)$ with initial values $s_0 = 0.1, 0.2, 0.3, \cdots, 0.9$ for $r = 2.8$. At what iteration number do you find the converged result for any choice of the initial value? Make observations about the pairs of s_0 producing identical results at iteration $n = 1$.

7 Non-linear Motion and Chaos – Applications

7.1 INTRODUCTION

In the previous chapter we introduced some of the basic theoretical concepts required for understanding and analysing non-linear and chaotic motions. It was all done by considering simple mechanical systems, such as the one-dimensional harmonic and anharmonic oscillators. In this chapter we aim to discuss applications of those concepts by considering examples in engineering, physics, astrophysics, weather science, biological science, neuroscience and neural networks and artificial intelligence.

7.2 NON-LINEAR ELECTRICAL CIRCUITS

An electrical circuit is **linear** if it is characterised by Ohm's law which describes the linear relationship $V = IR$ between voltage V and current I across a wire with electrical resistance R. Implicit in this statement is the assumption that the circuit's components, *e.g.* resistors, capacitors, inductors, *etc* do not change their values while current and voltage measurements are made. A **non-linear** electrical circuit element shows departure from Ohm's law. The simplest non-linear circuit element is the diode. An electrical circuit is considered **dynamic** iff it contains at least one capacitor and at least one inductor or resistor. If an electrical circuit contains one or more non-linear elements, equations describing such a dynamical system are not guaranteed to be solved exactly even in principle. A non-linear circuit is said to exhibit chaotic behaviour if small differences in its initial conditions lead to very different outcomes, *i.e.* when its behaviour is unpredictable and irregular.

Before proceeding, it is useful to draw analogies between parameters of a dynamical electrical circuit and a dynamical mechanical system. Considering the harmonic oscillator as a benchmark dynamic mechanical system, we can establish analogies between the two systems as listed in table 7.1.

7.2.1 DRIVEN LC SERIES CIRCUIT

Let us consider a simple LC series circuit with a time-dependent voltage source $V_s(t)$, as illustrated in figure 7.1. Applying Kirchoff's loop rule, the oscillatory motion of the circuit can be written as

$$V_L + V_C = V_s(t), \tag{7.1}$$

$$L\frac{\mathrm{d}^2Q}{\mathrm{d}t^2} + \frac{Q}{C} = V_0\cos(\Omega t), \tag{7.2}$$

where we have expressed $V_s(t) = V_0\cos(\Omega t)$, and V_L and V_C are the voltage drops across the inductor and capacitor, respectively. This is similar to the equation of motion of a driven harmonic oscillator, as discussed in section 6.9. Note that Ω is the frequency of the applied voltage and that the frequency of the oscillations in the LC circuit is $\omega = \sqrt{\frac{1}{LC}}$.

DOI: 10.1201/9781003383314-7

Table 7.1
Analogies between a dynamic electrical circuit and a dynamic mechanical system.

Electrical circuit	Mechanical system
charge Q	displacement x
current $I = \frac{dQ}{dt}$	velocity $v = \frac{dx}{dt}$
voltage drop*	force
inductance L	mass m
inverse capacitance $\frac{1}{C}$	spring constant Λ
resistance R	damping coefficient b
* voltage drop across resistor $V_R = IR$; capacitor $V_C = \frac{Q}{C}$; inductor $V_L = L\frac{dI}{dt} = L\frac{d^2Q}{dt^2}$	

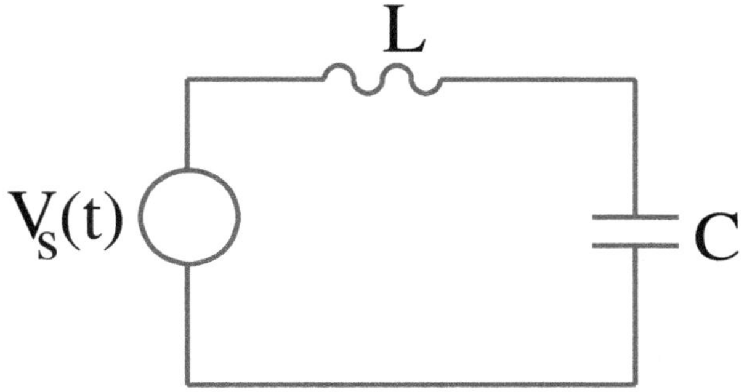

Figure 7.1 An LC series circuit driven by an alternating voltage source. In the figure the variables L and C represent the induction and capacitance, respectively, and $V_s(t)$ represents the source voltage at time t.

Following the discussion in section 6.8, we can express the second-order differential equation in (7.2) as the following two coupled first-order differential equations

$$\frac{dQ}{dt} = I, \tag{7.3}$$

$$\frac{dI}{dt} = -\frac{Q}{LC} + \frac{V_0}{L}\cos(\Omega t). \tag{7.4}$$

These equations can be solved numerically by using any of the algorithms described in Appendix F. General features of solutions will be similar to those discussed in section 6.9 and figures 6.11–6.13. We will therefore not present any results, but encourage the reader to attempt relevant problems at the end of this chapter.

7.2.2 DRIVEN LCR SERIES CIRCUIT

Let us now consider an LCR series circuit, driven by a periodic potential $V_s(t)$, as shown in figure 7.2. The equation of motion can be written, following Kirchoff's loop rule, as

$$V_L + V_C + V_R = V_s(t), \tag{7.5}$$

$$L\frac{d^2Q}{dt^2} + \frac{Q}{C} + R\frac{dQ}{dt} = V_0\cos(\Omega t). \tag{7.6}$$

This second-order differential equation can be expressed as the following two coupled first-order differential equations

$$\frac{dQ}{dt} = I, \tag{7.7}$$

$$\frac{dI}{dt} = -\frac{R}{L}I - \frac{Q}{LC} + \frac{V_0}{L}\cos(\Omega t), \tag{7.8}$$

and can be solved numerically by using any of the algorithms described in Appendix F. General features of solutions will be those of a periodically driven damped harmonic oscillator. Rather than presenting any plots here, we will set a problem for the reader at the end of this chapter. We will, however, remind the reader that the impedance of the driven LCR circuit, with the maximum value of the current $I_{\max}$, is

$$Z = \frac{|V_0|}{|I_{\max}|}, \tag{7.9}$$

$$= \sqrt{R^2 + (\Omega L - \frac{1}{\Omega C})^2}. \tag{7.10}$$

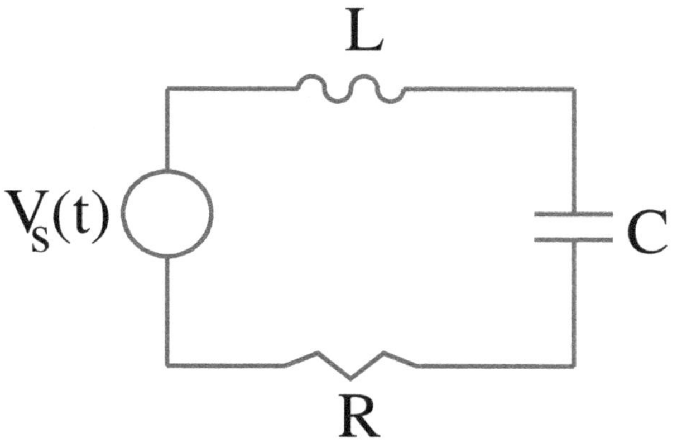

Figure 7.2 An LCR series circuit driven by an alternating voltage source $V_s(t)$. The variables L, C and R are the inductance, capacitance and resistance, respectively.

7.2.3 CHAOS IN ELECTRICAL CIRCUITS

While there is no generally accepted definition of a chaotic phenomenon, it was mentioned in the previous chapter that chaotic dynamical behaviour exhibits deterministic as well as non-deterministic features. A distinctive property of chaotic motion is that it is sensitive to initial conditions. A quantitative measure of chaotic behaviour is generally presented by the Lyapunov exponent λ, as described in section 6.12.2.

An electric circuit that could give rise to chaotic behaviour must include at least one locally active non-linear element, powered by a battery, with a current *vs* voltage non-linear function whose slope must be negative somewhere on the curve. The Chua circuit, invented in 1983, is one of the simplest electronic circuits that exhibits chaos and many well-known bifurcation phenomena (Chua, 1984; Zhong and Ayrom, 1984; Kennedy, 1993). Several generalised versions of the Chua circuit have also been realised. As shown in figure 7.3, the original Chua circuit contains five circuit elements: a linear inductor L, a linear resistor R, two capacitors C_1 and C_2 and a *locally active non-linear element* (called Chua diode). The Chua diode is a voltage-controlled resistor N_R, with a 3-segment piece-wise-linear odd-symmetric driving-point characteristic, as shown in figure 7.4. The non-linear current *vs* voltage relationship $I_{N_R} = f(V_{N_R})$ across the Chua diode is defined analytically as

$$\begin{aligned} I_{N_R} &= f(V_{N_R}), \\ &= \begin{cases} G_b V_{N_R} + (G_b - G_a)E & \text{if} \quad V_{N_R} < -E \\ G_a V_{N_R} & \text{if} \quad -E \le V_{N_R} \le E \\ G_b V_{N_R} + (G_a - G_b)E & \text{if} \quad V_{N_R} > E, \end{cases} \end{aligned} \tag{7.11}$$

where $E > 0$, $G_a < 0$ and $G_b < 0$. With this choice, the circuit possesses three equilibrium points: one at the origin with locally negative slope or *conductance* G_a and two in the outer regions with locally negative conductance G_b. All the three equilibrium points are *unstable*.

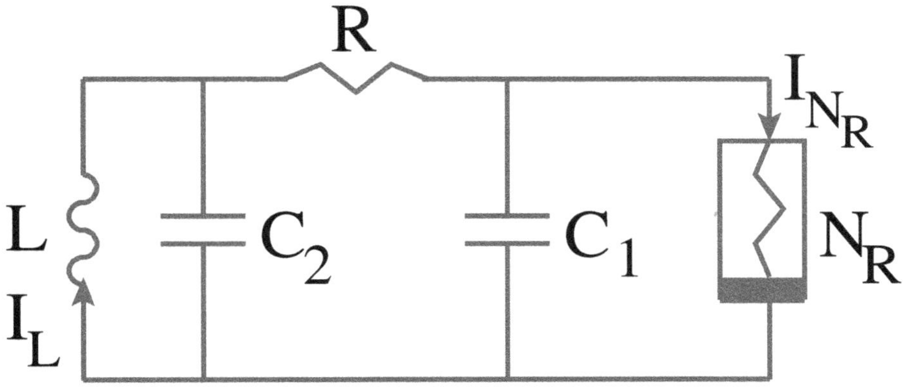

Figure 7.3 Chua's electrical circuit. It contains five circuit elements: a linear inductor L, a linear resistor R, two capacitors C_1 and C_2, and the Chua diode (a locally active non-linear element). N_R is a voltage-controlled non-linear resistor.

The voltages V_{C_1} and V_{C_2} across the two capacitors and the current I_L through the inductor can be chosen as *state variables* for the Chua circuit. Using these, the Chua circuit may be described by three ordinary first-order differential equations, known as Chua equations. Referring to figures 7.3 and 7.4, these equations are (Kennedy 1993)

$$\frac{dI_L}{dt} = -\frac{V_{C_2}}{L}, \tag{7.12}$$

$$\frac{dV_{C_2}}{dt} = \frac{I_L}{C_2} - \frac{G}{C_2}(V_{C_2} - V_{C_1}), \tag{7.13}$$

$$\begin{aligned} \frac{dV_{C_1}}{dt} &= \frac{G}{C_1}(V_{C_2} - V_{C_1}) - \frac{1}{C_1} f(V_{C_1}), \\ &= \begin{cases} \frac{G}{C_1}V_{C_2} - \frac{G'_b}{C_1}V_{C_1} - \left(\frac{G_b - G_a}{C_1}\right)E & \text{if} \quad V_{C_1} < -E \\ \frac{G}{C_1}V_{C_2} - \frac{G'_a}{C_1}V_{C_1} & \text{if} \quad -E \le V_{C_1} \le E \\ \frac{G}{C_1}V_{C_2} - \frac{G'_b}{C_1}V_{C_1} - \left(\frac{G_a - G_b}{C_1}\right)E & \text{if} \quad V_{C_1} > E, \end{cases} \end{aligned} \tag{7.14}$$

where $G = 1/R$, $G'_a = G + G_a$ and $G'_b = G + G_b$.

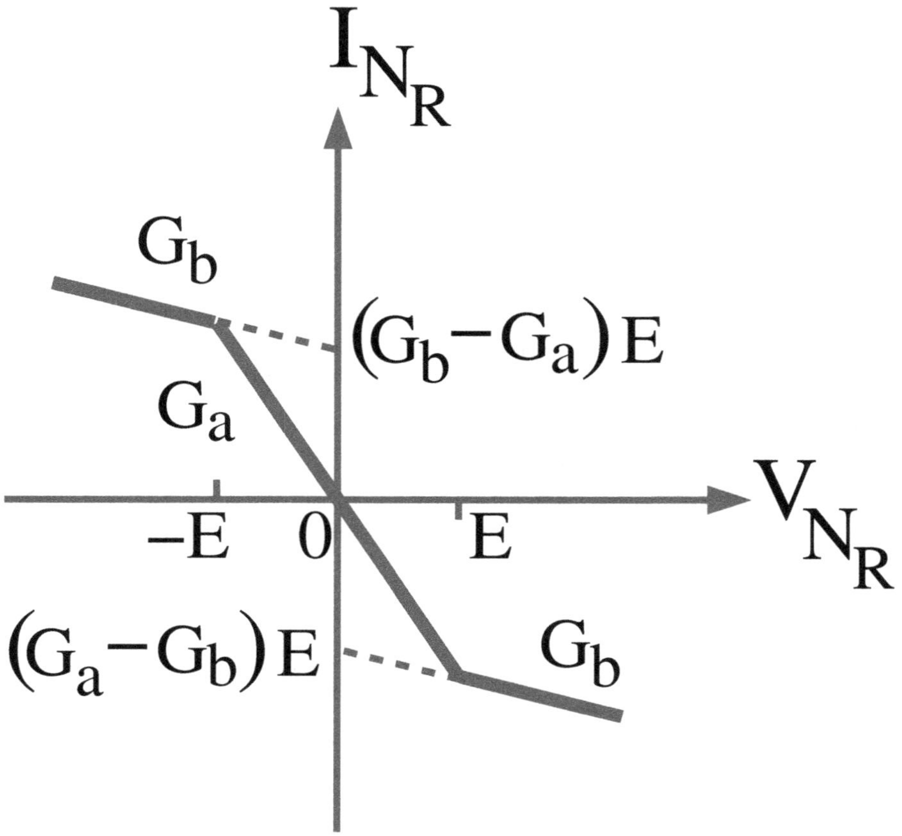

Figure 7.4 The driving-point characteristic of the non-linear resistor N_R in Chua's electric circuit. There are breakpoints at $\pm E$ and (negative) slopes G_a and G_b in the inner and outer regions, respectively. I_{N_R} and V_{N_R} are, respectively, the current and voltage across the non-linear resistor N_R.

Dynamics of Chua's circuit has been studied using specialised non-linear dynamics simulation packages, such as INSITE (Parker and Chua, 1987) and SPICE (Kennedy, 1993). We will not present solutions of Chua's equations here, but would refer the reader to the excellent tutorial by Kennedy (1993) where simulated period-doubling sequence and chaotic dynamics for Chua's oscillator has been studied in detail. Experimental verification of chaos from Chua's circuit was provided in the work of Zhong and Ayrom (1985).

7.2.4 A SIMPLE CIRCUIT WITH CHAOTIC BEHAVIOUR

Simple electronic circuits, designed as electrical analogues of the Duffing-Homes mathematical oscillator, have been built to demonstrate and study non-linear effects and chaos (see, *e.g.* Jones and Trefan, 2001; Tamaševičiūte *et al.* 2008). Here we discuss the circuit suggested by Tamaševičiūte *et al.* (2008).

The dynamics of a periodically forced Duffing-Homes oscillator is governed by the second-order non-autonomous differential equation

$$\frac{d^2x}{dt^2} + b\frac{dx}{dt} - x + x^3 = A\sin(\omega t), \tag{7.15}$$

where b is the damping coefficient, and A and ω are, respectively, the amplitude and frequency of the external driving force. This is equivalent to two first-order non-autonomous differential equations

$$\frac{dx}{dt} = y, \tag{7.16}$$

$$\frac{dy}{dt} = -by + x - x^3 + A\sin(\omega t). \tag{7.17}$$

These equations can be solved numerically, subject to initial conditions for x and y, using a variety of algorithms as discussed in the previous chapter.

The Duffing-Holmes equation (7.15) can be expressed in the Duffing form in equation (6.39), with the considerations $h_1 = b$, $h_2 = -1$ and $h_3 = 1$. As $h_2 < 0, h_3 > 0$, the system is governed by a double-well potential. This can be mathematically shown by obtaining the expression for the potential energy associated with the external force $f(x) = x - x^3$

$$W(x) = -\int f(x)\mathrm{d}x = -\frac{x^2}{2} + \frac{x^4}{4}. \tag{7.18}$$

The plot in figure 7.5 shows the double-well nonparabolic potential energy curve $W(x)$.

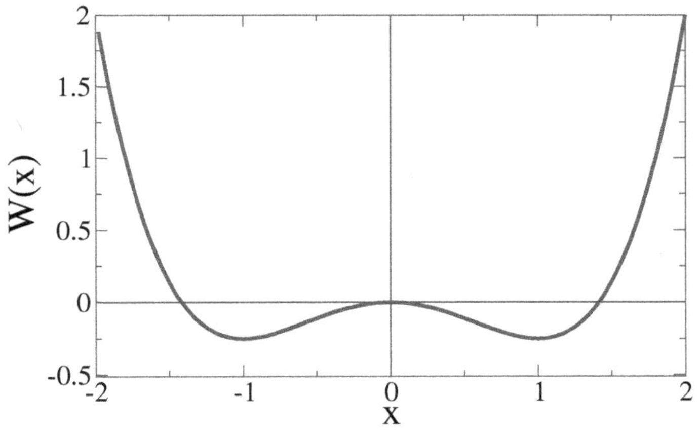

Figure 7.5 Double-well nonparabolic potential energy $W(x)$ derived from the non-linear force $f(x) = x - x^3$.

Using a chosen set of values for the parameters b, A and ω, numerical results for the dynamics of the Duffing-Homes oscillator were presented in figures 6.20-6.22. It was noted that this oscillator performs motion in each of the two wells as well as around the two wells and the system can exhibit chaotic behaviour.

Tamaševičiūte *et al.* described a simple analogue electrical circuit for simulating the Duffing-Homes oscillator equation in (7.15). As shown in figure 7.6, it is an externally driven damped LCR circuit containing a single operational amplifier, two diodes and four resistors. All the LCR elements are linear. The non-linearity is provided by the positive feedback loop consisting of the resistor R3, and the two diodes D1 and D2. The operational amplifier OA plays two roles: the buffer for the external sinusoidal force and the amplifying stage for the positive non-linear feedback.

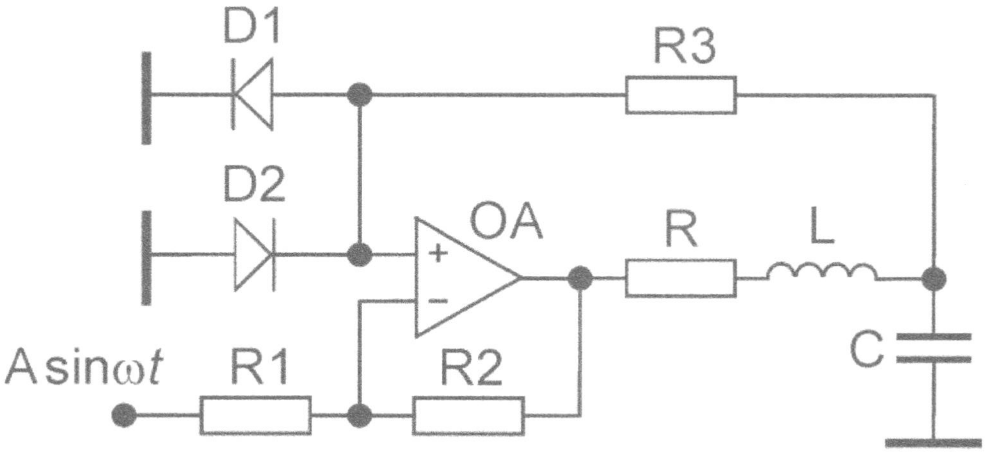

Figure 7.6 An electrical circuit diagram of the Duffing-Homes oscillator. It contains a single amplifier (OA), two diodes (D1 and D2) and four resistors (R, R1, R2 and R3), and all the LCR elements are linear. Here, L is an inductor, C is a capacitor, A is the amplitude of the sinusoidal signal of frequency ω. From Tamaševičiūte *et al.* (2008).

Using Kirchhoff's laws, the circuit can be described by the following two first-order differential equations

$$C\frac{dV_C}{dt} = I_L, \tag{7.19}$$

$$L\frac{dI_L}{dt} = F_E(V_C) - I_L R + A\sin(\omega t), \tag{7.20}$$

where V_C is the voltage across the capacitor C and I_L is the current through the inductor L. The non-linear function $F_E(V_C)$ can be expressed by using a three-segment piecewise linear approximation:

$$F_E(V_C) = \begin{cases} -(V_C + kV^*), & \text{if} \quad V_C < -V^*, \\ (k-1)V_C, & \text{if} \quad -V^* \le V_C \le V^*, \\ -(V_C - kV^*), & \text{if} \quad V_C > V^*, \end{cases} \tag{7.21}$$

where $k = R_2/R_1 + 1$ is the gain of the amplifying stage and V^* is the voltage drop across an opened diode. For silicon diodes $V^* \approx 0.5$ V at 0.1 mA. For convenience, the choice $k = 2$ is made. It is assumed that $R_{d0} >> R_3 >> R_{d1}$, where R_{d0} and R_{d1} are the resistances of the diode in the closed and opened stages, respectively. It is also assumed that $R_3 >> \sqrt{L/C}$.

Equations (7.19)-(7.21) can be conveniently re-expressed as

$$\frac{dx}{dt'} = y, \tag{7.22}$$

$$\frac{dy}{dt'} = F_E(x) - by + a\sin(\omega' t), \tag{7.23}$$

where

$$F_E(x) = \begin{cases} -(x+1), & \text{if} \quad x < -0.5, \\ x, & \text{if} \quad -0.5 \leq x \leq 0.5, \\ -(x-1), & \text{if} \quad x > 0.5, \end{cases} \tag{7.24}$$

and the following dimensionless variables and parameters have been introduced

$$x = \frac{V_C}{2V*}, y = \frac{\rho I_L}{2V*}, t' = \frac{t}{\sqrt{LC}}, \omega' = \omega\sqrt{LC}, a = \frac{A}{2V*}, b = \frac{R}{\rho}, \rho = \sqrt{\frac{L}{C}}. \tag{7.25}$$

The corresponding non-parabolic potential energy term has the form

$$W_E(x) = -\int F_E(x)\mathrm{d}x = \frac{1}{2}\begin{cases} (x+1)^2 - 0.5, & \text{if} \quad x < -0.5, \\ -x^2, & \text{if} \quad -0.5 \leq x \leq 0.5, \\ (x-1)^2 - 0.5, & \text{if} \quad x > 0.5. \end{cases} \tag{7.26}$$

Figure 7.7 shows a plot of $W_E(x)$.

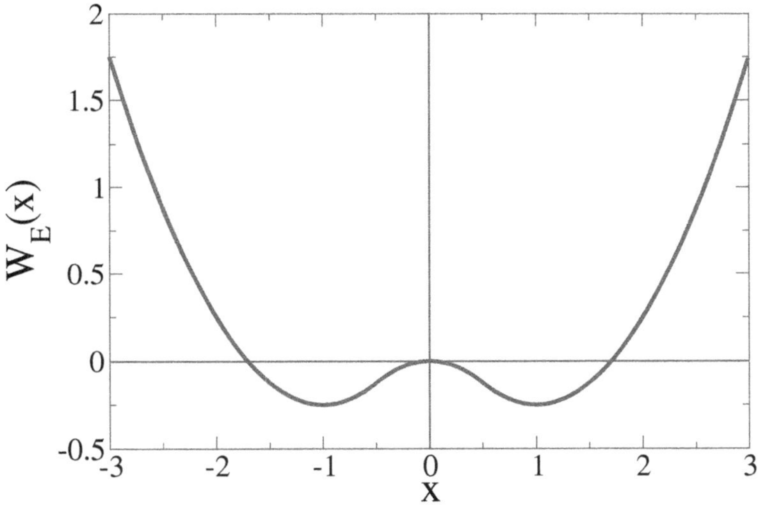

Figure 7.7 Plot of the two-well potential energy $W_E(x)$ derived from the non-linear force expression in equation (7.26) for which the force is expressed as $F_E(x) = -(x+1)$ when $x < -0.5$, $F_E(x) = x$ when $-0.5 \leq x \leq 0.5$ and $F_E(x) = -(x-1)$ when $x > 0.5$.

Equations (7.22)-(7.24) can be numerically solved by using any of the algorithms described in Appendix F. However, we will present results obtained by Tamaševičiūte *et al.* (2008) who simulated the circuit using the package "Electronics Workbench Professional", which is based on a sophisticated software called SPICE. The parameters used were: $L = 19$ mH, $C = 470$ nF (resonance frequency $f_0 \approx 1.7$ kHz, $\rho \approx 200\Omega$), $R_1 = R_2 = R_3 = 10$ kΩ, $R = 20\Omega$, and $f = \omega/2\pi = 1.5$ kHz. It was noted that the vastly different dynamical behaviours of the circuit can result by varying the damping coefficient b, and the driving amplitude A and frequency ω.

Phase space orbits and Poincaré maps presented in figures 7.8–7.12 clearly indicate that the dynamics changes in an irregular manner as the magnitude of the amplitude A is set to several values between 120 mV and 240 mV. The circuit undergoes period-doubling when A increases from

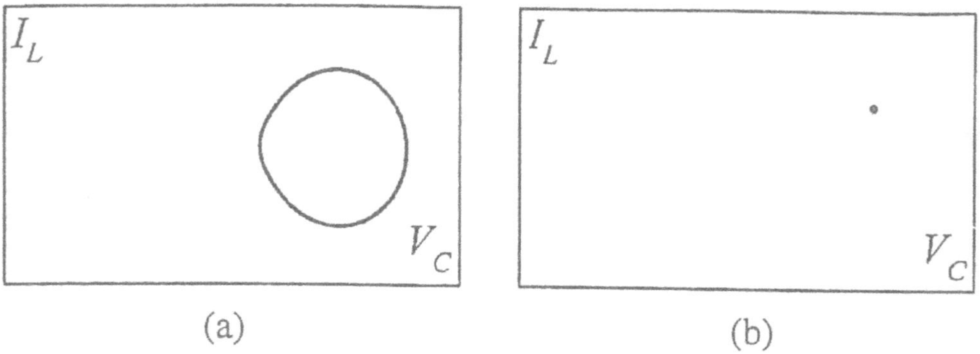

Figure 7.8 Simulated phase space orbit and Poincaré map for the Duffing-Homes oscillator described by equations (7.19)-(7.21) and figure 7.6 with the following parameters: $L = 19$ mH, $C = 470$ nF, $R_1 = R_2 = R_3 = 10$ kΩ, $R = 20Ω$, $f = \omega/2\pi = 1.5$ kHz and the driving force amplitude $A = 120$ mV. From Tamaševičiūte *et al.* (2008).

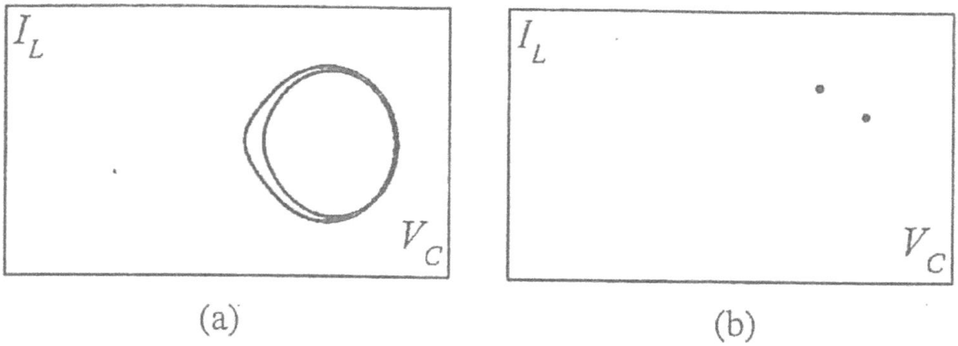

Figure 7.9 Simulated phase space orbit and Poincaré map for the Duffing-Homes oscillator described by equations (7.19)-(7.21) and figure 7.6 with the following parameters: $L = 19$ mH, $C = 470$ nF, $R_1 = R_2 = R_3 = 10$ kΩ, $R = 20Ω$, $f = \omega/2\pi = 1.5$ kHz and the driving force amplitude $A = 140$ mV. From Tamaševičiūte *et al.* (2008).

120 mV to 140 mV. For $A = 160$ mV and $A = 240$ mV it shows chaotic behaviour. The simulation results in figure 7.11 predict that, interspersing between the two chaotic scenarios shown in figures 7.10 and 7.12, period-5 "two-heart" oscillations are set in for $A = 196$ mV. Consistent with the phase orbit and Poincaré map plots, figure 7.13 shows chaotic nature of the time variation of the voltage across the capacitor for $A = 240$ mV.

The experimentally obtained results, taken from a hardware electrical circuit, are shown in figures 7.14–7.19. These results are quite consistent with the predictions of the simulation studies. These studies clearly vindicate that the non-linear behaviour of the Duffing-Homes oscillator can, at least in principle, be controlled so as to obtain periodic as well as non-periodic or chaotic behaviours by making suitable choices for the damping and external voltage parameters.

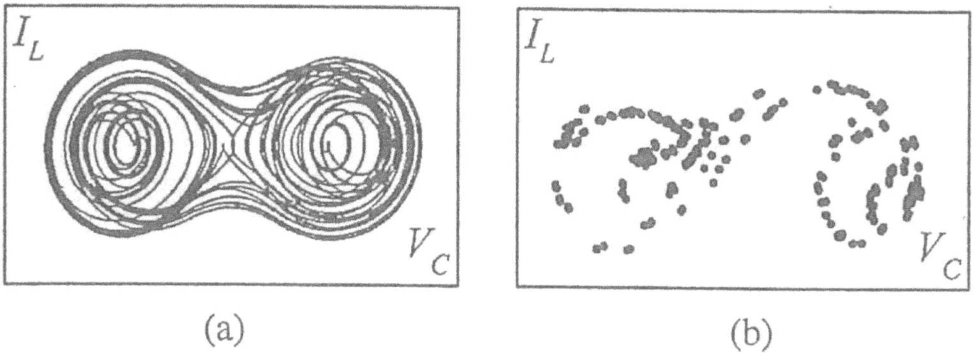

Figure 7.10 Duffing-Homes oscillator described by equations (7.19)-(7.21) and figure 7.6 with the following parameters: $L = 19$ mH, $C = 470$ nF, $R_1 = R_2 = R_3 = 10$ kΩ, $R = 20\Omega$, $f = \omega/2\pi = 1.5$ kHz and the driving force amplitude $A = 160$ mV. From Tamaševičiūte *et al.* (2008).

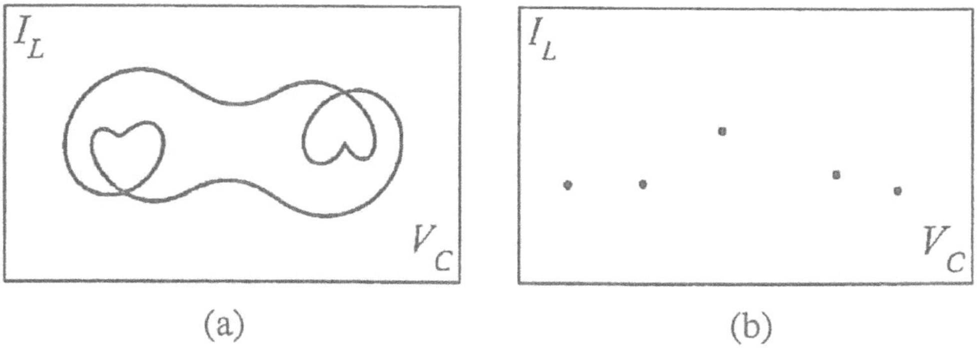

Figure 7.11 Simulated phase space orbit and Poincaré map for the Duffing-Homes oscillator described by equations (7.19)-(7.21) and figure 7.6 with the following parameters: $L = 19$ mH, $C = 470$ nF, $R_1 = R_2 = R_3 =$ 10 kΩ, $R = 20\Omega$, $f = \omega/2\pi = 1.5$ kHz and the driving force amplitude $A = 196$ mV. From Tamaševičiūte *et al.* (2008).

7.3 CELESTIAL MECHANICS

In section 3.3.8.3 we noted that the two-body problem of a planet moving in the Newtonian gravitational field of a fixed star can be solved analytically. However, except for special cases of high symmetry, a n-body Newtonian gravitational problem cannot be analytically solved for $n > 2$. Since the time of the pioneering work by Poincaré, it has been concluded that such systems demonstrate complex non-linear behaviour. Numerical methods, such as those described in Appendix F, are usually adopted to solve many-body celestial orbit problems. The topic of celestial mechanics is huge, both from observational and modelling viewpoints, as can be appreciated from a recent article by Horner *et al.* (2020). Such detailed discussions are far beyond the scope of this book. In this section we will apply the Euler-Cromer algorithm to study celestial orbital motion by choosing two examples of three-body problems, as discussed in Wild (1980).

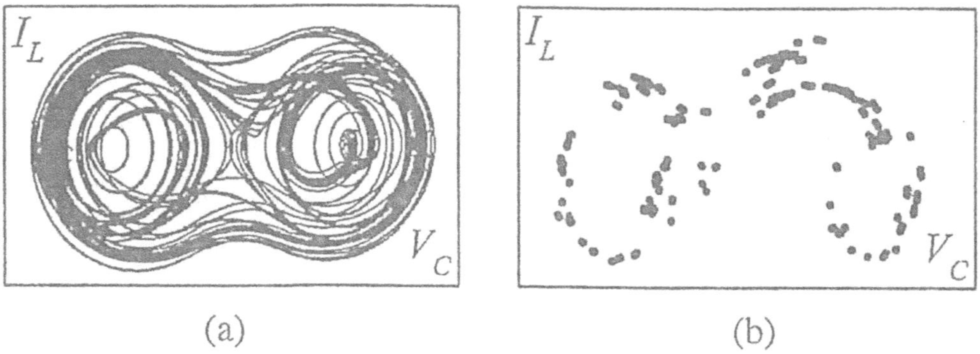

Figure 7.12 Simulated phase space orbit and Poincaré map for the Duffing-Homes oscillator described by equations (7.19)-(7.21) and figure 7.6 with the following parameters: $L = 19$ mH, $C = 470$ nF, $R_1 = R_2 = R_3 = 10$ kΩ, $R = 20\Omega$, $f = \omega/2\pi = 1.5$ kHz and the driving force amplitude $A = 240$ mV. From Tamaševičiūte *et al.* (2008).

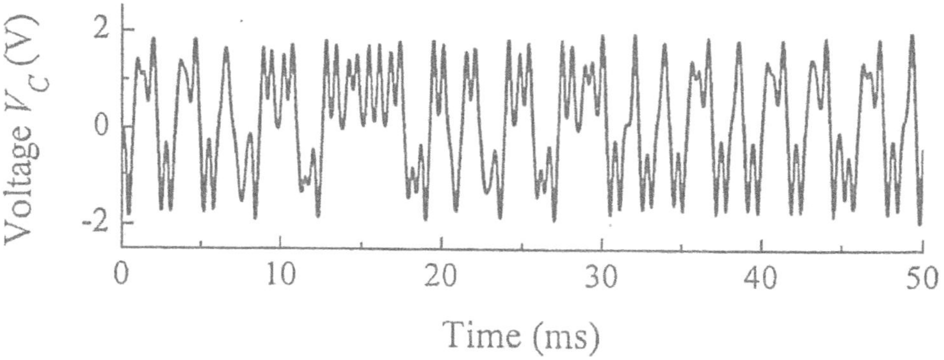

Figure 7.13 Chaotic time variation of voltage across the capacitor V_C in the Duffing-Holmes electrical circuit in figure 7.6 and described by equations (7.19)-(7.21) with $A = 240$ mV. From Tamaševičiūte *et al.* (2008).

7.3.1 MOTION OF TWO PLANETS AROUND A STAR

An analysis of a system of two planets and a star should be made using a three-dimensional model. This is because the orbits of the planets are not necessarily in the same plane. However, for simplicity, we will consider a two-dimensional model, with two planets orbiting around a star in the same plane. Figure 7.20 shows such a mini planetary model, where we consider a star of mass M at the origin and two planets of masses m_1 and m_2 lying at vectors $\boldsymbol{r}_1 = (x_1, y_1)$ and $\boldsymbol{r}_2 = (x_2, y_2)$ in the x-y plane.

The equations of motion of the planets are

$$m_1 \frac{\mathrm{d}^2 \boldsymbol{r}_1}{\mathrm{d}t^2} = -\frac{GMm_1}{r_1^3}\boldsymbol{r}_1 - \frac{Gm_2m_1}{r_{12}^3}\boldsymbol{r}_{12}, \tag{7.27}$$

$$m_2 \frac{\mathrm{d}^2 \boldsymbol{r}_2}{\mathrm{d}t^2} = -\frac{GMm_2}{r_2^3}\boldsymbol{r}_2 - \frac{Gm_1m_2}{r_{21}^3}\boldsymbol{r}_{21}, \tag{7.28}$$

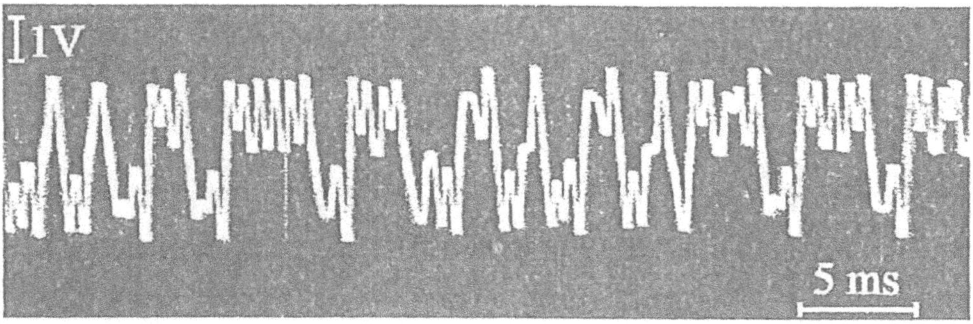

Figure 7.14 Experimental result for chaotic time variation of voltage across the capacitor V_C in the Duffing-Homes electrical circuit shown in figure 7.6 and described by equations (7.19)-(7.21) with $A = 200$ mV. From Tamaševičiūte *et al.* (2008).

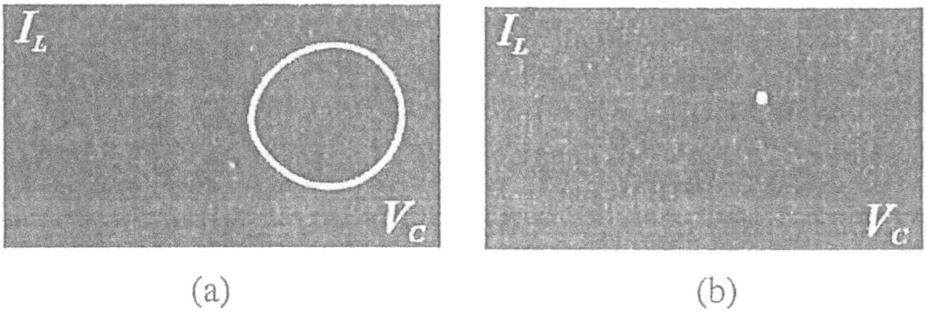

Figure 7.15 Experimental phase space orbits and Poincaré map for electrical circuit shown in figure 7.6 and described by equations (7.19)-(7.21) with $A = 120$ mV. From Tamaševičiūte *et al.* (2008).

where $r_j = \sqrt{(x_j^2 + y_j^2)}$, $\boldsymbol{r}_{ij} = -\boldsymbol{r}_{ji} = \boldsymbol{r}_i - \boldsymbol{r}_j$, $r_{ij} = \sqrt{\boldsymbol{r}_{ij} \cdot \boldsymbol{r}_{ij}}$ and G is the Gravitational constant.

We rewrite the above equations as

$$\frac{\mathrm{d}^2 x_1}{\mathrm{d}t^2} = -\frac{GM}{r_1^3} x_1 - \frac{Gm_2}{r_{12}^3} x_{12}, \tag{7.29}$$

$$\frac{\mathrm{d}^2 y_1}{\mathrm{d}t^2} = -\frac{GM}{r_1^3} y_1 - \frac{Gm_2}{r_{12}^3} y_{12}, \tag{7.30}$$

$$\frac{\mathrm{d}^2 x_2}{\mathrm{d}t^2} = -\frac{GM}{r_2^3} x_2 - \frac{Gm_1}{r_{21}^3} x_{21}, \tag{7.31}$$

$$\frac{\mathrm{d}^2 y_2}{\mathrm{d}t^2} = -\frac{GM}{r_2^3} y_2 - \frac{Gm_1}{r_{21}^3} y_{21}, \tag{7.32}$$

where $x_{12} = -x_{21} = x_1 - x_2$ and $y_{12} = -y_{21} = y_1 - y_2$. Each of these second-order differential equations can be numerically solved, subject to initial conditions.

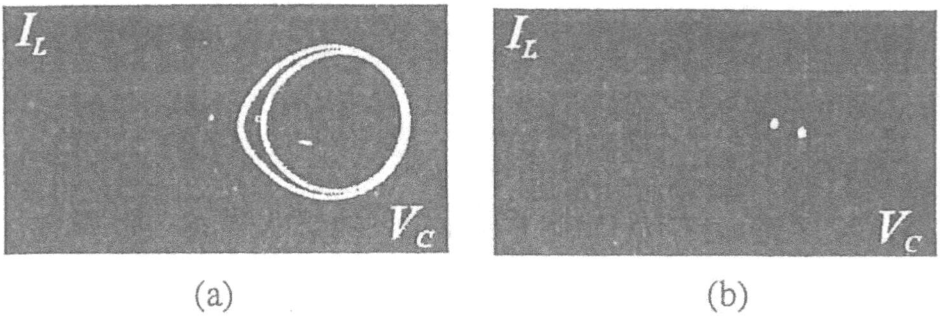

Figure 7.16 Experimental phase space orbits and Poincaré map for electrical circuit shown in figure 7.6 and described by equations (7.19)-(7.21) with $A = 140$ mV. From Tamaševičiūte *et al.* (2008).

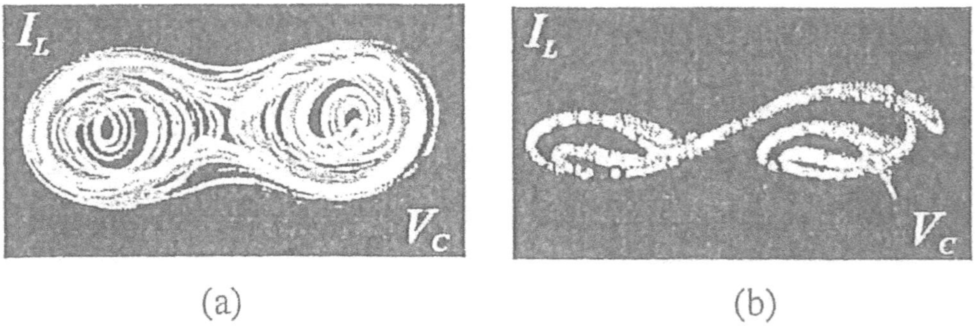

Figure 7.17 Experimental phase space orbits and Poincaré map for electrical circuit shown in figure 7.6 and described by equations (7.19)-(7.21) with $A = 150$ mV. From Tamaševičiūte *et al.* (2008).

For using the Euler-Cromer procedure, described in equations (F.18)-(F.21), we split each of the above equations as two coupled first order equations. Let us consider equation (7.29). Rewriting this as

$$\frac{\mathrm{d}^2 x_1}{\mathrm{d}t^2} = g_1(G, M, m_1, m_2, x_1, x_2, y_1, y_2) \tag{7.33}$$

we further express it as

$$\frac{\mathrm{d}x_1}{\mathrm{d}t} = v_{x_1}, \tag{7.34}$$

$$\frac{\mathrm{d}v_{x_1}}{\mathrm{d}t} = g_1(G, M, m_1, m_2, x_1, x_2, y_1, y_2). \tag{7.35}$$

The Euler-Cromer iterative steps are

$$v_{x_1}|_{n+1} = v_{x_1}|_n + g_1|_n \Delta t, \tag{7.36}$$

$$x_1|_{n+1} = x_1|_n + v_{x_1}|_{n+1} \Delta t, \tag{7.37}$$

where n is the iteration number and the time step Δt is sufficiently small. The other three equations can be similarly modelled for computation.

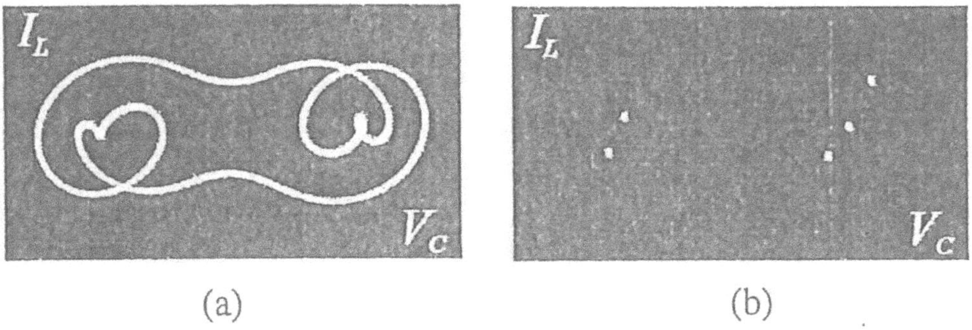

Figure 7.18 Experimental phase space orbits and Poincaré map for electrical circuit shown in figure 7.6 and described by equations (7.19)-(7.21) with $A = 160$ mV. From Tamaševičiūte *et al.* (2008).

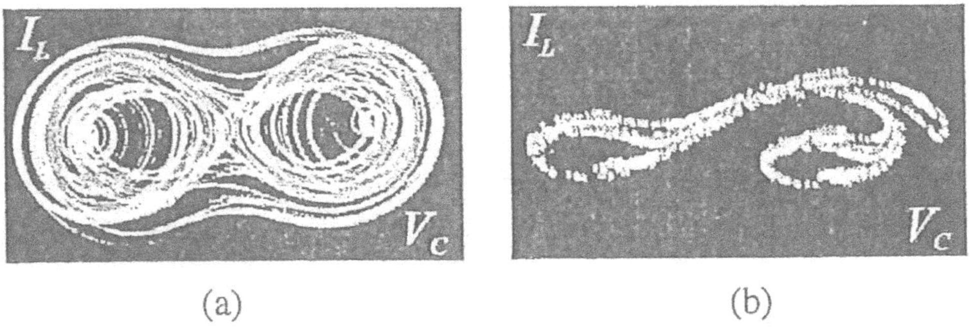

Figure 7.19 Experimental phase space orbits and Poincaré map for electrical circuit shown in figure 7.6 and described by equations (7.19)-(7.21) with $A = 200$ mV. From Tamaševičiūte *et al.* (2008).

The numerical results presented in figure 7.21 were obtained by using the Sun (mass $M = 1.989 \times 10^{30}$ kg) for the star at $(0,0)$, the Earth (mass $m_1 = 5.972 \times 10^{24}$ kg) for planet 1 at (x_1, y_1) and the Jupiter (mass $m_2 = 1.898 \times 10^{27}$ kg) for planet 2 at (x_2, y_2). Using astronomical units, we take $GM = 4\pi^2$. The initial conditions for the Earth and Jupiter were set respectively as $x_1 = 1.0$ AU, $y_1 = 0.0$, $x_2 = 5.2$ AU, $y_2 = 0.0$ for the positions and $v_{x_1} = \frac{dx_1}{dt} = 0.0$, $v_{y_1} = \frac{dy_1}{dt} = \sqrt{GM/x_1} = 2\pi$ AU/yr, $v_{x_2} = \frac{dx_2}{dt} = 0.0$ and $v_{y_2} = \frac{dy_2}{dt} = \sqrt{\frac{GM}{x_2}} = 2.755$ AU/yr. The choices for the velocity initial conditions were made by using the general condition for planetary circular orbits. Complete and stable orbits were obtained by making 2000 iterations with a time step of $\Delta t = 0.01$ yr. Almost circular orbits, revealed from the numerical data, for both the Sun and Jupitar are indeed expected, as due to the very large distance between the two planets means that there is minimal perturbation from one on the motion of the other. It should be mentioned that the choice for the initial conditions adopted in the simulation described above were made for convenience. If desired, another set of initial conditions could be set. For example, one may choose x_1, y_1, v_1 and v_2 such that $\sqrt{(x_1^2 + y_1^2)} = 1.0$ AU and $\sqrt{(v_1^2 + v_2^2)} = 2\pi$ AU/yr.

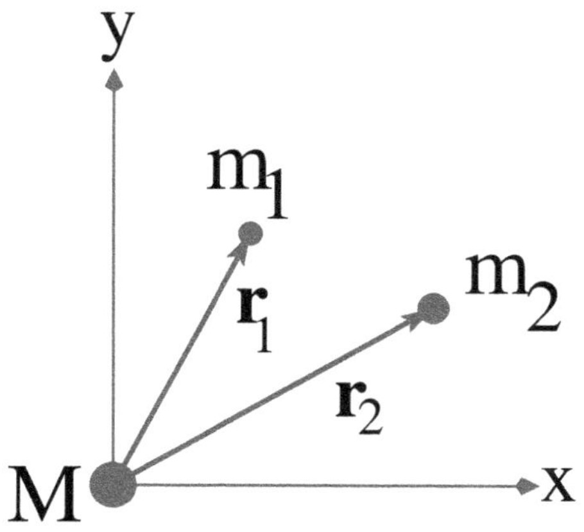

Figure 7.20 A mini planetary system, with two planets orbiting in a plane around a star. The planets of masses m_1 and m_2 are located, respectively, at position vectors $r_1 = (x_1, y_1)$ and $r_2 = (x_2, y_2)$ from a stationary star of mass M at the origin. All three bodies have been considered to lie in the x-y plane.

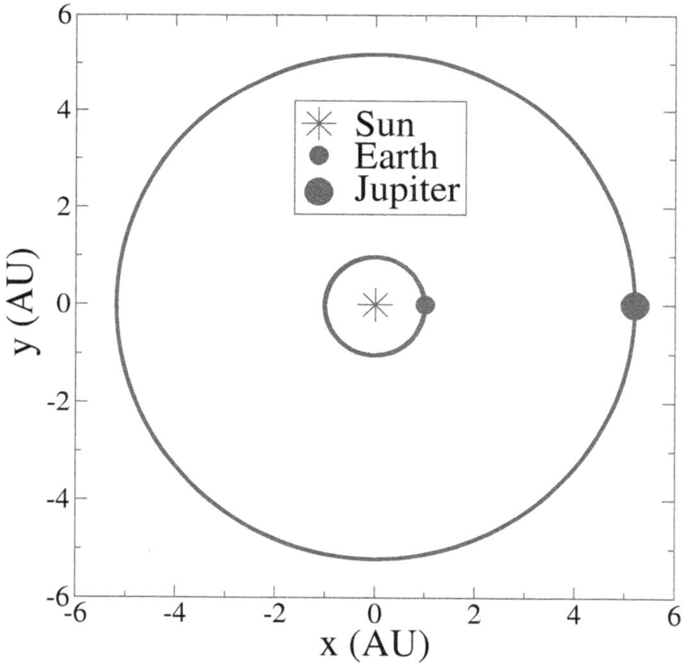

Figure 7.21 Computationally modelled two-dimensional orbits of the Earth and the Jupiter around the Sun in the x-y plane. The star symbol shows the location of the Sun and the smaller and bigger filled dots show the initial locations of the Earth and the Jupiter, respectively. AU = Astronomical Unit.

7.3.2 EULER'S THREE-BODY PROBLEM – MOTION OF A PLANET AROUND TWO STARS

In the previous example we analysed the orbital motions of two planets around a fixed star. Now we will consider the orbital motion of a planet around two fixed stars. We will assume that the two stars do not influence one another. This simple three-body system has been studied since the time of Euler in the 1760s, and hence the problem is known as Euler's three-body problem.

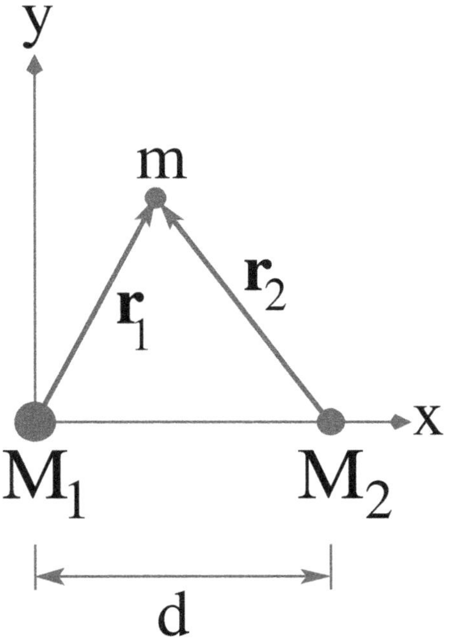

Figure 7.22 Geometry of Euler's three-body system. A planet of mass m moves around two fixed stars of masses M_1 and M_2. All three bodies have been considered to lie in the x-y plane and the planet is at position vectors $\boldsymbol{r}_1$ and $\boldsymbol{r}_2$ from the stars of masses M_1 and M_2, respectively. The stars are positioned along the x-axis and the distance between them is d.

Let us consider all three bodies to lie in the (x-y) plane, as shown in figure 7.22. We consider star 1 of mass M_1 at $(0,0)$, star 2 of mass M_2 at $(d,0)$ and the planet of mass m at $\boldsymbol{r} = (x,y)$. Let the planet be at vector $\boldsymbol{r}_1$ from star 1 and at vector $\boldsymbol{r}_2$ from star 2. Note that these considerations make $\boldsymbol{r} = \boldsymbol{r}_1$. The equation of motion of the planet can be expressed as

$$m\frac{\mathrm{d}^2\boldsymbol{r}}{\mathrm{d}t^2} = -\frac{GM_1m}{r_1^3}\boldsymbol{r}_1 - \frac{GM_2m}{r_2^3}\boldsymbol{r}_2, \tag{7.38}$$

$$= -\frac{GM_1m}{r_1^3}(x,y) - \frac{GM_2m}{r_2^3}(x-d,y), \tag{7.39}$$

where $r_1 = r = \sqrt{(x^2+y^2)}$, $r_2 = \sqrt{(x-d)^2+y^2}$. These can be expressed in terms of the x and y components as

$$\frac{\mathrm{d}^2x}{\mathrm{d}t^2} = -\frac{GM_1}{r_1^3}x - \frac{GM_2}{r_2^3}(x-d) = g_1, \tag{7.40}$$

$$\frac{\mathrm{d}^2y}{\mathrm{d}t^2} = -\frac{GM_1}{r_1^3}y - \frac{GM_2}{r_2^3}y = g_2, \tag{7.41}$$

where g_1 and g_2 have been introduced as short hand notations.

Equations (7.40) and (7.41) can be expressed as coupled first-order differential equations

$$\frac{dx}{dt} = v_x, \tag{7.42}$$

$$\frac{dv_x}{dt} = g_1, \tag{7.43}$$

$$\frac{dy}{dt} = v_y, \tag{7.44}$$

$$\frac{dv_y}{dt} = g_2. \tag{7.45}$$

and the Euler-Cromer iterative steps for solving these equations are

$$v_x|_{n+1} = v_x|n + g_1|_n\Delta t, \tag{7.46}$$

$$x_{n+1} = x_n + v_x|_{n+1}\Delta t, \tag{7.47}$$

$$v_y|_{n+1} = v_y|n + g_2|_n\Delta t, \tag{7.48}$$

$$y_{n+1} = y_n + v_y|_{n+1}\Delta t, \tag{7.49}$$

where n is the iteration number and Δt is a sufficiently small time step.

In general, the stability and shape of the orbits can be sensitive to the relative positions of the three objects and the initial conditions of the planet $\left(e.g.\ \text{its position } \boldsymbol{r} = (x,y) \text{ and velocity } \boldsymbol{v} = (v_x, v_y) = (\frac{dx}{dt}, \frac{dy}{dt}) \text{ at } t = 0\right)$. Notice that the mass m of the planet does not appear in its equation of motion.

We will present results of two different numerical simulations, considering $M_1 = M_2 = M = M_{\text{Sun}}$ (*i.e.* both stars of identical mass) and adopting Astronomical Units.

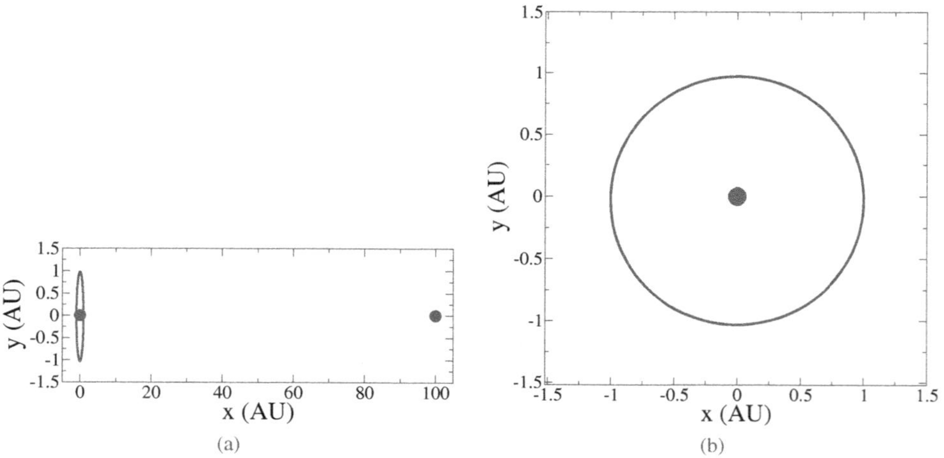

Figure 7.23 Numerical simulation of the motion of a planet around two stars of identical mass (mass of the Sun). The positions of the two stars are marked with the symbol of a filled circle. Using Astronomical Units (AU), the initial position and velocity vectors of the planet are set as $\boldsymbol{r} = (1.0, 0.0)$ and $\boldsymbol{v} = (0.0, 2\pi)$, and the distance between the stars is $d = 100$. The Euler-Cromer algorithm was used with a time step of $\Delta t = 0.01$ yr. It is clear from the diagram in panel (a) that the planet revolves around only around the star nearest to it. The diagram in panel (b) shows that the planet performs a circular orbit around the nearest star.

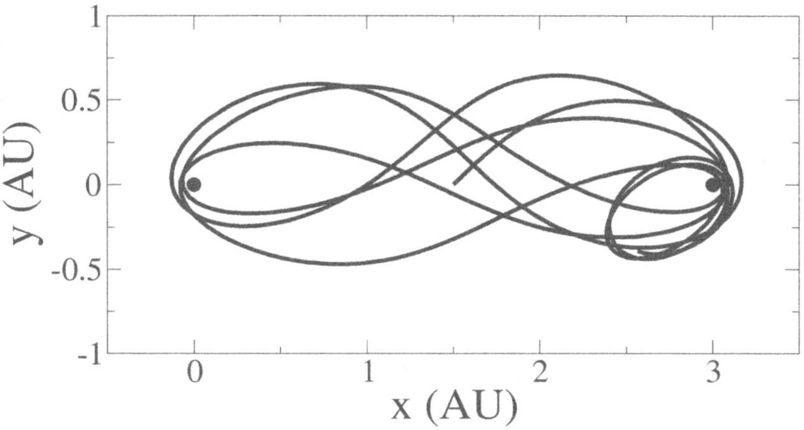

Figure 7.24 Numerical simulation of the motion of a planet around two stars of identical mass (each assigned mass of the Sun). The positions of the two stars are marked with the symbol of a filled circle. Using Astronomical Units (AU), initial position and velocity vectors of the planet are $\boldsymbol{r} = (1.5, 0.0)$ and $\boldsymbol{v} = (3.0, 3.0)$, and the distance between the stars is $d = 3.0$. The Euler-Cromer algorithm was used with a time step of $\Delta t = 0.01$ yr. The planet executes orbital motion around both stars, generating Lissajous curves in the shape of "figure of eight".

In the first simulation, using Astronomical Units (AU), we consider planet's initial position as $\boldsymbol{r} = (1.0, 0.0)$ and its velocity vector as $\boldsymbol{v} = (0.0, 2\pi)$, and set large distance $d = 100.0$ between the two stars. Each star was assigned the mass of the Sun ($M_1 = M_2 = M_{\text{Sun}}$). A time step $\Delta t = 0.01$ yr was used for $n = 200$ iterations. The result is shown in figure 7.23. Clearly, the planet makes a circular orbit only about the star close to it. This is because the other star is too far to have any significant influence.

In the second simulation, using Astronomical Units (AU), we consider the two stars quite close to each other ($d = 3.0$) and initially place the planet half way between the two stars $\left(\boldsymbol{r} = (1.5, 0.0)\right)$ and give it a velocity diagonally in the x-y plane $\left(\boldsymbol{v} = (3.0, 3.0)\right)$. As seen in figure 7.24, the planet executes orbits around both stars, generating Lissajous curves in the shape of "figure of eight".

7.4 NON-LINEAR WEATHER DYNAMICS

In this section we will present a brief discussion of the application of non-linear dynamics in meteorology.

Weather and climate predictions are classified as the predictions of the first and second kind, respectively, (Lorentz, 1975; Palmer, 1993). Whereas predictions of the first kind are initial value problems, predictions of the second kind do not depend on initial conditions. Weather depends on changes in short-term parameters related to atmosphere and oceans. In contrast, climate depends on changes in long-term parameters such as CO_2 content *etc.* In general, a physically sound theoretical model, a very large number of realistic parameters, powerful computers and robust computational algorithms are required for making reliable weather forecasts in different geographical regimes. While this is far beyond the scope of this book, here we will discuss model non-linear equations, developed by Lorentz (1963), for understanding chaotic behaviour of atmospheric weather.

Lorentz considered the problem of non-static and non-periodic fluid flow. His approach was based on the work of Saltzmann (1962) who investigated convection in the Earth's atmosphere, caused by thermal inequalities mainly due to solar heating. Saltzmann approximated the atmosphere as a uniformly extended fluid of depth h, with free upper boundary and rigid lower boundary, and a temperature difference ΔT between its upper and lower ends. All motions were considered parallel to the x-z plane, with no variations along the y-axis. The gravitational force on the fluid was directed along the z-axis. With these considerations, the fluid velocity components are expressed as

$$v_x = \frac{\partial \psi}{\partial z}; \qquad v_z = -\frac{\partial \psi}{\partial x}, \tag{7.50}$$

where ψ is a stream function for the two-dimensional motion. The fluid density, although assumed to be constant, is considered to vary with temperature through the gravity term. This is the Oberbeck-Boussinesq approximation (Oberbeck, 1879; Boussinesq, 1903). With the above considerations, and assuming both ψ and $\nabla^2\psi$ to vanish at both boundaries, Saltzmann obtained the following hydrodynamic equations

$$\frac{\partial(\nabla^2\psi)}{\partial t} = -\frac{\partial(\psi,\nabla^2\psi)}{\partial(x,z)} + \nu\nabla^4\psi + g\alpha\frac{\partial\theta}{\partial x}, \tag{7.51}$$

$$\frac{\partial\theta}{\partial t} = -\frac{\partial(\psi,\theta)}{\partial(x,z)} + \frac{\Delta T}{h}\frac{\partial\psi}{\partial x} + \kappa\nabla^2\theta, \tag{7.52}$$

where θ is the departure of the temperature from its static state value and the first term on the right-hand side is written in the form of a Jacobian operator, *viz*

$$\frac{\partial(A,B)}{\partial(x,z)} = \frac{\partial A}{\partial x}\frac{\partial B}{\partial z} - \frac{\partial B}{\partial x}\frac{\partial A}{\partial z}. \tag{7.53}$$

The constants g, α, ν and κ are the acceleration due to gravity, the coefficient of thermal expansion, the kinematic viscosity and the thermal conductivity, respectively. By expressing ψ and θ in double Fourier series in x and z, with the coefficients as functions of time t alone, Saltzmann reduced the partial differential equations in (7.51) and (7.52) to a set of ordinary differential equations. Lorentz made some drastic simplifications to that work by Saltzmann and presented the following three coupled non-linear first-order differential equations

$$\frac{\mathrm{d}X}{\mathrm{d}t} = -\sigma X + \sigma Y, \tag{7.54}$$

$$\frac{\mathrm{d}Y}{\mathrm{d}t} = rX - Y - XZ, \tag{7.55}$$

$$\frac{\mathrm{d}Z}{\mathrm{d}t} = XY - bZ. \tag{7.56}$$

Here t is dimensionless time, $X(t)$, $Y(t)$ and $Z(t)$ are dimensionless variables as functions of time alone, and the parameters σ, r and b are defined below:

$$\sigma = \nu/\kappa \qquad \text{Prandtl number}, \tag{7.57}$$

$$\begin{aligned} R &= \frac{g\alpha h^3\Delta T}{\nu\kappa} && \text{Rayleigh number}, \\ r &= \frac{R}{R_c} = \frac{R}{\frac{\pi^4}{a^2}(1+a^2)^3} && \text{reduced Rayleigh number}, \end{aligned} \tag{7.58}$$

and

$$b = \frac{4}{(1+a^2)} \qquad \text{width/height ratio of } x\text{–}z \text{ plane.} \tag{7.59}$$

The minimum value of the critical Rayleigh number R_c occurs when $a^2 = \frac{1}{2}$. In equations (7.54)-(7.56), X represents the intensity of the convective motion, Y represents the temperature difference between the ascending and descending currents and Z represents the distortion of the vertical temperature profile from linearity. Similar signs of X and Y denote that warm fluid is rising and cold fluid is descending. A positive value of Z indicates that the strongest gradients occur near the boundaries.

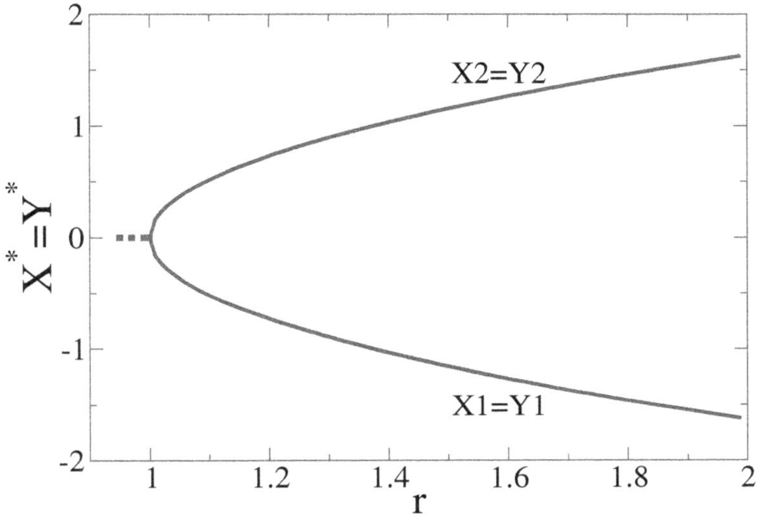

Figure 7.25 Pitchfork bifurcation of the solutions for the fixed point variables X^* and Y^* with respect to the parameter r in the Lorentz model of weather pattern with the choice $b = 8/3$. The two solutions for X^* are $X1$ and $X2$, and the two solutions for Y^* are $Y1$ and $Y2$.

Using additional mathematical analysis, Lorentz suggested that the criterion for the onset of convection is $r = 1$, and therefore, in general $r > 1$. Steady convention is stable for $\sigma < (b+1)$ and unstable for $\sigma > (b+1)$. Fixed points in the (X,Y,Z) space can be determined by setting $\frac{dX}{dt} = \frac{dY}{dt} = \frac{dZ}{dt}=0$ in equations (7.54)-(7.56). It can be established that $(0,0,0)$ is a fixed point for all values of the parameters. For $r > 1$, there are additional fixed points at $\left(\pm\sqrt{b(r-1)}, \pm\sqrt{b(r-1)}, r-1\right)$. These fixed points coalesce with $(0,0,0)$ in the limit $r \to 1^+$ in the form of a pitchfork bifurcation, as seen in figure 7.25. It is also interesting to note that the Lorentz equations show the symmetry $(X,Y) = (-X,-Y)$, meaning that for a solution $\left(X(t), Y(t), Z(t)\right)$ there also exists the solution $\left(-X(t), -Y(t), Z(t)\right)$.

Solutions of the Lorentz equations sensitively depend upon the choice of the initial condition $\left(X(0), Y(0), Z(0)\right)$ and the constants σ, r and b. Reasonable values for the Earth's atmosphere are $\sigma = 10$, $r = 28$ and $b = 8/3$. Using these values, results of numerical calculations, based on the Euler algorithm with time step $\Delta t = 0.01$ over 5000 iterations are presented in figures 7.26 and 7.27. The initial condition was set to $\left(X(0), Y(0), Z(0)\right) = (0,1,0)$.

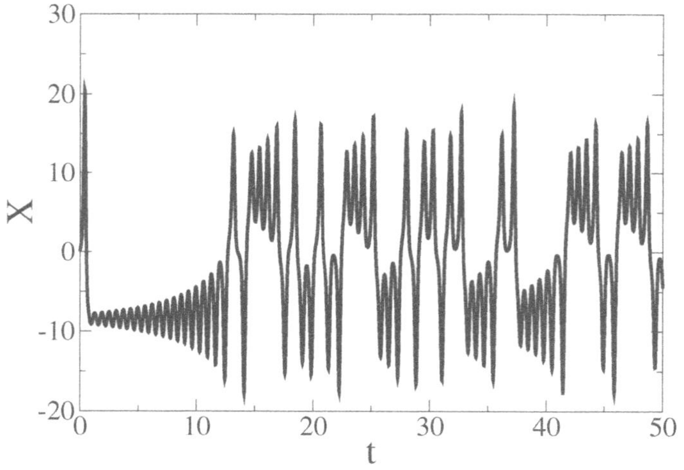

Figure 7.26 Time variation of the convective overturning variable X in the Lorentz equation (see equations (7.54)-(7.56)). Here t is dimensionless time and the parameters used are: the Prandtl number $\sigma = 10$, the reduced Rayleigh number $r = 28$ and the ratio of the width to height of the x-z plane $b = 8/3$. The step in the dimensionless time is $\Delta t = 0.01$ and the number of iterations in the Euler algorithm is 5000. The initial condition was set to $(X(0), Y(0), Z(0)) = (0, 1, 0)$.

Let us look at the time evolution of the variable X, shown in figure 7.26. For a few iterations the state of rest shows instability[1]. After a few initial iterations, the solution settles into a bounded range. However, there is no clear periodicity in the solution. Also, the solution settles as a different bounded long-term behaviour when a slightly different initial condition is chosen, such as $(0, 1.01, 0)$ rather than $(0, 1, 0)$. Similar observations can be made for the time evolution of the terms Y and Z. Therefore, we can conclude that the solutions of the Lorentz equations are irregular with the characteristics of being chaotic.

In the (X-Y-Z) space, each point represents the state of instantaneous 'weather' and a trajectory represents the evolution of weather with time. The numerically obtained time evolution of weather is shown as the three-dimensional plot in figure 7.27. The orbital trajectories form the shape of butterfly wings. Such a figure is called a **strange attractor**[2], and more specifically the **Lorentz attractor**. The orbits flip back and forth between the two wings of the butterfly diagram, but they never intersect each other. Starting at the initial location $(0, 1, 0)$, the trajectory moves to the right wing of the butterfly, then switches towards the centre of the left wing. Having made a few outwards spirals in the left wing, the trajectory swings back to the right wing, and after making a few outward spirals it switches back to the left wing, and so on. The number of spirals made on either side is unpredictably different for each switch from the left to the right wing. In other words, there is no fixed pattern for such flips and the total number of orbits in one wing is not the same as in the other wing. This is the characteristics of randomness.

The irregular and unpredictable, or chaotic, solutions of the Lorentz equations lead to the conclusion that, based on parameters and initial conditions, it is not possible to predict atmospheric weather pattern (*i.e.* make forecasts of the first kind) indefinitely into the future. This qualitative

[1] see Lorentz (1963) for an explanation of the initial instability.

[2] A strange attractor displays sensitive dependence on initial conditions, in that two points initially close to each other at some time are exponentially separated with time.

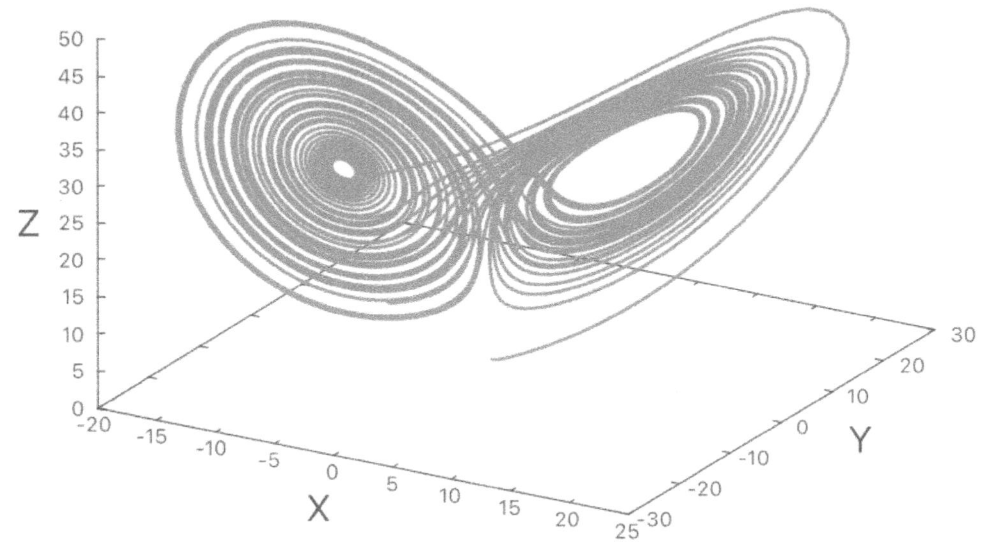

Figure 7.27 Numerically obtained Lorentz orbital trajectories in the X-Y-Z space. Here dimensionless variables $X(t)$, $Y(t)$ and $Z(t)$ represent the time variation of the convective overturning, the horizontal temperature and the vertical temperature, respectively. The parameters used are: the Prandtl number (ratio of viscosity to thermal conductivity) $\sigma = 10$, the reduced Rayleigh number (temperature difference between the top and bottom ends) $r = 28$ and width to height ratio of the x-z plane slice $b = 8/3$. The Euler algorithm is used to solve equations (7.54)-(7.56)) for 5000 iterations using a step of $\Delta t = 0.01$ in the dimensionless time. The initial condition was set to $\left(X(0), Y(0), Z(0)\right) = (0, 1, 0)$.

description of the chaotic property using the Lorentz model weather pattern is in agreement with results obtained from realistic models involving sophisticate algorithms and a very large number of parameters. A detailed coverage of the development of this topic is beyond the scope of this book and interested readers should read Palmer (1991, 1993, 1999, 2000) for further discussion.

7.5 NON-LINEAR DYNAMICAL MODEL FOR ELECTROCARDIOGRAM SIGNALS

In this section we will discuss the application of non-linear dynamics to a biophysical system. We will do this by discussing a mathematical modelling of heart physiology, represented by electrocardiogram (ECG) signals. Generally, the ECG records the electrical activity of the heart as the potential differences between electrodes attached to arms, legs and chest. Standard ECG equipment uses 3, 5, or 12 leads (*i.e.* cables from the electrodes to the electrical recorder).

In order to understand the relationship between an ECG trace and the electrical activity of the heart we first discuss the conduction system of the heart. This, in turn, requires knowing the anatomy and physiology of the heart, which we will briefly describe below.

Anatomy and physiology of the cardiac system

The structure of human heart is shown in figure 7.28. The heart is enclosed in a sac, called the pericardium. Inside the pericardium, the heart has a middle muscular layer called the myocardium, and an inner lining called the endocardium. The myocardium, or cardiac muscle, is one of three major types of muscle in the body. The myocardium encloses four cavities (or heart chambers) – two atria and two ventricles. These cavities are: left atrium (LA), left ventricle (LV), right atrium (RA) and right ventricle (RV). There are four valves in the heart – tricuspid valve, mitral valve, aortic valve and pulmonary valve. The RA and RV are separated by the tricuspid valve. The LA and LV are separated by the mitral valve. Nerve cells, called the bundle of His (after His, who discovered these in 1893), extend from the AV node to the septum (a wall of tissue that separates the left and right sides of the heart). Purkinje fibres, discovered by Purkyně in 1839, extend from the septum to the ventricles. The bundle of His and Purkinje fibres are together known as the His-Purkinje (or, HP) complex or node.

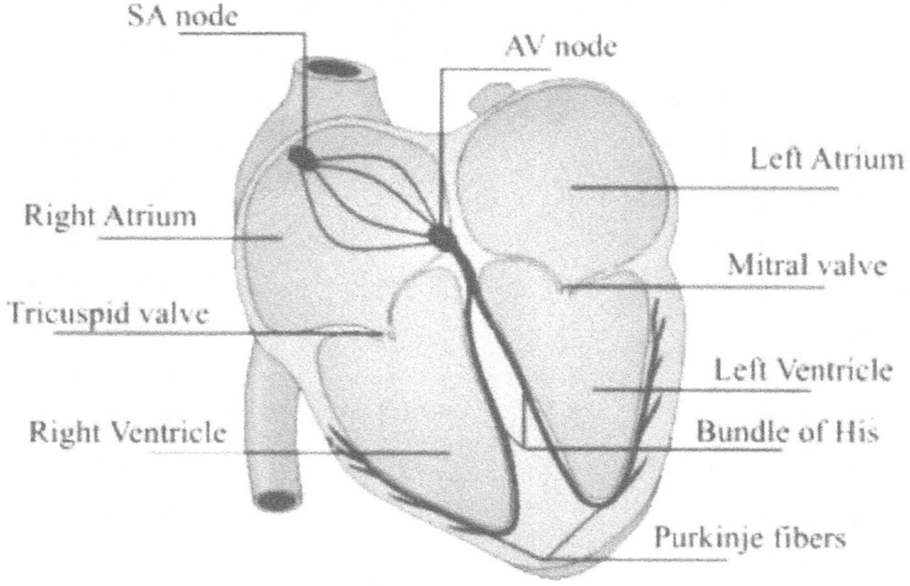

Figure 7.28 The structure of human heart. There are four cavities: two atria (right atrium and left atrium) and two ventricles (right ventricle and left ventricle). The electrical impulse in the cardiac system involves a network formed by the sinoatrial node (SA), the atrioventricular node (AV) and the His-Purkinje complex (HP). From Cheffer *et al.* (2021).

Deoxygenated blood from the head, neck and the rest of the body is received by the RA. It then flows through the tricuspid valve to RV. The RV then pumps blood through the pulmonary valve into the lungs, where it is oxygenated. The LA receives the oxygenated blood via the pulmonary veins and transfers it into the LV through the mitral valve. The LV then delivers the oxygenated blood to the whole body through the aorta, the largest artery in the body.

Heart's electrical system

The blood flow through the heart is controlled by contractions caused by its electrical conduction system. The cardiac conduction system is comprised of three specialised cells and nodes, *viz* sinoatrial node (SA), atrioventricular (AV) node and the complex made of the bundle of His and Purkinje fibres (the HP complex). The SA node is a small mass of specialised tissue and is located in the upper part of the RA chamber (see figure 7.28). The AV node is located close to the tricuspid valve near the central area of the heart.

The SA node sets the heart rhythm (sinus rhythm) and generates an electrical stimulus, 60 to 100 times in a normal heart. This generates an electrical potential across the cell membrane, which is known as the **cardiac action potential**. The process is spontaneous **depolarisation** of the SA node. The impulse travels to the AV node and after slowing down for a very short period of time continues down the conduction path via the His bundle (or bundle of His) to the Purkinje fibres. The Purkinje fibres deliver the electrical signal to the ventricles, making them contract. As the ventricles contract, blood flows from the RV to the pulmonary arteries and from the LV to the aorta. In turn, aorta sends blood to the rest of the body.

The cardiac muscle is comprised of billions of cells (basic biological units of living beings) containing electrolytes. Each of these cells is enclosed by a fatty membrane and surrounded by extra-cellular fluid. There are two major types of cardiac cells: myocardial contractile cells and myocardial conducting cells. The myocardial contractile and conducting cells constitute, respectively, the bulk (approximately 99%) and a very small fraction (approximately 1%) of the cells in the atria and ventricles. The contractile cells synchonise contractions necessary to pump blood through the body. The conducting cells trigger the contractions of the contractile cells to propel the blood. Neighbouring cardiac cells communicate with each other through inter-cellular ion movement. The generation of cardiac action potential requires the presence of ion gradient between the intra- and extra-cellular regions. The most important ions are K^+, Na^+ and Ca^{2+}.

Three coupled oscillator model for cardiac rhythm

An ECG trace records the heart's electrical activity in the form of waves. A normal ECG cardiac cycle is characterised by three components: P wave, QRS complex and T wave, as schematically illustrated in figure 7.29. The P wave represents the impulse generated by the SA node, the QRS complex represents the depolarisation of ventricles via the HP node and the T wave reflects ventricular repolarisation. The R–R interval in the ECG trace is used to measure the ventricular cardiac cycle and to calculate heart rate. This information suggests that a mathematical modelling of the cardiac rhythm can be performed by considering a network formed by the SA node, the AV node and the HP complex.

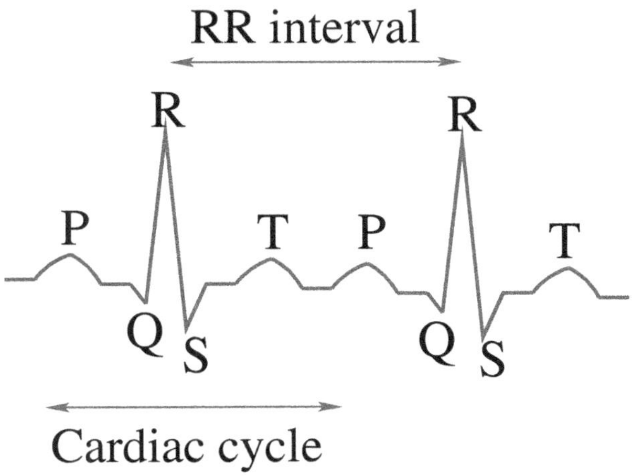

Figure 7.29 Important components of a normal ECG trace: P wave, QRS complex and T wave. The P wave is generated by the SA node, the QRS complex is formed by ventricular contraction via the HP complex and the T wave reflects ventricular repolarisation. The R–R interval in the ECG trace is used to measure the ventricular cardiac cycle and to calculate heart rate.

Grudzinski and Zebrowski (2004) have shown that the concept of a modified van der Pol relaxation oscillator can be successfully used to model the important physiological behaviour of the SA and AV nodes. Their model expresses the oscillatory motion of these nodes using the differential equation

$$\frac{d^2x}{dt^2} = -\alpha(x-v_1)(x-v_2)\frac{dx}{dt} - \frac{x(x+d_1)(x+d_2)}{d_1d_2}, \tag{7.60}$$

where α defines the shape of the heart stimulus pulse, v_1 and v_2 determine the signal amplitude, and $(0,0)$, $(-d_1,0)$ and $(d_2,0)$ are the equilibrium points in the $(x,\dot{x})$ phase space. The parameters are restricted as $\alpha, d_1, d_2 > 0$ and $v_1v_2 < 0$. As discussed in section 6.8, this second-order differential equation can be expressed as two coupled first-order differential equations and numerically solved by employing one of the algorithms described in Appendix F.

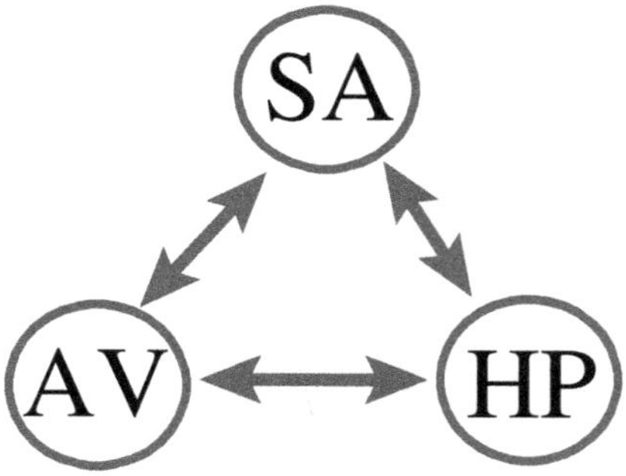

Figure 7.30 Conceptual model for cardiac dynamics, using the SA (sinoatrial) node, AV (atrioventricular) node and the HP node (representing the complex made of the bundle of His and the Purkinje fibres) as three coupled non-linear oscillators with bi-directional couplings. For normal rhythm only uni-directional coupling is considered, such that the electrical impulse is conducted from the SA node to the AV node and then from the AV node to the HP node.

Based on the Grudzinski-Zebrowski modification of the van der Pol's relaxation oscillator description, Gois and Savi (2009) presented a mathematical treatment for generating synthetic ECG signals by treating SA, AV and HP as three coupled non-linear oscillators with delayed symmetrical and uni-directional connections. Cheffer *et al.* (2021) presented an extension of the Gois-Savi model by considering independent coupling parameters between these oscillators for asymmetrical and bi-directional connections. They considered F_{SA}, F_{AV} and F_{HP} as external stimuli for the SA, AV and HP oscillators, respectively. The time-delayed terms in the model represent the transmitting time spent among each of the oscillators. The basic conceptual model of the cardiac dynamics using these oscillators is schematically illustrated in figure 7.30. The system dynamics is governed by the following coupled first-order differential equations for each of the three oscillators

$$\begin{aligned}
\text{SA}: \quad \frac{dx_{SA}}{dt} &= y_{SA}, \\
\frac{dy_{SA}}{dt} &= F_{SA}(t) - \alpha_{SA}(x_{SA} - v_{SA_1})(x_{SA} - v_{SA_2})y_{SA} \\
&\quad - \frac{x_{SA}(x_{SA} + d_{SA_1})(x_{SA} + d_{SA_2})}{d_{SA_1}d_{SA_2}} \\
&\quad - k_{(AV-SA)}x_{SA} + k^{\tau}_{(AV-SA)}x_{AV}^{\tau_{(AV-SA)}} \\
&\quad - k_{(HP-SA)}x_{SA} + k^{\tau}_{(HP-SA)}x_{HP}^{\tau_{(HP-SA)}},
\end{aligned} \tag{7.61}$$

$$\text{AV}: \quad \frac{dx_{AV}}{dt} = y_{AV},$$

$$\begin{aligned}
\frac{\mathrm{d}y_{\mathrm{AV}}}{\mathrm{d}t} &= F_{\mathrm{AV}}(t) - \alpha_{\mathrm{AV}}(x_{\mathrm{AV}} - v_{\mathrm{AV}_1})(x_{\mathrm{AV}} - v_{\mathrm{AV}_2})y_{\mathrm{AV}} \\
&\quad - \frac{x_{\mathrm{AV}}(x_{\mathrm{AV}} + d_{\mathrm{AV}_1})(x_{\mathrm{AV}} + d_{\mathrm{AV}_2})}{d_{\mathrm{AV}_1} d_{\mathrm{AV}_2}} \\
&\quad - k_{(\mathrm{SA-AV})} x_{\mathrm{AV}} + k^{\tau}_{(\mathrm{SA-AV})} x_{\mathrm{SA}}^{\tau(\mathrm{SA-AV})} \\
&\quad - k_{(\mathrm{HP-AV})} x_{\mathrm{AV}} + k^{\tau}_{(\mathrm{HP-AV})} x_{\mathrm{HP}}^{\tau(\mathrm{HP-AV})},
\end{aligned} \tag{7.62}$$

$$\begin{aligned}
\mathrm{HP}: \quad \frac{\mathrm{d}x_{\mathrm{HP}}}{\mathrm{d}t} &= y_{\mathrm{HP}}, \\
\frac{\mathrm{d}y_{\mathrm{HP}}}{\mathrm{d}t} &= F_{\mathrm{HP}}(t) - \alpha_{\mathrm{HP}}(x_{\mathrm{HP}} - v_{\mathrm{HP}_1})(x_{\mathrm{HP}} - v_{\mathrm{HP}_2})y_{\mathrm{HP}} \\
&\quad - \frac{x_{\mathrm{HP}}(x_{\mathrm{HP}} + d_{\mathrm{HP}_1})(x_{\mathrm{HP}} + d_{\mathrm{HP}_2})}{d_{\mathrm{HP}_1} d_{\mathrm{HP}_2}} \\
&\quad - k_{(\mathrm{SA-HP})} x_{\mathrm{HP}} + k^{\tau}_{(\mathrm{SA-HP})} x_{\mathrm{SA}}^{\tau(\mathrm{SA-HP})} \\
&\quad - k_{(\mathrm{AV-HP})} x_{\mathrm{HP}} + k^{\tau}_{(\mathrm{AV-HP})} x_{\mathrm{AV}}^{\tau(\mathrm{AV-HP})}.
\end{aligned} \tag{7.63}$$

Here k_{m-n} and k^{τ}_{m-n} are coupling coefficients between nodes m and n, and $x_i^{\tau_{m-n}} = x_i(t - \tau_{m-n})$ are state variables with a time delay τ_{m-n}. The spatiotemporal external excitations are expressed as $F_m(t) = \rho_m \sin(\omega_m t)$.

The ECG signal can be represented by forming a linear combination of the state variables

$$ECG \equiv X = X_{\mathrm{SA}} + X_{\mathrm{AV}} + X_{\mathrm{HP}}, \tag{7.64}$$

with

$$\begin{aligned}
X_{\mathrm{SA}} &= \frac{\beta_0}{3} + \beta_1 x_{\mathrm{SA}}, \\
X_{\mathrm{AV}} &= \frac{\beta_0}{3} + \beta_2 x_{\mathrm{AV}}, \\
X_{\mathrm{HP}} &= \frac{\beta_0}{3} + \beta_3 x_{\mathrm{HP}},
\end{aligned} \tag{7.65}$$

where β_0, β_1, β_2 and β_3 are adjustable parameters. The parameters β_1, β_2 and β_3 represent the relative weight factors associated with the contributions from the SA, AV and HP nodes, respectively. The temporal rate of change of the ECG signal, therefore, is

$$\frac{\mathrm{d}(ECG)}{\mathrm{d}t} \equiv \dot{X} = Y = \beta_1 y_{\mathrm{SA}} + \beta_2 y_{\mathrm{AV}} + \beta_3 x_{\mathrm{HP}}. \tag{7.66}$$

In the above discussion, all the variables are dimensionless. Cheffer *et al.* defined a dimensional time variable $\bar{t} = \beta_t t$, where β_t is estimated by the ratio between experimental and numerical R–R intervals.

We will present and analyse numerically obtained results for the normal rhythm, atrial flutter rhythm and ventricular flutter rhythm, using the Euler-Cromer algorithm[3] with a constant time step of 0.005 s. The parameters used for each rhythm type, presented in the respective figure captions, are taken from the work of Cheffer *et al.* The initial conditions for starting the numerical iteration were set, following the work of Cheffer *et al.* , as $x_{\mathrm{SA}}(0) = -0.1, x_{\mathrm{AV}}(0) = -0.6, x_{\mathrm{HP}}(0) = -3.3, y_{\mathrm{SA}}(0) = 0.025, y_{\mathrm{AV}}(0) = 0.1, x_{\mathrm{HP}}(0) = 10.0/25.0$.

---

[3]Cheffer *et al.* used a fourth-order Runge-Kutta algorithm to solve equations (7.61)–(7.63).

Normal rhythm

Normal cardiac rhythm is maintained by uni-directional couplings between the three oscillators (electrical impulse conducted from SA to AV, followed by that from AV to HP) and does not require any external stimulus (*i.e.* $F_{SA}(t) = F_{AV}(t) = F_{HP}(t) = 0$). The results presented in figure 7.31 show that the ECG traces from all three oscillators are periodic. The results also show that the amplitudes of the traces from the SA and AV oscillators are smaller than a quarter of that from the HP oscillator. Figure 7.32 confirms that each period in the total ECG trace is characterised with the main features, *viz* the P, QRS and T waves, of the experimental ECG trace from a normal heart.

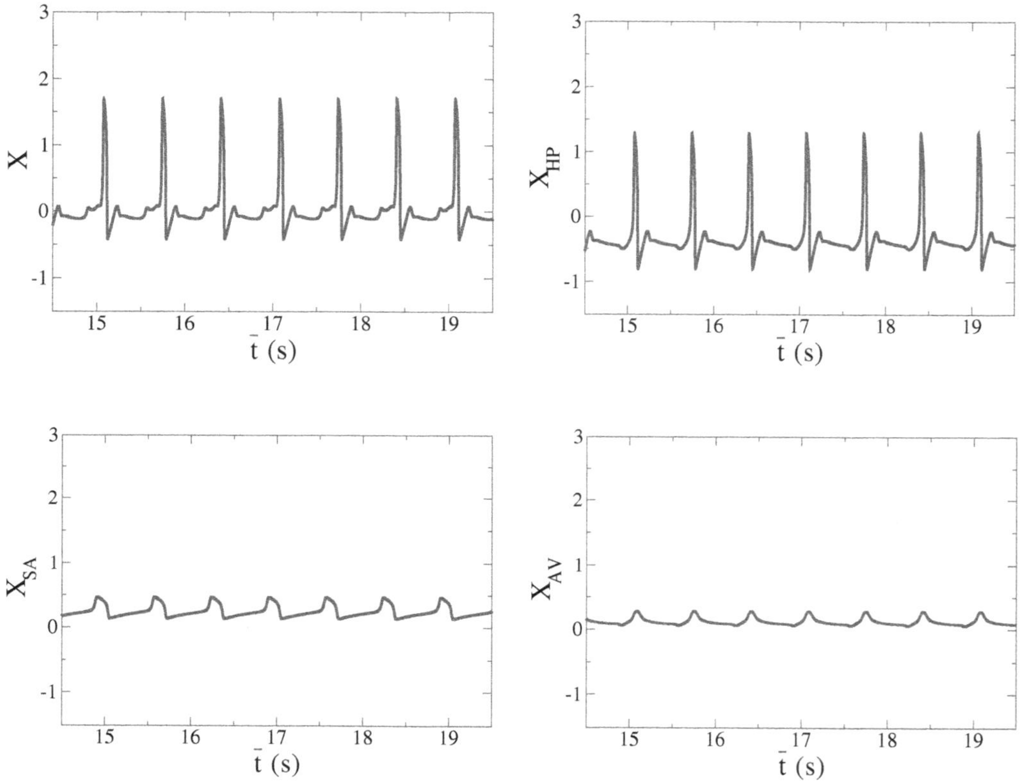

Figure 7.31 Computed temporal variation of ECG features with normal rhythm. Equations (7.61)–(7.63) were numerically solved using the Euler-Cromer algorithm for 50,000 iterations with a time step of 0.005 s. The variable X represents the total ECG feature, with the components X_{SA}, X_{AV} and X_{HP} contributed by the SA, AV and HP nodes, respectively. The variable $\bar{t}$ has been defined as $\bar{t} = \beta_t t - 0.5$, with t as time and $\beta_t = 0.1048$. No external stimulus was imposed, *i.e.* $F_{SA} = F_{AV} = F_{HP} = 0$. The parameters used are: $\beta_0 = 1.0$, $\beta_1 = 0.06$, $\beta_2 = 0.1$, $\beta_3 = 0.3$, $\alpha_{SA} = 3.0$, $v_{SA_1} = 1.0$, $v_{SA_2} = -1.9$, $d_{SA_1} = 1.9$, $d_{SA_2} = 0.55$, $\alpha_{AV} = 3.0$, $v_{AV_1} = 0.5$, $v_{AV_2} = -0.5$, $d_{AV_1} = 4.0$, $d_{AV_2} = 0.67$, $\alpha_{HP} = 7.0$, $v_{HP_1} = 1.65$, $v_{HP_2} = -2.0$, $d_{HP_1} = 7.0$, $d_{HP_2} = 0.67$, $k_{SA-AV} = k^{\tau}_{SA-AV} = 3.0$, $\tau_{SA-AV} = 0.8$, $k_{AV-HP} = k^{\tau}_{AV-HP} = 55.0$, $\tau_{AV-HP} = 0.1$.

Figure 7.33 presents the phase space orbits. Each oscillator generates a periodic orbit. Notice that the total ECG phase orbit is a knotted periodic loop. This is because while the SA and AV nodes execute normal loop periods, the HP node makes a knotted periodic loop. The computed results show that the HP node makes almost the entire contribution to the area in the phase space.

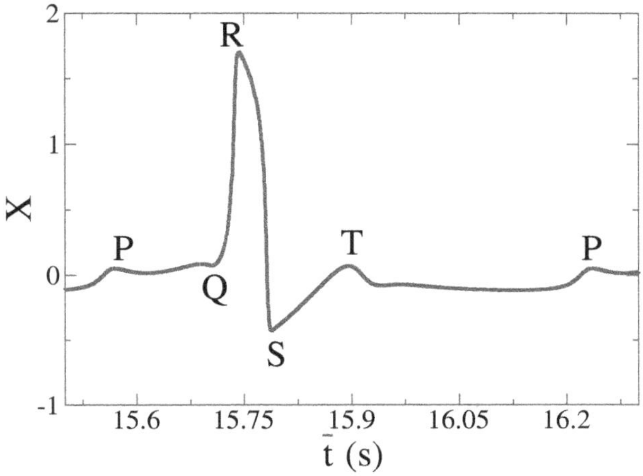

Figure 7.32 Computational confirmation of the main features (*viz* the P, QRS and T waves) in the computed temporal variation of ECG with normal rhythm.

Atrial flutter rhythm

For normal heart rhythm, the atrial and ventricular rates are usually the same. Atrial flutter is an irregular heart rhythm, characterised with the atrial rate being much faster than the ventricular rate. The notation atrial flutter n:1 is normally used to imply that for every n atrial flutter wave, there is one QRS complex. An ECG trace of an atrial flutter shows a sawtooth pattern for the P wave and an irregular R–R interval.

Figure 7.34 shows the numerical results for an ECG trace of atrial flutter. No external stimulus was applied (*i.e.* $F_{SA}(t) = F_{AV}(t) = F_{HP}(t) = 0$). Compared to the normal rhythm, the damping factor α_{AV} for the AV oscillator has been increased by a factor of 2.5 and the SA–AV and AV–HP coupling constants have been significantly reduced. The SA node shows a sawtooth pattern, but the AV and HP nodes produce regular features. The resultant ECG trace is characterised with a sawtooth pattern for the P wave but unclearly identifiable Q, S and T features. The atrial flutter P waves are called flutter waves (or f waves). Figure 7.35 shows the computational results for the phase space orbits. The knot in the periodic loop for the total ECG phase orbit with the atrial flutter rhythm is more structured than that for the normal case.

Ventricular flutter rhythm

Another type of irregular heart rhythm is the ventricular flutter, caused by high frequency ventricular contraction. Cheffer *et al.* modelled this by changing the parameters for the HP oscillator. In particular, they used much reduced values for the AV–HP coupling constants.

The numerical results are presented in figures 7.36 and 7.37. The QRS structure of the ECG trace is maintained, but there is a band of closely lying phase space orbits. This is in sharp contrast to the simple knotted orbital loop structure for the normal and atrial flutter rhythms. In other words, the phase space plot confirms an irregular characteristic of the ventricular flutter rhythm.

Further considerations and key message

The computational studies, discussed above, of the ECG features for the normal, atrial flutter and ventricular flutter rhythms have been extended to the cases of atrial fibrillation and ventricular fibrillation in the publications by Gois and Savi (2009) and Cheffer *et al.* (2021). Further considerations and analysis of results have been presented in the publication by Kuate and Fotsin (2022), where experimental investigations have also been made to confirm theoretical predictions.

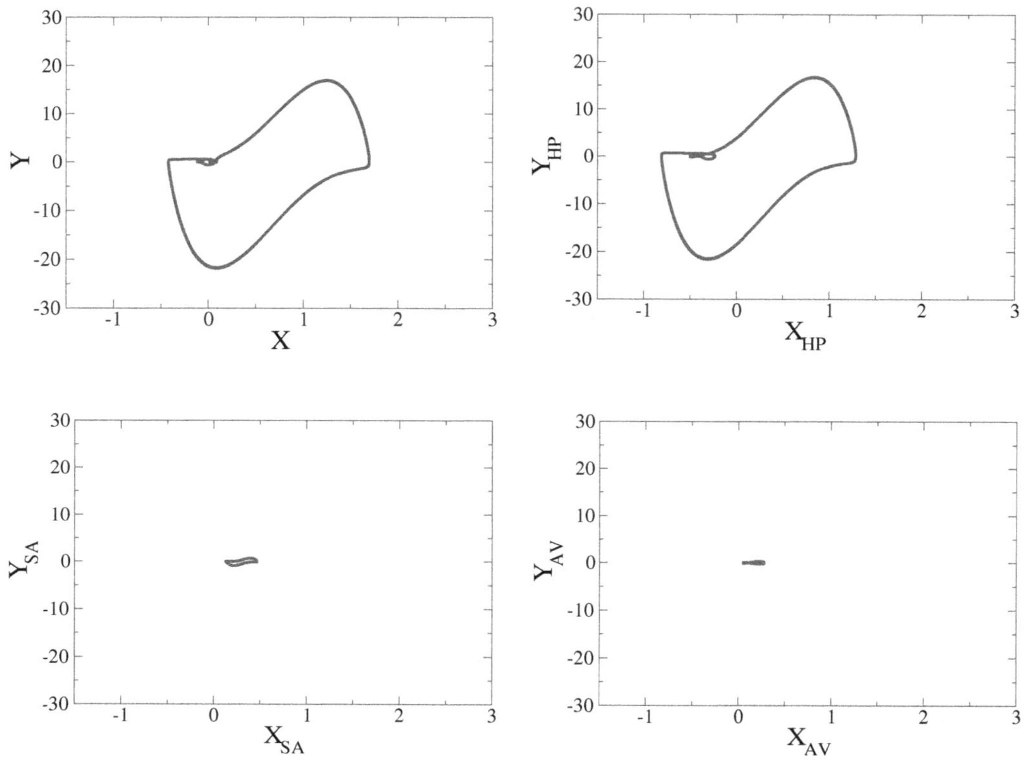

Figure 7.33 Computed phase-space orbits for ECG with normal rhythm. The variable X represents the total ECG feature, with the components X_{SA}, X_{AV} and X_{HP} contributed by the SA, AV and HP nodes, respectively. The variable Y is the time rate of change of X, *etc.* Equations (7.61)–(7.63) were numerically solved using the Euler-Cromer algorithm with a constant time step of 0.005 s. The variable $\bar{t}$ has been defined as $\bar{t} = \beta_t t - 0.5$, with t as time and $\beta_t = 0.1048$. No external stimulus was imposed, *i.e.* $F_{SA} = F_{AV} = F_{HP} = 0$. The parameters used are: $\beta_0 = 1.0$, $\beta_1 = 0.06$, $\beta_2 = 0.1$, $\beta_3 = 0.3$, $\alpha_{SA} = 3.0$, $v_{SA_1} = 1.0$, $v_{SA_2} = -1.9$, $d_{SA_1} = 1.9$, $d_{SA_2} = 0.55$, $\alpha_{AV} = 3.0$, $v_{AV_1} = 0.5$, $v_{AV_2} = -0.5$, $d_{AV_1} = 4.0$, $d_{AV_2} = 0.67$, $\alpha_{HP} = 7.0$, $v_{HP_1} = 1.65$, $v_{HP_2} = -2.0$, $d_{HP_1} = 7.0$, $d_{HP_2} = 0.67$, $k_{SA-AV} = k^{\tau}_{SA-AV} = 3.0$, $\tau_{SA-AV} = 0.8$, $k_{AV-HP} = k^{\tau}_{AV-HP} = 55.0$, $\tau_{AV-HP} = 0.1$.

It is important to mention that in the above described mathematical and numerical studies no serious attempt has really been made to reproduce any real ECG trace. The key takeaway message from these numerical studies is that by suitably adjusting parameters for the van der Pol type SA, AV and HP non-linearly coupled oscillators it is possible to explain experimentally observed ECG traces and identify the relative importance of each of the three oscillators.

7.6 NON-LINEAR NEURONAL DYNAMICS

7.6.1 INTRODUCTION

Our next example of the application of the theory of non-linear dynamics will be on the topic of computational neuroscience. Traditionally classified as a subdivision of biology, neuroscience has developed into a multidisciplinary subject, combining concepts and methods from biophysics, chemistry, engineering, computer science, physiology, psychology, cognitive science and medical

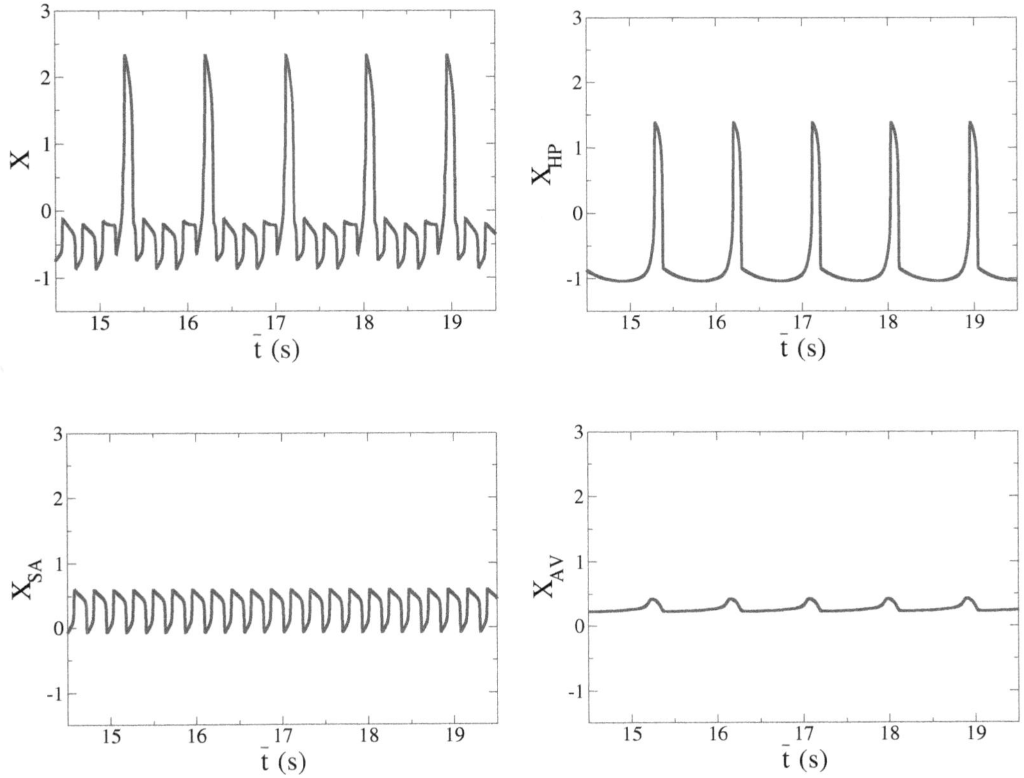

Figure 7.34 Computed temporal variation of ECG features with atrial flutter rhythm. Equations (7.61)–(7.63) were numerically solved using the Euler-Cromer algorithm for 50,000 iterations with a time step of 0.005 s. The variable X represents the total ECG feature, with the components X_{SA}, X_{AV} and X_{HP} contributed by the SA, AV and HP nodes, respectively. The variable $\bar{t}$ has been defined as $\bar{t} = \beta_t t - 0.5$, with t as time and $\beta_t = 0.0809$. No external stimulus was imposed, *i.e.* $F_{SA} = F_{AV} = F_{HP} = 0$. The parameters used are: $\beta_0 = 1.0$, $\beta_1 = 0.06$, $\beta_2 = 0.1$, $\beta_3 = 0.3$, $\alpha_{SA} = 3.0$, $v_{SA_1} = 1.65$, $v_{SA_2} = -4.2$, $d_{SA_1} = 1.9$, $d_{SA_2} = 0.55$, $\alpha_{AV} = 7.0$, $v_{AV_1} = 0.5$, $v_{AV_2} = -0.5$, $d_{AV_1} = 4.0$, $d_{AV_2} = 0.67$, $\alpha_{HP} = 7.0$, $v_{HP_1} = 1.65$, $v_{HP_2} = -2.0$, $d_{HP_1} = 7.0$, $d_{HP_2} = 0.67$, $k_{SA-AV} = 0.66$, $k^{\tau}_{SA-AV} = 0.02$, $\tau_{SA-AV} = 0.66$, $k_{AV-HP} = 14.0$, $k^{\tau}_{AV-HP} = 60.0$, $\tau_{AV-HP} = 0.1$.

science. The most fundamental entity in neuroscience is a nerve cell, called **neuron**. A neuron body, of diameter in the μm range and length in the mm$-$m range, is covered by a membrane. The neural membrane is an impenetrable double layer of lipids, which is characterised by an electric potential difference between its inner and outer layers. The inner layer has a negative potential against the outer layer. This potential difference is called the membrane potential. The membrane potential can vary with time, but typically lies in the range -70 mV to -90 mV at rest. Human brain contains many billions of neurons, which are intricately linked together and carry information in the form of electrical signal. Electrical signal in the nervous system is generated when neurons are **spiked** by the development of **action potential**, *i.e.* an electric pulse, inside the neuron membrane. We will briefly discuss the anatomy of neurons before presenting theoretical models and computational results for the non-linear dynamics of spiking neurons.

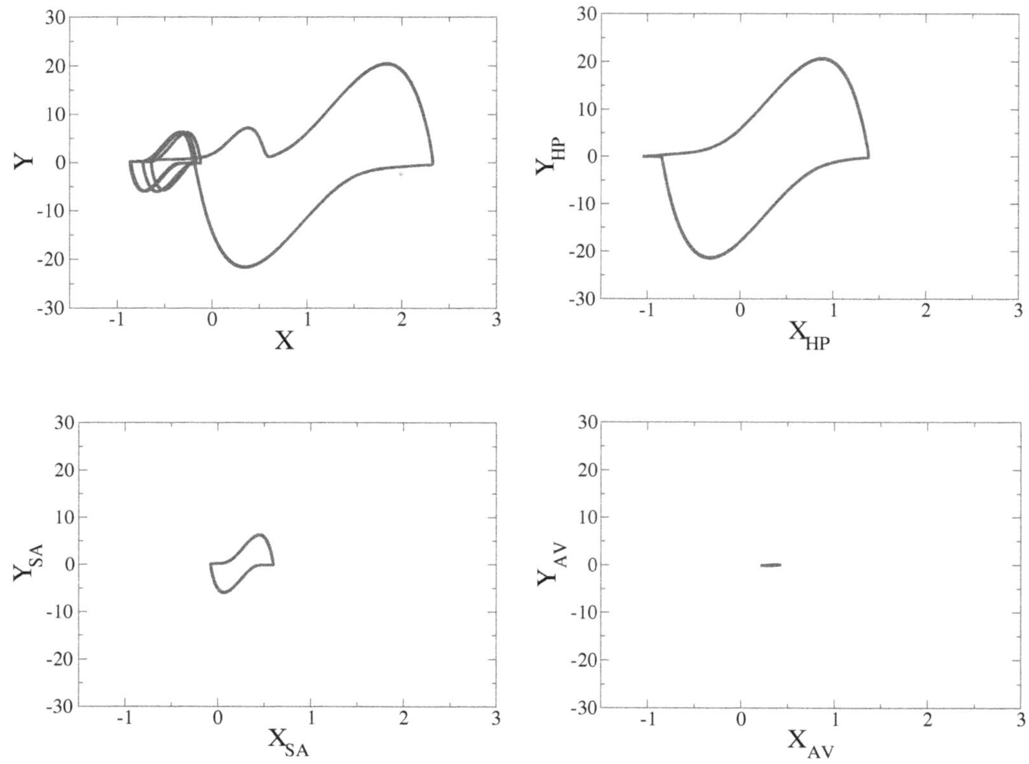

Figure 7.35 Computed temporal variation of ECG features with atrial flutter rhythm. Equations (7.61)–(7.63) were numerically solved using the Euler-Cromer algorithm for 50,000 iterations with a time step of 0.005 s. The variable X represents the total ECG feature, with the components X_{SA}, X_{AV} and X_{HP} contributed by the SA, AV and HP nodes, respectively. The variable $\bar{t}$ has been defined as $\bar{t} = \beta_t t - 0.5$, with t as time and $\beta_t = 0.0809$. No external stimulus was imposed, *i.e.* $F_{\text{SA}} = F_{\text{AV}} = F_{\text{HP}} = 0$. The parameters used are: $\beta_0 = 1.0$, $\beta_1 = 0.06$, $\beta_2 = 0.1$, $\beta_3 = 0.3$, $\alpha_{\text{SA}} = 3.0$, $v_{\text{SA}_1} = 1.65$, $v_{\text{SA}_2} = -4.2$, $d_{\text{SA}_1} = 1.9$, $d_{\text{SA}_2} = 0.55$, $\alpha_{\text{AV}} = 7.0$, $v_{\text{AV}_1} = 0.5$, $v_{\text{AV}_2} = -0.5$, $d_{\text{AV}_1} = 4.0$, $d_{\text{AV}_2} = 0.67$, $\alpha_{\text{HP}} = 7.0$, $v_{\text{HP}_1} = 1.65$, $v_{\text{HP}_2} = -2.0$, $d_{\text{HP}_1} = 7.0$, $d_{\text{HP}_2} = 0.67$, $k_{\text{SA-AV}} = 0.66$, $k^{\tau}_{\text{SA-AV}} = 0.02$, $\tau_{\text{SA-AV}} = 0.66$, $k_{\text{AV-HP}} = 14.0$, $k^{\tau}_{\text{AV-HP}} = 60.0$, $\tau_{\text{AV-HP}} = 0.1$.

7.6.2 ANATOMY AND PHYSIOLOGY OF NEURON

Figure 7.38 shows a confocal microscopic image of mouse neurons, obtained from Srivastava (2024). It reveals that a neuron has three main parts: **dendrites**, a **soma** (or cell body) and an **axon**. The anatomy of a neuron is analogous to the structure of a tree, in that the dendrites represent branches of the tree, the soma is the tree trunk and the axon represents the roots of the tree. Dendrites have leaf-like structures, called **spines**. The soma houses the nucleus, containing neuron's DNA, from where proteins are transported to the axon and dendrites. Dendrites receive input in the form of a short electrical signal, called spike or action potential, from other neurons. The axon provides a channel for the action potential to be sent to another neuron past a **synapse** (joint, or connection, between two neurons). A typical axon makes a few thousand synapses with other neurons. The neuron that sends (receives) the signal is called the presynaptic (postsynaptic) neuron. The distance between neighbouring presynaptic and postsynaptic neurons is about 2-4 nm. The passage of action potential between neurons, known as synaptic transmission, or neurotransmission, is the

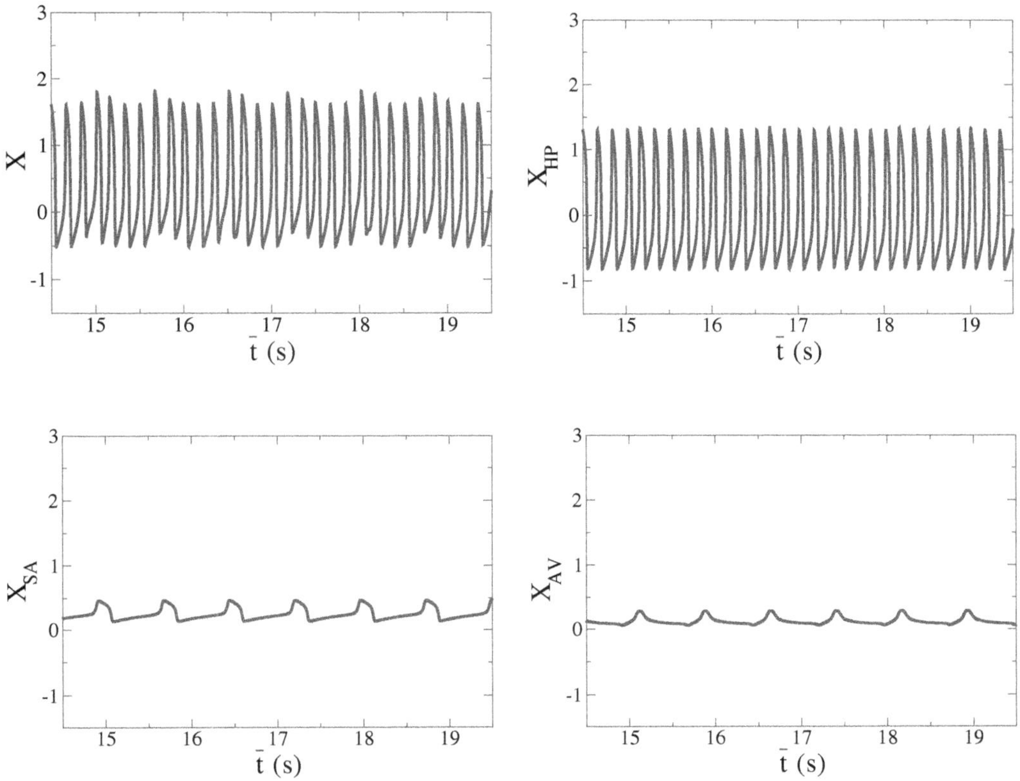

Figure 7.36 Computed temporal variation of ECG features with ventricular flutter rhythm. The variable X represents the total ECG feature, with the components X_{SA}, X_{AV} and X_{HP} contributed by the SA, AV and HP nodes, respectively. The variable $\bar{t}$ has been defined as $\bar{t} = \beta_t t - 0.44$, with t as time and $\beta_t = 0.12$. No external stimulus was imposed, *i.e.* $F_{SA} = F_{AV} = F_{HP} = 0$. The parameters used are: $\beta_0 = 1.0$, $\beta_1 = 0.06$, $\beta_2 = 0.1$, $\beta_3 = 0.3$, $\alpha_{SA} = 3.0$, $v_{SA_1} = 1.0$, $v_{SA_2} = -1.9$, $d_{SA_1} = 1.9$, $d_{SA_2} = 0.55$, $\alpha_{AV} = 3.0$, $v_{AV_1} = 0.5$, $v_{AV_2} = -0.5$, $d_{AV_1} = 4.0$, $d_{AV_2} = 0.67$, $\alpha_{HP} = 7.0$, $v_{HP_1} = 1.65$, $v_{HP_2} = -2.0$, $d_{HP_1} = 7.0$, $d_{HP_2} = 0.67$, $k_{SA-AV} = 3.0$, $k^{\tau}_{SA-AV} = 3.0$, $\tau_{SA-AV} = 0.8$, $k_{AV-HP} = 45.0$, $k^{\tau}_{AV-HP} = 30.0$, $\tau_{AV-HP} = 0.1$.

fundamental process underlying brain function. Figure 7.39 presents a simple schematic drawing of a neuron, labelling the three important elements of its anatomy, *viz* dendrites, soma and axon.

7.6.3 SPIKING NEURON MODELS

Many different dynamical models of neurons have been proposed over the past one hundred years. Izhikevich (2004, 2007) has elucidated the biological plausibility and computational efficiency of some of the most useful dynamical models of neurons. Basically, dynamical models for spiking neurons can be grouped in two categories: phenomenological models and conduction-based models.

A. Phenomenological models

We will discuss three phenomenological models of spiking neurons.

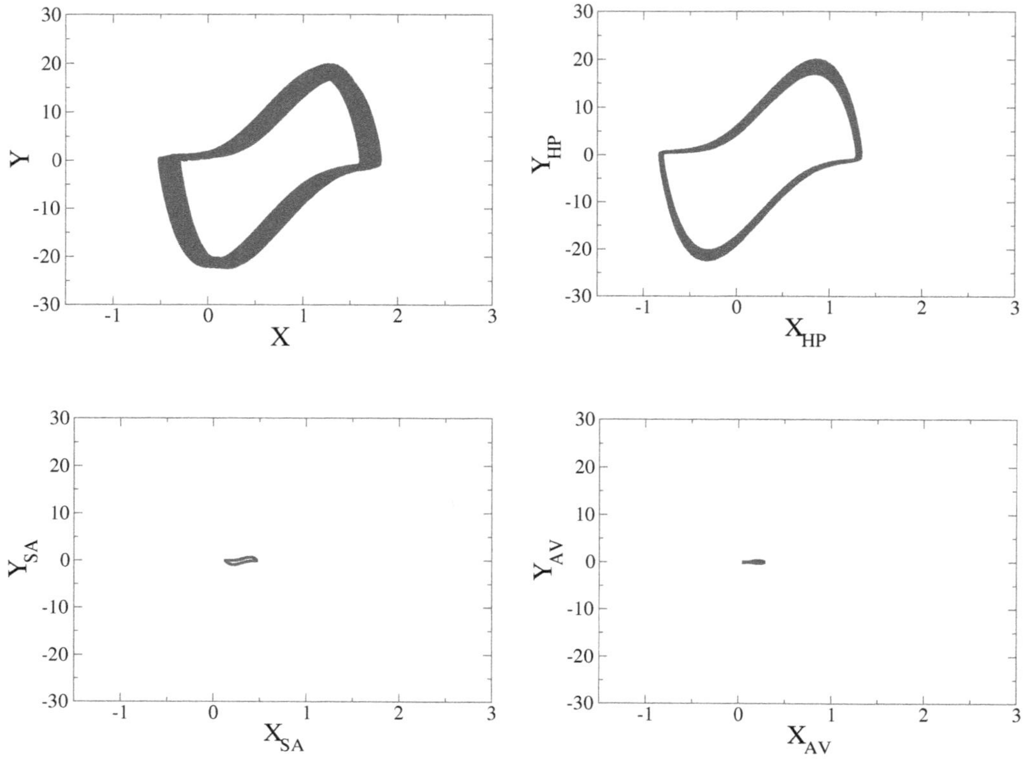

Figure 7.37 Computed phase-space orbits for ECG with ventricular flutter rhythm. The variable X represents the total ECG feature, with the components X_{SA}, X_{AV} and X_{HP} contributed by the SA, AV and HP nodes, respectively. The variable Y is the time rate of change of X. The variable $\bar{t}$ has been defined as $\bar{t} = \beta_t t - 0.44$, with t as time and $\beta_t = 0.12$. No external stimulus was imposed, *i.e.* $F_{SA} = F_{AV} = F_{HP} = 0$. The parameters used are: $\beta_0 = 1.0$, $\beta_1 = 0.06$, $\beta_2 = 0.1$, $\beta_3 = 0.3$, $\alpha_{SA} = 3.0$, $v_{SA_1} = 1.0$, $v_{SA_2} = -1.9$, $d_{SA_1} = 1.9$, $d_{SA_2} = 0.55$, $\alpha_{AV} = 3.0$, $v_{AV_1} = 0.5$, $v_{AV_2} = -0.5$, $d_{AV_1} = 4.0$, $d_{AV_2} = 0.67$, $\alpha_{HP} = 7.0$, $v_{HP_1} = 1.65$, $v_{HP_2} = -2.0$, $d_{HP_1} = 7.0$, $d_{HP_2} = 0.67$, $k_{SA-AV} = 3.0$, $k^{\tau}_{SA-AV} = 3.0$, $\tau_{SA-AV} = 0.8$, $k_{AV-HP} = 45.0$, $k^{\tau}_{AV-HP} = 30.0$, $\tau_{AV-HP} = 0.1$.

7.6.3.1 Linear, or leaky, integrate-and-fire (LIF) model

The dynamical behaviour of a neuron can be represented by an RC circuit with a threshold, consisting of a parallel capacitor and resistor, as illustrated in figure 7.40. Here C is membrane capacitance, R is the membrane resistance, V is the membrane potential, V_{rest} is the resting membrane potential (also known as the Nernst equilibrium potential), V_{th} is a threshold potential and I_{inj} is a current injected to the membrane. The integrate-and-fire (IF) model assumes that when the voltage across the neuron membrane reaches the threshold V_{th}, the neuron 'fires' a spike (an action potential) and the voltage is reset back to the resting potential V_{rest}. The 'integrate' word is used because the membrane voltage is summed (integrated) over input currents.

Using Kirchhoff's law, the rate of change of the capacitative current across the membrane can be expressed by the linear (one-dimensional) ordinary differential equation

$$C\frac{\mathrm{d}V}{\mathrm{d}t} = -\frac{1}{R}(V - V_{rest}) + I_{inj} = -\text{leak current} + \text{injected current}, \tag{7.67}$$

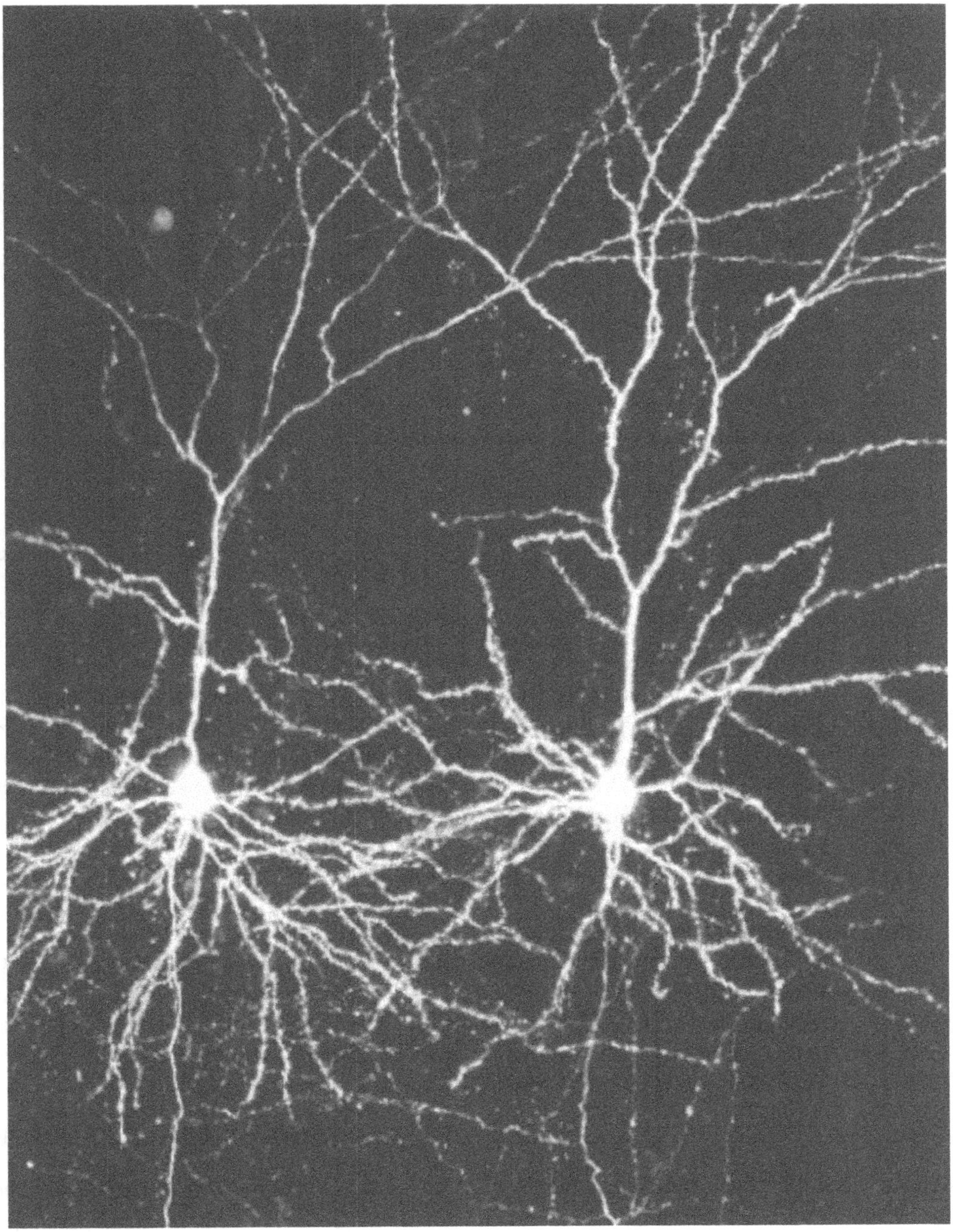

Figure 7.38 Confocal microscopic image of mouse neurons. It reveals three main parts of neuron anatomy: dendrites, soma and axon. Courtesy of Professor Deepak Srivastava (King's College, London).

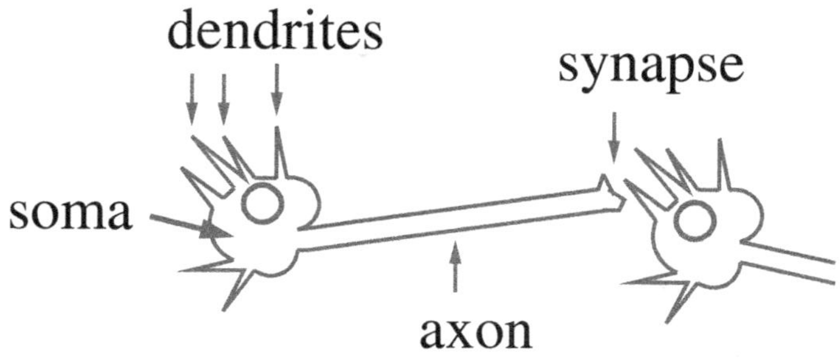

Figure 7.39 A simple schematic sketch of a neuron, labelling the three main parts of its anatomy (dendtries, soma and axon) and the synapse (joint between two neurons).

where $\frac{dV}{dt}$ is the time derivative of the membrane potential. The source of the injected current I_{inj} could be an electrode or other synapses.

Let us express equation (7.67) in the form

$$\frac{dV}{dt} = -\frac{1}{\tau}(V - V_{\text{rest}}) + \frac{1}{C}I_{\text{inj}}, \tag{7.68}$$

or

$$\frac{dV}{dt} = f(V) + \frac{1}{C}I_{\text{inj}}, \tag{7.69}$$

where $\tau = RC$ is the time constant of the RC circuit. As $f(V) \propto V$, the leaky model in equations (7.67) and (7.68) can also be called the linear model. The leaky (linear) integrate-and-fire (LIF) model was proposed by Lapicque in 1907 (Lapicque, 1907) and applied widely during the 1960s and 1970s (see, *e.g.* Stein, 1967; Knight , 1972). We will rewrite equation (7.67) with explicit mention of the spike generation and reset mechanism (we have considered $V_{\text{reset}} = V_{\text{rest}}$)

$$C\frac{dV}{dt} = -\frac{1}{R}(V - V_{\text{rest}}) + I_{\text{inj}}, \quad \text{if } V \geq V_{\text{th}}, \quad V \to V_{\text{rest}}. \tag{7.70}$$

Note that the reset potential V_{reset} need not be the same as the resting potential V_{rest}.

We will employ the Euler algorithm to present and analyse the time evolution of the potential V under different conditions of the injected current I_{inj}. As remarked in previous sections, the number of iterations and the time step should be chosen sensibly. It is advisable to test the computational results against the analytical solution of equation (7.67) for a constant current input $I_{\text{inj}}(t) = I_c$ and without imposition of voltage spike and reset, which is

$$V(t) = V_{\text{rest}} + I_c R + \left(V(t=0) - V_{\text{rest}} - I_c R\right) e^{-t/\tau}. \tag{7.71}$$

We have stated the values of system variables and computational parameters in the captions to the figures where results are presented.

Figure 7.41 shows the time variation of the membrane potential when a current step of 2 μA for $t \geq 2$ ms is applied. Panel (a) shows the current step. Panel (b) shows that, starting at the rest value of -70 mV, the voltage starts to exponentially rise after $t = 2$ ms and saturates at -50 mV in the absence of any resetting. This means that the neuron reacts to the spike at the current input. Panel (c) shows that resetting to -70 mV at a threshold of -55 mV leads to a sawtooth like time variation of the potential. Each time the potential reaches the threshold value, the neuron fires and the voltage hyperpolarises back to its resting value.

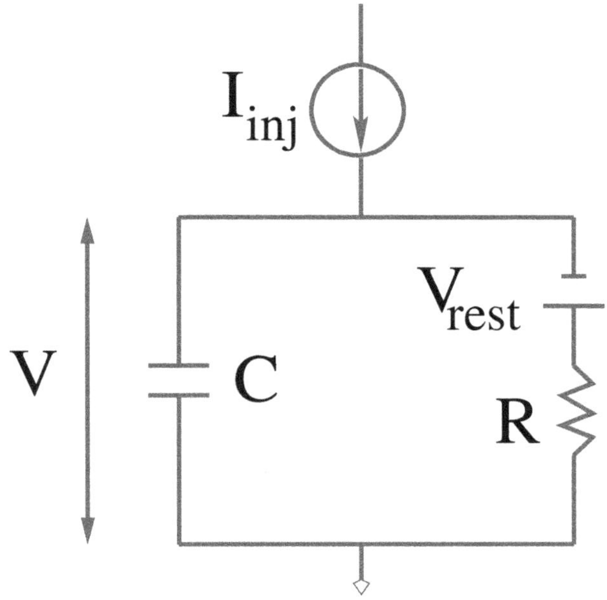

Figure 7.40 An electrical circuit representation of a dynamical neural membrane. Here C is the membrane capacitance, R is the membrane resistance, V is the membrane potential, V_{rest} is the resting membrane potential (also known as the Nernst equilibrium potential), V_{th} is a threshold potential and I_{inj} is a current injected to the membrane.

Figure 7.42 shows the time variation of the membrane potential when a current pulse of height 2 μA for a duration of 5 ms is applied. Panel (a) shows the current pulse. Panel (b) shows that, staring at the rest value of -70 mV at $t = 10$ ms, the voltage rises to -62 mV at $t \approx 14.9$ ms and then decays very close to its resting value at around $t \approx 60$ ms. Panel (c) shows that upon the imposition of the threshold at -65 mV, the potential acquires a double-peak feature. A very sharp feature develops between 10 ms and 13 ms, and the neuron fires a potential of slightly shorter height which decays to its resting point at around $t \approx 60$ ms.

Now consider two pulses of current, each of height 2 μA and time duration 5 ms starting at $t = 10$ ms and $t = 60$ ms, as shown in figure 7.43. As expected, the potential development due to the second pulse starts when the potential profile due to the first pulse reaches the rest value. A close inspection of panel (b) shows that while the potential due to the first pulse peaks at -62 mV, the potential due to the second pulse peaks at a slightly higher value. As seen in panel (c), the effect of resetting the potential at the threshold is to generate a double peak structure around each of the two current peaks.

Complicated voltage propagation can be expected when the injected current takes the form of a number of pulses with random heights and widths. Denoting a pulse as (A,B), where A and B are its height and width (or time duration), figure 7.44 shows computational results for six pulses (5 μA, 15 ms), (10 μA, 6 ms), (10 μA, 2 ms), (5 μA, 4 ms), (3 μA, 5 ms) and (7 μA, 2.5 ms). Panel (a) shows the six current pulses. Panel (b) shows six voltage peaks corresponding to the six pulses. With the imposition of the voltage rest at the threshold $V_{\text{th}} = -65$ mV, the voltage output shows a complicated pattern, as seen in panel (c).

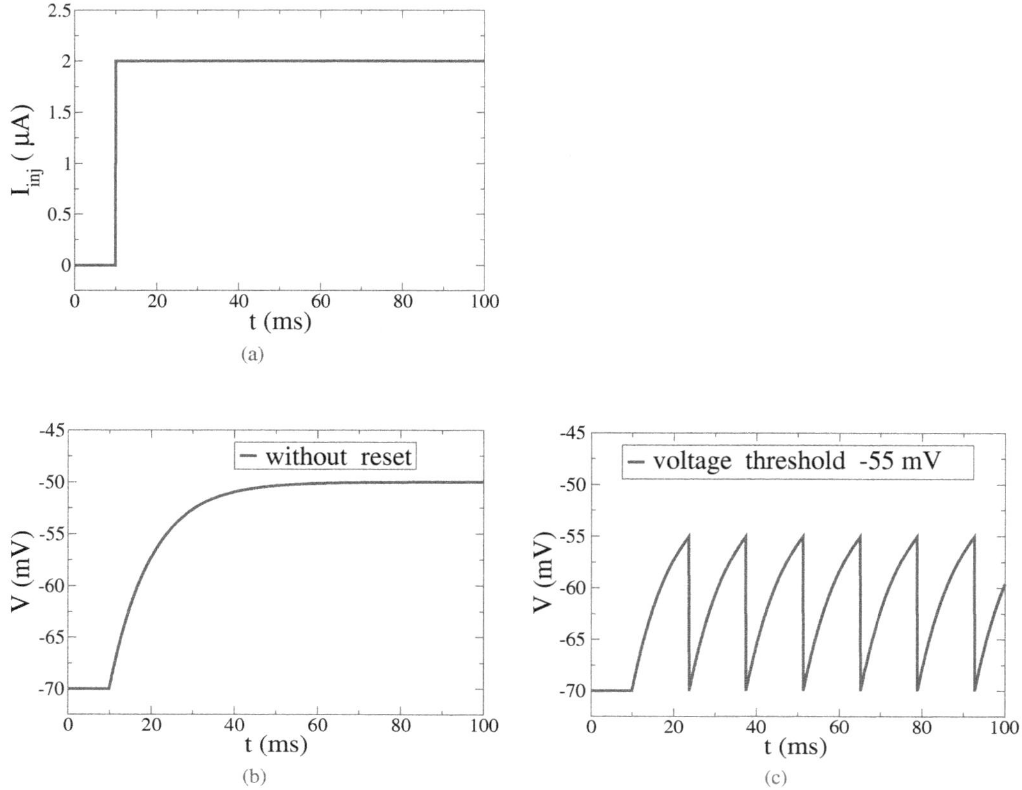

Figure 7.41 Computed temporal variation of membrane potential V in a LIF neuron subjected to a step input current $I_{\text{inj}} = 2\ \mu$A, shown in panel (a). The membrane resting potential is $V_{\text{rest}} = -70$ mV, the membrane resistance is $R = 10$ kΩ, the membrane time constant is $\tau = 10$ ms and the voltage threshold is -55 mV. The Euler algorithm was simulated for 1000 iterations with a time step of $\Delta t = 0.1$ ms. Panel (b) shows that in the absence of any reset the potential rises exponentially before saturating a constant value after the time lapse of approximately 50 s. Panel (c) shows a sawtooth type variation of the potential when a voltage threshold of -55 mV is imposed.

7.6.3.2 Quadratic integrate-and-fire (QIF) model

The **quadratic integrate-and-fire** (QIF) model considers f in equation (7.69) as a quadratic function of the membrane potential, *viz* $f(V) \propto V^2$, and the spike generation and reset mechanism are maintained as in the LIF model (see, *e.g.* Latham *et al.* 2000; Hansel and Mato, 2001). Equation (7.70) can then be replaced with (Sterratt *et al.* 2023)

$$C\frac{\mathrm{d}V}{\mathrm{d}t} = -\frac{(V - V_{\text{rest}})(V_{\text{th}} - V)}{R(V_{\text{th}} - V_{\text{rest}})} + I_{\text{inj}}, \quad \text{if } V \geq V_{\text{th}}, \quad V \to V_{\text{rest}}. \tag{7.72}$$

Note that in the above equation the quadratic function is zero both at the resting potential V_{rest} and the threshold potential V_{th}. Also, note once again that we have considered the reset potential as the rest potential: $V_{\text{reset}} = V_{\text{rest}}$

Obviously, we will expect differences in the time evolution of the membrane potential obtained from the applications of the LIF and QIF models. Results from the applications of the LIF and QIF

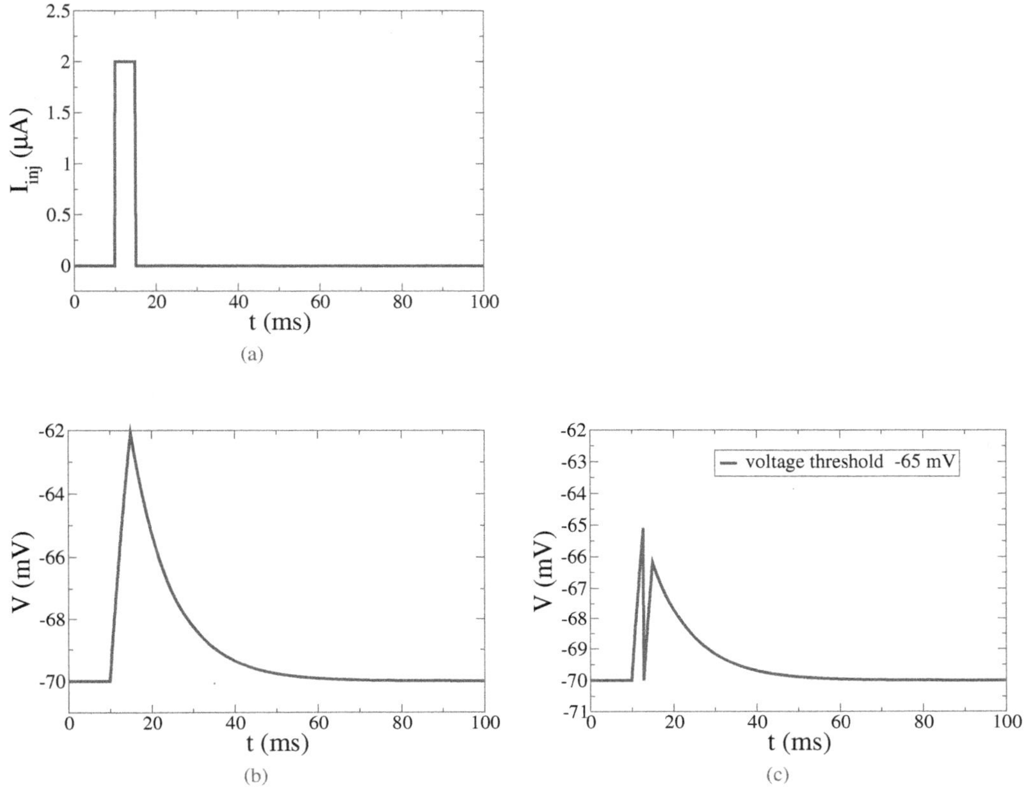

Figure 7.42 Computed temporal variation of the membrane potential V in a LIF neuron subjected to a pulse of input current $I_{\text{inj}} = 2\ \mu$A for a duration of 5 ms, as shown in panel (a). The membrane resting potential is $V_{\text{rest}} = -70$ mV, membrane resistance is $R = 10$ kΩ, the membrane time constant is $\tau = 10$ ms and the voltage threshold is $V_{\text{th}} = -65$ mV. The Euler algorithm was simulated for 1000 iterations with a time step of Δt=0.1 ms. As seen in panel (b), at the onset of the pulse, the voltage shoots up to -62 mV in about 14.9 ms and then decays to value very close to its rest value after about 60 ms. Panel (c) shows the results with the threshold voltage of -65 mV. Soon after the voltage peak, the potential acquires a second peak, of a slightly shorter height than the first peak, before deaying to its rest value.

models, using identical computational parameters, are presented in figure 7.45. As expected, the membrane potential rises (depolarises) much faster in the QIF neuron compared to the LIF neuron. This can be seen in both panel (b) (without voltage reset) and as well as panel (c) (with voltage reset). Consistent with this difference is the observation in panel (c) that during the first 95 ms there are 11 voltage peaks in the QIF neuron, compared to 7 peaks in the LIF neuron.

Equilibrium and stability:

The equilibrium of the membrane potential dynamics occur when $\mathrm{d}V/\mathrm{d}t = 0$. In other words, the equilibrium corresponds to the membrane potential for which the steady-state I_{inj}-V curve passes through zero. Writing equations (7.67) and (7.72) in the form $C\frac{\mathrm{d}V}{\mathrm{d}t} = G(V) + I_{\text{inj}}$, this happens when $G(V) + I_{\text{inj}} = 0$. The equilibrium is stable when the I_{inj}-V curve changes sign from "negative" to "positive" as V increases. The following expressions for V_{eq} can be obtained

$$\text{LIF model}: \quad V_{\text{eq}} = V_{\text{rest}} + RI_{\text{inj}}, \tag{7.73}$$

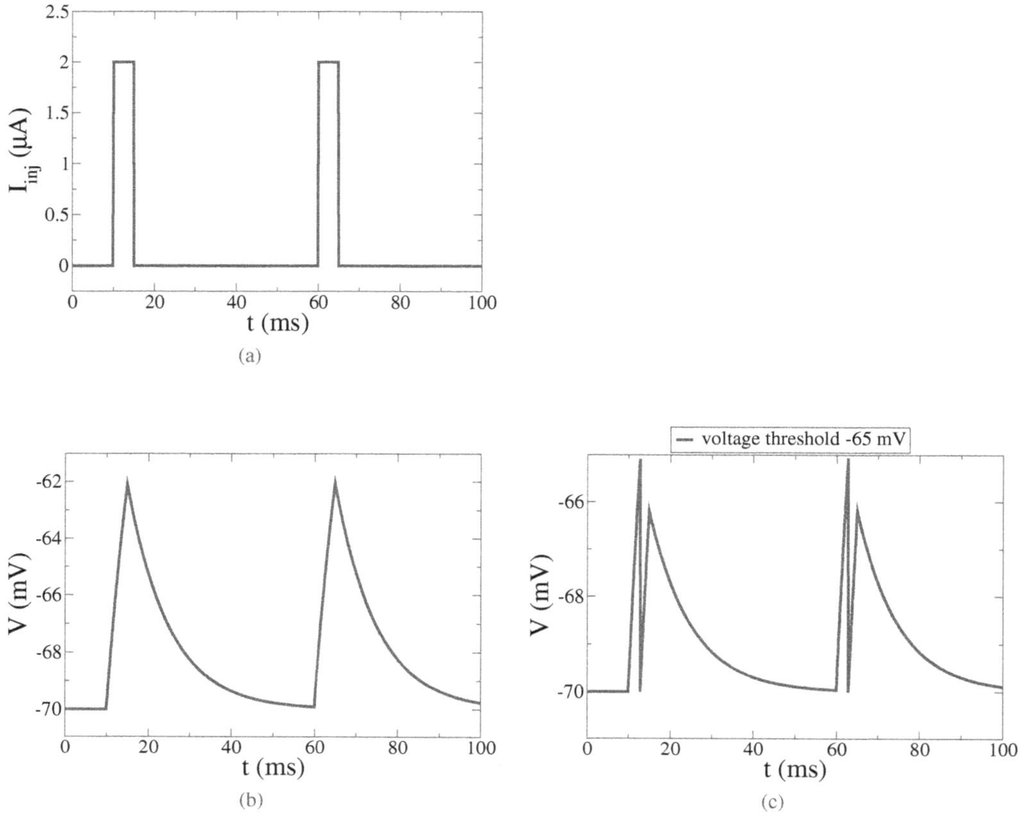

Figure 7.43 Computed temporal variation of the membrane potential V in a LIF neuron subjected to two pulses of input current $I_{\text{inj}} = 2\ \mu\text{A}$ for a duration of 5 ms, as shown in panel (a). The membrane resting potential is $V_{\text{rest}} = -70$ mV, membrane resistance is $R = 10$ kΩ, the membrane time constant is $\tau = 10$ ms and the voltage threshold is $V_{\text{th}} = -65$ mV. The Euler algorithm was simulated for 1000 iterations with a time step of Δt=0.1 ms. As the time gap between the two pulses is quite large, the potential development due to the second pulse starts when the potential profile due to the first pulse reaches the rest value. A close inspection of panel (b) shows that while the potential due to the first pulse peaks at -62 mV, the potential due to the second pulse peaks at a slightly higher value. Imposition of the voltage threshold results in the development of two almost identical double-peak structures, as seen in panel (c).

$$\text{QIF model}: \quad V_{\text{eq}} = \frac{1}{2}(V_{\text{th}} + V_{\text{rest}})\left[1 \pm \sqrt{1 - 4\frac{V_{\text{th}}V_{\text{rest}} + RI_{\text{inj}}(V_{\text{th}} - V_{\text{rest}})}{(V_{\text{th}} + V_{\text{rest}})^2}}\right]. \quad (7.74)$$

With $I_{\text{inj}} = 0$, the equilibrium in the LIF model is V_{rest} and there are two equilibria in the QIF model: V_{rest} and V_{th}. These locations change when a non-zero current I_{inj} is present.

Under the voltage-clamp condition, *i.e.* when $\frac{dV}{dt} = 0$, the injected current I_{inj} equals the net membrane current which we will label as I. Figure 7.46 shows the current-voltage (I-V) plot in the LIF and QIF models. Clearly, the equilibrium in the LIF model is stable [see panel (a)]. In the QIF model [panel (b)], the equilibrium on the left is stable and on the right is unstable. Negative values of the I-V curve correspond to a net inward current which causes the membrane to depolarise.

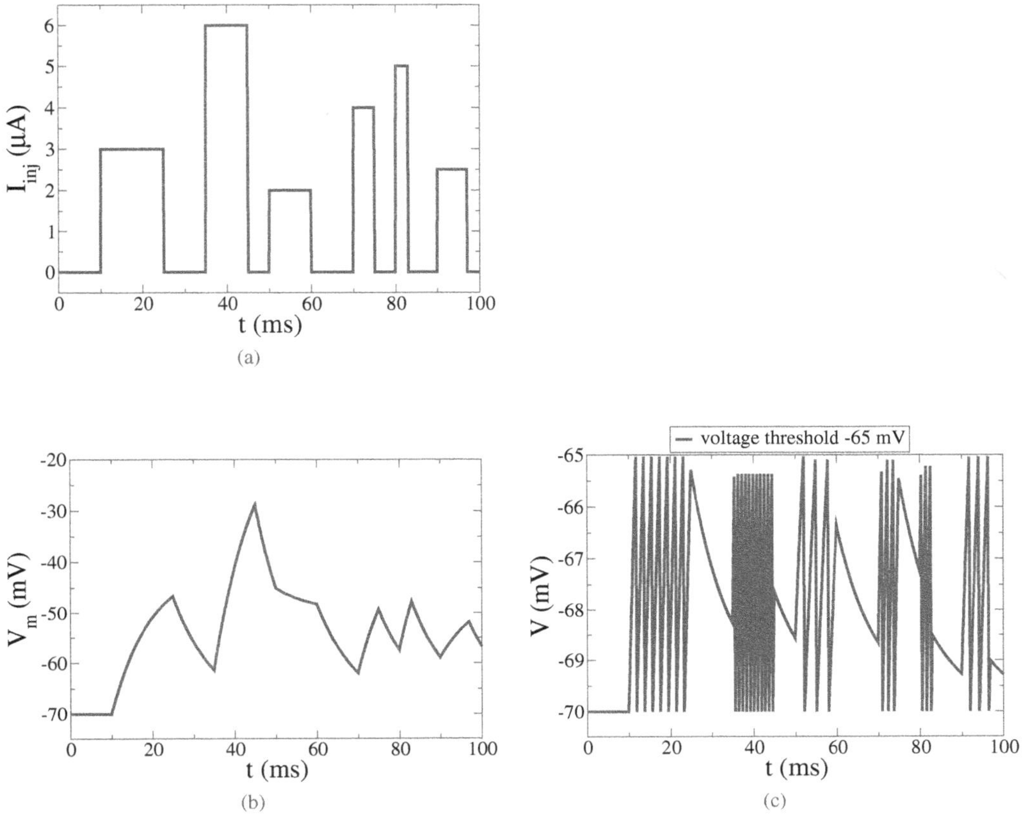

Figure 7.44 Computed temporal variation of the membrane potential V in a LIF neuron subjected to six pulses with random heights and widths. The membrane resting potential is $V_{\rm rest} = -70$ mV, membrane resistance is $R = 10$ kΩ, the membrane time constant is $\tau = 10$ ms and the voltage threshold is $V_{\rm th} = -65$ mV. The Euler algorithm was simulated for 1000 iterations with a time step of Δt=0.1 ms. Panel (a) shows the six current pulses. Panel (b) shows six voltage peaks corresponding to the six pulses. With the imposition of the voltage rest at the threshold $V_{\rm th} = -65$ mV, the voltage output shows a complicated pattern, as shown in panel (c).

Positive values of the I-V curve correspond to a net outward current, which causes the membrane to hyperpolarise.

The presence of two equilibria (the resting state and the excited state) in the QIF model indicates bi-stability. Figure 7.47 shows plots of the capacitive current across the membrane, *viz* $C\frac{dV}{dt}$, against the membrane potential V in the LIF model (panels on the left) as well as the QIF model (panels on the right). In the LIF model, the larger the injected current $I_{\rm inj}$, the larger the membrane potential where the capacitive current changes sign from positive to negative. It is interesting to note the change in the capacitative current with the membrane potential in the QIF model. For $I_{\rm inj} = 0$, the capacitive current becomes zero at the two $V_{\rm eq}$ points: $V_{\rm rest}$ and $V_{\rm th}$, and it takes negative values between $V_{\rm rest}$ and $V_{\rm th}$. For $I_{\rm inj} = 0.3\mu$A the capacitive current takes negative values in a shorter range of V, but the bi-stability is retained. However, for $I_{\rm inj} \geq 0.37\mu$A there are no negative values of the capacitive current, meaning that bi-stability is lost. The special value $I_{\rm inj} = 0.37\mu$A, at which the two equilibria coalesce, is called the bifurcation value.

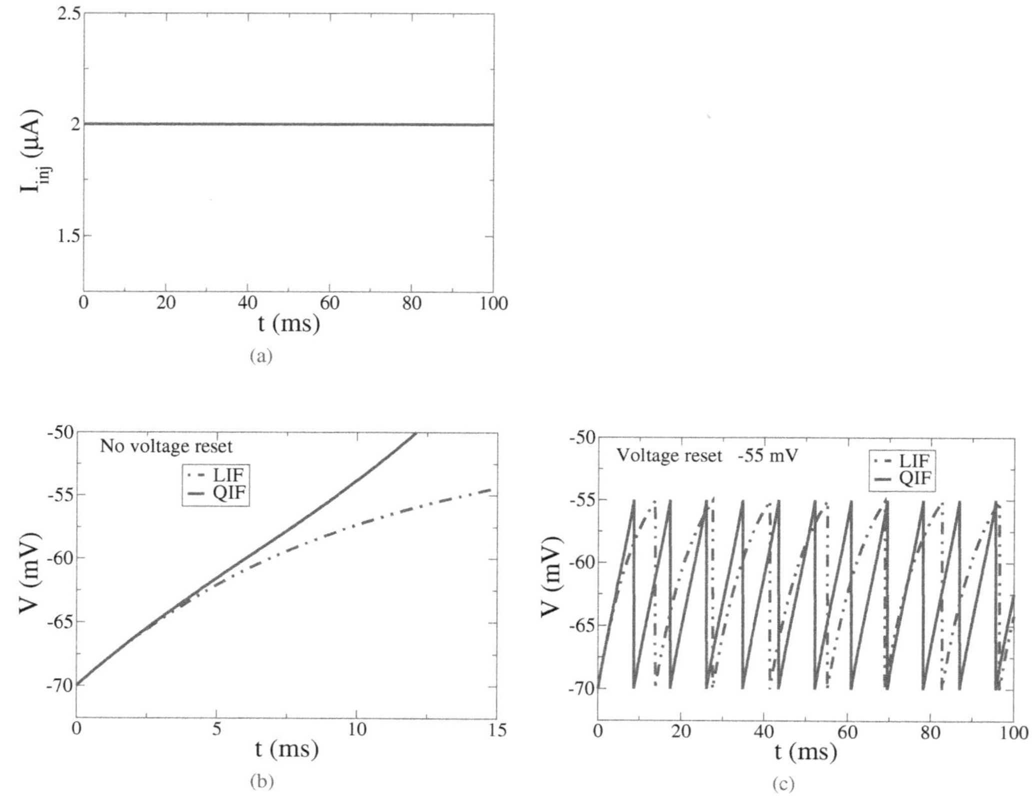

Figure 7.45 Computed temporal variation of membrane potential V in LIF and QIF neurons subjected to a constant input current $I_{\text{inj}} = 2\ \mu$A. The voltage threshold is $V_{\text{th}} = -55$ mV, the membrane resting potential is $V_{\text{rest}} = -70$ mV, the membrane resistance is $R = 10$ kΩ and the membrane time constant is $\tau = 10$ ms. The Euler algorithm was simulated for 1000 iterations with a time step of Δt=0.1 ms.

7.6.3.3 Izhikevich model

Although the one-dimensional (*i.e.* one-variable) LIF and QIF models illustrate a number of important properties of neurons, they are too simplistic to describe realistic neuron dynamics. In addition to 'firing', neurons show 'recovery' behaviour after firing a spike. In order to capture both firing and recovery behaviours, Izhikevich proposed a two-dimensional (*i.e.* two-variable) model for neuron dynamics (Izhikevich, 2003, 2004, 2007). In this model the capacitive current across the membrane is described by the following two first-order differential equations

$$\begin{aligned} C\frac{\mathrm{d}V}{\mathrm{d}t} &= -\frac{(V-V_{\text{rest}})(V_{\text{th}}-V)}{R(V_{\text{th}}-V_{\text{rest}})} - W + I_{\text{inj}} \\ \frac{\mathrm{d}W}{\mathrm{d}t} &= a[b(V-V_{\text{rest}})-W] \quad \text{if } V \geq V_{\text{peak}}, \text{ then } V \leftarrow c, W \leftarrow W+d, \end{aligned} \tag{7.75}$$

where W is the recovery current (which has a slower rate of change than the potential V), V_{peak} is the spike cutoff and a, b, c, d are four independent parameters. W is an amplifying variable when $b < 0$ and a resonant variable when $b > 0$. The constant a is the recovery time constant. The parameter d describes the balance between the outward and inward currents during the spike and affects the

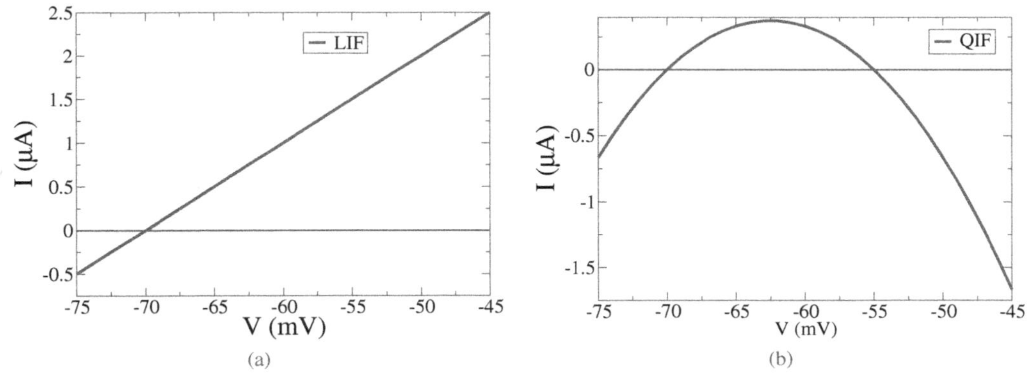

Figure 7.46 The current-voltage plot for the dynamical equilibrium of the membrane potential in the LIF model [panel (a)] and the QIF model [panel (b)]. The parameters used are $V_{\text{rest}} = -70$ mV, $V_{\text{th}} = -55$ mV and $R = 10$ kΩ.

after-spike behaviour. Notice that in the absence of the recovery variable, *i.e.* when $W = 0$, the Izhikevich model reduces to the QIF model.

Figure 7.48 shows the time evolution of the membrane potential V and the recovery current W when I_{inj} is a step current of height 70 μA. The Euler algorithm was employed with a constant time step of $\Delta t = 1$ ms. For the choice of parameters stated in the figure caption, the potential shows periodic firing behaviour. After each potential firing, the recovery current reduces and becomes negative, but with a slower rate than the potential.

As mentioned before, under the voltage-clamp condition, *i.e.* when the membrane potential is constant ($\frac{dV}{dt} = 0$), the injected current I_{inj} equals the net membrane current which we will label as I. Two types of current-voltage relations can be described: instantaneous relation $I_0(V)$ *vs* V and asymptotic steady-state relation $I_\infty(V)$ *vs* V. These are obtained, respectively, by setting $\frac{dV}{dt} = 0$ with $W = 0$ and $\frac{dV}{dt} = 0$ with $\frac{dW}{dt} = 0$. The instantaneous current-voltage relation reflects transmembrane processes taking place on the time scale of the action potential. The steady-scale current-voltage relation shows the asymptotic values of all transmembrane processes. Using equation (7.75), we can express the instantaneous and steady-state currents as

$$I_0(V) = -k(V - V_{\text{rest}})(V - V_{\text{th}}), \tag{7.76}$$

$$I_\infty(V) = -k(V - V_{\text{rest}})(V - V_{\text{th}}) + b(V - V_{\text{rest}}), \tag{7.77}$$

where

$$k = \frac{1}{R(V_{\text{th}} - V_{\text{rest}})}. \tag{7.78}$$

The set of points where $\frac{dV}{dt} = 0$ are called the V-**nullclines**, and the set of points where $\frac{dW}{dt} = 0$ are known as the W-nullclines. An intersection of the V nullcline and the W nullcline is an equilibrium point $(V_{\text{eq}}, W_{\text{eq}})$ in the V-W plane. Figure 7.49 shows the phase portrait in the V-W plane. Also shown are the V-nullcline and W-nullcline. Starting with the rest point, the trajectory follows the spiking behaviour until V_{peak} is reached, after which the dynamics of the variable W leads to the reset point. Following the reset, slow afterhyperpolarisation (AHP) follows. Thereafter, the firing behaviour follows again. Note that, in contrast to the integrate-and-fire (IF) model, in the Izhikevich model, the voltage reset occurs at the peak V_{peak} (not at the threshold V_{th}) of the spike.

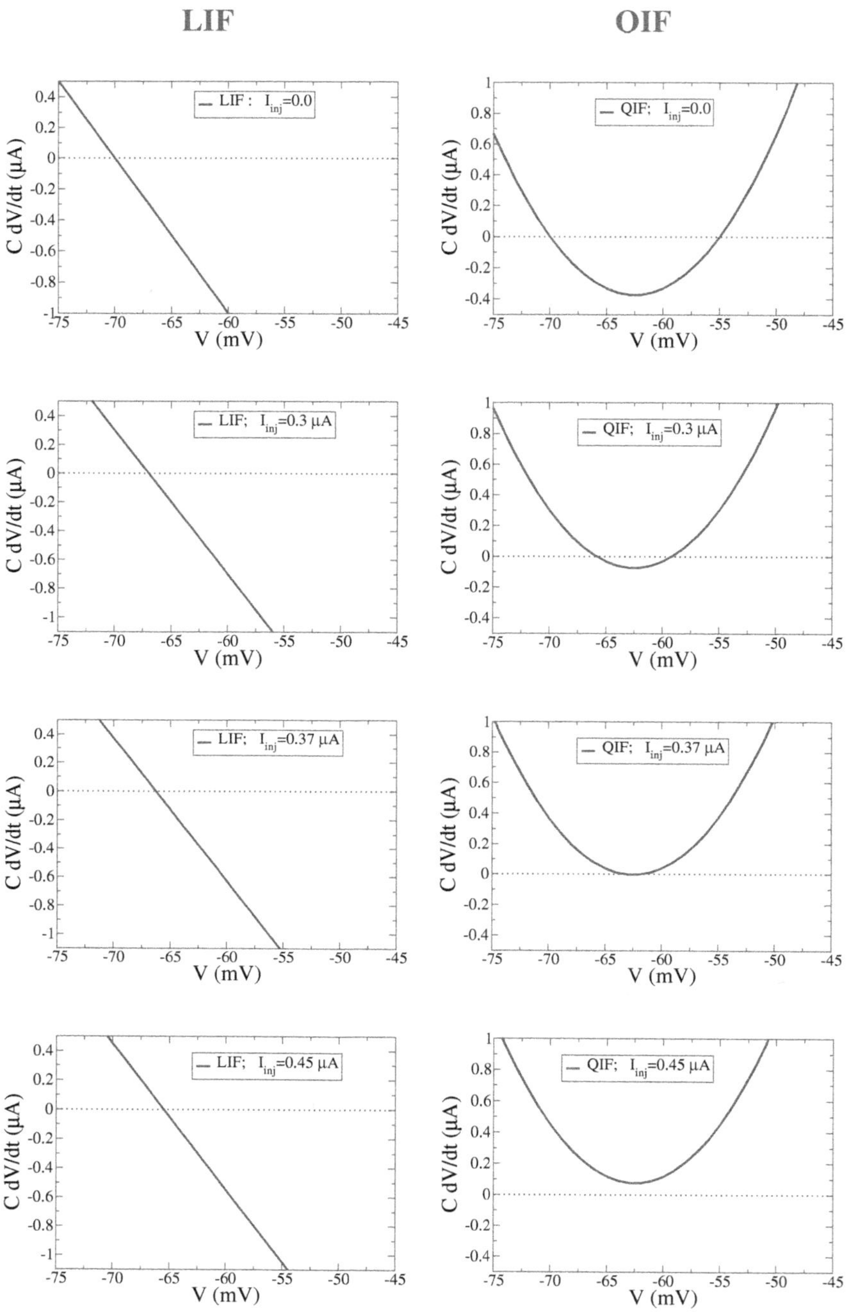

Figure 7.47 Plot of the capacitative current across the membrane, $C\frac{dV}{dt}$, against the membrane potential V in the LIF model (panels on the left) and the QIF model (panels on the right). The parameters used are $V_{\rm rest} = -70$ mV, $V_{\rm th} = -55$ mV and $R = 10$ kΩ.

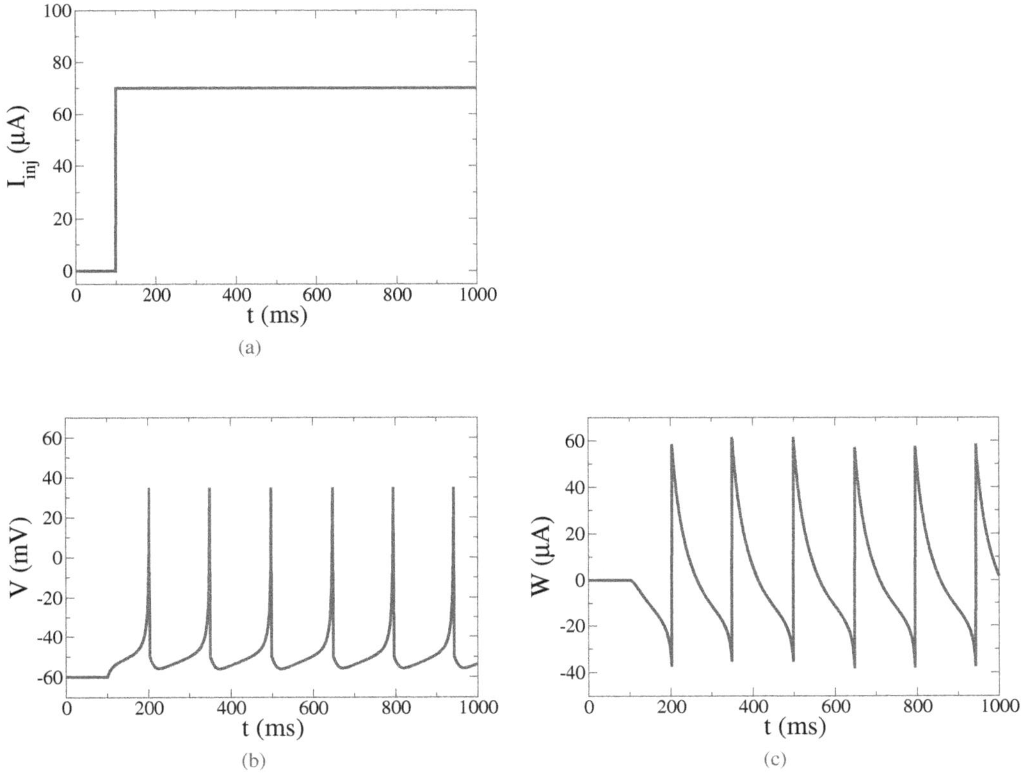

Figure 7.48 Time evolution of the membrane potential V and recovery current W in the Izhikevich model for neuron dynamics. The Euler algorithm was employed with a constant time step of $\Delta t = 1$ ms. The parameters used are: $I_{\text{inj}} = 70\ \mu$A, $C = 100.0\ \mu$F, $R = 0.07$ kΩ, $V_{\text{rest}} = -60$ mV, $V_{\text{th}} = -40$ mV, $V_{\text{peak}} = 35.0$ mV, $a = 0.03$, $b = -2.0$, $c = -50.0$ mV and $d = 100.0\ \mu$A in equation (7.75).

The Izhikevich model is considered to be the simplest model for neuron dynamics that is capable of demonstrating spiking, bursting and other properties of biological neurons, making it suitable for simulations of large-scale neural networks.

7.6.3.4 Hindmarsh-Rose model

Another two-dimensional phenomenological model of neuron dynamics is due to Hindmarsh and Rose (1982, 1984). This model incorporates the real neuron property that each action potential is separated by a long inter-spike interval. The time evolution of the capacitative neural membrane potential x is described by the following two first-order differential equations

$$\begin{aligned} \frac{dx}{dt} &= y - a_1 x^3 + a_2 x^2 + J_{\text{inj}}, \\ \frac{dy}{dt} &= a_3 - a_4 x^2 - y. \end{aligned} \tag{7.79}$$

Here y represents a recovery variable and a_1, a_2, a_3, a_4 are adjustable parameters. Hindmarsh and Rose wrote the above equations using dimensionless units for the variables, making J_{inj} a unitless injected current variable. These equations differ from the corresponding equations in the Izhikevich model. The expression for the time rate of change of the membrane potential contains quadratic and

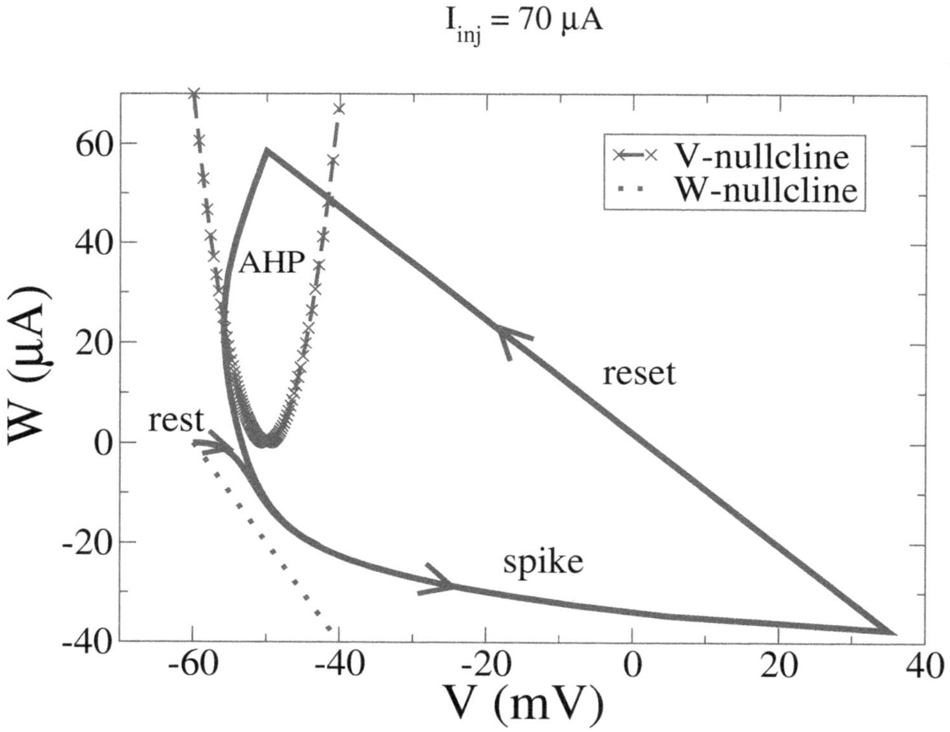

Figure 7.49 Phase portrait in the Izhikevich model for neuron dynamics. V is membrane potential and W is the recovery current. The parameters used are: $I_{\text{inj}} = 70\ \mu$A, $C = 100.0\ \mu$F, $R = 0.07$ kΩ, $V_{\text{rest}} = -60$ mV, $V_{\text{th}} = -40$ mV, $V_{\text{peak}} = 35.0$ mV, $a = 0.03$, $b = -2.0$, $c = -50.0$ mV and $d = 100.0\ \mu$A. The two equations in (7.75) were solved by using the Euler algorithm for 1000 iterations with the time step of 1 ms. Afterhyperpolarisation in indicated as AHP.

cubic potential terms in the Hindmarsh-Rose model, while it contains linear and quadratic potential terms in the Izhikevich model. Also, the time evolution of the recovery current varies as a quadratic function of the potential in the Hindmarsh-Rose model, against a linear function in the Izhikevich model.

Following the discussion presented in the section dealing with the Izhikevich model, and using equation (7.79), the instantaneous and steady-state current-voltage relations in the Hindmarsh-Rose model can be expressed as

$$J_0(x) = a_1x^3 - a_2x^2, \tag{7.80}$$

$$J_\infty(x) = -a_3 - (a_2 - a_4)x^2 + a_1x^3. \tag{7.81}$$

We will present numerical results in figures 7.50 and 7.51, using the Euler-Cromer algorithm with a constant dimensionless time step of $\Delta t = 0.01$ and the parameters used by Hindmarsh and Rose (1984), *viz.* $a_1 = 1.0$, $a_2 = 3.0$, $a_3 = 1.0$ and $a_4 = 5.0$. A current J_{inj} in the form of a short pulse of height 0.2 and duration 5 units is injected at $t = 20$, as shown in figure 7.50(a). Panels (b) and (c) in figure 7.50 show, respectively, the time evolutions of the membrane current variable x and the recovery variable y. Figure 7.51 shows the variation of the steady-state current variable $J_\infty(x)$ [panel (a)] and the phase-space portrait [panel (b)].

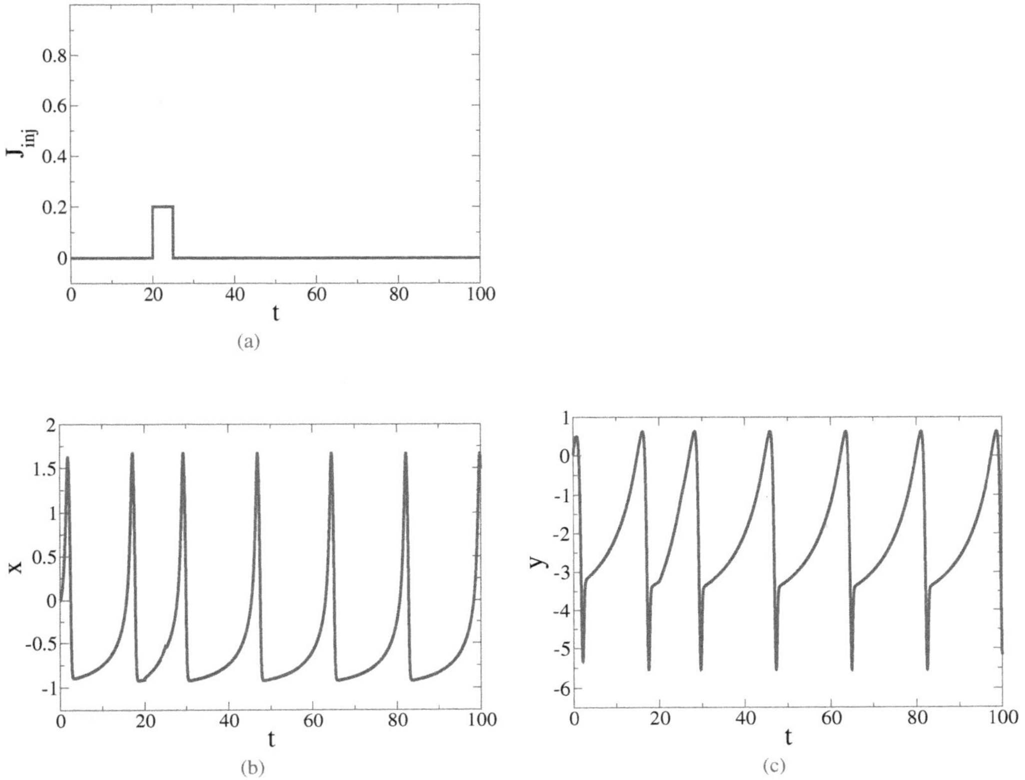

Figure 7.50 (a) Current pulse J_{inj}, (b) time evolution of the membrane potential x and (c) time evolution of the recovery current y in the Hindmarsh-Rose model for neuron dynamics. The Euler-Cromer algorithm was sued with a constant dimensionless time step of $\Delta t = 0.01$. The parameters used in equation (7.81) are: current pulse J_{inj} of height 0.2 and width 5; $a_1 = 1.0$, $a_2 = 3.0$, $a_3 = 1.0$ and $a_4 = 5.0$. A current J_{inj} in the form of a short pulse of height 0.2 and duration 5 units is injected at $t = 20$.

As mentioned before, equilibrium values of the variable x are obtained by solving $\frac{dx}{dt} = \frac{dy}{dt} = 0$, *i.e.* as the intersection points of the x-nullcline and the y-nullcline. These points are also known as the roots of the equation $J_\infty(x) = 0$. Three such points are expected, as $\frac{dx}{dt}$ is a cubic polynomial in x. From the plot in panel (a) of figure 7.51, the equilibrium values of x are -1.6, -1 and 0.6. The same three points can be noted as the intersection of the x-nullcline and y-nullcline plots in panel (b). Stability of these equilibria has been discussed by Hindmarsh and Rose (1984) using analytical arguments. Their analysis suggests that -1.6 is a stable point, -1 is a saddle point and 0.6 is an unstable point.

The phase orbit in figure 7.51 is a limit cycle. In it, the part A $\rightarrow$ B $\rightarrow$ C is an action potential phase and the part C $\rightarrow$ A is a recovery phase. After crossing the $\frac{dx}{dt} = 0$ curve at C, the limit cycle is constrained to move along the narrow path between the two nullclimes. As the phase path is close to both the nullclines, it progresses from C to A much slower compared to the its motion along A $\rightarrow$ B $\rightarrow$ C.

Having discussed phenomenological models, we now turn to a discussion of conduction-based models of neuronal dynamics.

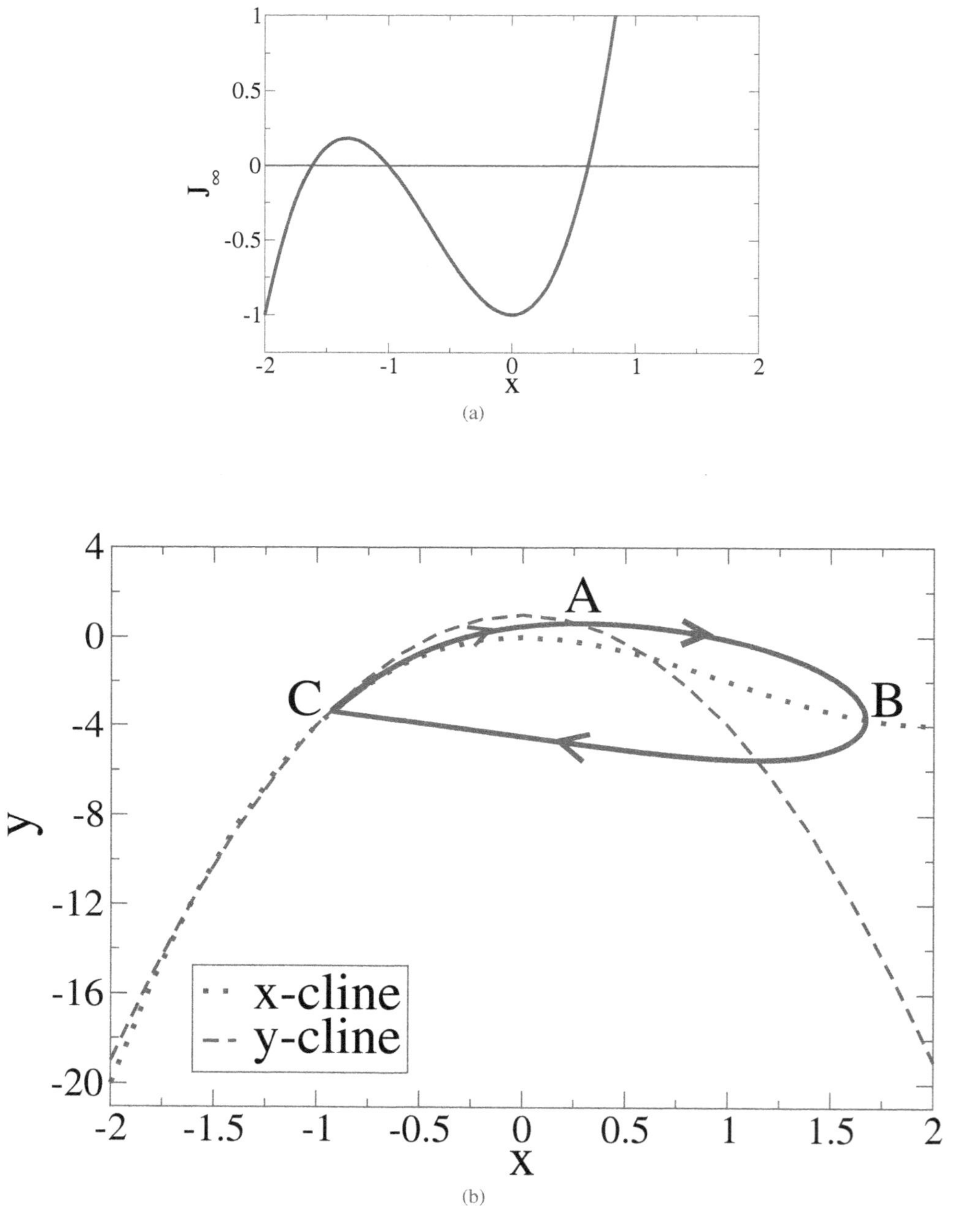

Figure 7.51 Steady-state current-voltage curve $J_\infty(x)$ and the x-y phase portrait in the Hindmarsh-Rose model for neuron dynamics. The parameters used in equation (7.81) are: current pulse J_{inj} of height 0.2 and width 5, and $a_1 = 1.0$, $a_2 = 3.0$, $a_3 = 1.0$ and $a_4 = 5.0$. The Euler-Cromer algorithm was used with a constant dimensionless time step of $\Delta t = 0.01$. A current J_{inj} in the form of a short pulse of height 0.2 and duration 5 units is injected at $t = 20$.

B. Conduction-based models

Ionic currents

Conduction-based models of neuron dynamics attempt to express the first term on the right hand side in equations (7.67), (7.70) and (7.72) as arising from ionic currents in the cell membrane. Important transmembrane currents are generated by four ionic species: sodium (Na^+), potassium (K^+), calcium (Ca^{2+}) and chlorine (Cl^-). In a typical mammalian neuron, concentration of K^+ is much higher inside the membrane cell and concentrations of Na^+ and Cl^- are higher outside the cell. These differences in the ionic concentrations on the two sides of the membrane are maintained by active ion pumps (composed of multiple proteins which are embedded within the membrane). Net ionic currents can be expressed, respectively, as

$$\begin{aligned} I_{\mathrm{K}} &= g_{\mathrm{K}}(V-E_{\mathrm{K}}), \\ I_{\mathrm{Na}} &= g_{\mathrm{Na}}(V-E_{\mathrm{Na}}), \\ I_{\mathrm{Ca}} &= g_{\mathrm{Ca}}(V-E_{\mathrm{Ca}}), \\ I_{\mathrm{Cl}} &= g_{\mathrm{Cl}}(V-E_{\mathrm{Cl}}), \end{aligned} \tag{7.82}$$

where, respectively, E_{K}, $\mathrm{E_{Na}}$, $\mathrm{E_{Ca}}$ and $\mathrm{E_{Cl}}$ are the Nerst equilibrium potentials of K, Na, Ca and Cl, and g_{K}, g_{Na}, g_{Ca} and g_{Cl} are the contributions of these ions towards the total membrane conductance. The corresponding ionic resistances are $R_{\mathrm{K}} = 1/g_{\mathrm{K}}$, $R_{\mathrm{Na}} = 1/g_{\mathrm{Na}}$, $R_{\mathrm{Ca}} = 1/g_{\mathrm{Ca}}$ and $R_{\mathrm{Cl}} = 1/g_{\mathrm{Cl}}$. The ionic conductances are functions of the membrane voltage V and time t.

7.6.3.5 Hodgkin-Huxley model

One of the most successful models for neuron non-linear dynamics is due to Hodgkin and Huxley (1952). Based on experimental observations, Hodgkin and Huxley established that there are three major types of currents flowing through a squid axon. These are: voltage-gated persistent K^+ current with four activation gates, voltage-gated transient Na^+ current with three activation gates and one inactivation gate, and Ohmic leak current carried by Cl^- ions. Figure 7.52 shows an electrical circuit representation of a cell membrane. With these considerations, the capacitative membrane current is expressed as the following four ordinary differential equations

$$\begin{aligned} C\frac{\mathrm{d}V}{\mathrm{d}t} &= -\frac{1}{R_{\mathrm{L}}}(V-E_{\mathrm{L}}) - \frac{1}{R_{\mathrm{K}}}(V-E_{\mathrm{K}}) - \frac{1}{R_{\mathrm{Na}}}(V-E_{\mathrm{Na}}) + I_{\mathrm{inj}}, \\ \frac{\mathrm{d}n}{\mathrm{d}t} &= (1-n)\alpha_n(V) + n\beta_n(V), \\ \frac{\mathrm{d}m}{\mathrm{d}t} &= (1-m)\alpha_m(V) + m\beta_m(V), \\ \frac{\mathrm{d}h}{\mathrm{d}t} &= (1-h)\alpha_h(V) + h\beta_h(V), \end{aligned} \tag{7.83}$$

where

$$\frac{1}{R_{\mathrm{L}}} = g_{\mathrm{L}}, \quad \frac{1}{R_{\mathrm{K}}} = n^4 g_{\mathrm{K}} \quad \text{and} \quad \frac{1}{R_{\mathrm{Na}}} = m^3 h g_{\mathrm{Na}}, \tag{7.84}$$

and

$$\alpha_n = \frac{0.01(10-V)}{\exp\left(\frac{10-V}{10}\right)-1}, \quad \alpha_m = \frac{0.1(25-V)}{\exp\left(\frac{25-V}{10}\right)-1}, \quad \alpha_h = 0.07\exp\left(\frac{-V}{20}\right), \tag{7.85}$$

$$\beta_n = 0.125\exp\left(\frac{-V}{80}\right), \quad \beta_m = 4\exp\left(\frac{-V}{18}\right), \quad \beta_h = \frac{1}{\exp\left(\frac{30-V}{10}\right)+1}. \tag{7.86}$$

Typical values of the parameters are (Izhikevich 2007):

$$E_{\mathrm{K}} = -12\,\mathrm{mV}, \quad E_{\mathrm{Na}} = 120\,\mathrm{mV}, \quad E_{\mathrm{L}} = 10.6\,\mathrm{mV} \tag{7.87}$$

and

$$g_{\mathrm{K}} = 36\ \mathrm{mS/cm^2}, \quad g_{\mathrm{Na}} = 120\ \mathrm{mS/cm^2}, \quad g_{\mathrm{L}} = 0.3\ \mathrm{mS/cm^2}. \tag{7.88}$$

The resting potential in the Hodgkin-Huxley model is zero ($V_{\mathrm{rest}} = 0$).

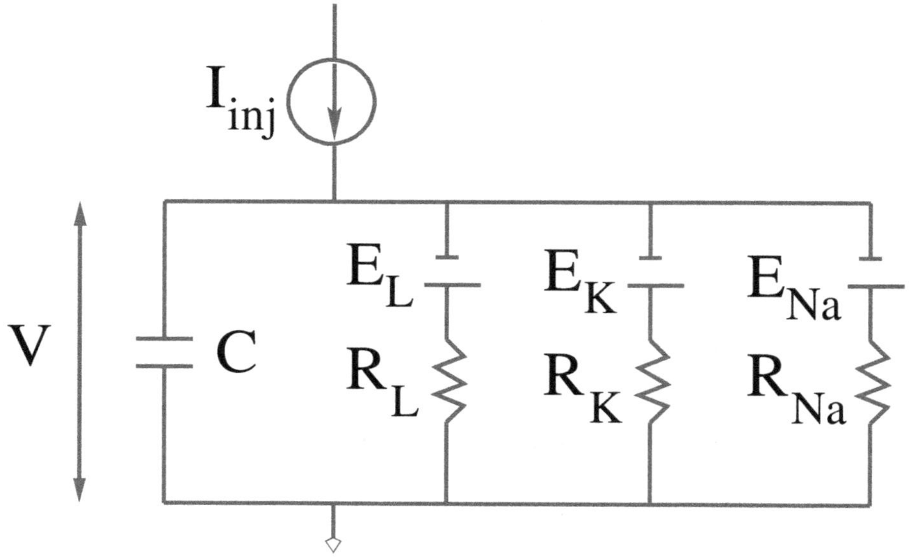

Figure 7.52 An electrical circuit representation of the Hodgkin-Huxley model of a neural membrane subjected to an external current. The model assumes that there are three major types of current flowing through an axon. Here the symbols are as follows: C is the capacitance and V is the potential across the membrane; I_{inj} is the current pulse injected into the neuron cell; $E_{\mathrm{K}}, E_{\mathrm{Na}}, E_{\mathrm{Ca}}$ and $E_{\mathrm{L}} \equiv E_{\mathrm{Cl}}$ are the Nerst equilibrium potentials, and $R_{\mathrm{K}}, R_{\mathrm{Na}}, R_{\mathrm{Ca}}, R_{\mathrm{L}} \equiv R_{\mathrm{Cl}}$ are the ionic resistances of K, Na, Ca and Cl, respectively.

In above, n, m and h are gating variables, whose rate of activation (opening) and inactivation (closing, or blocking) are described by the functions $\alpha(V)$ and $\beta(V)$, respectively. The variable n describes the activation of the K^+ channel, m describes the activation of the Na^+ channel and h describes the inactivation (blocking) of the Na^+ channel. The time evolution of these variables can be re-expressed as

$$\frac{\mathrm{d}n}{\mathrm{d}t} = (n_\infty - n)/\tau_n, \quad \frac{\mathrm{d}m}{\mathrm{d}t} = (m_\infty - m)/\tau_m, \quad \frac{\mathrm{d}h}{\mathrm{d}t} = (h_\infty - h)/\tau_h, \tag{7.89}$$

where

$$n_\infty = \alpha_n/(\alpha_n + \beta_n), \quad m_\infty = \alpha_m/(\alpha_m + \beta_m), \quad h_\infty = \alpha_h/(\alpha_h + \beta_h), \tag{7.90}$$

and

$$\tau_n = 1/(\alpha_n + \beta_n), \quad \tau_m = 1/(\alpha_m + \beta_m), \quad \tau_h = 1/(\alpha_h + \beta_h). \tag{7.91}$$

Equation (7.83) can be solved for n, m and h either analytically or computationally using an algorithm such as the Euler method. For $t \geq 0$, the analytical solution for $n(t)$ is

$$n(t) = n_\infty(v_0) + \big(n(0) - n_\infty(V_0)\big)\exp[-t/\tau_n(V_0)], \tag{7.92}$$

where $n_\infty(V_0)$ and $\tau_n(V_0)$ are values at $t = 0$. Analogous solutions can be written down for $m(t)$ and $h(t)$. It may be useful to note that the equilibrium values of the variables n, m and h are: $n_\infty(V = 0) = 0.318$, $m_\infty(V = 0) = 0.053$ and $h_\infty(V = 0) = 0.596$.

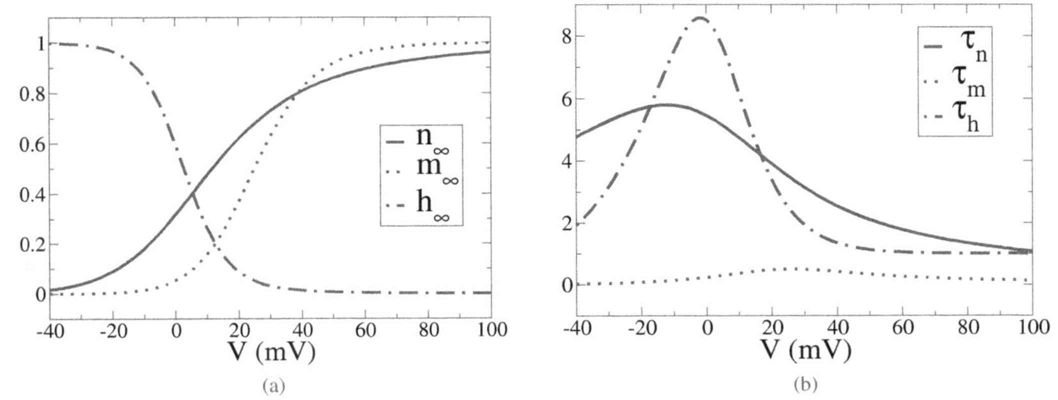

Figure 7.53 The activation functions $n_\infty(V)$ and $m_\infty(V)$, the inactivation function $h_\infty(V)$ and the time constants functions $\tau_n(V)$, $\tau_m(V)$ and $\tau_h(V)$ in the Hodgkin-Huxley neuron model.

The steady-state values of the activation functions $n_\infty(V)$ and $m_\infty(V)$ and the inactivation function $h_\infty(V)$ are shown in panel (a) of figure 7.53. The time constant functions $\tau_n(V)$, $\tau_m(V)$ and $\tau_h(V)$ are shown in panel (b) of figure 7.53. The values $n_\infty(0)$, $m_\infty(0)$ and $h_\infty(0)$ for $V = V_{\text{rest}} = 0$ correspond to the stable resting state when $I_{\text{inj}} = 0$. As $\tau_m(V)$ is much smaller than $\tau_n(V)$ and $\tau_h(V)$, the activation function $m(V)$ changes faster than the functions $n(V)$ and $h(V)$. Also, as $\tau_n(V)$ and $\tau_h(V)$ are relatively large near V_{rest}, recovery of the variables n and h is slower.

Full graphical visualisation of the solutions of the Hodgkin-Huxley equations requires the four-dimensional phase space with coordinates (V, n, m, h). As this is normally not possible, results of reduced-dimensional systems have been discussed in several works, including FitzHugh (1960), Morris and Lecar (1981) and Tsumoto *et al.* (2006). We will discuss the Morris-Lecar model in the next section. However, it is instructive to investigate the behaviour of the Hodgkin-Huxley model by making two-dimensional plots of its solutions.

We solved the Hodgkin-Huxley equations by using the Euler-Cromer algorithm with time step $\Delta t = 0.01$ ms and the initial conditions $V = 0$, $n = 0$, $m = 0$, $h = 0$. The parameters are listed in equations (7.87) and (7.88). The choice of the initial conditions is arbitrary, and it will be left as an exercise for the reader to try $V(t = 0) = -6$ mV, $n(t = 0) = n_\infty(V = 0) = 0.318$, $m(t = 0) = m_\infty(V = 0) = 0.053$ and $h(t = 0) = h_\infty(V = 0) = 0.596$.

Figure 7.54 shows the temporal variation of the membrane potential V, the potassium activation variable n, the sodium activation variable m and the sodium inactivation variable h when the neuron is subjected to three different choices for the injected current I_{inj}. As seen in panels (a)-(c), in absence of any injection of external current, the membrane potential depolarises and then repolarises, and after 20 ms it settles at its resting level ($V = 0$). As seen in panels (d)-(f), application of the short electric pulse at $t = 2$ ms of height 10 μA and duration 5 ms, results in small changes in the temporal variation of the gate variables n, m and h, and the membrane potential V. In particular, there develops a small kink in the polarisation of the potential at the start of the upstroke. As seen in panels (g)-(i), when a constant current of 10 μA is applied, each of the variables n, m, h and V shows three peaks during the simulation period of 40 ms. This behaviour can be generalised: when a high external current is injected, the neuron cell fires repetitively. In general, injection of a pulse of greater amplitude and shorter duration produces the action potential of shorter amplitude and which rises more rapidly and with greater depolarisation. This will be left as an exercise for the reader to verify.

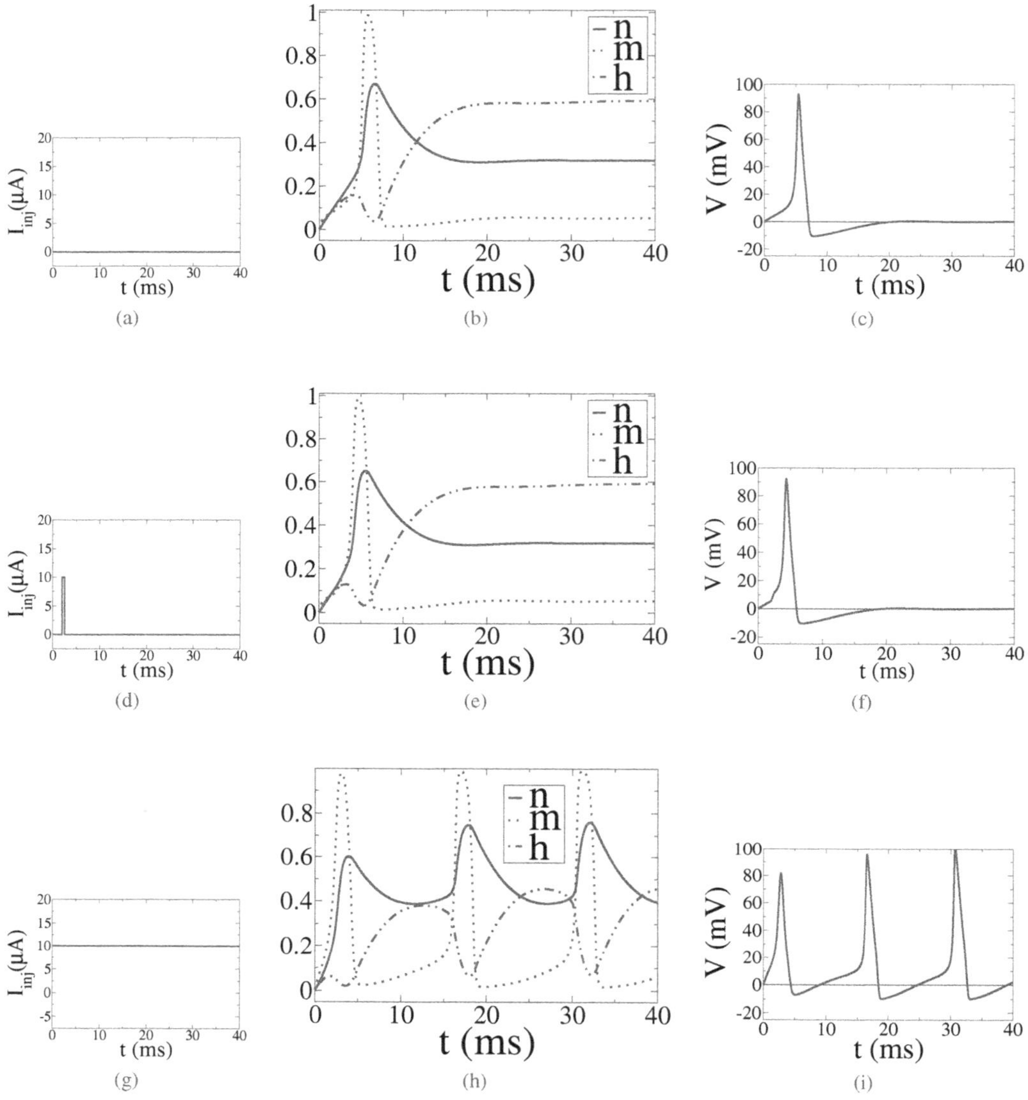

Figure 7.54 Temporal variation of the membrane potential V and the variables n, m, h in the Hodgkin-Huxley model of neuron dynamics. The initial conditions for the numerical simulation, using the Euler algorithm, are set as $V = 0$, $n = 0$, $m = 0$, $h = 0$. The parameters are listed in equations (7.87) and (7.88). The membrane capacitance was taken as $C = 1\mu\text{F/cm}^2$. The time step for the simulation is $\Delta t = 0.01$ ms.

In figure 7.55 we show the phase portrait plot, *viz* the orbit in the (V-$\frac{dV}{dt}$) plane, for the short current pulse $I_{\text{inj}} = 10\ \mu$A injected at 2 ms for 0.5 ms. Clearly, at the end of the simulation, which was carried out for 40 ms, the membrane potential settles at its resting value of 0 mV. This indicates that the resting potential is a stable equilibrium point.

Figure 7.56 presents the solutions of the Hodgkin-Huxley equations in the n-V, m-V and h-V planes. The choices for I_{inj} were the same as in figure 7.54, *i.e.* results in the absence of current are shown in panels (a)-(c), results for the short pulse of 10 μA are shown in panels (d)-(f) and the

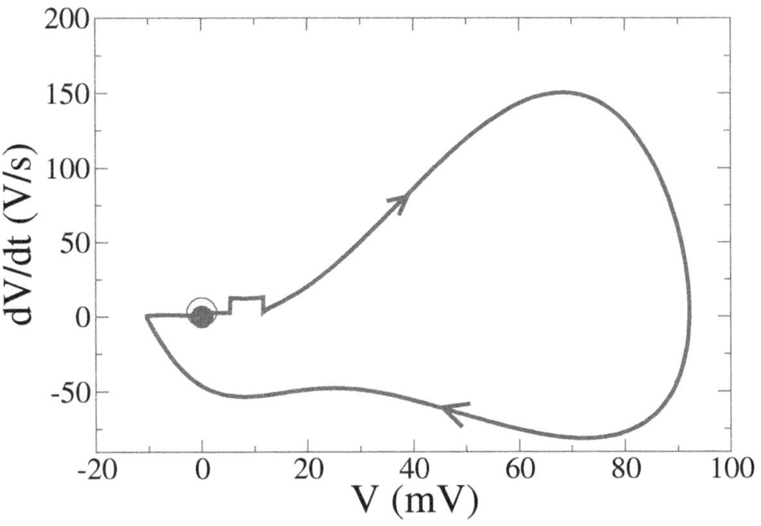

Figure 7.55 Phase portrait for the membrane potential in the Hodgkin-Huxley model when it is subjected to a short current pulse of $I_{\text{inj}} = 10\ \mu$A of duration 0.5 ms at $t = 2$ ms. The initial conditions for the numerical simulation, using the Euler algorithm, are set as $V = 0$, $n = 0$, $m = 0$, $h = 0$. The parameters are listed in equations (7.87) and (7.88). The membrane capacitance was taken as $C = 1\mu\text{F/cm}^2$. The time step for the simulation is $\Delta t = 0.01$ ms. The orbit progresses clockwise with the open and filled circles indicating, respectively, the start and end of the simulation.

results for a constant current of 10 μA are shown in panels (g)-(i). The membrane capacitance was taken as $C = 1\ \mu\text{F/cm}^2$. The plots in the n-V, m-V and h-V planes follow the temporal behaviours of the variables n, m and h with V. The results clearly suggest that the roles of the variables n and h are mutually coordinated: when the sodium channels close, the potassium channels open. The projections of the phase orbits in these planes show different behaviours for different choices of I_{inj}. It can be noticed that when the constant current of 10 μA is injected, limit cycle trajectories are generated in each of the n-V, m-V and h-V planes. This indicates that application of a constant current can play an important role in the non-linear dynamics of neurons. We will investigate this in some detail for a two-dimensional version of the Hodgkin-Huxley model in the next section, in the form of the Morris-Lecar model.

7.6.3.6 Morris-Lecar model

The Morris-Lecar model describes the two-dimensional conduction-based non-linear dynamics of neurons. Morris and Lecar (1981) developed this model for investigating voltage oscillations in Barnacle giant muscle fibre. This fibre possesses a simple conductance system consisting of voltage- and time-dependent Ca^{++} and K^{+} channels. By employing mathematical techniques of non-linear dynamics, the Morris-Lecar model concluded that subject to changeable parameters, the system can exhibit varied oscillatory and bistable behaviours. Using mathematical, computational and neuroscientific languages, the Morris-Lecar model can be considered as a two-dimensional reduction of the four-dimensional Hodgkin-Huxley model, as explained below.

As the Ca^{++} current changes much faster than the K^{+} current, the Morris-Lecar model assumes that the former is always in equilibrium. Based on this assumption, the capacitative membrane

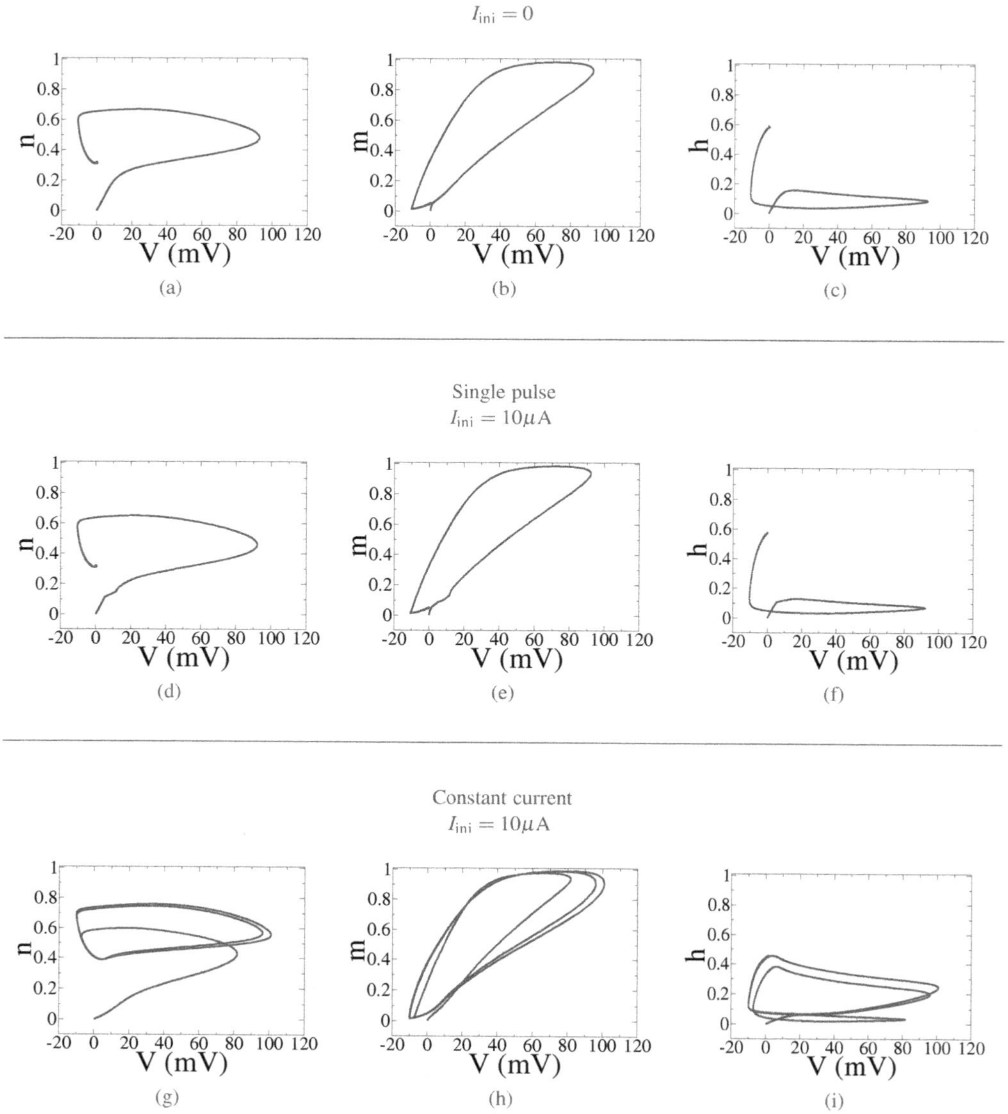

Figure 7.56 Solutions of the Hodgkin-Huxley equations plotted in the n-V, m-V and h-V planes. The results for no external current ($I_{\text{inj}} = 0$) are shown in panels (a)-(c). Panels (d)-(f) show the results when the membrane is subjected to a short current pulse of $I_{\text{inj}} = 10\ \mu$A of duration 0.5 ms at $t = 2$ ms. Panels (g)-(i) show the results when the membrane is subjected to a constant current pulse of $I_{\text{inj}} = 10\ \mu$A. The initial conditions for the numerical simulation are $V = 0$, $n = 0$, $m = 0$, $h = 0$. The parameters are listed in equations (7.87) and (7.88).

current is expressed as the following two ordinary differential equations (Morris and Lecar, 1981; Tsumoto *et al.* 2006)

$$C\frac{\mathrm{d}V}{\mathrm{d}t} = -g_{\mathrm{L}}(V - E_{\mathrm{L}}) - g_{\mathrm{K}}N(V - E_{\mathrm{K}}) - g_{\mathrm{Ca}}M(V - E_{\mathrm{Ca}}) + I_{\mathrm{inj}},$$

$$\frac{dN}{dt} = \phi\frac{(N_\infty(V)-N)}{\tau_N(V)}, \tag{7.93}$$

where

$$\begin{aligned} M_\infty(V) &= \frac{1}{2}\Big[1+\tanh\big(\frac{(V-V_1)}{V_2}\big)\Big], \\ N_\infty(V) &= \frac{1}{2}\Big[1+\tanh\big(\frac{(V-V_3)}{V_4}\big)\Big], \\ \tau_N^{-1}(V) &= \cosh\Big(\frac{(V-V_3)}{V_4}\Big). \end{aligned} \tag{7.94}$$

In the above equations, C is the membrane capacitance, V is the membrane potential, g_L, g_K, g_{Ca} are respectively the leak membrane, potassium and calcium conductances, and E_L, E_K, E_{Ca} are the corresponding equilibrium potentials. M and N are analogous to the Hodgkin-Huxley parameters m and n, current injected to the membrane is I_{inj}, and ϕ is a temperature factor. V_1 is the potential at which the calcium current is half-activated (*i.e.* when $M_\infty = 0.5$) and V_2 is a measure of the steepness of the activation voltage. V_3 is potential at which the potassium current is half-activated (*i.e.* when $N_\infty = 0.5$) and V_4 is a measure of the slope of the activation voltage. The system described will be referred as the M-L neuron.

The V-nullcline, the curve defined by $dV/dt = 0$, is described by the function

$$N(V) = \frac{I_{inj} - g_L(V-E_L) - g_{Ca}M_\infty(V-E_{Ca})}{g_K(V-E_K)} \tag{7.95}$$

and the N-nullcline, the curve defined by $dN/dt = 0$, is described by the function

$$N(V) \equiv N_\infty(V) = \frac{1}{2}\Big[1+\tanh\big(\frac{(V-V_3)}{V_4}\big)\Big]. \tag{7.96}$$

The Morris-Lecar model is capable of describing different types of non-linear dynamics, dependent on the choice of parameters. Note that the parameters C, g_L, g_K, E_L, E_K, E_{Ca} are fixed for the Barnacle fibre. Therefore, the dynamical behaviour of the M-L neuron should be investigated by considering different choices for five parameters: ϕ, g_{Ca}, V_3, V_4 and I_{inj}. Of these, I_{inj} is the dominant parameter. Three sets of these five parameters have been used to discuss different types of bifurcation analysis of the M-L neuron. These sets, taken from literature (Ermentrout and Terman, 2010; Azizi and Mugabi, 2020), are listed in table 7.2. For each of these sets I_{inj} acts as the main bifurcation parameter and ϕ, g_{Ca}, V_3, V_4 as secondary parameters.

Detailed discussion of mathematical analysis of different types of bifurcation can be found in several books, including Strogatz (1994), Wiggins (2003) and Izhikevich (2007). Here we will make the bifurcation analysis by examining phase space portraits. The Morris-Lecar equations were solved by employing the Euler-Cromer algorithm with a time step of $\Delta t = 0.01$ ms. Initial values were set as $V(t=0) = 0$ and $N(t=0) = 0$.

Results and analysis for Set 1 (Hopf)

Results for Set 1 (Hopf) are displayed in figures 7.57-7.59. In the presence of an external current, the time evolutions of V and N show different behaviours. Accordingly, phase portraits also show different behaviours. Regular changes are noted as I_{inj} increases up to a certain value and beyond that a different pattern is noted.

$I_{inj} = 0$ *(No external current)*: As seen in figure 7.57, in the absence of any external current ($I_{inj} = 0$), the potential rises to just under 40 mV in approximately 5.5 ms and then drops down to about -69.2 mV at around 37 ms of the simulation. As the simulation progresses further, the potential rises

Table 7.2
Three sets of parameters for Morris-Lecar neuronal dynamics.

Parameter	Set 1 (Hopf)	Set 2 (SNLC)	Set 3 (Homoclinic)
ϕ	0.04	0.067	0.23
g_{Ca} (mS/cm^2)	4.4	4	4
V_3 (mV)	2	12	12
V_4 (mV)	30	17.4	17.4
E_{Ca} (mV)	120	120	120
E_{K} (mV)	-84	-84	-84
E_{L} (mV)	-60	-60	-60
g_{K} (mS/cm^2)	8	8	8
g_{L} (mS/cm^2)	2	2	2
V_1 (mV)	-1.2	-1.2	-1.2
V_2 (mV)	18	18	18
C (μF/cm^2)	20	20	20

gently and settles down to approximately -60.9 mV for $t \geq 123$ ms. The activation variable N rises to approximately 0.417 in 19.4 ms and then gradually drops down to approximately 0.02 for $t \geq 20$ ms. The portrait in the $(V, \frac{dV}{dt})$ phase space settles at the phase point where $V = \frac{dV}{dt} = 0$. Similarly, the phase orbit in the N-V plane settles at the intersections of the V- and N-nullclines, *viz.* at the stable equilibrium point.

$I_{\text{inj}} = 88\ \mu A$: As seen in figure 7.58, the minimum values of V and N are higher than those for $I_{\text{inj}} = 0$. Small oscillations in the temporal variations of V and N are seen as time increases. The portrait in the $(V, \frac{dV}{dt})$ phase space exhibits a small periodic region towards the end of the simulation. In agreement with these features, towards the end of the simulation, the portrait in the N-V plane also shows a small periodic orbit around the intersection of the nullclines, the stable equilibrium point.

$I_{\text{inj}} = 89\ \mu A$: The results in figure 7.59 show that the temporal behaviours of V and N change suddendly as I_{inj} is increased from 88 μA to 89 μA. Both V and N exhibit oscillations with large amplitudes for the entire simulation period. Consequently, we see development of limit cycles in both the $(V, \frac{dV}{dt})$ phase space and the N-V plane. Importantly, the equilibrium state has lost stability and the limit cycle in the N-V plane is established around the stable equilibrium point. This qualitative change in the phase orbit from a stable equilibrium to a limit cycle is a signature of **Hopf bifurcation** (see p 249 in Strogatz, 1994).

Results and analysis for Set 2 (SNLC)

Phase portraits for Set 2 (**saddle-node bifurcation on limit cycle**, or SNLC) are displayed in figures 7.60-7.62.

$I_{\text{inj}} = 0$ *(No external current)*: As seen in figure 7.60, in the absence of external current, there are three equilibrium points (at the intersections of the V- and N-nullclines), one stable and two unstable, which are quite close by. Both V and N stabilise at their resting values. The phase path in the $(V, \frac{dV}{dt})$ phase space stops at the stable equilibrium point. The phase orbit in the N-V plane settles at the stable equilibrium point.

$I_{\text{inj}} \approx 40\ \mu A$: As seen in figure 7.61, when a finite external current $I_{\text{inj}} = 40\ \mu$A is applied, the stable equilibrium point and both unstable equilibria move towards each other. It is found that for $I_{\text{inj}} = 45$ μA, there are two equilibrium points. The potential V acquires a single dip before flattening out and

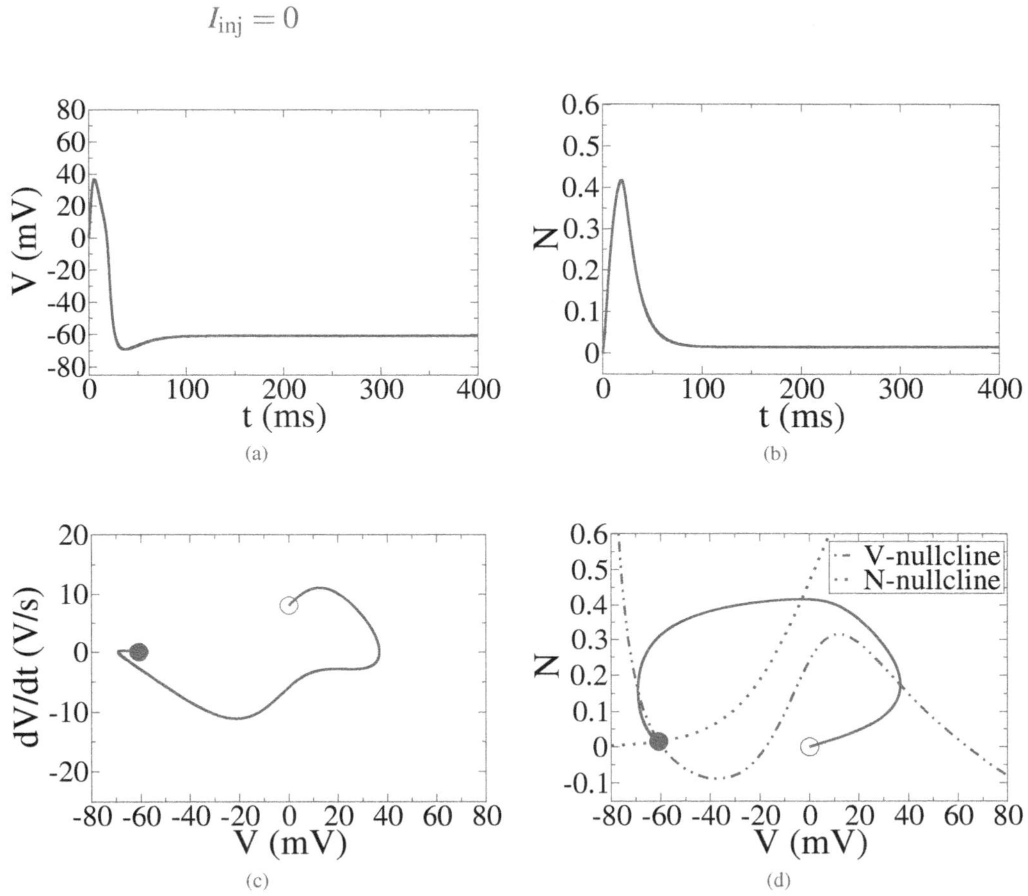

Figure 7.57 Numerical solutions of the Morris-Lecar equations for Set 1 (Hopf) using the parameters listed in table 7.2 and in the absence of external current ($I_{\text{inj}} = 0$). Plotted are the time variation of the membrane potential V, the time variation of the potassium gating variable N, the phase portrait in the $(V, \frac{dV}{dt})$ phase space and the phase orbit in the N-V plane. The open and filled circles indicate, respectively, the start and finish of the simulation.

these two equilibrium points coalesce. The phase path in the $(V, \frac{dV}{dt})$ phase space traverses past the stable equilibrium point. A similar behaviour is seen in the N-V plane.

$I_{\text{inj}} = 50\ \mu A$: As seen in figure 7.62, when the external current is increased to $I_{\text{inj}} = 50\ \mu$A, the two coalesced equilibrium points have annihilated each other but the third (unstable) equilibrium point has remained unmoved. Both the potential V and the variable N have become oscillatory for the entire simulation time. This has led to the development of limit cycles in both the $(V, \frac{dV}{dt})$ phase space and the N-V plane. The change of the behaviour from $I_{\text{inj}} = 40\ \mu$A to $I_{\text{inj}} = 50\ \mu$A is a signature of fold or **saddle-node bifurcation**, or saddle-node bifurcation on limit cycle (SNLC), or saddle-node bifurcation on invariant circle (SNIC).

Results and analysis for Set 3 (Homoclinic)

Results for Set 3 (Homoclinic) are displayed in figures 7.63-7.65.

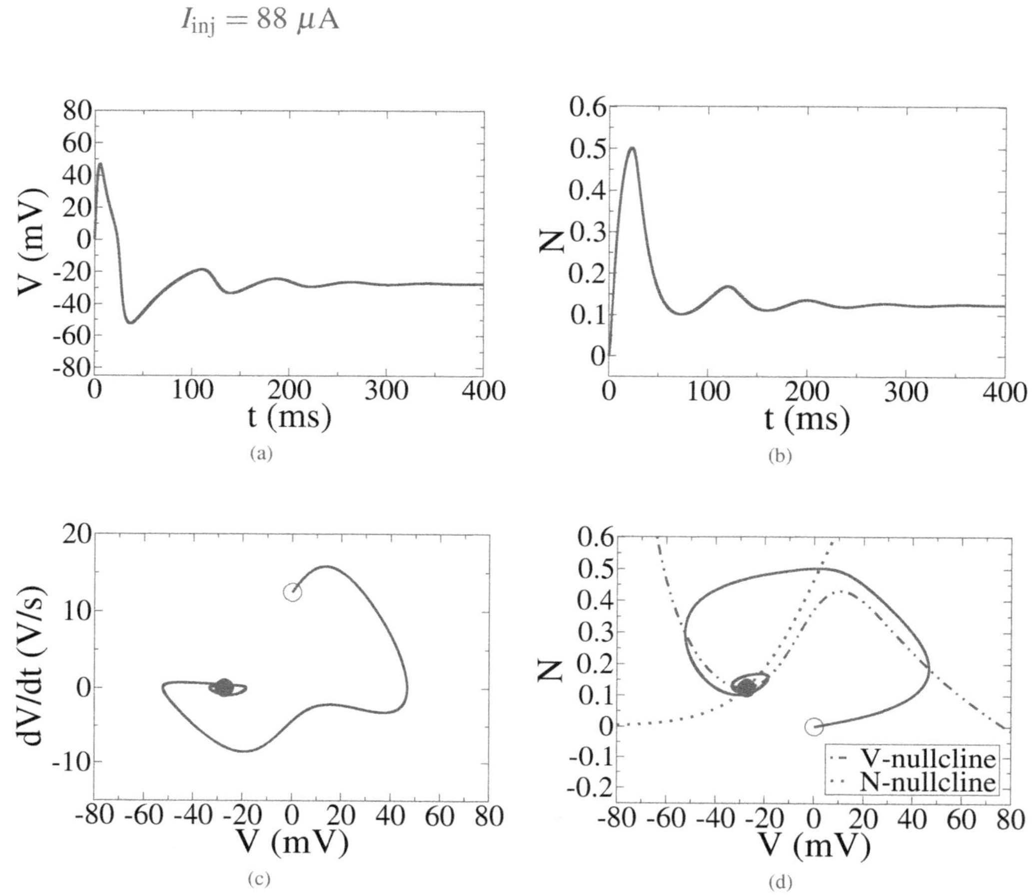

Figure 7.58 Numerical solutions of the Morris-Lecar equations for Set 1 (Hopf) using the parameters listed in table 7.2 and in the absence of external current $I_{\text{inj}} = 88\ \mu$A. Plotted are the time variation of the membrane potential V, the time variation of the potassium gating variable N, the phase portrait in the $(V, \frac{dV}{dt})$ space and the phase orbit in the N-V plane. The open and filled circles indicate, respectively, the start and finish of the simulation.

The only difference in the parameter lists for Set 2 and Set 3 is the temperature factor ϕ. While this factor does not influence the nullclines, it does so for the temporal variation of N, and hence, the phase paths. Therefore, the nullcline curves for Set 3 (Homocline) are the same as for Set 2 (SNLC), indicating three equilibrium points.

$I_{\text{inj}} = 0$ *(No external current)*: As seen in figure 7.63, in the absence of external current, both V and N stabilise at their resting values. Compared to Set 2, N stabilises faster for Set 3 due to the larger value of ϕ. The phase path in the $(V, \frac{dV}{dt})$ phase space stops at the stable equilibrium point. Similar to Set 2, the phase orbit for Set 3 in the N-V plane settles at the stable equilibrium point.

$I_{\text{inj}} = 40\ \mu A$: Changes in the dynamics occur as an increased amount of external current is applied. The plots in figure 7.64 for $I_{\text{inj}} = 40\ \mu$A show that both V and N become oscillatory and limit cycles have been established in the $(V, \frac{dV}{dt})$ phase space and the N-V plane. The limit cycle in the N-V plane is accompanied by a saddle point.

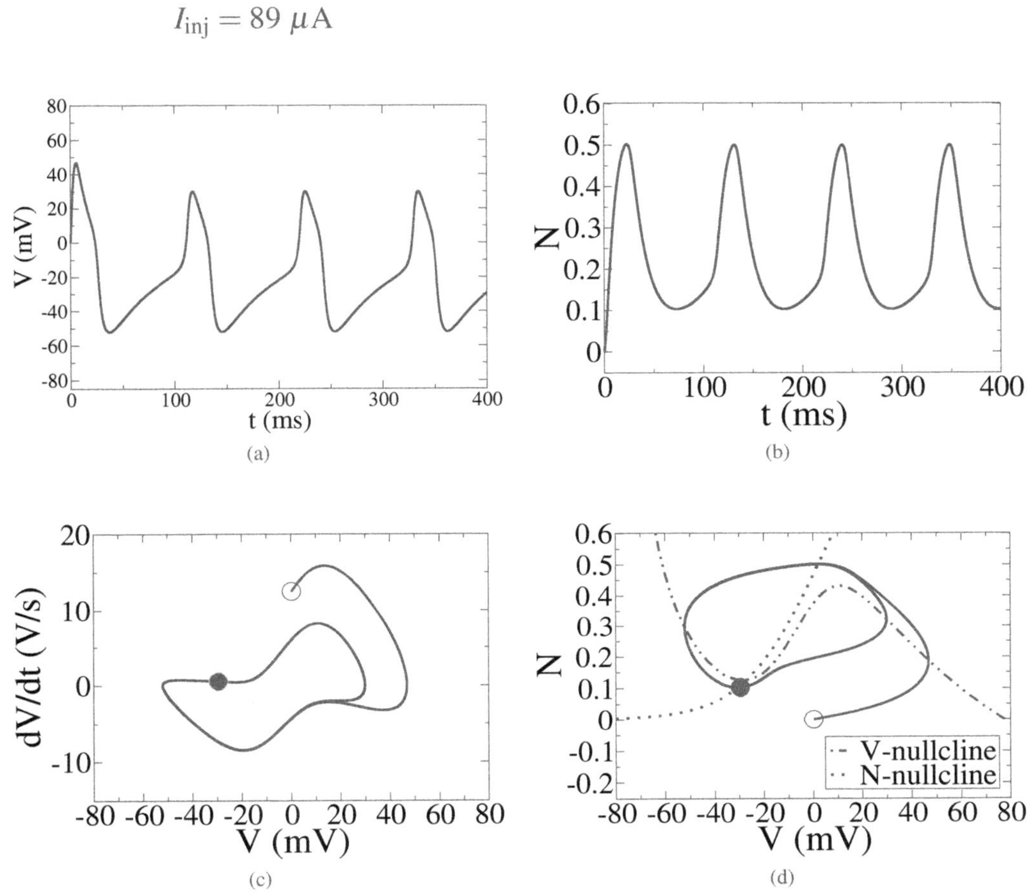

Figure 7.59 Numerical solutions of the Morris-Lecar equations for Set 1 (Hopf) using the parameters listed in table 7.2 and in the absence of external current $I_{\text{inj}} = 89\ \mu$A. Plotted are the time variation of the membrane potential V, the time variation of the potassium gating variable N, the phase portrait in the $(V, \frac{dV}{dt})$ phase space and the phase orbit in the N-V plane. The open and filled circles indicate, respectively, the start and finish of the simulation.

$I_{\text{inj}} = 50\ \mu A$: The plots for $I_{\text{inj}} = 50\ \mu$A in figure 7.65 indicate that the oscillations in the temporal development of V and N die after about 200 ms. The limit cycle in the $(V, \frac{dV}{dt})$ phase space is reduced to a dot-like infinitesimally small phase-space region. In the N-V plane, the limit cycle (stable equilibrium) and a saddle point (unstable equilibrium) have intersected each other. This behaviour could be interpreted as collision between the saddle point and the limit cycle.

The change observed in the dynamical behaviour in increasing the external current from $I_{\text{inj}} = 40\ \mu$A to $I_{\text{inj}} = 50\ \mu$A, *viz* the intersection of the stable equilibrium and an unstable equilibrium with each other, is an example of **homoclinic bifurcation**.

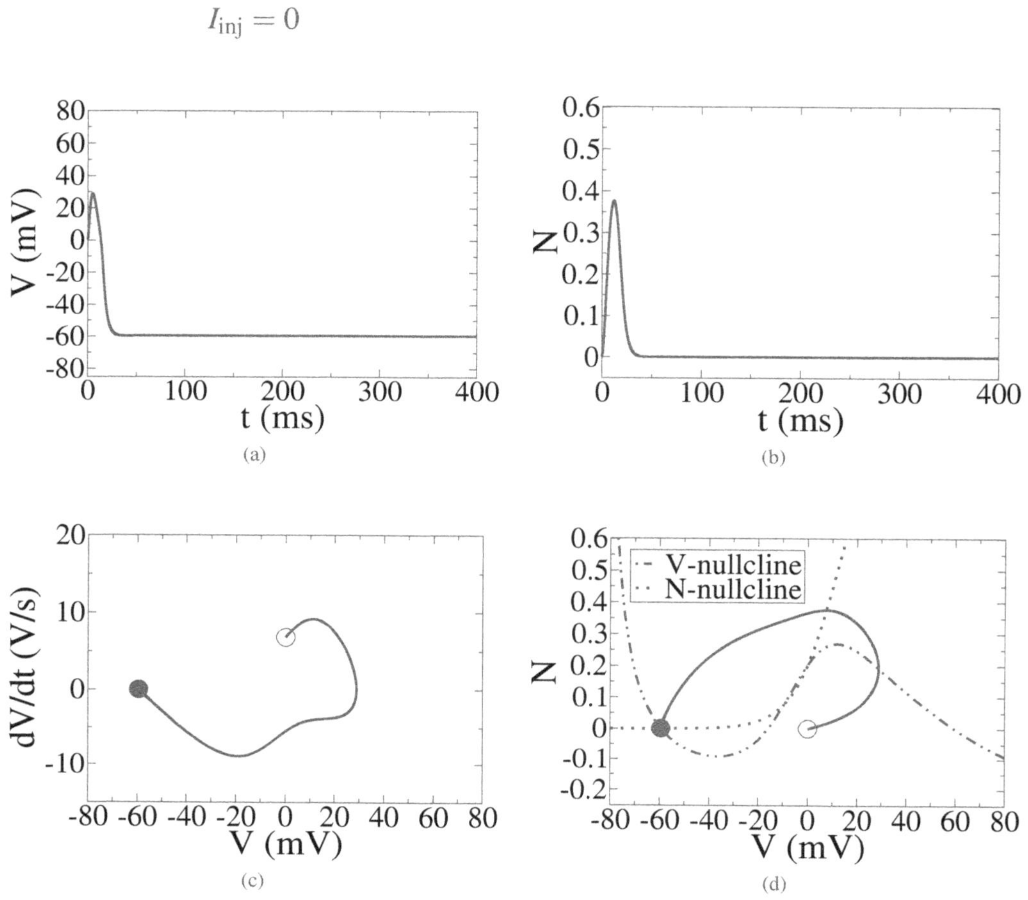

Figure 7.60 Numerical solutions of the Morris-Lecar equations for Set 2 (**saddle-node bifurcation on limit cycle**, or SNLC) of the parameters listed in table 7.2 and in the absence of external current ($I_{inj} = 0$). Plotted are the time variation of the membrane potential V, the time variation of the potassium gating variable N, the phase portrait in the $(V, \frac{dV}{dt})$ phase space and the phase orbit in the N-V plane. The open and filled circles indicate, respectively, the start and finish of the simulation.

7.7 NEURAL NETWORKS AND ARTIFICIAL INTELLIGENCE

7.7.1 INTRODUCTION

In the previous section we discussed several models of neuronal (or, neural) dynamics. Groups of interconnected neurons, called neural circuits, work together to perform specific functions in brain. There are a few different types of neural circuits. In order to understand how the brain processes information, it is important to understand how the highly complicated biological neural network (BNN) works and how it can be modified/tailored to achieve desired results. This is a highly formidable task. However, computational efforts using artificial neural networks (ANN), suitable mathematical and statistical models, and 'deep learning' processes can help in the study of neuroscience. Deep learning is a technique for implementing machine (computer) learning. Machine learning is the practice of using computer algorithms to analyse data, learn from it and make

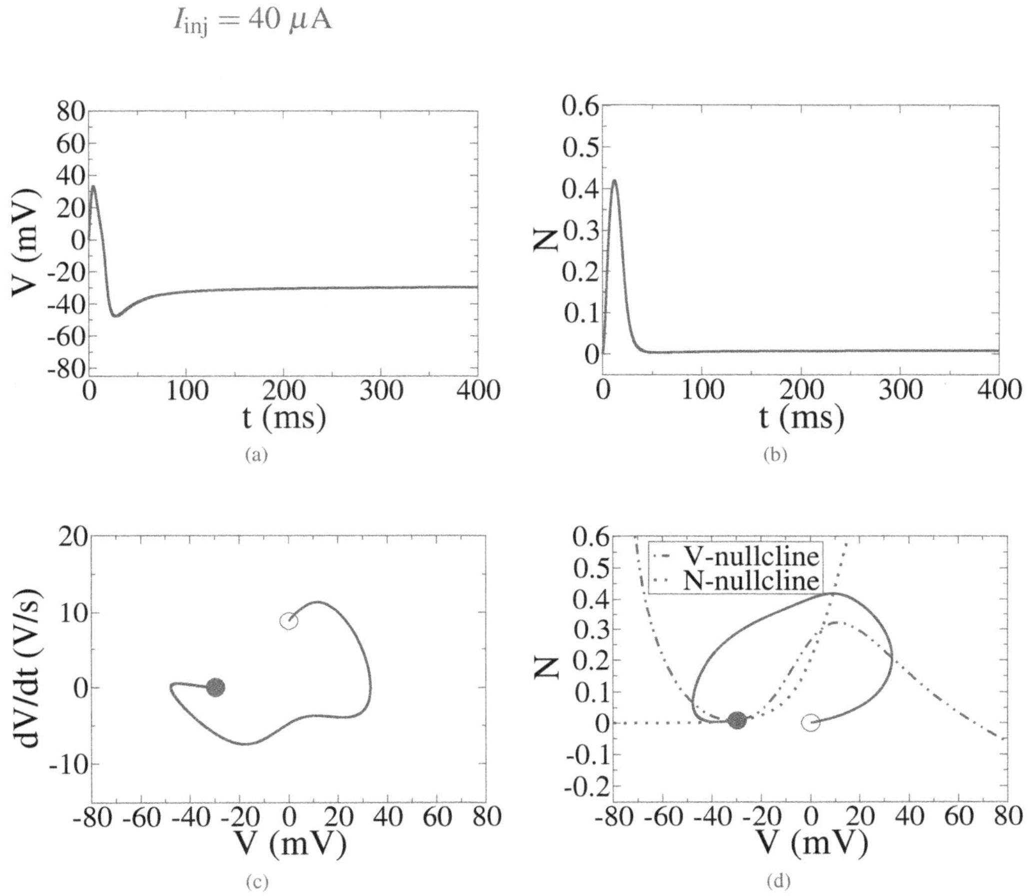

Figure 7.61 Numerical solutions of the Morris-Lecar equations for Set 2 (SNLC) of the parameters listed in table 7.2 and in the presence of external current $I_{inj} = 40\ \mu A$. Plotted are the time variation of the membrane potential V, the time variation of the potassium gating variable N, the phase portrait in the $(V, \frac{dV}{dt})$ space and the phase orbit in the N-V plane. The open and filled circles indicate, respectively, the start and finish of the simulation.

prediction. Artificial intelligence (AI) is the simulation of human intelligence using machine learning. Using the language of "biological computation", this means that ANNs provide an approach for building neural networks, and understanding and optimising their *collective* behaviour using AI.

It should be added here that the role of neural networks is not limited to the field of neuroscience or biological computation. In general, AI and chatboxes based on it, such as ChatGPT, are now routinely used as computer tools in many diverse areas in scientific, medical, agricultural, humanities and social sciences, where learning, problem solving, decision making and forecasting are required. Many useful review articles keep appearing in the literature and the reader should follow useful references provided in such publications. As an example, the review articles by Yang and Wang (2020), Surianarayanan *et al.* (2023) and Badrulhisham *et al.* (2024) discuss the use and limitations of machine learning in neuroscience. The articles by Hauer (2022) and Qian *et al.* (2024) discuss the importance and limitations of AI ethics in work and society.

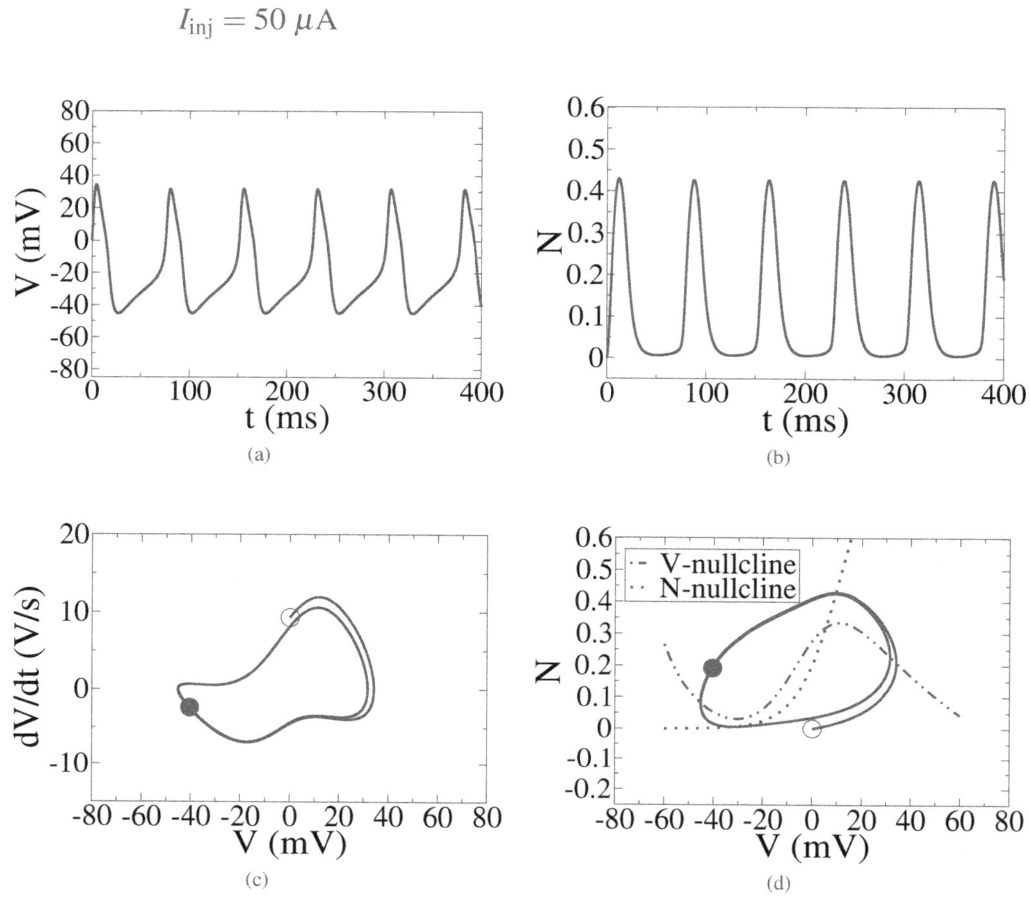

Figure 7.62 Numerical solutions of the Morris-Lecar equations for Set 2 (SNLC) of the parameters listed in table 7.2 and in the presence of external current $I_{inj} = 50\ \mu$A. Plotted are the time variation of the membrane potential V, the time variation of the potassium gating variable N, the phase portrait in the $(V, \frac{dV}{dt})$ phase space and the phase orbit in the N-V plane. The open and filled circles indicate, respectively, the start and finish of the simulation.

In the context of applications of non-linear dynamics discussed in this chapter, we will present a brief description of ANNs and their working principle.

7.7.2 ARTIFICIAL NEURAL NETWORKS

Artificial neural networks (ANNs) are generally described as layered structures, where 'layer' refers to a collection of 'units', or artificial neurons. There are several types of ANN, which can be differentiated based on their **feed-forward** and **feed-back** capabilities. In a feed-forward ANN there is an input layer, an output layer, one or more 'hidden' layers of artificial neurons and neural information is sent from the input layer towards the output layer. In the feed-back ANN, the output is fed back into the network to achieve optimised output result. Types of feed-forward ANN include: **single-layer feed-forward network**, **multi-layer feed-forward network**, **convolution network**, **recurrent network**, *etc*. The single-layer ANN only has one layer of neurons, which is also known

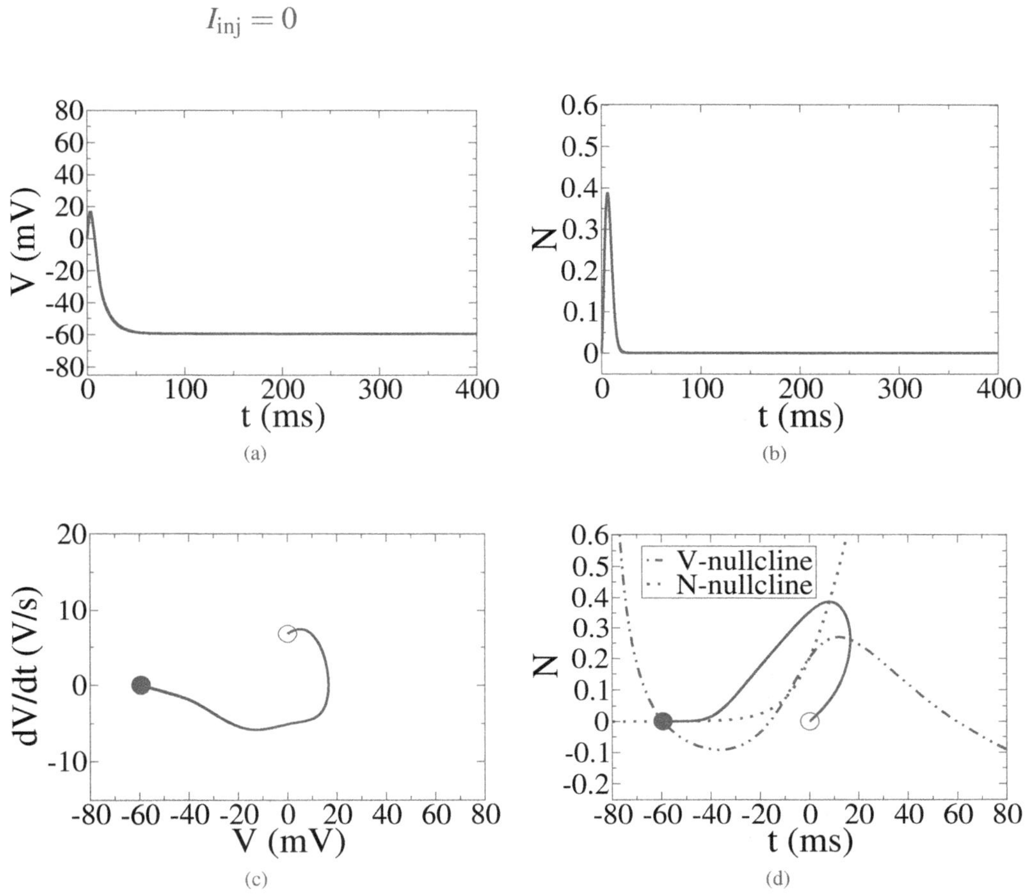

Figure 7.63 Numerical solutions of the Morris-Lecar equations for Set 3 (Homoclinic) of the parameters listed in table 7.2 and in the absence of external current ($I_{\text{inj}} = 0$). Plotted are the time variation of the membrane potential V, the time variation of the potassium gating variable N, the phase portrait in the $(V, \frac{dV}{dt})$ phase space and the phase orbit in the N-V plane. The open and filled circles indicate, respectively, the start and finish of the simulation.

as the 'output layer'. A single-layer ANN is also called a **perceptron**, after Rosenblatt (1958) who introduced the concept of ANN. A multi-layer ANN has an 'input' layer, an 'output layer' and one or more 'hidden layers' in between. A multi-layer feed-forward ANN can be referred to as a **multi-layer perceptron**. A N-layer perceptron has the output layer and (N-1) hidden layers. Figure 7.66 shows sketches of a perceptron (a single-layer ANN) and a two-layer perceptron (a two-layer ANN)[4].

7.7.3 WORKING PRINCIPLE OF ARTIFICIAL NEURAL NETWORKS

The working principle of an ANN can be described easily if we simply consider a single neuron in the perceptron and assume that it receives inter-synaptic *linear input* (in the form of signals, such

[4]The input layer is not counted as a layer of the ANN.

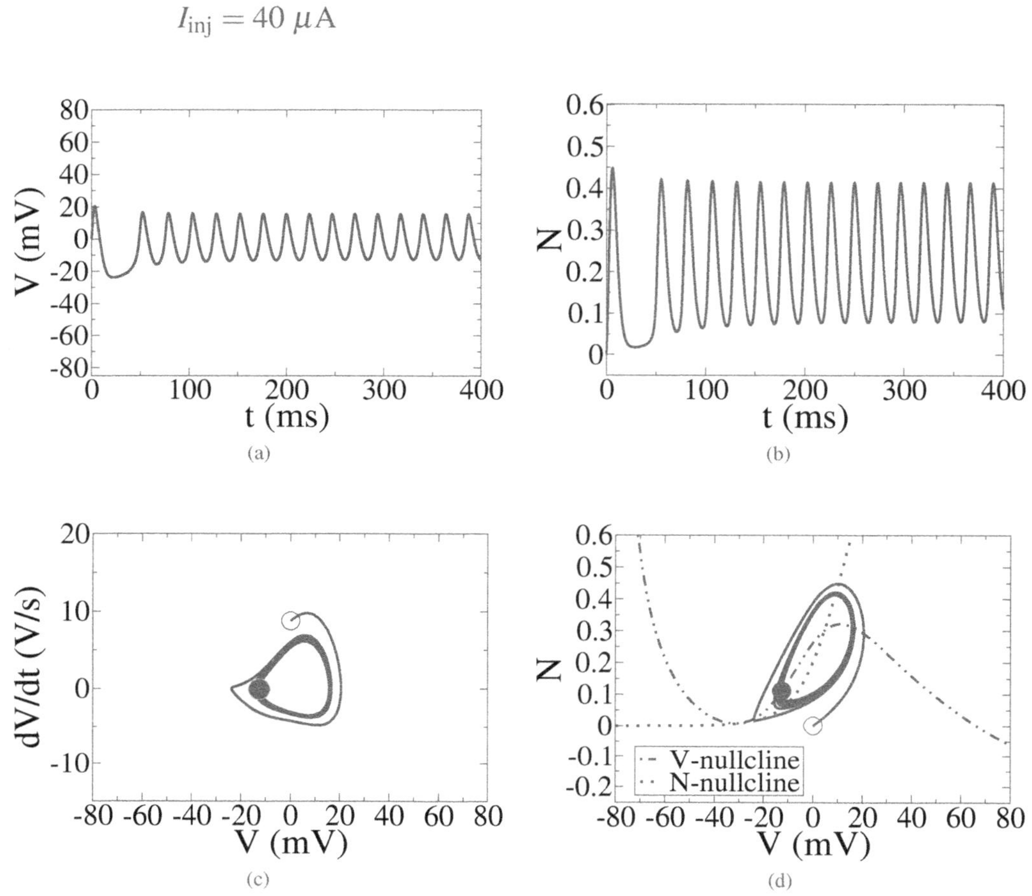

Figure 7.64 Numerical solutions of the Morris-Lecar equations for Set 3 (Homoclinic) of the parameters listed in table 7.2 and in the presence of external current $I_{\rm inj} = 40\ \mu$A. Plotted are the time variation of the membrane potential V, the time variation of the potassium gating variable N, the phase portrait in the $(V, \frac{dV}{dt})$ phase space and the phase orbit in the N-V plane. The open and filled circles indicate, respectively, the start and finish of the simulation.

as spikes) $\{x_i\}$ with their weights $\{w_i\}$ from a number of neurons. The artificial neuron computes the sum of weighted inputs, $\{w_i x_i\}$ and adds a bias b (if it is injected to the neuron) and then applies a non-linear function f, called **activation function**, for producing a *non-linear output*. This mathematical model is schematically shown in figure 7.67. The activation function f can be taken as the binary step function, defining the output y as

$$y(\boldsymbol{x}) = f(\boldsymbol{x}) = \begin{cases} 0 \text{ if } [\sum_j w_j x_j + b] \leq 0, \\ 1 \text{ if } [\sum_j w_j x_j + b] > 0. \end{cases} \tag{7.97}$$

Note that we have used the vector quantity $\boldsymbol{x}$ to denote all the inputs $x_1, x_2, \cdots$. As a useful alternative to adopting the binary step function, a continuous form of the activation function f can be

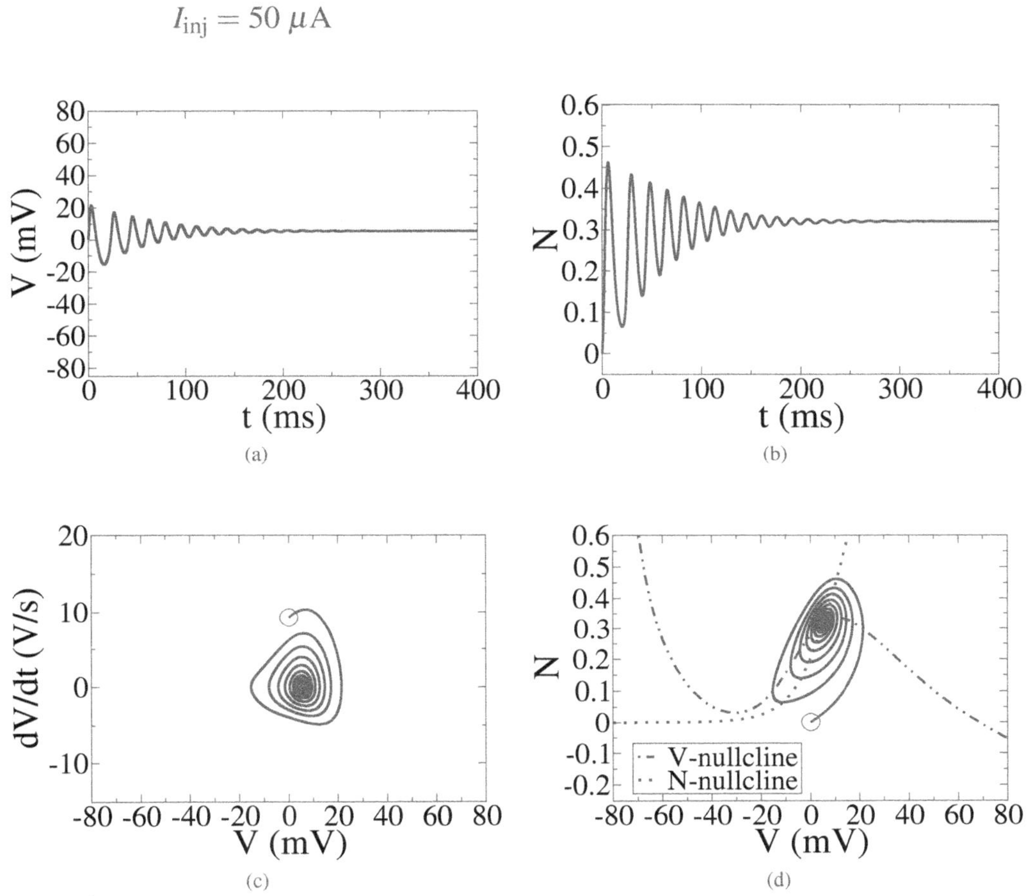

Figure 7.65 Numerical solutions of the Morris-Lecar equations for Set 3 (Homoclinic) of the parameters listed in table 7.2 and in the presence of external current $I_{inj} = 50\ \mu$A. Plotted are the time variation of the membrane potential V, the time variation of the potassium gating variable N, the phase portrait in the $(V, \frac{dV}{dt})$ phase space and the phase orbit in the N-V plane. The open and filled circles indicate, respectively, the start and finish of the simulation.

considered as the sigmoid function. The output, then, is expressed as

$$\begin{aligned} y(\boldsymbol{x}) &= \sigma\big(\sum_j w_j x_j + b\big), \\ &= \frac{1}{1+\exp\big[-\big(\sum_j w_j x_j + b\big)\big]}. \end{aligned} \tag{7.98}$$

The above description can be easily extended to the case when the perceptron contains several units (artificial neurons). The output will then be written as a vector quantity $\boldsymbol{y}$ to denote contributions from all neurons. Considering m inputs for each of n neurons in the output layer, we can generalise equation (7.98) to read the total output from the perceptron as

$$\boldsymbol{y}(\boldsymbol{x}) = \sum_{i=1}^{n} y_i(\boldsymbol{x}) = \sigma\big(\big(\sum_{i=1}^{n}\big(\sum_{j=1}^{m} w_{ij}x_j + b_i\big)\big) = \sigma(w\boldsymbol{x} + \boldsymbol{b}), \tag{7.99}$$

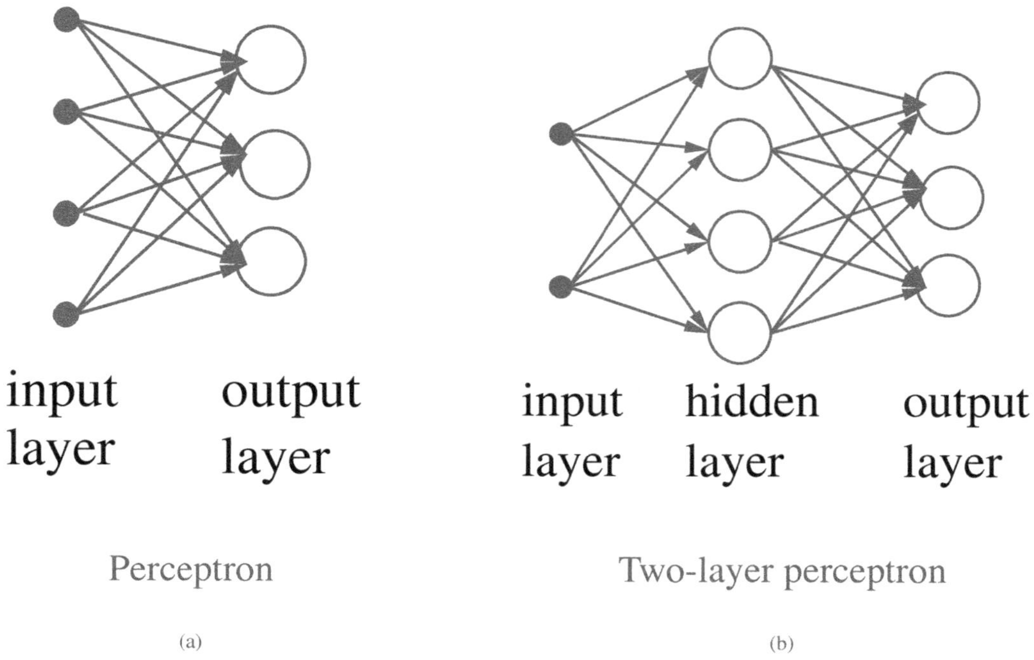

Figure 7.66 Examples of simple feed-forward artificial neural networks: (a) the perceptron, *i.e.* a single-layer feed-forward artificial neural network and (b) a two-layer perceptron. The perceptron has the input layer with 4 units and the output layer with 3 units. The two-layer perceptron has the input layer with 2 units, one hidden layer with 4 units and the output layer with 3 units. Open circles represent units (*i.e.* artificial neurons) and the much smaller filled circles comprise the input layer.

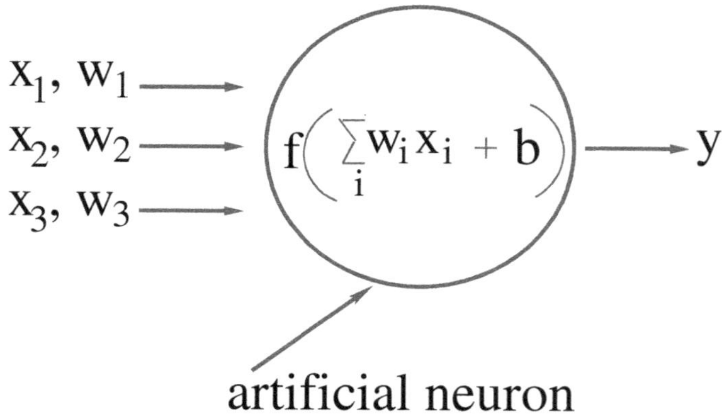

Figure 7.67 A mathematical model of a neuron in an artificial neural network. The neuron receives the weighted inputs $\{w_i x_i\}$ from other neurons, computes the sum $\sum_i w_i x_i$, adds the bias b and then applies an activation function $f = f(\sum_i w_i x_i + b)$ for producing the output y.

where now $\boldsymbol{y} = \{y_i\} = (y_1, y_2, \cdots, y_n)$ is a vector quantity, $w = \{w_{ij}\}$ is a $n \times m$ matrix with w_{ij} denoting the weight of the synapse connecting the jth neuron to the ith neuron and $\boldsymbol{b} = \{b_i\}$ represents biases received by the neurons.

The mathematical description for the one-layer perceptron can be extended to write down the output from a multi-layer perceptron. The output from the first layer becomes the input for the second layer, and so on. Let us assign a layer-index to the input, output, bias and weight factor, so that for the ith-layer we have $\boldsymbol{x}^{(i)}$, $\boldsymbol{b}^{(i)}$ and $w^{(i)}$. We express the output from an N-layer ANN as

$$\begin{aligned} \boldsymbol{x}^{(1)} &= \boldsymbol{x}, \quad \boldsymbol{b}^{(1)} = \boldsymbol{b}, \quad w^{(1)} = w, \\ \boldsymbol{x}^{(l)} &= f\big(w^{(l)}\boldsymbol{x}^{(l-1)} + \boldsymbol{b}^{(l)}\big), \quad 1 < l < N, \\ \boldsymbol{y} &= f\big(w^{(N)}\boldsymbol{x}^{(N-1)} + \boldsymbol{b}^{(N)}\big). \end{aligned} \tag{7.100}$$

In addition to the forward-propagation algorithm described above, it is important also to consider the back-propagation of signals in ANN. We will not discuss this here but refer the reader to books dedicated to the theory of neural computation, such as Hertz *et al.* (2018), Aggarwal (2023) and Herrmann *et al.* (2025).

7.7.4 MACHINE LEARNING: THEORETICAL CONCEPTS

An application of the working principle of an ANN requires inputting $\boldsymbol{x}^{(l)}$ with the corresponding inter-synaptic weight factor $w^{(l)}$ for all units in every lth neuron layer. A chosen ANN architecture seeks to produce a 'target output' $\boldsymbol{y}_{\text{target}}$. Specific values of the elements of the matrix $w^{(l)}$ are randomly assigned at initialisation. The resulting output $\boldsymbol{y}$ is usually very different from the target output. In order to reduce the difference between the output $\boldsymbol{y}$ and the target output $\boldsymbol{y}_{\text{target}}$, the ANN is 'trained' to optimise the matrix w. The objective function to minimise is the so-called loss function, or the square error function

$$E = \frac{1}{N}\sum_l E(\boldsymbol{y}^{(l)}, \boldsymbol{y}^{(l)}_{\text{target}}), \tag{7.101}$$

where $E(\boldsymbol{y}^{(l)}, \boldsymbol{y}^{(l)}_{\text{target}})$ is the difference between the target output $\boldsymbol{y}^{(l)}_{\text{target}}$ and the actual output $\boldsymbol{y}^{(l)}$. It is usually too expensive to compute the loss function for the entire ANN system. In practice, the loss minimisation is attempted by sampling a small number of randomly selected subset.

For a single-layer ANN (the perceptron), we write the square error function as

$$E = \frac{1}{2}||\boldsymbol{y} - \boldsymbol{y}_{\text{target}}||^2 = \frac{1}{2}\sum_i (y_i - y_{\text{target},i})^2. \tag{7.102}$$

This function should be minimised with respect to w. Widrow and Hoff (1960) suggested adjusting w iteratively until E is minimised. Iterative approaches, such as the Newton-Raphson algorithm, the gradient descent algorithm and conjugate gradients algorithms, are applied to find the optimum weight matrix w. A detailed discussion of these can be found in Hetrtz *et al.* (2018).

In the quasi-**Newton-Raphson method**, the matrix w is iteratively solved using the equation

$$w^r = w^{(r-1)} - \eta \mathscr{G} \nabla_w E, \tag{7.103}$$

where η is a small step size called the **learning rate** and $\mathscr{G} \approx J^{-1}$ is an approximation to the inverse of the Jacobian matrix J with elements

$$J_{ij} = \frac{\partial^2 E}{\partial x_i \partial x_j}. \tag{7.104}$$

Several inverse-Jacobian updating procedures have been proposed. We will not discuss this point here but refer the reader to the works of Broyden (1965), Srivastava (1984) and Eyert (1996) for detailed discussion.

In the **gradient descent method** the weight matrix is adjusted by moving a small distance in the w-space down the gradient of E. Differentiation of E with respect to an element w_{ij} gives

$$\frac{\partial E}{\partial w_{ij}} = (y_i - y_{\text{target,i}})x_j, \tag{7.105}$$

from which we can express the error in w_{ij} as

$$\Delta w_{ij} \propto -\frac{\partial E}{\partial w_{ij}} = \eta\,(y_{\text{target,i}} - y_i)x_j, \tag{7.106}$$

where η is a small step size, called the **learning rate**. The expression for the error weight matrix Δw_{ij} in equation (7.106), which depends simultaneously on the input x_j to the jth layer and the output y_i from the ith layer, is of the form of the Hebbian learning (Hebb, 1949). A weight matrix element w_{ij} at the rth iteration is updated as

$$\begin{aligned} w_{ij}^{r} &= w_{ij}^{(r-1)} + \Delta w_{ij}, \\ &= w_{ij}^{(r-1)} + \eta\,(y_{\text{target,i}} - y_i^{(r-1)})x_j. \end{aligned} \tag{7.107}$$

If a known value of η is used, the procedure is known as the **supervised learning** rule for optimising the weight factors w_{ij}. The iterative procedure is stopped when the value of the square error E drops below an acceptable level.

FURTHER READING

Electrical circuits

1. Chua L O 1984 *IEEE Transactions on Circuits and Systems* **CAS-31** 69
2. Homes P 1979 *Phil Trans Roy Soc London: A. Math & Phys Sciences* **292** 419
3. Kennedy M P 1993 *IEEE Trans Circuits and Systems – I: Fundamental Theory and Applications* **40** 657
4. Parker T S and Chua L O 1987 *Proc IEEE* **75** 1081
5. Zhong G Q and Ayrom F 1985 *Int J Circuit Theory Appl* **13** 93

Astrophysics

1. Wild J 1980 *Am J Phys* **48** 297
2. Horner J, Vervoort P, Kane S R, Ceja A, Waltham D, Gilmore J and Turner S K 2020 *The Astronomical Journal* **159** 10

Non-linear weather and climate dynamics

1. Lorentz E N 1963 *J Atmos Sci* **42** 433
2. Palmer T N 1991 *The New Scientist Guide to Chaos* ed Nina Hall (Penguin Books: London); 1993 *Weather* **48** 314; 1999 *J Clim* **12** 575; 2000 *Rep Prog Phys* **63** 71
3. Saltzmann B 1962 *J Atmos Sci* **19** 329
4. Strogatz S H 1994 *Nonlinear Dynamics and Chaos* (Westview: Cambridge, MA)

Non-linear dynamical model for electrocardiogram signals

1. Cheffer A, Savi M A, Pereira T L and de Paula A S 2021 *Appl Math Modelling* **96** 152
2. Gois S R F S M and Savi M A 2009 *Chaos, Solitons and Fractals* **41** 2553
3. Grudziǹski K and Żebrowski J J 2004 *Physica* A **336** 153
4. Kuate G C G and Fotsin H B 2022 *Phys Scr* **97** 045205

5. McSharry P E *et al.* 2003 *IEEE Trans on Biomed Engg* **50** 289

Non-linear neural dynamics

1. Azizi T and Mugabi R 2020 *Applied Mathematics* **11** 203
2. Ermentrout G B and Terman D H 2010 *Mathematical Foundations of Neuroscience* Vol 35 (Springer Science & Business Media: Berlin)
3. Hansel D and Mato G 2001 *Phys Rev Lett* **86** 4175
4. Izhikevich E M 2003 *IEEE Trans Neural Networks* **14** 1569; 2004 *IEEE Trans Neural Networks* **15** 1063; 2007 *Dynamical Systems in Neuroscience* (MIT Press: Cambridge, Massachusetts)
5. Knight B W 1972 *J Gen Physio* **59** 734
6. Lapicque L 1907 *J Physiol Paris* **9** 620
7. Latham P E *et al.* 2000 *J Neurophysiology* **83** 808
8. Morris C and Lecar H 1981 *Biophys J* **35** 193
9. Stein R B 1967 *Proc R Soc Lond* **167** 64
10. Sterratt D, Graham B, Gillies A and Willshaw D 2023 *Principles of Computational Modelling in Neuroscience* 2nd edn (Cambridge University Press: Cambridge)
11. Strogatz S H 1994 *Nonlinear Dynamics and Chaos* (Westview: Cambridge, MA)
12. Tsumoto K *et al.* 2006 Neurocomputing **69** 293
13. Wiggins S 2003 *Introduction to Applied Nonlinear Dynamical Systems and Chaos* 2nd edn (Springer: New York)

Neural networks and artificial intelligence

1. Hebb D O 1949 *The Organization of Behavior: A Neuropsychological Theory* (Wiley: New York)
2. Rosenblatt F 1958 *Psychol Rev* **65** 386
3. Widrow B and Hoff Jr M E 1960 *IRE WESCON Conv Rec* **4** 96
4. Hertz J, Krogh A and Palmer R G 2018 *Introduction to the Theory of Neural Computation* (CRC: Boca Raton)
5. Yang G R and Wang X-J 2020 *Neuron* **107** 1048
6. Aggarwal C C 2023 *Neural Networks and Deep Learning* 2nd edn (Springer: Switzerland)
7. Herrmann L, Jokeit M, Weeger O and Kollmannsberger S 2025 *Deep Learning in Computational Mechanics – An Introductory Course* 2nd edn (Springer: Switzerland)

PROBLEMS

Section: *Non-linear Electrical Circuits*

1. Consider an LC circuit with $L = 2 \times 10^{-3}$ H and $C = 1 \times 10^{-6}$ F driven by a time dependent voltage of amplitude $V_0 = 1$ volt and angular frequency Ω. Write a computer program using the Euler-Cromer algorithm to obtain numerical solutions for $f = \frac{\Omega}{2\pi} = 1000$ Hz. Use initial charge and current as $Q_0 = 7.5 \times 10^{-3}$ F and $I_0 = 0.0$. Plot variation of charge Q with time t, variation of current I with time t, and phase orbits in the (Q-I) plane.
2. Turn the LC circuit in the previous question into an LCR circuit by inserting a resistance $R = 5$ ohm is series with L and C. Using the Euler-Cromer algorithm, obtain numerical solutions for $f = \frac{\Omega}{2\pi} = 1000$ Hz. Use initial charge and current as $Q_0 = 7.5 \times 10^{-3}$ F and $I_0 = 0.0$. Plot variation of charge Q with time t, variation of current I with time t, and phase orbits in the (Q-I) plane. Comment on the changes in the results compared to those for the previous question.

3. Write a computer program based on the Euler-Cromer algorithm to investigate the chaotic dynamics of the Duffin-Holmes oscillator. Compare your results with those presented in figures 7.8–7.12, which were obtained by using the SPICE package.

Section: *Celestial dynamics*

1. Write a computer program, using either the Euler-Cromer algorithm or the Runge-Kutta algorithm, to reproduce the Earth and Jupiter orbits around the Sun. Does Jupiter have any effect on Earth's orbit?
2. Write a computer program using the Euler-Cromer algorithm to analyse the orbits of two planets in figure 7.20 by considering $m_1/M = 0.001$ and $m_2/M = 0.01$. Discuss the qualitative effect of the mutual interaction between the two planets.
3. Consider Euler's three-body problem discussed in section 7.3.2 with $M_1 = M_2 = M_{\text{Sun}}$ and $d = 0.2$ AU. Discuss the orbital motion of the planet when it is assigned the initial conditions, in Astronomical Units, $\boldsymbol{r} = (1.0, 0.0)$ and $\boldsymbol{v} = (0.0, 2\pi)$.
4. Consider Euler's three-body problem discussed in section 7.3.2 with $M_1 = M_2 = M_{\text{Sun}}$ and inter-star separation $d = 2.01$ AU. Set the initial conditions, in Astronomical Units, $\boldsymbol{r} = (1.0, 1.0)$ and $\boldsymbol{v} = (0.0, 2)$. Discuss the planet's orbit.

Section: *Non-linear weather and climate dynamics*

1. Plot the time variation of the Lorentz variables $Y(t)$ and $Z(t)$ with the parameters used for figure 7.25. Discuss similarities and differences with the plot for $X(t)$.
2. Use the Euler algorithm and the parameters in figure 7.26 to plot trajectories in the X-Y and X-Z planes. Make observations and relate these to the plot in figure 7.26.
3. Using either the Euler algorithm or any other algorithm, compute 3D Lorentz trajectories with the same values of σ, r and b as in figure 7.26 but slightly different initial conditions for $\left(X(0), Y(0), Z(0)\right)$. Discuss differences that arise in the solutions.

Section: *Non-linear model for electrocardiogram signals*

1. In figures 7.34 – 7.37 we studied the ECG features with atrial flutter and ventricular flutter rhythms. Write a computer program to simulate the ECG with atrial fibrillation and ventricular fibrillation rhythms. Follow the research paper by Cheffer *et al.* (2021) and use the parameters therein. Discuss your results and compare with the results presented in that paper.

Section: *Non-linear neural dynamics*

1. Derive the analytical solution in equation (7.71). Write a computer program using either the Euler algorithm or a Runge-Kutta algorithm and plot the time variation of the potential. Compare the numerical results with the analytical results.
2. Write a computer code using the Euler algorithm to obtain time variation of the membrane potential in the presence of two sets of current pulses : (a) six pulses of equal height but random widths, and (b) six pulses of equal width but randomly different heights. Use the membrane resting potential -70 mV, membrane resistance 10 kΩ, time constant 10 ms and the potential threshold -65 mV and -59 mV for the first and second set of pulses, respectively. Comment on differences between the results for the two sets.
3. Show, either algebraically or using a computer simulation, that, for $b = 0$, equation (7.75) is reduced to a form of the quadratic integrate-and-fire model.

4. Plot and compare the instantaneous and steady-state currents for the Izhikevich and Hindmarsh-Rose models of neural dynamics. Use the parameters listed in the captions for figures 7.47 and 7.50.
5. Using a computational simulation of the Hodgkin-Huxley model, show that injection of a pulse of greater amplitude and shorter duration produces the action potential of shorter amplitude and which rises more rapidly and with greater depolarisation.
6. What determines the shape of the nullclines in the Morris-Lecar model of neuronal dynamics?
7. Plot the V- and N-nullclines for the three sets (Hopf, SNLC and Homoclinic cases) in the Morris-Lecar model of neuronal dynamics with different amounts of injected current I_{inj} and discuss the number and nature of stable solutions.

Section: *Neural networks and artificial intelligence*

1. *Linear regression:* With reference to artificial intelligence and machine learning, consider an error function $E = (y - y_{\text{target}})^2$, where $y = wx + b = 0.8x + 0.1$ provides sample data for the variable x in the range $0.1 \leq x \leq 1.0$ and y_{target} is a target function. Derive an expression for the linear regression algorithm for w similar to that in equation (7.107) and using a computer program with the learning rate $\eta = 0.1$ show that optimum linear expression $y_{\text{target}} = x$ can be reached after a few iterative steps. Try other values of the learning rate, such as $\eta = 0.05$ and $\eta = 0.2$, and comment on the change in the rate of convergence of the results.
2. *Quadratic regression:* With reference to artificial intelligence and machine learning, consider an error function $E = (y - y_{\text{target}})^2$, where $y = wx^2 + b = 0.8x^2 + 0.1$ provides sample data for variables x and y_{target} is a target function. Derive an expression for the quadratic regression algorithm for w that modifies equation (7.107). Using a computer program with the learning rate $\eta = 0.1$ show that optimum linear expression $y_{\text{target}} = x^2$ can be reached after a few iterative steps. Try other values of the learning rate, such as $\eta = 0.05$ and $\eta = 0.2$, and comment on the change in the rate of convergence of the results.

Part V

Elements of classical field theory

8 Lagrangian and Hamiltonian of a Classical Field

8.1 INTRODUCTION

Classical field theory (CFT) may be considered as a many-body version of the classical mechanics (CM) of particles and objects as we have discussed in previous chapters. In other words, CFT may be considered as an extension of classical mechanics (CM) for the analysis of systems with many particles, *i.e.* systems with a large number of degrees of freedom, with or without the consideration of special relativity (SR). CFT can be applied to various branches of theoretical physics, such as condensed matter (including the topic considered in part III), gravitation, electromagnetism and electrodynamics. Table 8.1 shows the relationship between CM and CFT.

Table 8.1
Relationship between classical mechanics (CM) of a system of N particles, special relativity (SR) and classical field theory (CFT).

few particles (small N)	many-body system ($N \to \infty$)	$N \to \infty$ and SR
CM	non-relativistic CFT	relativistic CFT

So, what is a 'classical field'?. A 'field' is existence of happening of an event, or simply presence of a variable in space and time. 'Classical' obviously means description based on firm principles discussed in previous chapters of this book. Note that in dealing with CM we used particle positions and their time dependence, *viz* $\{\boldsymbol{R}_i(t)\}$. When dealing with classical field theory we invoke the concept of a **wave field** which is described by its amplitude at all points in space and the time dependence of the amplitude. We will use the symbol $\phi(\boldsymbol{r},t)$ for the amplitude of a wave field at point $\boldsymbol{r}$ and time t. It is important to note that $\phi(\boldsymbol{r},t)$ is not necessarily a scalar quantity. As ϕ is defined at all points in space, we can express the Lagrangian L as an integral

$$L = \int \mathscr{L} \mathrm{d}^3 r, \tag{8.1}$$

where $\mathscr{L}$ is the 'Lagrangian density'. The Lagrangian density $\mathscr{L}$ is a **functional**[1] that, due to continuous dependence of ϕ on $\boldsymbol{r}(t)$, can in general be expressed as $\mathscr{L} = \mathscr{L}(\phi, \boldsymbol{\nabla}\phi, \nabla^2\phi, ..., \frac{\mathrm{d}\phi}{\mathrm{d}t}, t)$. However, for most physical systems of interest we can drop higher derivatives of ϕ and write

$$\mathscr{L}(\boldsymbol{r}) = \mathscr{L}(\phi, \boldsymbol{\nabla}\phi, \frac{\mathrm{d}\phi}{\mathrm{d}t}, t). \tag{8.2}$$

We may also refer to $\phi(\boldsymbol{r},t)$ as a field coordinate. If we do this, then it is easy to see the transition from CM to CFT as described in table 8.2.

[1] A functional is a function involving another function or functions.

DOI: 10.1201/9781003383314-8

Table 8.2
Change of variables and quantities for making a transition from particle classical mechanics to classical field theory.

Classical mechanics	Classical field theory
Particle position variable $\boldsymbol{R}(t)$	Wave field amplitude $\phi(\boldsymbol{r},t)$
Particle velocity $\frac{\mathrm{d}\boldsymbol{R}(t)}{\mathrm{d}t}$	Wave field amplitude velocity $\frac{\mathrm{d}\phi(\boldsymbol{r},t)}{\mathrm{d}t}$
Lagrangian $L(\boldsymbol{R},\frac{\mathrm{d}\boldsymbol{R}}{\mathrm{d}t},t)$	Lagrangian density $\mathscr{L}(\phi,\nabla\phi,\frac{\mathrm{d}\phi}{\mathrm{d}t},t)$

8.2 LAGRANGIAN CLASSICAL FIELD THEORY

In order to develop the Lagrangian classical field theory we apply the Hamilton's variational principle described in section 2.3.6. The action integral associated with the Lagrangian density $\mathscr{L}$ in equation (8.2) is

$$J=\int L\mathrm{d}t=\int\int \mathscr{L}(\phi,\nabla\phi,\frac{\mathrm{d}\phi}{\mathrm{d}t},t)\mathrm{d}t\mathrm{d}^3r. \tag{8.3}$$

Following the discussion in Appendix C, the Euler-Lagrange variational principle is then

$$\delta J=\delta\int_{t_1}^{t_2} L\mathrm{d}t=\delta\int_{t_1}^{t_2}\int\mathscr{L}\mathrm{d}t\mathrm{d}^3r=\int_{t_1}^{t_2}\int\delta\mathscr{L}\mathrm{d}t\mathrm{d}^3r=0, \tag{8.4}$$

subject to $\delta\phi(\boldsymbol{r},t_1)=\delta\phi(\boldsymbol{r},t_2)=0$. Further, following the discussion in Appendix C, the variation $\delta\mathscr{L}$ can be written as

$$\begin{aligned}\delta\mathscr{L} &= \frac{\partial\mathscr{L}}{\partial\phi}\delta\phi+\sum_\alpha\frac{\partial\mathscr{L}}{\partial(\frac{\partial\phi}{\partial r_\alpha})}\delta(\frac{\partial\phi}{\partial r_\alpha})+\frac{\partial\mathscr{L}}{\partial(\frac{\partial\phi}{\partial t})}\delta(\frac{\partial\phi}{\partial t})\\ &= \frac{\partial\mathscr{L}}{\partial\phi}\delta\phi+\sum_\alpha\frac{\partial\mathscr{L}}{\partial(\frac{\partial\phi}{\partial r_\alpha})}\frac{\partial(\delta\phi)}{\partial r_\alpha}+\frac{\partial\mathscr{L}}{\partial(\frac{\partial\phi}{\partial t})}\frac{\partial(\delta\phi)}{\partial t},\end{aligned} \tag{8.5}$$

where $r_\alpha=x,$ or $y,$ or z. Thus, Hamilton's variational principle in the context of the classical field theory reads

$$\int_{t_1}^{t_2}\int\Big[\frac{\partial\mathscr{L}}{\partial\phi}\delta\phi+\sum_\alpha\frac{\partial\mathscr{L}}{\partial(\frac{\partial\phi}{\partial r_\alpha})}\frac{\partial(\delta\phi)}{\partial r_\alpha}+\frac{\partial\mathscr{L}}{\partial(\frac{\partial\phi}{\partial t})}\frac{\partial(\delta\phi)}{\partial t}\Big]\mathrm{d}t\mathrm{d}^3r=0. \tag{8.6}$$

To proceed, we integrate by parts the second term with respect to the space coordinates and the third term with respect to the time. The second term in equation (8.6) becomes

$$\begin{aligned}\text{second term} &= \int_{t_1}^{t_2}\Big[\sum_\alpha\frac{\partial\mathscr{L}}{\partial(\frac{\partial\phi}{\partial r_\alpha})}(\delta\phi)|_{r_\alpha\to-\infty}^{r_\alpha\to+\infty}-\int\sum_\alpha\frac{\partial}{\partial r_\alpha}\Big[\frac{\partial\mathscr{L}}{\partial(\frac{\partial\phi}{\partial r_\alpha})}\Big]\delta\phi\mathrm{d}^3r\Big]\mathrm{d}t,\\ &= -\int_{t_1}^{t_2}\int\sum_\alpha\frac{\partial}{\partial r_\alpha}\Big[\frac{\partial\mathscr{L}}{\partial(\frac{\partial\phi}{\partial r_\alpha})}\Big]\delta\phi\mathrm{d}t\mathrm{d}^3r.\end{aligned} \tag{8.7}$$

The first integral in the first line of the above equation, which corresponds to the 'end' or 'surface' contribution, vanishes because the variable ϕ falls off to zero at infinitely large distances (or can be eliminated by applying the periodic boundary condition discussed in section 4.5 for a large but finite size of the system). With the same consideration, the third term in equation (8.6) can be reduced to

$$\text{third term}=-\int_{t_1}^{t_2}\int\frac{\partial}{\partial t}\Big(\frac{\partial\mathscr{L}}{\partial(\frac{\partial\phi}{\partial t})}\Big)\delta\phi\mathrm{d}t\mathrm{d}^3r. \tag{8.8}$$

With equations (8.7) and (8.8), and treating $\delta\phi$ as an arbitrary variation, equation (8.6) requires the following to be satisfied

$$\frac{\partial \mathscr{L}}{\partial \phi} = \sum_{\alpha} \frac{\partial}{\partial r_\alpha}\left[\frac{\partial \mathscr{L}}{\partial(\frac{\partial \phi}{\partial r_\alpha})}\right] + \frac{\partial}{\partial t}\left(\frac{\partial \mathscr{L}}{\partial(\frac{\partial \phi}{\partial t})}\right). \tag{8.9}$$

This differential equation is the statement of the Lagrangian classical field theory.

The Lagrangian classical field theory has been widely applied in different branches of theoretical physics. Here we will present a few examples, including electromagnetism and vibrational/oscillatory behaviour of atoms in crystalline structures and of elastic continuum structures.

8.2.1 ELECTROMAGNETISM

We begin by writing Maxwell's equations in vacuum

$$\begin{aligned} \nabla\cdot\boldsymbol{E} = 0, \quad \nabla\times\boldsymbol{E} = -\frac{\partial \boldsymbol{B}}{\partial t}, \\ \nabla\cdot\boldsymbol{B} = 0, \quad \nabla\times\boldsymbol{B} = \frac{1}{c^2}\frac{\partial \boldsymbol{E}}{\partial t}, \end{aligned} \tag{8.10}$$

where $\boldsymbol{E}$ and $\boldsymbol{B}$ respectively are the electric and magnetic fields, and c is the speed of light in vacuum. And we express the vector fields $\boldsymbol{E}$ and $\boldsymbol{B}$ in terms of a scalar potential Φ and a vector potential $\boldsymbol{A}$ in the form[2]

$$\begin{aligned} \boldsymbol{E} &= -\nabla\Phi - \frac{\partial \boldsymbol{A}}{\partial t}, \\ \boldsymbol{B} &= \nabla\times\boldsymbol{A}. \end{aligned} \tag{8.11}$$

We express the Lagrangian density of the electromagnetic field as

$$\mathscr{L} = \frac{1}{2}\left(\frac{\partial \boldsymbol{A}}{\partial t} + \nabla\Phi\right)^2 - \frac{1}{2}c^2\left(\nabla\times\boldsymbol{A}\right)^2. \tag{8.12}$$

This Lagrangian density is a function of four variables Φ, A_x, A_y and A_z, *i.e.* $\mathscr{L} = \mathscr{L}(\Phi,\boldsymbol{A}) = \mathscr{L}(\Phi,A_x,A_y,A_z)$. As $\mathscr{L}$ is not an explicit function of Φ and $\boldsymbol{A}$, equation (8.9) can be fully expressed in the form of the following equations

$$\phi = \Phi: \qquad 0 = \sum_{\alpha} \frac{\partial}{\partial r_\alpha}\left[\frac{\partial \mathscr{L}}{\partial(\frac{\partial \Phi}{\partial r_\alpha})}\right], \quad \alpha = x,y,z, \tag{8.13}$$

$$\phi = \boldsymbol{A}: \quad 0 = \sum_{\alpha} \frac{\partial}{\partial r_\alpha}\left[\frac{\partial \mathscr{L}}{\partial(\frac{\partial A_\beta}{\partial r_\alpha})}\right] + \frac{\partial}{\partial t}\left(\frac{\partial \mathscr{L}}{\partial(\frac{\partial A_\beta}{\partial t})}\right), \quad \alpha,\beta = x,y,z. \tag{8.14}$$

It will be left as an exercise to show that equations (8.13) and (8.14) can be expressed respectively as

$$0 = \nabla\cdot\left(\frac{\partial \boldsymbol{A}}{\partial t} + \nabla\Phi\right), \tag{8.15}$$

$$0 = -c^2\nabla\times\nabla\times\boldsymbol{A} + \frac{\partial}{\partial t}\left(\frac{\partial \boldsymbol{A}}{\partial t} + \nabla\Phi\right). \tag{8.16}$$

Using the definitions in equation (8.11), equations (8.15) and (8.16) read as $\nabla\cdot\boldsymbol{E} = 0$ and $\nabla\times B = \frac{1}{c^2}\frac{\partial \boldsymbol{E}}{\partial t}$, respectively. These are two of Maxwell's equations written in equation (8.10).

---

[2]The symbols Φ and $\boldsymbol{A}$ used in this sub-section should not be confused with similar symbols used in the sub-sections not dealing with electromagnetism.

8.2.2 MONATOMIC LINEAR CHAIN

In section 4.4 we discussed the harmonic oscillations of an infinitely large number N of atoms arranged regularly in the form of a monatomic linear chain. Figure 4.4 shows the atomic arrangement. Each atom, of mass m, is separated from its neighbouring atoms by distance a. Only nearest neighbour linear forces are considered, with the harmonic force constant Λ. At a given instance nth atom is displaced from its equilibrium by u_n. The Lagrangian of this system is

$$L = \frac{1}{2}m\sum_n\left(\frac{\mathrm{d}u_n}{\mathrm{d}t}\right)^2 - \frac{1}{2}\Lambda\sum_n(u_{n+1}-u_n)^2. \tag{8.17}$$

Due to the translations symmetry of the chain, there is just one atom per unit cell. This allows us to define the Lagrangian density $\mathscr{L}$ as the Lagrangian per unit cell: $L=\sum_n\mathscr{L}$. The expression for $\mathscr{L}$ is thus,

$$\mathscr{L} = \frac{1}{2}m\left(\frac{\mathrm{d}u_n}{\mathrm{d}t}\right)^2 - \frac{1}{2}\Lambda(u_{n+1}-u_n)^2. \tag{8.18}$$

Considering $\phi = u_n$, equation (8.9) becomes

$$\begin{aligned} \frac{\mathscr{L}}{\mathrm{d}u_n} &= \frac{\mathrm{d}}{\mathrm{d}t}\left(\frac{\mathrm{d}\mathscr{L}}{\mathrm{d}(\frac{\mathrm{d}u_n}{\mathrm{d}t})}\right), \\ -\Lambda(2u_n-u_{n-1}-u_{n+1}) &= m\frac{\mathrm{d}^2u_n}{\mathrm{d}t^2}, \end{aligned} \tag{8.19}$$

where we have noted that the nth atom is spring attached to its neighbours, *viz* to the $(n-1)$st and $(n+1)$st atoms. This is the equation of motion obtained in equation (4.23) using the Newtonian mechanics.

8.2.3 DIATOMIC LINEAR CHAIN

The Lagrangian of the diatomic linear chain discussed in section 4.5.2 is

$$\begin{aligned} L &= \frac{1}{2}m\sum_n\left(\frac{\mathrm{d}u_{2n}}{\mathrm{d}t}\right)^2 + \frac{1}{2}M\sum_n\left(\frac{\mathrm{d}u'_{2n+1}}{\mathrm{d}t}\right)^2 \\ &\quad -\frac{1}{2}\Lambda\sum_n(u'_{2n+1}-u_{2n})^2 - \frac{1}{2}\Lambda\sum_n(u'_{2n-1}-u_{2n})^2. \end{aligned} \tag{8.20}$$

The Lagrangian density (Lagrangian per unit cell length) is

$$\begin{aligned} \mathscr{L} &= \frac{1}{2}m\left(\frac{\mathrm{d}u_{2n}}{\mathrm{d}t}\right)^2 + \frac{1}{2}M\left(\frac{\mathrm{d}u'_{2n+1}}{\mathrm{d}t}\right)^2 \\ &\quad -\frac{1}{2}\Lambda(u'_{2n+1}-u_{2n})^2 - \frac{1}{2}\Lambda(u'_{2n-1}-u_{2n})^2. \end{aligned} \tag{8.21}$$

In this case the field variable ϕ takes two values: u_{2n} and u'_{2n+1}. Equation (8.9) therefore turns into two equations:

$$\begin{aligned} \frac{\mathscr{L}}{\mathrm{d}u_{2n}} &= \frac{\mathrm{d}}{\mathrm{d}t}\left(\frac{\mathrm{d}\mathscr{L}}{\mathrm{d}(\frac{\mathrm{d}u_{2n}}{\mathrm{d}t})}\right), \\ -\Lambda(2u_{2n}-u'_{2n-1}-u'_{2n+1}) &= m\frac{\mathrm{d}^2u_{2n}}{\mathrm{d}t^2}, \end{aligned} \tag{8.22}$$

and

$$\frac{\mathscr{L}}{\mathrm{d}u'_{2n+1}} = \frac{\mathrm{d}}{\mathrm{d}t}\left(\frac{\mathrm{d}\mathscr{L}}{\mathrm{d}(\frac{\mathrm{d}u'_{2n+1}}{\mathrm{d}t})}\right),$$

$$-\Lambda(2u'_{2n+1} - u_{2n} - u_{2n+2}) = m\frac{d^2u'_{2n+1}}{dt^2}. \tag{8.23}$$

These are the same as equations (4.31) and (4.32) which were derived from the Newtonian approach.

8.2.4 THREE-DIMENSIONAL CRYSTALS

Let us now extend the discussion to three-dimensional crystals. The periodic boundary condition discussed in section 4.4 for a periodic linear chain will be applied to a large three-dimensional crystal containing $N \to \infty$ number of unit cells. Let, from a chosen origin, $\boldsymbol{l}$ denote lth unit cell and $\boldsymbol{b}$ denote bth atom of mass m_b in the lth unit cell. Let $\boldsymbol{u}(\boldsymbol{l},\boldsymbol{b})$ denote the displacement from equilibrium of bth atom in lth unit cell. Let us also consider $\Phi(\boldsymbol{lb};\boldsymbol{l'b'})$ as the harmonic force constant matrix[3] such that $\Phi_{\alpha\beta}(\boldsymbol{lb};\boldsymbol{l'b'})$ represents force along α direction on atom $(\boldsymbol{l},\boldsymbol{b})$ when atom $(\boldsymbol{l}',\boldsymbol{b}')$ is displaced along β direction. With the above considerations, the dynamical Lagrangian of the crystal is

$$L = \frac{1}{2}\sum_{\boldsymbol{lb}} m_b \sum_{\alpha}\left(\frac{\partial u_\alpha(\boldsymbol{lb})}{\partial t}\right)^2 - \frac{1}{2}\sum_{\boldsymbol{lb},\boldsymbol{l'b'}}\sum_{\alpha,\beta}\Phi_{\alpha\beta}(\boldsymbol{lb};\boldsymbol{l'b'})u_\alpha(\boldsymbol{lb})u_\beta(\boldsymbol{l'b'}). \tag{8.24}$$

As we have considered an infinite number of unit cells, we can write

$$\sum_{\boldsymbol{l}} \to \int d\boldsymbol{l} \equiv \int d^3l = N_0\Omega, \tag{8.25}$$

where N_0 is number of unit cells and Ω is the volume of a unit cell so that $N_0\Omega$ is the crystal volume. We then express $L = \sum_{\boldsymbol{l}}\mathscr{L} \equiv \int \mathscr{L} d^3l$, with the Lagrangian density as

$$\mathscr{L} = \frac{1}{2}\sum_{\boldsymbol{b}} m_b \sum_{\alpha}\left(\frac{\partial u_\alpha(\boldsymbol{0b})}{\partial t}\right)^2 - \frac{1}{2}\sum_{\boldsymbol{b},\boldsymbol{l'b'}}\sum_{\alpha,\beta}\Phi_{\alpha\beta}(\boldsymbol{0b};\boldsymbol{l'b'})u_\alpha(\boldsymbol{0b})u_\beta(\boldsymbol{l'b'}). \tag{8.26}$$

Note that in the above equation $u_\alpha(\boldsymbol{0b})$ denotes the αth component of the displacement of the bth basis atom in the 0th unit cell.

As $\mathscr{L} = \mathscr{L}(\boldsymbol{u}, \frac{d\boldsymbol{u}}{dt})$, the Lagrange equations of motion are

$$\frac{\partial \mathscr{L}}{\partial u_\alpha(\boldsymbol{0b})} = \frac{d}{dt}\left(\frac{\partial \mathscr{L}}{\partial\left(\frac{\partial u_\alpha}{\partial t}\right)}\right), \quad \alpha = x, y, z. \tag{8.27}$$

With equation (8.26) these can be expressed as

$$m_b\frac{d^2(u_\alpha(\boldsymbol{0b}))}{dt^2} = -\sum_{\boldsymbol{l'b'}\beta}\Phi_{\alpha\beta}(\boldsymbol{0b};\boldsymbol{l'b'})u_\beta(\boldsymbol{l'b'}), \quad \alpha,\beta = x,y,z. \tag{8.28}$$

Extending the discussion presented in section 4.4, we seek a solution in the form

$$u_\alpha(\boldsymbol{lb}) = \frac{1}{\sqrt{m_b}}\sum_{\boldsymbol{q}s} U_\alpha(\boldsymbol{q}s;\boldsymbol{b})\exp[i(\boldsymbol{q}\cdot\boldsymbol{x}(\boldsymbol{lb}) - \omega t)], \tag{8.29}$$

where $\boldsymbol{U}(\boldsymbol{q}s)$ is amplitude of a wave of vector $\boldsymbol{q}$ and polarisation branch s and $\boldsymbol{x}(\boldsymbol{lb}) \equiv \boldsymbol{l} + \boldsymbol{b}$ is the equilibrium position vector of the bth atom in the lth unit cell. With this choice, equation (8.28) can be expressed as

$$\omega^2(\boldsymbol{q}s)e_\alpha(\boldsymbol{b};\boldsymbol{q}s) = \sum_{\boldsymbol{b}'\beta} C_{\alpha\beta}(\boldsymbol{bb}'|\boldsymbol{q})e_\beta(\boldsymbol{b}';\boldsymbol{q}s), \tag{8.30}$$

[3] Again, Φ in this sub-section has a different meaning than the same symbol used in sub-section 8.2.1. This is unfortunate but unavoidable in order to retain symbols normally used in these topics in physics.

where $\{C_{\alpha\beta}\}$ is known as the C-type dynamical matrix and $\boldsymbol{e}$ is an eigenvector whose components satisfy orthonormality and completeness relations (Born and Huang, 1954; Maradudin *et al.* , 1971; Srivastava, 2022). The elements of the matrix C are expressed as

$$C_{\alpha\beta}(\boldsymbol{bb}'|\boldsymbol{q}) = \frac{1}{\sqrt{m_b m_{b'}}} \sum_{\boldsymbol{l}'} \Phi_{\alpha\beta}(\boldsymbol{0b};\boldsymbol{l}'\boldsymbol{b}') \exp[-i\boldsymbol{q}\cdot(\boldsymbol{x}(\boldsymbol{0b}) - x(\boldsymbol{l}'\boldsymbol{b}'))]. \tag{8.31}$$

This Lagrangian-based lattice dynamical description of three-dimensional crystals is an alternative to the Newtonian-based theory described in section 5.4. Whereas in section 5.4 we obtained expressions for forces acting on each basis atom in a unit cell, using the approach presented in this section we have to obtain an expression for the inter-atomic force constant matrix $\{\Phi_{\alpha\beta}\}$.

Determining the force constant matrix elements $\{\Phi_{\alpha\beta}(\boldsymbol{0b};\boldsymbol{l}'\boldsymbol{b}')\}$ is in general a demanding task. This requires obtaining the second derivative of crystal potential energy. Elements of the matrix $\{\Phi\}$ obey certain symmetry rules. Discussion of these issues is beyond the scope of this book but details are available in several text books, including Born and Huang (1954), Maradudin *et al.* (1971) and Srivastava (1999, 2022).

We proceed with a brief discussion for the cubic diamond structure. With consideration only of first nearest neighbour interaction for the tetrahedrally coordinated atomic structure in the cubic diamond structure, a general form of the inter-atomic force constant matrix $\{\Phi_{\alpha\beta}\}$ between any two neighbouring atoms is (Herman, 1959)

$$\Phi = \begin{pmatrix} A & B & B \\ B & A & B \\ B & B & A \end{pmatrix}, \tag{8.32}$$

where A and B are two adjustable parameters. These parameters are the same as the two general force constants defined in equation (5.57). With this choice the dynamical matrix for the cubic diamond structure can be set up. The resulting 6×6 dynamical matrix will be the same as presented in equations (5.58)-(5.60). Some results have already been presented and discussed in equations (5.61)-(5.63) and figure 5.11.

8.2.5 THREE-DIMENSIONAL HARMONIC ELASTIC CONTINUUM

In dealing with the classical theory of three-dimensional crystals we considered linear displacements of atoms from their equilibrium sites while maintaining harmonic spring connectivity with their neighbours. The theory of elasticity ignores consideration of the microscopic atomic structure of a crystal in favour of a macroscopic picture of a continuum. We consider a solid characterised by mass density $\rho = \sum_{\boldsymbol{lb}} m_b/N_0\Omega$, replace an atomic position $\boldsymbol{x}(\boldsymbol{lb})$ with a continuous position vector $\boldsymbol{r}$, replace an atomic displacement vector $\boldsymbol{u}(\boldsymbol{lb})$ with a continuous displacement vector $\boldsymbol{u}(\boldsymbol{r})$ at point $\boldsymbol{r}$ and replace the force constants matrix $\{\Phi_{\alpha\beta}(\boldsymbol{lb};\boldsymbol{l}'\boldsymbol{b}')\}$ with a fourth-order **elastic modulus tensor** $\{c_{ijkl}\}$. This consideration is appropriate in the limit of very long-wavelength wave propagation.

The kinetic energy of the vibrating elastic solid can be written as

$$\mathscr{T} = \frac{1}{2}\int \rho \sum_i \left(\frac{\partial u_i(\boldsymbol{r})}{\partial t}\right)^2 \mathrm{d}^3 r. \tag{8.33}$$

Within the harmonic approximation, the potential energy can be written as (Landau and Lifshitz, 1986; Ashcroft and Mermin, 1976; Hori *et al.* , 2018)

$$\mathscr{V}_2 = \frac{1}{2}\int \left[\sum_{ijkl} \varepsilon_{ij} c_{ijkl} \varepsilon_{kl}\right] \mathrm{d}^3 r, \tag{8.34}$$

where

$$\varepsilon_{ij} = \frac{1}{2}\left(\frac{\partial u_i(\boldsymbol{r})}{\partial r_j} + \frac{\partial u_j(\boldsymbol{r})}{\partial r_i}\right) \tag{8.35}$$

is the **strain tensor**.

The Lagrangian density corresponding to the Lagrangian $L = \int (\mathscr{T} - \mathscr{V}_2)\mathrm{d}^3r$ is

$$\mathscr{L} = \frac{1}{2}\rho\sum_i\left(\frac{\partial u_i(\boldsymbol{r})}{\partial t}\right)^2 - \frac{1}{2}\left[\sum_{ijkl}\varepsilon_{ij}c_{ijkl}\varepsilon_{kl}\right]. \tag{8.36}$$

Clearly, $\mathscr{L}$ is a functional of $\frac{\mathrm{d}\boldsymbol{u}}{\mathrm{d}t}$ and $\{\nabla_j\boldsymbol{u}\}$. Therefore, expressing $\mathscr{L} = \mathscr{L}(\frac{\mathrm{d}\boldsymbol{u}}{\mathrm{d}t}, \{\nabla_j\boldsymbol{u}\})$, we interpret the Lagrange equation in (8.9) as

$$0 = \sum_j \frac{\partial}{\partial r_j}\left[\frac{\partial \mathscr{L}}{\partial\left(\frac{\partial u_i}{\partial r_j}\right)}\right] + \frac{\partial}{\partial t}\left[\frac{\partial \mathscr{L}}{\partial\left(\frac{\partial u_i}{\partial t}\right)}\right], \quad i, j = 1,2,3. \tag{8.37}$$

Using equations (8.35) and (8.36), the above equation can be expressed as

$$\rho\frac{\partial^2 u_i}{\partial t^2} = \sum_{jkl} c_{ijkl}\frac{\partial^2 u_k}{\partial r_j \partial r_l}, \quad i,j,k,l = 1,2,3. \tag{8.38}$$

Due to the invariance under the transpositions $i \leftrightarrow j$, $k \leftrightarrow l$, and $ij \leftrightarrow kl$, the tensor c_{ijkl} has a maximum of 21 (for triclinic crystal system) independent components. The number of independent components is smaller for crystals of higher symmetry. In particular, for cubic crystals there are only three independent components. It is usual to adopt the following simplified notation

$$11 \to 1; \quad 22 \to 2; \quad 33 \to 3; \quad 23,32 \to 4; \quad 31,13 \to 5; \quad 12,21 \to 6. \tag{8.39}$$

In this contracted notation the three independent second-order elastic constants for cubic crystals are c_{11}, c_{12} and c_{44}. In terms of these elastic constants, the bulk modulus B of cubic crystals is

$$B = \frac{1}{3}(c_{11} + 2c_{12}), \tag{8.40}$$

and the shear modulus η is

$$\begin{aligned} \eta &= c_{44} \quad \text{for shear on the (100) plane,} \\ &= \frac{1}{2}(c_{11} - c_{12}) \quad \text{for shear on the (110) plane,} \end{aligned} \tag{8.41}$$

For *isotropic systems* the elastic constants can be related to Lamé's constants λ and μ:

$$c_{11} = \lambda + 2\mu; \quad c_{12} = \lambda; \quad c_{44} = \mu, \tag{8.42}$$

and the bulk and shear moduli are

$$B = \frac{1}{3}(c_{11} + 2c_{12}) = \lambda + \frac{2}{3}\mu, \tag{8.43}$$

$$\eta = \frac{1}{2}(c_{11} - c_{12}) = c_{44} = \mu. \tag{8.44}$$

Seeking a solution in the form

$$\boldsymbol{u} = \boldsymbol{U}\exp[\mathrm{i}(\boldsymbol{q}\cdot\boldsymbol{r} - \omega t)], \tag{8.45}$$

equation (8.38) can be expressed as the *Green-Christoffel equations*

$$\Big(\sum_{jkl} c_{ijkl}q_j q_l - \rho\omega^2\delta_{ik}\Big)U_i = 0, \quad i,j,k,l = 1,2,3, \tag{8.46}$$

non-trivial solutions of which can be obtained by solving the 3×3 determinant

$$\Big|\sum_{jl} c_{ijkl}q_j q_l - \rho\omega^2\delta_{ik}\Big| = 0 \quad i,j,k,l = 1,2,3. \tag{8.47}$$

Equations (8.47) can be analytically solved for the dispersion relation $\omega = \omega(q)$ and the sound speed $v = \omega/q$ in cubic and isotropic elastic continuum media. More detail can be found in the references listed for further reading, particularly in Srivastava (2022) and de Launay (1956).

8.2.6 ELASTIC LINEAR CHAIN

For a one-dimensional elastic chain equation (8.38) reduces to the simple form

$$\rho\frac{\partial^2 u}{\partial t^2} = Y\frac{\partial^2 u}{\partial x^2}, \tag{8.48}$$

where ρ is mass density, x is a continuous space variable along the chain and Y is Young's modulus.

The above equation can be written as

$$\frac{\partial^2 u}{\partial x^2} = \frac{1}{v^2}\frac{\partial^2 u}{\partial t^2}, \tag{8.49}$$

where

$$v = \sqrt{\frac{Y}{\rho}} \tag{8.50}$$

is the speed of sound along the chain.

It is instructive to relate equation (8.49) of the linear elasticity theory to equation (8.19), or equation (4.23), of the harmonic lattice dynamical theory. Let us consider the chain in figure 4.4 with a hugely dense atomic population such that the interatomic separation a becomes a small distance $b \to 0$ but the chain length is kept unchanged to $L = Nb$. With this consideration, we write the Lagrangian of the vibrating chain as

$$\begin{aligned} L &= \frac{1}{2}m\sum_n\left(\frac{\partial u_n}{\partial t}\right)^2 - \frac{1}{2}\Lambda\sum_n(u_{n+1}-u_n)^2, \\ &= \frac{1}{2}\frac{m}{b}\sum_n b\left(\frac{\partial u_n}{\partial t}\right)^2 - \frac{1}{2}\Lambda b\sum_n b\left(\frac{u_{n+1}-u_n}{b}\right)^2. \end{aligned} \tag{8.51}$$

Now the discrete atomic chain structure can be treated as an elastic continuum by making the following correspondences

$$\begin{aligned} &b \to \mathrm{d}x, \quad \sum_n(\cdots)b \to \int(\cdots)\mathrm{d}x, \quad \frac{u_{n+1}-u_n}{b} \to \frac{\partial u}{\partial x}, \\ &\frac{m}{b} \to \rho\ \text{(density)}, \quad \Lambda\mathrm{b} \to \mathrm{Y}\ \text{(Young's modulus)}. \end{aligned} \tag{8.52}$$

It will be left as an exercise to show that the resulting Lagrange equation of motion can be reduced to the form given in equation (8.48). The dispersion relation for this wave equation is $\omega = vq$, where q is the wave number along the continuum chain.

Some useful information from the above discussion is that the Young's modulus Y is related to the inter-atomic force constant Λ in the form $\Lambda = Y/b$, *i.e.* the force constant is the elastic modulus per unit length. Note that it was mentioned in section 4.5.1 that the linear dispersion relation ω *vs* q for the linear chain requires the long wavelength condition $qa << 1$. The method of long wavelengths has been useful to determine some force constant parameters from experimentally measured elastic constants. This is useful as force constants cannot be measured experimentally.

8.2.7 ANHARMONIC ELASTIC CONTINUUM

In section 8.2.5 we wrote an expression for the Lagrangian of a harmonic elastic continuum. That procedure can be extended to write an expression for an anharmonic elastic continuum. The lowest order anharmonicity, *viz* the cubic anharmonicity, makes the most significant contribution to changes in physical properties. The third-order elastic energy term can be expressed as

$$\mathscr{V}_3 = \frac{1}{3!}\sum_{\substack{lmn\\ijk}} A_{ijk}^{lmn}\frac{\partial u_l}{\partial r_i}\frac{\partial u_m}{\partial r_j}\frac{\partial u_n}{\partial r_k}. \tag{8.53}$$

The quantity $\left\{A_{ijk}^{lmn}\right\}$ is a sixth rank tensor, which is a function of second- and third-order elastic constants. Extending the notation in section 8.2.5 we can write (Ziman, 1960; Drabble, 1966; Srivastava, 2022)

$$\begin{aligned}
\mathscr{V}_3 \;=\; & \int d^3r \sum_{i\neq j\neq k}\Big(\frac{1}{2}c_{11}u_{ii}\sum_p u_{pi}u_{pi} + \frac{1}{2}c_{12}u_{ii}\sum_p u_{pj}u_{pj} \\
& + c_{44}u_{ij}\sum_p u_{pi}u_{pj} + \frac{1}{6}C_{111}u_{ii}^3 + C_{123}u_{ii}u_{jj}u_{kk} \\
& + \frac{1}{2}C_{112}u_{ii}^2(u_{jj}+u_{kk}) + \frac{1}{2}C_{144}u_{ii}(u_{jk}+u_{kj})^2 \\
& + C_{456}(u_{ij}+u_{ji})(u_{jk}+u_{kj})(u_{ki}+u_{ik}) \\
& + \frac{1}{2}C_{166}(u_{ii}+u_{jj})(u_{ij}+u_{ji})^2\Big), \qquad i,j,k,p = 1,2,3,
\end{aligned} \tag{8.54}$$

where c_{ij} are the second-order elastic constants, $u_{ij} = \frac{\partial u_i}{\partial u_j}$ *etc* and C_{ijk} are the third-order elastic constants in Brugger's notation (Brugger, 1964).

For isotropic medium some of the elastic constants are inter-related, as follows

$$\begin{aligned}
C_{111} &= C_{123} + 6C_{144} + 8C_{456} \\
C_{112} &= C_{123} + 2C_{144} \\
C_{166} &= C_{144} + 2C_{456} \\
2c_{44} &= c_{11} - c_{12}.
\end{aligned} \tag{8.55}$$

Adopting the notation of Landau and Lifshitz (1976), we can write (Hamilton and Parrott, 1969; Srivastava, 2022)

$$\begin{aligned}
\mathscr{V}_3 \;=\; & \int d^3r \sum_{ijk}\left[\frac{\mathscr{C}}{3}\left(\frac{\partial u_i}{\partial r_i}\right)^3 + \frac{\lambda+\mathscr{B}}{2}\frac{\partial u_i}{\partial r_i}\left(\frac{\partial u_j}{\partial r_k}\right)^2 + \frac{\mathscr{B}}{2}\frac{\partial u_i}{\partial r_i}\frac{\partial u_j}{\partial r_k}\frac{\partial u_k}{\partial r_j}\right. \\
& \left. + \left(\mu + \frac{\mathscr{A}}{4}\right)\frac{\partial u_i}{\partial r_j}\frac{\partial u_k}{\partial r_i}\frac{\partial u_k}{\partial r_j} + \frac{\mathscr{A}}{12}\frac{\partial u_i}{\partial r_j}\frac{\partial u_j}{\partial r_k}\frac{\partial u_k}{\partial r_i}\right].
\end{aligned} \tag{8.56}$$

where $\mathscr{A}$, $\mathscr{B}$ and $\mathscr{C}$ are related to the third-order elastic constants C_{123}, C_{144} and C_{456} as follows

$$\mathscr{A} = 4C_{456} \qquad \mathscr{B} = C_{144} \qquad \mathscr{C} = \frac{1}{2}C_{123}, \tag{8.57}$$

and λ and μ are the Lamé constants defined in equation (8.42). We will extend the discussion of $\mathscr{V}_3$ in the next section and the next chapter.

8.3 HAMILTONIAN CLASSICAL FIELD THEORY

In chapter 3 we described how to derive an expression for the Hamiltonian of a system from its Lagrangian. It was then shown how the modified Hamilton's variational principle can be applied to derive Hamilton's equations of motion. Here we will adopt that approach to derive Hamiltonian density and Hamilton's equations of motion for the systems discussed in the previous section.

In order to proceed, we remind ourselves from section 8.1 that the Lagrangian of a classical field can be written as $L = \int \mathscr{L} \mathrm{d}^3 r$, with the Lagrangian density expressed as $\mathscr{L}(\boldsymbol{r}) = \mathscr{L}(\phi, \boldsymbol{\nabla}\phi, \frac{\mathrm{d}\phi}{\mathrm{d}t}, t)$, where $\phi = \phi(\boldsymbol{r}, t)$ is a *field coordinate variable* which has continuous dependence on $\boldsymbol{r}(t)$ at point $\boldsymbol{r}$ and time t. Let us define a *field momentum variable* $\xi(\boldsymbol{r})$ canonically conjugate to $\phi(\boldsymbol{r})$

$$\xi(\boldsymbol{r}) = \frac{\partial \mathscr{L}}{\partial\left(\frac{\partial \phi(\boldsymbol{r})}{\partial t}\right)}, \tag{8.58}$$

and using this define the *Hamiltonian density* as

$$\mathscr{H} = \xi(\boldsymbol{r})\frac{\partial \phi}{\partial t} - \mathscr{L}(\boldsymbol{r}). \tag{8.59}$$

The Hamiltonian of the system is then

$$H = \int \mathscr{H}(\boldsymbol{r}) \mathrm{d}^3 r. \tag{8.60}$$

Applying the modified Hamilton's principle in section 3.2.3 and following the discussion in section 8.2, we obtain the following field equations

$$\frac{\partial \phi}{\partial t} = \frac{\partial \mathscr{H}}{\partial \xi} - \sum_i \frac{\partial}{\partial r_i} \frac{\partial \mathscr{H}}{\partial\left(\frac{\partial \xi}{\partial r_i}\right)}, \tag{8.61}$$

$$\frac{\partial \xi}{\partial t} = -\frac{\partial \mathscr{H}}{\partial \phi} + \sum_i \frac{\partial}{\partial r_i} \frac{\partial \mathscr{H}}{\partial\left(\frac{\partial \phi}{\partial r_i}\right)}. \tag{8.62}$$

Note that equations (8.61) and (8.62) are a generalised version of the Hamilton's equations of motion in equations (3.14) and (3.15).

8.3.1 MONATOMIC LINEAR CHAIN

The Lagrangian density $\mathscr{L}$ for the monatomic linear chain is given in equation (8.17). It is a function of the generalised coordinate u_n. The corresponding generalised momentum is

$$p_n = \frac{\partial \mathscr{L}}{\partial\left(\frac{\partial q_n}{\partial t}\right)} = m\frac{\partial u_n}{\partial t}. \tag{8.63}$$

Using this equation, we obtain the following expression for the *Hamiltonian density* $\mathscr{H}$ for the chain

$$\begin{aligned} \mathscr{H} &= p_n \frac{\partial u_n}{\partial t} - \mathscr{L}, \\ &= \frac{p_n^2}{2m} + \frac{1}{2}\Lambda(u_{n+1} - u_n)^2. \end{aligned} \tag{8.64}$$

Hamilton's equations of motion then read

$$\frac{\partial u_n}{\partial t} = \frac{\partial \mathscr{H}}{\partial p_n} = \frac{p_n}{m}, \tag{8.65}$$

and

$$\frac{\partial p_n}{\partial t} = -\frac{\partial p_n}{\partial u_n} = \Lambda(2u_n - u_{n-1} - u_{n+1}). \tag{8.66}$$

Using equations (8.65) and (8.66), the equation of motion for the nth atom in the chain can be written as

$$m\frac{\partial^2 u_n}{\partial t^2} = -\Lambda(2u_n - u_{n+1} - u_{n-1}), \tag{8.67}$$

which is the same as equation (8.19) derived from the Lagrangian formulation.

Derivation of equations of motion using the Hamiltonian formulation of the dynamics of the diatomic linear chain will be set as an exercise.

8.3.2 THREE-DIMENSIONAL CRYSTALS

When considering three-dimensional crystals, we start with the expression for the Lagrangian density in equation (8.26) and define momenta conjugate to generalised coordinates $\{u_\alpha(\boldsymbol{lb})\}$

$$p_\alpha(\boldsymbol{lb}) = \frac{\partial \mathscr{L}}{\partial\left(\frac{\partial u_\alpha(\boldsymbol{lb})}{\partial t}\right)} = m_b\frac{\partial u_\alpha(\boldsymbol{0b})}{\partial t}, \tag{8.68}$$

where $\boldsymbol{u}(\boldsymbol{lb})$ is the displacement vector for bth atom of mass m_b in the lth unit cell. Using this definition, we express the Hamiltonian density as

$$\begin{aligned}\mathscr{H} &= \sum_{b\alpha} p_\alpha(\boldsymbol{0b})\frac{\partial u_\alpha(\boldsymbol{0b})}{\partial t} - \frac{1}{2}\sum_{b} m_b \sum_{\alpha}\left(\frac{\partial u_\alpha(\boldsymbol{0b})}{\partial t}\right)^2 \\ &\quad + \frac{1}{2}\sum_{b,\boldsymbol{l}'\boldsymbol{b}'}\sum_{\alpha,\beta}\Phi_{\alpha\beta}(\boldsymbol{0b};\boldsymbol{l}'\boldsymbol{b}')u_\alpha(\boldsymbol{0b})u_\beta(\boldsymbol{l}'\boldsymbol{b}'), \\ &= \sum_{b\alpha}\frac{p_\alpha^2(\boldsymbol{0b})}{2m_b} + \frac{1}{2}\sum_{b,\boldsymbol{l}'\boldsymbol{b}'}\sum_{\alpha,\beta}\Phi_{\alpha\beta}(\boldsymbol{0b};\boldsymbol{l}'\boldsymbol{b}')u_\alpha(\boldsymbol{0b})u_\beta(\boldsymbol{l}'\boldsymbol{b}').\end{aligned} \tag{8.69}$$

We now write down Hamilton's equations of motion as

$$\frac{\partial u_\alpha(\boldsymbol{0b})}{\partial t} = \frac{\partial \mathscr{H}}{\partial p_\alpha(\boldsymbol{0b})} = \frac{p_\alpha(\boldsymbol{ob})}{m_b}, \tag{8.70}$$

and

$$\frac{\partial p_\alpha(\boldsymbol{0b})}{\partial t} = -\frac{\partial \mathscr{H}}{\partial u_\alpha(\boldsymbol{0b})} = -\sum_{\boldsymbol{l}'\boldsymbol{b}'}\sum_{\alpha,\beta}\Phi_{\alpha\beta}(\boldsymbol{0b};\boldsymbol{l}'\boldsymbol{b}')u_\beta(\boldsymbol{l}'\boldsymbol{b}'). \tag{8.71}$$

When combined together, equations (8.70) and (8.71) represent equation (8.28) which was obtained by using the Lagrangian classical field theory.

8.3.3 ELASTIC LINEAR CHAIN

From the discussion in and below equation (8.51) we write the Lagrangian density of an elastic linear chain as

$$\mathscr{L} = \frac{1}{2}\rho\left(\frac{\partial u}{\partial t}\right)^2 - \frac{1}{2}Y\left(\frac{\partial u}{\partial x}\right)^2, \tag{8.72}$$

where ρ is mass density, $u = u(x)$ is displacement of medium at x and Y is Young's modulus. We treat u as a generalised coordinate and define its conjugate momentum as

$$p = \frac{\mathscr{L}}{\partial\left(\frac{\partial u}{\partial t}\right)} = \rho\frac{\partial u}{\partial t}. \tag{8.73}$$

The Hamiltonian density is

$$\mathscr{H} = p\frac{\partial u}{\partial t} - \mathscr{L} = \frac{1}{2\rho}p^2 + \frac{1}{2}Y\left(\frac{\partial u}{\partial x}\right)^2. \tag{8.74}$$

and Hamilton's equations of motion are

$$\frac{\partial u}{\partial t} = \frac{\partial \mathscr{H}}{\partial p} = \frac{p}{\rho} \tag{8.75}$$

and

$$\begin{aligned} \frac{\partial p}{\partial t} &= -\frac{\partial \mathscr{H}}{\partial u} + \frac{\partial}{\partial x}\frac{\partial \mathscr{H}}{\partial\left(\frac{\partial u}{\partial x}\right)}, \\ &= 0 + Y\frac{\partial^2 u}{\partial x^2}. \end{aligned} \tag{8.76}$$

Using equations (8.75) and (8.76) together, we obtain the wave equation

$$\rho\frac{\partial^2 u}{\partial t^2} = Y\frac{\partial^2 u}{\partial x^2}. \tag{8.77}$$

which was previously derived from the application of the classical Lagrangian field theory.

8.3.4 THREE-DIMENSIONAL HARMONIC ELASTIC CONTINUUM

In section 8.2.5 we developed the Lagrangian formulation of the elastic harmonic continuum. The Lagrangian density is given in equation (8.36). Using that, the conjugate momentum variable corresponding to the coordinate variable $u_i(\boldsymbol{r})$ is

$$p_i(\boldsymbol{r}) = \rho\frac{\partial u_i(\boldsymbol{r})}{\partial t}, \tag{8.78}$$

where ρ is mass density. Using this the expression for the Hamiltonian density is

$$\begin{aligned} \mathscr{H} &= \sum_i p_i(\boldsymbol{r})\frac{\partial u_i(\boldsymbol{r})}{\partial t} - \mathscr{L}, \\ &= \frac{1}{2\rho}\sum_i p_i^2(\boldsymbol{r}) + \frac{1}{2}\sum_{ijkl}\varepsilon_{ij}c_{ijkl}\varepsilon_{kl}, \end{aligned} \tag{8.79}$$

where ε_{ij} is a symmetric gradient of $\boldsymbol{u}(\boldsymbol{r})$ and $\{c_{ijkl}\}$ is the elastic modulus tensor.

Following the procedure described in section 8.3.3 we can straightforwardly derive the wave equation in equation (8.38).

8.3.5 ELECTROMAGNETISM – HAMILTONIAN FORMALISM

We start with the Lagrangian density of the electromagnetic field[4]

$$\mathscr{L} = \frac{1}{2}\left(\frac{\partial \boldsymbol{A}}{\partial t} + \nabla\Phi\right)^2 - \frac{1}{2}c^2\left(\nabla \times \boldsymbol{A}\right)^2 \tag{8.80}$$

[4]Reminder – The symbols Φ and $\boldsymbol{A}$ used in this sub-section should not be confused with similar symbols used in the sub-sections not dealing with electromagnetism.

as expressed in equation (8.12). Here $\nabla\Phi$ is the gradient of the scalar potential and A_x, A_y and A_z are the components of the vector potential. As the scalar potential Φ does not explicitly appear in the expression for $\mathscr{L}$, it is no longer a field variable and the momentum canonically conjugate to it must be zero. The momentum components conjugate to the field variables A_x, A_y and A_z are

$$p_\alpha = \frac{\partial A_\alpha}{\partial t} + \frac{\partial \Phi}{\partial r_\alpha}, \quad \alpha = x, y, z. \tag{8.81}$$

The Hamiltonian density of the electromagnetic field can now be expressed as

$$\begin{aligned} \mathscr{H} &= \boldsymbol{p}\cdot\frac{\partial \boldsymbol{A}}{\partial t} - \mathscr{L}, \\ &= \frac{1}{2}p^2 - \boldsymbol{p}\cdot\nabla\Phi + \frac{1}{2}c^2(\nabla\times\boldsymbol{A})^2. \end{aligned} \tag{8.82}$$

Now Hamilton's equations of motion can be obtained to read

$$\frac{\partial \boldsymbol{A}}{\partial t} = \boldsymbol{p} - \nabla\Phi, \tag{8.83}$$

and

$$\frac{\partial \boldsymbol{p}}{\partial t} = -c^2\nabla\times(\nabla\times\boldsymbol{A}). \tag{8.84}$$

Equation (8.83) confirms the definition of $\boldsymbol{p}$ in (8.81) and equation (8.84) is the same as equation (8.15).

FURTHER READING

1. Ashcroft N W and Mermin N D 1976 *Solid State Physics* (Sounders College: Philadelphia)
2. Born M and Huang K 1954 *Dynamical Theory of Crystal Lattices* (Oxford University Press: Oxford)
3. Brugger K 1964 *Phys Rev* **133** A1611
4. de Launay J 1956 *Solid State Physics* vol 2, ed F Seitz and D Turnbull (Academic: New York)
5. Herman F J 1959 *J Phys Chem Solids* **8** 405
6. Hori M, Wijerathne L, Riaz R and Ichimura T 2018 *Journal of JSCE* **6** 1
7. Landau L D and Lifshitz E M 1986 *Theory of Elasticity* 3rd edn (Butterworth and Heinemann: Oxford)
8. Maradudin A A *et al.* 1971 *Solid State Physics* Supplement 3 (Academic: New York)
9. Parrott J E 1969 *Solid State Theory - methods and applications* (Ed. P T Landsberg) (Wiley - Interscience: London)
10. Srivastava G P 1999 *Theoretical Modelling of Semiconductor Surfaces* (World Scientific: Singapore)
11. Srivastava G P 2022 *The Physics of Phonons* 2nd edn (CRC: Oxon)

PROBLEMS

1. Using the Lagrangian in equation (8.51) derive the wave equation (8.48).
2. Derive the equations of motion for a diatomic linear chain in equations (8.22) and (8.23) by using the Hamiltonian formalism of classical field theory.

3. Consider the anharmonic Hamiltonian of a one-dimensional continuum in the form

$$H = \int \mathrm{d}x \left[\frac{p^2}{2\rho} + \frac{A}{2} u^2(x) + \frac{B}{2} u^4(x) + \frac{Y}{2} \left(\frac{\partial u}{\partial x} \right)^2 \right],$$

where u and p are displacement and momenta canonical variables, Y is Young's modulus, and A and B are constants. Using classical field theory, derive the equation of motion for the displacement field $u(x)$.

4. (a) Consider the Lagrangian density for the Klein-Gordon field, also called scalar field,

$$\mathscr{L} = \frac{1}{2} \left(\frac{\partial \phi}{\partial t} \right)^2 - \frac{1}{2} \left(\frac{\partial \phi}{\partial x} \right)^2 - \frac{1}{2} m^2 \phi^2(x),$$

where m is the mass of the particle and $phi(x)$ is the field variable. Using this Lagrangian density, derive the Klein-Gordon equation of motion.

(b) Obtain an expression for the Hamiltonian density corresponding to the above Lagrangian density in (a). Obtain Hamilton's equations of motion and derive the Klein-Gordon wave equation.

Part VI

Beyond classical mechanics

9 Elements of Quantum Field Theory

9.1 INTRODUCTION

Under appropriate conditions classical mechanical equations can be turned into quantum mechanical equations. Let us, for example, consider the discussions in previous chapters of this book. We have tacitly assumed dynamics of objects of masses and sizes larger than those of an atom. What happens when we consider the dynamics of an object of atomic or subatomic size? In such case description of the position variable $\boldsymbol{r}$ of the object becomes uncertain in a certain range. We specify the *state* of the object using a *state function*[1] $\psi(\boldsymbol{r},t)$ such that $|\psi(\boldsymbol{r},t)|^2$ defines the probability that the object is located at $\boldsymbol{r}$ at time t. The dynamics of the object is described by the time dependence of this state function. This consideration, with $\psi(\boldsymbol{r},t)$ usually called the **wave function** of the object, is at the heart of quantum mechanics. We will later make another statement regarding classical *vs* quantum nature of a system by revisiting the action integral discussed in chapter 2. Excellent descriptions of analogy between classical mechanics and quantum mechanics exist in the literature, including the articles by Dirac (1945) and Sebens (2015), which readers may wish to peruse. In this chapter we will provide a short description of the transition from classical mechanics to quantum mechanics followed by a discussion of quantum field theory.

9.2 BASIC CONCEPTS IN QUANTUM MECHANICS

In chapter 3 we noted that in a (q,p) phase space the canonically conjugate coordinate-momentum pair of variables q and p of a classical dynamical system can both be simultaneously and accurately determined. In the quantum regime, these variables are subject to the **uncertainty principle**, according to which it is impossible to measure them both simultaneously with arbitrary accuracy. In fact, in the quantum language canonical variables, and quantities expressed by them, are known as observables and are represented by self-adjoint (Hermitian) **operators**. In other words, we speak of the coordinate operator $\hat{Q}$ and the conjugate momentum operator $\hat{P}$ in quantum physics rather than simply treat them as the coordinate variable q and the momentum variable p as in classical physics. It is usual to write the quantum operator for a three-dimensional coordinate $\boldsymbol{q}$ as $\hat{\boldsymbol{Q}} \equiv \hat{Q}$ and for a three-dimensional momentum $\boldsymbol{p}$ as $\hat{\boldsymbol{P}}$. The expression for the momentum operator is $\hat{\boldsymbol{P}} = -i\hbar\boldsymbol{\nabla}$.

Quantum mechanical measurement of an observable is equivalent to solving the eigenvalue problem satisfied by the Hermitian operator representing the observable. For example, the linear momentum of a freely moving object is the eigenvalue of the operator $\hat{\boldsymbol{P}}$ which satisfies the eigenvalue equation $\hat{\boldsymbol{P}}\psi(\boldsymbol{r}) = \boldsymbol{p}\psi(\boldsymbol{r})$. The corresponding eigenfunction is a planewave [*cf* equation (3.176)] $\psi(\boldsymbol{r}) = \psi_{\boldsymbol{k}}(\boldsymbol{r}) = A\exp(i\boldsymbol{k}\cdot\boldsymbol{r})$, where A is the wave amplitude and $\boldsymbol{k}$ is the wavevector. Operation of $\hat{\boldsymbol{P}}$ on this eigenfunction produces the eigenvalue expression $\boldsymbol{p} = \hbar\boldsymbol{k}$.

The **commutator** of two quantum operators $\hat{F}_1$ and $\hat{F}_2$ is defined as

$$[\hat{F}_1,\hat{F}_2] = \hat{F}_1\hat{F}_2 - \hat{F}_2\hat{F}_1. \tag{9.1}$$

A pair of compatible observables are represented by a pair of quantum mechanical operators which commute and have the same eigenfunction. Two commuting operators $\hat{F}_1$ and $\hat{F}_2$ satisfy the relations

[1] A state function is often regarded as a *state vector* in an infinite-dimensional **Hilbert space**.

DOI: 10.1201/9781003383314-9

$[\hat{F}_1, \hat{F}_2] = [\hat{F}_2, \hat{F}_1]$, so that $\hat{F}_1\hat{F}_2\psi = \hat{F}_2\hat{F}_1\psi$. According to the Heisenberg uncertainty principle, a pair of incompatible observables cannot be simultaneously measured. Referring to a pair of canonically conjugate observables as q and p, the uncertainty principle states that $\Delta q \Delta p \geq \hbar/2$, where $h = 2\pi\hbar$ is Planck's constant.

9.3 TRANSITION FROM CLASSICAL TO QUANTUM MECHANICS

There is a **correspondence principle** which says that the classical Poisson bracket of two variables f_1 and f_2 should be replaced with the **commutator** of the corresponding operators:

$$[f_1, f_2]_{\mathrm{CM}} \to \frac{1}{i\hbar}[\hat{F}_1, \hat{F}_2]_{\mathrm{QM}}. \tag{9.2}$$

Note that, using common practice, we have used lowercase letters for classical variables and uppercase letters with a hat for quantum operators. Table 9.1 lists a few examples of the analogy between Poisson bracket in classical mechanics and commutator bracket in quantum mechanics.

Table 9.1
Analogy between Poisson bracket in classical mechanics (CM) and commutator bracket in quantum mechanics (QM).

Poisson bracket in CM	commutator bracket in QM
$[q_i, p_j] = \delta_{ij}$	$\frac{1}{i\hbar}[\hat{Q}_i, \hat{P}_j] = \hat{I}\delta_{ij}$
$\Sigma_i[q_i, p_j] = 1$	$\frac{1}{i\hbar}\Sigma_i[\hat{Q}_i, \hat{P}_j] = \hat{I}$
$\frac{\mathrm{d}f}{\mathrm{d}t} = [f, H] + \frac{\partial f}{\partial t}$	$\frac{\mathrm{d}\hat{F}}{\mathrm{d}t} = -i\hbar[\hat{F}, \hat{H}] + \frac{\partial \hat{F}}{\partial t}$

In the third example in table 9.1 we have shown the analogy between the classical equation of motion for a variable f of a system governed by Hamiltonian H and the Heisenberg equation of motion for the corresponding quantum mechanical operator $\hat{F}$ governed by the Hamiltonian operator $\hat{H}$.

Consider a conservative classical system. For such a system the relation $p\dot{q} = 2T$ holds [*cf* equation (3.80)], where T is the kinetic energy. The change in the action integral for a repeated orbital motion can be expressed as

$$\begin{aligned} \Delta J &= \Delta \int_{t_1}^{t_2} p\dot{q}\mathrm{d}t = \Delta \int_{t_1}^{t_1+\tau} p\dot{q}\mathrm{d}t = \Delta \int_{t_1}^{t_1+\tau} p\mathrm{d}q \\ &= \Delta \oint p\mathrm{d}q, \end{aligned} \tag{9.3}$$

where $\oint p dq$ is the integral over a complete orbit of the motion. According to the old quantum theory[2] the integral $\oint p dq$ is quantised as a multiple of Plank's constant h:

$$J = nh, \quad n = \text{integer greater than zero}, \tag{9.4}$$

[2]The old quantum theory, developed by Planck, Einstein, Bohr, Sommerfeld and many others, predates Schrödinger's wave mechanics.

leading to

$$\Delta J = h. \tag{9.5}$$

Equation (9.5) is the Bohr-Sommerfeld quantum condition, *viz* periodic motion of a conservative system is limited to orbits for which **proper action** integral has discrete values as multiples of h. We can therefore say that Planck's constant h is a quantum of proper action for periodic motion of a conservative system. As a corollary of this, we can state that if the action is smaller (larger) than h, then the system is quantum (classical).

We will now show that the quantum mechanical Hamilton-Jacobi equation reduces to the classical Hamilton-Jacobi equation in the limit $h \to 0$. The quantum mechanical Hamilton-Jacobi equation is the Schödinger wave equation when the wave function is expressed in terms of the action integral. For the sake of brevity, we will consider a one-dimensional system of mass m for which the classical Hamiltonian function in the (q, p) phase space is

$$H = \frac{p^2}{2m} + \mathscr{V}(q). \tag{9.6}$$

The corresponding quantum Hamiltonian operator is[3]

$$\hat{H} = \frac{\hat{p}^2}{2m} + \mathscr{V}(q). \tag{9.7}$$

We will consider both time-dependent and time-independent forms of the Schödinger equation.

A: Let us consider the time-dependent Schödinger equation

$$\begin{aligned} \hat{H}\psi &= \hat{E}\psi, \\ -\frac{\hbar^2}{2m}\frac{\partial^2 \psi}{\partial q^2} + \mathscr{V}(q,t)\psi &= i\hbar\frac{\partial}{\partial t}\psi, \end{aligned} \tag{9.8}$$

where $\hat{P} = -i\hbar\frac{\partial}{\partial q}$ is the momentum operator, $\hat{E} = i\hbar\frac{\partial}{\partial t}$ is the energy operator and $\psi = \psi(q,t)$ is the eigenfunction of the Hamiltonian operator. We may try a solution of this equation in the form[4]

$$\psi = Ae^{iS/\hbar}, \tag{9.9}$$

where A is amplitude of the wavefunction and S is Hamilton's principal function (see sections 3.3.6 and 3.3.7). With this choice for ψ, equation (9.8) is known as quantum Hamilton-Jacobi equation. Note that we can write [see equation (3.165)]

$$S(q,t) = W(q) - Et, \tag{9.10}$$

where W is Hamilton's characteristic function and E is the energy eigenvalue.

With the substitution of equation (9.9) for ψ, equation (9.8) can be expressed as

$$\frac{1}{2m}\left(\frac{\partial S}{\partial q}\right)^2 - \frac{i\hbar}{2m}\frac{\partial^2 S}{\partial q^2} + \mathscr{V}(q,t) + \frac{\partial S}{\partial t} = 0. \tag{9.11}$$

In the limit $\hbar \to 0$, the quantum Hamilton-Jacobi equation in (9.11) reduces to

$$H\left(q, \frac{\partial S}{\partial q}, t\right) + \frac{\partial S}{\partial t} = 0, \tag{9.12}$$

---

[3] It is usual not to use a 'hat' over an operator representing a coordinate variable. The same applies for a function of a coordinate variable. For this reason we will not use a 'hat' over the potential energy term $\mathscr{V}$.

[4] Such a trial function is also the starting point for the WKB approximation in quantum mechanics. For this reason the WKB approximation is often called the semi-classical approximation.

which is the classical Hamilton-Jacobi equation in (3.157) with the recognition that $p = \frac{\partial S}{\partial q}$ is the momentum canonically conjugate to the coordinate q.

B: Let us consider the time-independent Schödinger equation

$$-\frac{\hbar^2}{2m}\frac{\partial^2 \psi}{\partial q^2} + \mathscr{V}(q)\psi = E\psi. \tag{9.13}$$

In this case we express ψ as

$$\begin{aligned} \psi(q,t) &= \psi(q)e^{-iEt/\hbar}, \\ &= Ae^{iW/\hbar}e^{-iEt/\hbar}. \end{aligned} \tag{9.14}$$

With the substitution of equation (9.14), equation (9.13) becomes

$$\frac{1}{2m}\left(\frac{\partial W}{\partial q}\right)^2 - \frac{i\hbar}{2m}\frac{\partial^2 W}{\partial q^2} + \mathscr{V}(q) = E. \tag{9.15}$$

In the limit $\hbar \to 0$, this form of the quantum Hamilton-Jacobi equation reduces to

$$\frac{1}{2m}\left(\frac{\partial W}{\partial q}\right)^2 + V(q) = E, \tag{9.16}$$

which can be expressed as the classical Hamilton-Jacobi equation

$$H\left(q, \frac{\partial W}{\partial q}\right) = E, \tag{9.17}$$

as stated in equation (3.166).

9.4 QUANTISATION OF BOSONIC QUANTUM FIELD

For developing classical field theory in the previous chapter we used ϕ and ξ as canonically conjugate physical variables. For developing quantum field theory we can proceed by regarding these as operators $\hat{\boldsymbol{\phi}}$ and $\hat{\boldsymbol{\xi}}$, with appropriate commutation relation between them (see, *e.g.* Schiff, 1968). This is the level of **first quantisation**. The Schödinger wave equation presents an example of this first quantisation level. However, an efficient approach for many-body systems can be developed by adopting the concept of **second quantisation**, a step further than the first quantisation. This is a scheme in which wave modes are quantised in the form of bundles of energy of **quasi-particles**. Such quasi-particles are regarded as the **field excitations** of many-body quantum systems [see Pines (1963) and references therein]. This requires developing and using the concepts of **number-** (or ***N***-) **representation** and **creation** and **annihilation operators**. The hierarchy in dealing with dynamics can therefore be indicated by considering a many-body system as

particles → waves → field excitations that are quasi-particles.

The second quantisation approach has been applied to describe quantum mechanics of many-body systems in various branches of physics. Good description of the approach can be found in many text books, including Mandl (1959), Pines (1963), Schiff (1968), Ziman (1969), Peskin and Schroeder (1995), Lancaster and Blundell (2014). Here we will consider a few examples in condensed matter physics.

Consider the examples of the vibrating masses discussed in sections 8.2 and 8.3. Using classical mechanics (particle picture), we described the lattice vibrations as normal modes. This is the first quantisation level. In the second quantisation picture we proceed to analyse the normal modes in terms of quasi-particles, or the elementary excitations, of the vibrational field known as **phonons** of characteristic frequencies and momenta. We can present a similar picture for second quantisation of elastic waves in an elastic continuum for which classical field theory was discussed in sections 8.2 and 8.3. We can also do the same for electromagnetic waves (*cf* sections 8.2.1 and 8.3.5) and develop the concept of **photons** as the quasi-particles, or the elementary excitations, of the electromagnetic field. Furthermore, we can extend the discussion of electron waves in a solid from the Schödinger equation to the second quantisation level and speak of **electronic excitations**.

Before proceeding further we note that the language of second quantisation has to be tailored to treat quantum particles either as bosons (characterised with integer spin) or as fermions (characterised with fractional spin).

9.4.1 BOSONIC CREATION AND ANNIHILATION OPERATORS

We start with discussion of creation and annihilation operators for a bosonic field. Let us consider a representation in which an operator $\boldsymbol{N}$ is diagonal and Hermitian (*i.e.* with real eigenvalues). We can express such an operator as

$$\hat{N} = \sum_q \hat{N}_q, \tag{9.18}$$

where q represents a state of the field[5]. Let us further express

$$\hat{N}_q = \hat{a}_q^+ \hat{a}_q. \tag{9.19}$$

The operators $\hat{a}_q^+$, $\hat{a}_q$ and $\hat{N}_q$ are called the creation, annihilation and number operators, respectively.

The states of a quantised quantum field, in the representation in which each $\hat{N}_q$ is diagonal, are the kets[6]

$$|n_1, n_2, \cdots, n_q, \cdots>,$$

where each n_q is an eigenvalue of $\hat{N}_q$ and must be a positive integer or zero.

Bosonic creation, annihilation and number operators obey the eigenvalue relations

$$\hat{a}_q^+ |n_1, n_2, \cdots, n_q, \cdots> = \sqrt{(n_q+1)}\, |n_1, n_2, \cdots, n_q+1, \cdots>, \tag{9.20}$$

$$\hat{a}_q |n_1, n_2, \cdots, n_q, \cdots> = \sqrt{n_q}\, |n_1, n_2, \cdots, n_q-1, \cdots>, \tag{9.21}$$

$$\hat{N}_q |n_1, n_2, \cdots, n_q, \cdots> = n_q\, |n_1, n_2, \cdots, n_q, \cdots>. \tag{9.22}$$

The operators $\hat{a}_q^+$ and $\hat{a}_q$ satisfy the commutation relations

$$[\hat{a}_q, \hat{a}_{q'}^+] \equiv \hat{a}_q \hat{a}_{q'}^+ - \hat{a}_{q'}^+ \hat{a}_q = \delta_{q,q'}, \tag{9.23}$$

$$[\hat{a}_q, \hat{a}_{q'}] = 0, \tag{9.24}$$

$$[\hat{a}_q^+, \hat{a}_{q'}^+] = 0. \tag{9.25}$$

We can specify the orthonormality of states as

$$< n_{q'} | n_q > = \delta_{q,q'}. \tag{9.26}$$

In the context of the second quantisation procedure, n_q represents the number of quanta of quasi-particles, or excitations, in the qth state. Using equations (9.20), (9.21) and (9.26), increase and decrease in number of such excitations in the qth state are obtained using the following expectation value results

$$< n_q + 1 | \hat{a}_q^+ | n_q > = \sqrt{(n_q+1)}, \tag{9.27}$$

$$< n_q - 1 | \hat{a}_q | n_q > = \sqrt{n_q}. \tag{9.28}$$

[5] Note that the notation q in this section is different from that in the previous section.

[6] A ket $|>$ represents an eigenstate and a bra $<|$ represents the complex conjugate of an eigenstate $|>$. The kets are regarded as state vectors in the **Fock space**, which is the Hibert space describing a quantum many-body system.

9.4.2 DIAGONAL FORM OF HAMILTONIAN FOR HARMONIC VIBRATIONS IN THREE-DIMENSIONAL CRYSTALS

We will use the second quantisation methodology to derive an expression for the diagonal form of the Hamiltonian for harmonic vibrations in three-dimensional crystals and obtain an expression for the corresponding energy. Using the notation described in section 8.3.2, and the expression for the Hamiltonian density in equation (8.69), we write the classical Hamiltonian function at the harmonic level as

$$H_{\text{harm}} = \sum_{\boldsymbol{lb}} \frac{\boldsymbol{p}(\boldsymbol{lb}) \cdot \boldsymbol{p}(\boldsymbol{lb})}{2m_b} + \frac{1}{2} \sum_{\boldsymbol{lb},\boldsymbol{l}'\boldsymbol{b}'} \sum_{\alpha,\beta} \Phi_{\alpha\beta}(\boldsymbol{lb};\boldsymbol{l}'\boldsymbol{b}') u_\alpha(\boldsymbol{lb}) u_\beta(\boldsymbol{l}'\boldsymbol{b}'), \tag{9.29}$$

where $\boldsymbol{u}(\boldsymbol{lb})$ is the displacement vector for bth atom of mass m_b in the lth unit cell, $\boldsymbol{p}(\boldsymbol{lb})$ is the momentum canonically conjugate to $\boldsymbol{u}(\boldsymbol{lb})$ and $\Phi_{\alpha\beta}(\boldsymbol{lb};\boldsymbol{l}'\boldsymbol{b}')$ represents linear force along α direction on atom $(\boldsymbol{l},\boldsymbol{b})$ when atom $(\boldsymbol{l}',\boldsymbol{b}')$ is displaced along β direction. We will assume that the one-dimensional cyclic, or periodic, boundary condition described in section 4.5 can be appropriately extended to three dimensions for both $\boldsymbol{u}(\boldsymbol{lb})$ and $\boldsymbol{p}(\boldsymbol{lb})$.

The first step towards diagonalisation of the Hamiltonian is to make a transformation from the canonically conjugate variables $\boldsymbol{u}(\boldsymbol{lb})$ and $\boldsymbol{p}(\boldsymbol{lb})$ to canonically conjugate variables $\boldsymbol{X}(\boldsymbol{qb})$ and $\boldsymbol{P}(\boldsymbol{qb})$, representing respectively *new coordinates* and *new momenta*, by making use of the following Fourier analyses

$$\begin{aligned} \boldsymbol{u}(\boldsymbol{lb}) &= \frac{1}{\sqrt{N_0\Omega}} \sum_{\boldsymbol{q}} \boldsymbol{X}(\boldsymbol{qb}) \mathrm{e}^{i\boldsymbol{q}\cdot\boldsymbol{l}}, \\ \boldsymbol{p}(\boldsymbol{lb}) &= \frac{1}{\sqrt{N_0\Omega}} \sum_{\boldsymbol{q}} \hat{\boldsymbol{P}}(\boldsymbol{qb}) \mathrm{e}^{-i\boldsymbol{q}\cdot\boldsymbol{l}}, \end{aligned} \tag{9.30}$$

where $N_0\Omega$ is crystal volume (N_0 unit cells, each of volume Ω). To proceed, we will regard the variables $\boldsymbol{X}(\boldsymbol{qb})$ and $\boldsymbol{P}(\boldsymbol{qb})$ as operators $\hat{\boldsymbol{X}}(\boldsymbol{qb}) = \boldsymbol{X}(\boldsymbol{qb})$ and $\hat{\boldsymbol{P}}(\boldsymbol{qb})$. Note that while operators representing $\boldsymbol{u}(\boldsymbol{lb})$ and $\boldsymbol{p}(\boldsymbol{lb})$ are Hermitian, the operators representing $\boldsymbol{X}(\boldsymbol{qb})$ and $\boldsymbol{P}(\boldsymbol{qb})$ are not, because

$$\begin{aligned} \boldsymbol{X}^*(\boldsymbol{qb}) &= \boldsymbol{X}(-\boldsymbol{qb}) = \frac{1}{\sqrt{N_0\Omega}} \sum_{\boldsymbol{q}} \boldsymbol{u}(\boldsymbol{lb}) \mathrm{e}^{i\boldsymbol{q}\cdot\boldsymbol{l}}, \\ \boldsymbol{P}^*(\boldsymbol{qb}) &= \boldsymbol{P}(-\boldsymbol{qb}) = \frac{1}{\sqrt{N_0\Omega}} \sum_{\boldsymbol{q}} \boldsymbol{p}(\boldsymbol{lb}) \mathrm{e}^{-i\boldsymbol{q}\cdot\boldsymbol{l}}. \end{aligned} \tag{9.31}$$

It can be shown that the operators $\boldsymbol{X}(\boldsymbol{qb})$ and $\hat{\boldsymbol{P}}(\boldsymbol{qb})$ satisfy the commutation relation

$$[\boldsymbol{X}(\boldsymbol{qb}), \hat{\boldsymbol{P}}(\boldsymbol{q}'\boldsymbol{b}')] = \hat{\boldsymbol{I}} i\hbar \delta_{\boldsymbol{qq}'} \delta_{\boldsymbol{bb}'}. \tag{9.32}$$

Before proceeding further, we note that lattice translational symmetry allows the harmonic force constant matrix to be expressed as

$$\Phi(\boldsymbol{lb},\boldsymbol{l}'\boldsymbol{b}') = \Phi(\boldsymbol{0b},(\boldsymbol{l}'-\boldsymbol{l})\boldsymbol{b}'). \tag{9.33}$$

We then express the force constant matrix in $(\boldsymbol{q},\boldsymbol{b})$ space as

$$\Phi(\boldsymbol{bb}'|\boldsymbol{q}) = \Phi(\boldsymbol{0b},\boldsymbol{hh}') \mathrm{e}^{-i\boldsymbol{q}\cdot\boldsymbol{h}}, \tag{9.34}$$

where $\boldsymbol{h} = \boldsymbol{l}' - \boldsymbol{l}$.

At this stage the Hermitian form of the harmonic Hamiltonian operator can therefore be expressed as

$$\hat{H}_{\text{harm}} = \frac{1}{2}\sum_{\boldsymbol{qb}} \frac{1}{m_b}\hat{\boldsymbol{P}}(\boldsymbol{qb})\cdot\hat{\boldsymbol{P}}^{\dagger}(\boldsymbol{qb}) + \frac{1}{2}\sum_{\boldsymbol{qbb'}} \boldsymbol{X}(\boldsymbol{qb})\cdot\Phi(\boldsymbol{bb'}|\boldsymbol{q})\cdot\boldsymbol{X}(\boldsymbol{qb'}), \tag{9.35}$$

where we have used the following property of Fourier series

$$\frac{1}{N_0\Omega}\sum_{\boldsymbol{h}} e^{i(\boldsymbol{Q}-\boldsymbol{Q}')\cdot\boldsymbol{h}} = \delta_{\boldsymbol{Q},\boldsymbol{Q}'}. \tag{9.36}$$

In the second step towards the diagonalisation of the crystal Hamiltonian we introduce creation operators $\hat{a}^{\dagger}_{\boldsymbol{q}s}$ and annihilation operators $\hat{a}_{\boldsymbol{q}s}$ by using the following relations

$$\hat{a}^{\dagger}_{\boldsymbol{q}s} = \frac{1}{\sqrt{2\hbar\omega(\boldsymbol{q}s)}}\hat{\boldsymbol{P}}^{\dagger}(\boldsymbol{q}s) + i\sqrt{\frac{\omega(\boldsymbol{q}s)}{2\hbar}}\boldsymbol{X}(\boldsymbol{q}s), \tag{9.37}$$

$$\hat{a}_{\boldsymbol{q}s} = \frac{1}{\sqrt{2\hbar\omega(\boldsymbol{q}s)}}\hat{\boldsymbol{P}}(\boldsymbol{q}s) - i\sqrt{\frac{\omega(\boldsymbol{q}s)}{2\hbar}}\boldsymbol{X}^{\dagger}(\boldsymbol{q}s). \tag{9.38}$$

In the above equations the operators $\hat{\boldsymbol{X}}(\boldsymbol{q}s) \equiv \boldsymbol{X}(\boldsymbol{q}s)$ and $\hat{P}(\boldsymbol{q}s)$ are obtained using the following transformations

$$\boldsymbol{X}(\boldsymbol{q}s) = \sum_{\boldsymbol{b}} \sqrt{m_b}\boldsymbol{e}^*(\boldsymbol{b}|\boldsymbol{q}s)\cdot\boldsymbol{X}(\boldsymbol{qb}), \tag{9.39}$$

$$\hat{\boldsymbol{P}}(\boldsymbol{q}s) = \sum_{\boldsymbol{b}} \frac{1}{\sqrt{m_b}}\boldsymbol{e}(\boldsymbol{b}|\boldsymbol{q}s)\cdot\hat{\boldsymbol{P}}(\boldsymbol{qb}), \tag{9.40}$$

where $\omega(\boldsymbol{q}s)$ are sth branch eigenfrequencies for the dynamical matrix described in section 8.2.4 whose eigenvectors satisfy the orthogonality relation

$$\sum_{\boldsymbol{b}} \boldsymbol{e}^*(\boldsymbol{b}|\boldsymbol{q}s)\cdot\boldsymbol{e}(\boldsymbol{b}|\boldsymbol{q}s')) = \delta_{ss'} \tag{9.41}$$

and the closure relation

$$\sum_{s} e^*_{\alpha}(\boldsymbol{b}|\boldsymbol{q}s)e_{\beta}(\boldsymbol{b}'|\boldsymbol{q}s) = \delta_{\alpha\beta}\delta_{\boldsymbol{bb}'}. \tag{9.42}$$

It can be easily verified that the operators $\hat{a}_{\boldsymbol{q}s}$ and $\hat{a}^{\dagger}_{\boldsymbol{q}s}$ satisfy the commutation relation

$$[\hat{a}_{\boldsymbol{q}s}, \hat{a}^{\dagger}_{\boldsymbol{q}'s'}] = \hat{\boldsymbol{I}}\delta_{\boldsymbol{q},\boldsymbol{q}'}\delta_{s,s'}. \tag{9.43}$$

With the help of the above transformations, the harmonic Hamiltonian operator can finally be expressed as

$$\begin{aligned}
\hat{H}_{\text{harm}} &= \frac{1}{4}\sum_{\boldsymbol{q}s}\hbar\omega(\boldsymbol{q}s)(\hat{a}_{\boldsymbol{q}s} + \hat{a}^{\dagger}_{-\boldsymbol{q}s})(\hat{a}^{\dagger}_{\boldsymbol{q}s} + \hat{a}_{-\boldsymbol{q}s}) \\
&\quad + \frac{1}{4}\sum_{\boldsymbol{q}s}\hbar\omega(\boldsymbol{q}s)(\hat{a}^{\dagger}_{\boldsymbol{q}s} - \hat{a}_{-\boldsymbol{q}s})(\hat{a}_{\boldsymbol{q}s} - \hat{a}^{\dagger}_{-\boldsymbol{q}s}), \\
&= \frac{1}{2}\sum_{\boldsymbol{q}s}\hbar\omega(\boldsymbol{q}s)(\hat{a}_{\boldsymbol{q}s}\hat{a}^{\dagger}_{\boldsymbol{q}s} + \hat{a}^{\dagger}_{\boldsymbol{q}s}\hat{a}_{\boldsymbol{q}s}), \\
&= \sum_{\boldsymbol{q}s}\hbar\omega(\boldsymbol{q}s)(\hat{a}^{\dagger}_{\boldsymbol{q}s}\hat{a}_{\boldsymbol{q}s} + \frac{1}{2}\hat{\boldsymbol{I}}).
\end{aligned} \tag{9.44}$$

In deriving steps in the above equation we have made use of the commutation relations in equation (9.43) and the fact that summation over values of $-\boldsymbol{q}$ merely duplicates the summation over values of $\boldsymbol{q}$. The expression in equation (9.44) represents the harmonic Hamiltonian operator in diagonal form.

We can summarise the steps taken for achieving the diagonal form of $\hat{H}_{\text{harm}}$. The particle picture in equation (9.29) was transformed into the wave picture (first quantisation) in equation (9.35), which was subsequently transformed into the quasi-particle picture (second quantisation) in equation (9.44).

The expression for the Hamiltonian operator in equation (9.44) can be used to obtain the vibrational energy for a harmonic crystal. Consistent with the development of the creation and annihilation operators $\hat{a}^{\dagger}_{\boldsymbol{q}s}$ and $\hat{a}_{\boldsymbol{q}s}$, we denote system's eigenvector in state $\boldsymbol{q}s$ as $|n_{\boldsymbol{q}s}>$, containing $n_{\boldsymbol{q}s}$ quasi-particles. Using equations (9.19) and (9.22), we can immediately write down the eigenvalue equation satisfied by $\hat{H}_{\text{harm}}$:

$$\begin{aligned}
\hat{H}_{\text{harm}}|n_{\boldsymbol{q}s}> &= \sum_{\boldsymbol{q}s}\hbar\omega(\boldsymbol{q}s)(\hat{a}^{\dagger}_{\boldsymbol{q}s}\hat{a}_{\boldsymbol{q}s}+\frac{1}{2}\hat{\boldsymbol{I}})|n_{\boldsymbol{q}s}>, \\
&= \sum_{\boldsymbol{q}s}\hbar\omega(\boldsymbol{q}s)(n_{\boldsymbol{q}s}+\frac{1}{2})|n_{\boldsymbol{q}s}>, \\
&= \sum_{\boldsymbol{q}s}\varepsilon_{\boldsymbol{q}s}|n_{\boldsymbol{q}s}>. \qquad (9.45)
\end{aligned}$$

The expression for crystal vibrational energy, the expectation value of $\hat{H}_{\text{harm}}$, is thus

$$\begin{aligned}
E &= \sum_{\boldsymbol{q}s}<n_{\boldsymbol{q}s}|\hat{H}_{\text{harm}}|n_{\boldsymbol{q}s}>, \\
&= \sum_{\boldsymbol{q}s}\varepsilon_{\boldsymbol{q}s}, \\
&= \sum_{\boldsymbol{q}s}\hbar\omega(\boldsymbol{q}s)(n_{\boldsymbol{q}s}+\frac{1}{2}). \qquad (9.46)
\end{aligned}$$

It is physically meaningful to talk about thermal average energy $\bar{E}$:

$$\bar{E} = \sum_{\boldsymbol{q}s}\bar{\varepsilon}_{\boldsymbol{q}s} = \sum_{\boldsymbol{q}s}\hbar\omega(\boldsymbol{q}s)(\bar{n}_{\boldsymbol{q}s}+\frac{1}{2}). \qquad (9.47)$$

The term $\frac{1}{2}\sum_{\boldsymbol{q}s}\hbar\omega(\boldsymbol{q}s)$ represents the zero-point energy and the term $\sum_{\boldsymbol{q}s}\hbar\omega(\boldsymbol{q}s)\bar{n}_{\boldsymbol{q}s}$ is the excitation energy. The excitations, or quasi-particles, are called **phonons**. A diagrammatic representation for a phonon is sketched in figure 9.1.

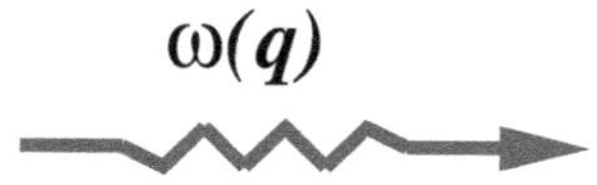

Figure 9.1 A diagrammatic representation for a phonon of wavevector $\boldsymbol{q}$ and frequency $\omega(\boldsymbol{q})$.

The thermal average of the phonon occupation number $\bar{n}_{\boldsymbol{q}s}$ at temperature T is given by the Bose-Einstein distribution function

$$\bar{n}_{\boldsymbol{q}s} = \frac{1}{e^{\hbar\omega(\boldsymbol{q}s)/k_{\text{B}}T}-1}, \qquad (9.48)$$

where k_{B} is Boltzmann's constant. We say that the average number of phonons of wave vector $\boldsymbol{q}$ in polarisation mode s at temperature T is $\bar{n}_{\boldsymbol{q}s}$, each carrying energy $\hbar\omega(\boldsymbol{q}s)$. Obviously, there are no phonons at absoluate zero. However, the number of phonons is not conserved but increases linearly with temperature at high temperatures (when $k_B T >> \hbar\omega(\boldsymbol{q}s)$) and decreases exponentially as $\exp(-\hbar\omega/k_B T)$ at low temperatures (when $k_B T << \hbar\omega(\boldsymbol{q}s)$).

9.4.3 DIAGONAL FORM OF HAMILTONIAN FOR ISOTROPIC HARMONIC ELASTIC CONTINUUM

In this sub-section we will use the second quantisation methodology to derive an expression for the diagonal form of Hamiltonian for isotropic harmonic elastic continuum. For this we use equations (8.33), (8.34) and (8.42) in section 8.2.5 and equation (8.79) in section 8.3.4 to write the classical Hamiltonian in the form

$$H = \int d^3r \Big[\sum_k \frac{p_k^2(\boldsymbol{r})}{2\rho} + \frac{\lambda}{2}\sum_k \Big(\frac{\partial u_k(\boldsymbol{r})}{\partial r_k}\Big)^2 + \frac{\mu}{4}\sum_{kl}\Big(\frac{\partial u_k(\boldsymbol{r})}{\partial r_l} + \frac{\partial u_l(\boldsymbol{r})}{\partial r_k}\Big)^2\Big], \tag{9.49}$$

where ρ is mass density, $\boldsymbol{u}(\boldsymbol{r})$ and $\boldsymbol{p}(\boldsymbol{r})$ are respectively displacement and momentum vectors of the continuum at point $\boldsymbol{r}$, and λ and μ are Lamè's constants of isotropic elasticity.

Following equations (9.30), (9.30), (9.39) and (9.40), we express $\boldsymbol{u}(\boldsymbol{r})$ and $\boldsymbol{p}(\boldsymbol{r})$ as

$$\begin{aligned} \boldsymbol{u}(\boldsymbol{r}) &= \frac{1}{\sqrt{\rho V_{\text{norm}}}}\sum_{\boldsymbol{q}s} \boldsymbol{e}_{\boldsymbol{q}s} X_{\boldsymbol{q}s} e^{i\boldsymbol{q}\cdot\boldsymbol{r}}, \\ \boldsymbol{p}(\boldsymbol{r}) &= \sqrt{\frac{\rho}{V_{\text{norm}}}}\sum_{\boldsymbol{q}s} \boldsymbol{e}_{\boldsymbol{q}s} P_{\boldsymbol{q}s} e^{-i\boldsymbol{q}\cdot\boldsymbol{r}}, \end{aligned} \tag{9.50}$$

where V_{norm} is normalisation volume and $\boldsymbol{e}_{\boldsymbol{q}s}$ is considered as polarisation unit vector. Using the relation

$$\frac{1}{V_{\text{norm}}}\int d^3r\, e^{(\boldsymbol{q}-\boldsymbol{q}')\cdot\boldsymbol{r}} = \delta_{\boldsymbol{q},\boldsymbol{q}'}, \tag{9.51}$$

and the closure and orthonormality conditions associated with the polarisation unit vectors $\boldsymbol{e}_{\boldsymbol{q}s}$ in equations (9.41) and (9.42), the three terms in equation (9.49) can then be expressed as follows.

$$\text{First term} = \int d^3r \sum_k \frac{p_k^2(\boldsymbol{r})}{2\rho} = \frac{1}{2}\sum_{\boldsymbol{q}s} P_{\boldsymbol{q}s}^* P_{\boldsymbol{q}s}, \tag{9.52}$$

$$\text{Second term} = \frac{\lambda}{2}\int d^3r \sum_k \Big(\frac{\partial u_k(\boldsymbol{r})}{\partial r_k}\Big)^2 = \frac{\lambda}{2\rho}\sum_{\boldsymbol{q}s} X_{\boldsymbol{q}s}^* X_{\boldsymbol{q}s}(\boldsymbol{e}_{\boldsymbol{q}s}\cdot\boldsymbol{q})^2, \tag{9.53}$$

$$\begin{aligned} \text{Third term} &= \frac{\mu}{4\rho}\int d^3r \sum_{kl}\Big(\frac{\partial u_k(\boldsymbol{r})}{\partial r_l} + \frac{\partial u_l(\boldsymbol{r})}{\partial r_k}\Big)^2, \\ &= \frac{\mu}{2\rho}\sum_{\boldsymbol{q}s} X_{\boldsymbol{q}s}^* X_{\boldsymbol{q}s}[q^2 + (\boldsymbol{e}_{\boldsymbol{q}s}\cdot\boldsymbol{q})^2]. \end{aligned} \tag{9.54}$$

Adding the three terms, we can express the quantum Hamiltonian operator as[7]

$$\hat{H} = \frac{1}{2}\sum_{\boldsymbol{q}s}\Big\{\hat{P}_{\boldsymbol{q}s}\hat{P}_{\boldsymbol{q}s}^\dagger + \frac{1}{\rho}\left[\mu q^2 + (\lambda+\mu)(\boldsymbol{e}\cdot\boldsymbol{q})^2\right]\hat{X}_{\boldsymbol{q}s}\hat{X}_{\boldsymbol{q}s}^\dagger\Big\}. \tag{9.55}$$

---

[7]Note that $\hat{X} = X$ and $\hat{X}^\dagger = X^\dagger$, as mentioned before.

Now, using equations (9.39) and (9.40), we express $\hat{X}_{\boldsymbol{q}s}$ and $\hat{P}_{\boldsymbol{q}s}$ in terms of creation and annihilation operators

$$\hat{X}_{\boldsymbol{q}s} = -i\sqrt{\frac{\hbar}{2\omega_{\boldsymbol{q}s}}}(\hat{a}^{\dagger}_{\boldsymbol{q}s} - \hat{a}_{-\boldsymbol{q}s}), \tag{9.56}$$

$$\hat{P}_{\boldsymbol{q}s} = \sqrt{\frac{\hbar\omega_{\boldsymbol{q}s}}{2}}(\hat{a}^{\dagger}_{\boldsymbol{q}s} + \hat{a}_{-\boldsymbol{q}s}). \tag{9.57}$$

With substitution of equations (9.56) and (9.57) in (9.55) the Hamiltonian operator for an isotropic elastic continuum reads

$$\begin{aligned} \hat{H} &= \frac{1}{2}\sum_{\boldsymbol{q}s}\Big\{\frac{\hbar\omega_{\boldsymbol{q}s}}{2}(\hat{a}_{\boldsymbol{q}s} + \hat{a}^{\dagger}_{-\boldsymbol{q}s})(\hat{a}^{\dagger}_{\boldsymbol{q}s} + \hat{a}_{-\boldsymbol{q}s}) \\ &\quad + \frac{\hbar}{2\rho\,\omega_{\boldsymbol{q}s}}[\mu q^2 + (\lambda+\mu))(\boldsymbol{e}\cdot\boldsymbol{q})^2](\hat{a}^{\dagger}_{\boldsymbol{q}s} - \hat{a}_{-\boldsymbol{q}s})(\hat{a}_{\boldsymbol{q}s} - \hat{a}^{\dagger}_{-\boldsymbol{q}s})\Big\}. \end{aligned} \tag{9.58}$$

For an isotropic elastic continuum the polarisation index takes two values: *longitudinal acoustic* (LA) and doubly-degenerate *transverse acoustic* (TA). For the longitudinal and transverse polarisations $\boldsymbol{e}\cdot\boldsymbol{q} = q$ and $\boldsymbol{e}\cdot\boldsymbol{q} = 0$, respectively. The Hamiltonian operator for the two polarisations can be expressed separately.

For LA :

$$\begin{aligned} \hat{H}_{\mathsf{LA}} &= \frac{1}{4}\sum_{\boldsymbol{q}}\Big\{\frac{\hbar\omega_{\boldsymbol{q}\mathsf{LA}}}{2}(\hat{a}_{\boldsymbol{q}\mathsf{LA}} + \hat{a}^{\dagger}_{-\boldsymbol{q}\mathsf{LA}})(\hat{a}^{\dagger}_{\boldsymbol{q}\mathsf{LA}} + \hat{a}_{-\boldsymbol{q}\mathsf{LA}}) \\ &\quad + \frac{\hbar}{\rho\,\omega_{\boldsymbol{q}\mathsf{LA}}}(\lambda+2\mu)q^2(\hat{a}^{\dagger}_{\boldsymbol{q}\mathsf{LA}} - \hat{a}_{-\boldsymbol{q}\mathsf{LA}})(\hat{a}_{\boldsymbol{q}\mathsf{LA}} - \hat{a}^{\dagger}_{-\boldsymbol{q}\mathsf{LA}})\Big\}. \end{aligned} \tag{9.59}$$

and

for each TA :

$$\begin{aligned} \hat{H}_{\mathsf{TA}} &= \frac{1}{4}\sum_{\boldsymbol{q}}\Big\{\frac{\hbar\omega_{\boldsymbol{q}\mathsf{TA}}}{2}(\hat{a}_{\boldsymbol{q}\mathsf{TA}} + \hat{a}^{\dagger}_{-\boldsymbol{q}\mathsf{TA}})(\hat{a}^{\dagger}_{\boldsymbol{q}\mathsf{TA}} + \hat{a}_{-\boldsymbol{q}\mathsf{TA}}) \\ &\quad + \frac{\hbar}{\rho\,\omega_{\boldsymbol{q}\mathsf{TA}}}\mu q^2(\hat{a}^{\dagger}_{\boldsymbol{q}\mathsf{TA}} - \hat{a}_{-\boldsymbol{q}\mathsf{TA}})(\hat{a}_{\boldsymbol{q}\mathsf{TA}} - \hat{a}^{\dagger}_{-\boldsymbol{q}\mathsf{TA}})\Big\}. \end{aligned} \tag{9.60}$$

Solutions of the Green-Christoffel determinantal equation (8.47) for the isotropic elastic continuum are (*cf* Srivastava, 2022)

$$\omega^2_{\boldsymbol{q}\mathsf{LA}} = \frac{(\lambda+2\mu)q^2}{\rho}, \tag{9.61}$$

and

$$\omega^2_{\boldsymbol{q}\mathsf{TA}} = \frac{\mu q^2}{\rho}. \tag{9.62}$$

Using these results, the Hamiltonian operator for polarisation s can be written in a compact manner as

$$\begin{aligned} \hat{H}_s &= \frac{1}{4}\sum_{\boldsymbol{q}}\hbar\omega_{\boldsymbol{q}s}\big[(\hat{a}_{\boldsymbol{q}s} + \hat{a}^{\dagger}_{-\boldsymbol{q}s})(\hat{a}^{\dagger}_{\boldsymbol{q}s} + \hat{a}_{-\boldsymbol{q}s}) + (\hat{a}^{\dagger}_{\boldsymbol{q}s} - \hat{a}_{-\boldsymbol{q}s})(\hat{a}_{\boldsymbol{q}s} - \hat{a}^{\dagger}_{-\boldsymbol{q}s})\big], \\ &= \frac{1}{2}\sum_{\boldsymbol{q}}\hbar\omega_{\boldsymbol{q}s}(\hat{a}_{\boldsymbol{q}s} + \hat{a}^{\dagger}_{\boldsymbol{q}s})(\hat{a}^{\dagger}_{\boldsymbol{q}s} + \hat{a}_{\boldsymbol{q}s}), \\ &= \sum_{\boldsymbol{q}}\hbar\omega_{\boldsymbol{q}s}(\hat{a}^{\dagger}_{\boldsymbol{q}s}\hat{a}_{\boldsymbol{q}s} + \frac{1}{2}\hat{I}), \end{aligned} \tag{9.63}$$

which is of the same form as equation (9.44). Hence, the eigenvalue equation satisfied by $\hat{H}_s$ is

$$\hat{H}_s|n_{qs}>=\sum_{qs}(n_{qs}+\frac{1}{2})\hbar\omega_{qs}|n_{qs}>, \tag{9.64}$$

which has the same form as equation (9.45) for a crystalline solid.

9.4.4 DIAGONAL FORM OF HAMILTONIAN FOR ELECTROMAGNETIC FIELD IN VACUUM

We start with the expression for the Hamiltonian density for the electromagnetic field in vacuum, expressed in equation (8.82)

$$\begin{aligned}\mathscr{H} &= \boldsymbol{p}\cdot\frac{\partial \boldsymbol{A}}{\partial t}-\mathscr{L}, \\ &= \frac{1}{2}\boldsymbol{p}^2-\boldsymbol{p}\cdot\boldsymbol{\nabla}\Phi+\frac{1}{2}c^2(\boldsymbol{\nabla}\times\boldsymbol{A})^2,\end{aligned} \tag{9.65}$$

where $\mathscr{L}$ is the Lagrangian density, $\boldsymbol{A}$ is the vector potential, Φ is the scalar potential[8] and $\boldsymbol{p}$ is the momentum conjugate to the generalised coordinate $\boldsymbol{A}$.

In the *Coulomb gauge*, also known as the transverse gauge (Berestetskii *et al.* , 1999),

$$\boldsymbol{\nabla}\cdot\boldsymbol{A}=0. \tag{9.66}$$

In this gauge, using the Maxwell equation $\boldsymbol{\nabla}\cdot\boldsymbol{E}=0$ in vacuum, we get

$$\begin{aligned}\boldsymbol{\nabla}\cdot\boldsymbol{E} &= 0, \\ -\boldsymbol{\nabla}\cdot\left(\frac{\partial \boldsymbol{A}}{\partial t}+\boldsymbol{\nabla}\Phi\right) &= 0, \\ \nabla^2\Phi &= 0.\end{aligned} \tag{9.67}$$

The Lagrangian does not contain Φ, therefore Φ is not a generalised coordinate. As Φ is a constant, $\boldsymbol{\nabla}\Phi=0$ and the only solution of the Laplace equation in (9.67) that is regular over all space is $\Phi=0$. With these considerations, the expression for the classical Hamiltonian density is

$$\mathscr{H}=\frac{1}{2}p^2+\frac{1}{2}c^2(\boldsymbol{\nabla}\times\boldsymbol{A})^2. \tag{9.68}$$

Within the Coulomb gauge, the Maxwell equation $\boldsymbol{\nabla}\times\boldsymbol{B}=\frac{1}{c^2}\frac{\partial \boldsymbol{E}}{\partial t}$ becomes the wave equation $\nabla^2\boldsymbol{A}=\frac{1}{c^2}\frac{\partial^2\boldsymbol{A}}{\partial t^2}$, whose solution can be expressed in planewaves

$$\boldsymbol{A}(\boldsymbol{r},t)=\frac{1}{\sqrt{V}_{\text{norm}}}\sum_{\boldsymbol{Q}s}\hat{\boldsymbol{e}}_{\boldsymbol{Q}s}X_{\boldsymbol{Q}s}e^{i(\boldsymbol{Q}\cdot\boldsymbol{r}-\omega t)}, \tag{9.69}$$

with the linear dispersion relation

$$\omega=\omega_{\boldsymbol{Q}s}=cQ. \tag{9.70}$$

In the above, c is the speed of light in vacuum, $\boldsymbol{Q}$ is a wave vector[9], s is a transverse polarisation index, $\boldsymbol{e}_{\boldsymbol{Q}s}$ is polarisation unit vector and V_{norm} is normalisation volume.

---

[8]The symbols Φ and $\boldsymbol{A}$ used in this sub-section should not be confused with similar symbols used in the sub-sections not dealing with electromagnetism.

[9]Note that the symbol $\boldsymbol{Q}$ used in this sub-section should not be confused with that used in section 9.2.

With the above consideration, the canonical momentum $\boldsymbol{p}(\boldsymbol{r},t)$, electric field $\boldsymbol{E}(\boldsymbol{r},t)$ and magnetic field $\boldsymbol{B}(\mathrm{r},\mathrm{t})$ can also be expressed using planewaves with transverse polarisations

$$\boldsymbol{p}(\boldsymbol{r},t) = -\boldsymbol{E}(\boldsymbol{r},t) = \frac{1}{\sqrt{V}_{\text{norm}}} \sum_{\boldsymbol{Q}s} \boldsymbol{e}_{\boldsymbol{Q}s} P_{\boldsymbol{Q}s} \mathrm{e}^{-i(\boldsymbol{Q}\cdot\boldsymbol{r}-\omega t)}, \tag{9.71}$$

$$\nabla \times \boldsymbol{A}(\boldsymbol{r},t) = \boldsymbol{B}(\boldsymbol{r},t) = \frac{i}{\sqrt{V}_{\text{norm}}} \sum_{\boldsymbol{Q}s} (\boldsymbol{Q} \times \boldsymbol{e}_{\boldsymbol{Q}s}) X_{\boldsymbol{Q}s} \mathrm{e}^{i(\boldsymbol{Q}\cdot\boldsymbol{r}-\omega t)}. \tag{9.72}$$

We can now proceed in the same manner as we did in sections 9.4.2 and 9.4.3. Using the following rule

$$\frac{1}{V_{\text{norm}}} \int \mathrm{d}^3 r\, \mathrm{e}^{i(\boldsymbol{Q}-\boldsymbol{Q}')\cdot\boldsymbol{r}} = \delta_{\boldsymbol{Q},\boldsymbol{Q}'} \tag{9.73}$$

and the closure and orthonormality considerations for $\hat{\boldsymbol{e}}_{\boldsymbol{Q}s}$ similar to those in equations (9.41) and (9.42), we can obtain the following expression for the classical Hamiltonian

$$H = \sum_{\boldsymbol{Q}s} \left[\frac{1}{2} P_{\boldsymbol{Q}s} P^*_{\boldsymbol{Q}s} + \frac{1}{2} \omega^2_{\boldsymbol{Q}s} X_{\boldsymbol{Q}s} X^*_{\boldsymbol{Q}s}\right]. \tag{9.74}$$

The expression for quantum Hamiltonian operator, then, is

$$\hat{H} = \sum_{\boldsymbol{Q}s} \left[\frac{1}{2} \hat{P}_{\boldsymbol{Q}s} \hat{P}^\dagger_{\boldsymbol{Q}s} + \frac{1}{2} \omega^2_{\boldsymbol{Q}s} \hat{X}_{\boldsymbol{Q}s} \hat{X}^*_{\boldsymbol{Q}s}\right]. \tag{9.75}$$

We now express $X_{\boldsymbol{Q}s}$ and $P_{\boldsymbol{Q}s}$ in terms of creation and annihilation operators $\hat{c}^\dagger_{\boldsymbol{Q}s}$ and $\hat{c}_{\boldsymbol{Q}s}$[10]

$$\hat{X}_{\boldsymbol{Q}s} = \sqrt{\frac{\hbar}{2\omega_{\boldsymbol{Q}s}}} (\hat{c}^\dagger_{\boldsymbol{Q}s} - \hat{c}_{-\boldsymbol{Q}s}), \tag{9.76}$$

$$\hat{P}_{\boldsymbol{Q}s} = \sqrt{\frac{\hbar\omega_{\boldsymbol{Q}s}}{2}} (\hat{c}^\dagger_{\boldsymbol{Q}s} + \hat{c}_{-\boldsymbol{Q}s}). \tag{9.77}$$

With these, and noting that sum over $-\boldsymbol{Q}$ duplicates sum over $\boldsymbol{Q}$, we express the Hamiltonian operator as

$$\begin{aligned} \hat{H} &= \frac{1}{4} \sum_{\boldsymbol{Q}s} \hbar\omega_{\boldsymbol{Q}s} \left[(\hat{c}^\dagger_{\boldsymbol{Q}s} + \hat{c}_{-\boldsymbol{Q}s})(\hat{c}_{\boldsymbol{Q}s} + \hat{c}^\dagger_{-\boldsymbol{Q}s}) + (\hat{c}^\dagger_{\boldsymbol{Q}s} - \hat{c}_{-\boldsymbol{Q}s})(\hat{c}_{\boldsymbol{Q}s} - \hat{c}^\dagger_{-\boldsymbol{Q}s})\right], \\ &= \frac{1}{2} \sum_{\boldsymbol{Q}s} \hbar\omega_{\boldsymbol{Q}s} (\hat{c}_{\boldsymbol{Q}s} \hat{c}^\dagger_{\boldsymbol{Q}s} + \hat{c}^\dagger_{\boldsymbol{Q}s} \hat{c}_{\boldsymbol{Q}s}). \end{aligned} \tag{9.78}$$

It can be easily verified that the operators $\hat{c}_{\boldsymbol{Q}s}$ and $\hat{c}^\dagger_{\boldsymbol{Q}s}$ satisfy the commutation relation

$$[\hat{c}_{\boldsymbol{Q}s}, \hat{c}^\dagger_{\boldsymbol{Q}s}] = \hat{I}, \tag{9.79}$$

using which, the expression for the Hamiltonian operator can be simplified as

$$\begin{aligned} \hat{H} &= \sum_{\boldsymbol{Q}s} \hbar\omega_{\boldsymbol{Q}s} (\hat{c}^\dagger_{\boldsymbol{Q}s} \hat{c}_{\boldsymbol{Q}s} + \frac{1}{2} \hat{I}), \\ &= \sum_{\boldsymbol{Q}s} \hbar\omega_{\boldsymbol{Q}s} (\hat{N}_{\boldsymbol{Q}s} + \frac{1}{2} \hat{I}), \end{aligned} \tag{9.80}$$

[10]The operator $\hat{c}_{\boldsymbol{Q}s}$ should not be confused with c, the speed of light.

where $\hat{N}_{\boldsymbol{Q}s}$ is the number operator. The Hamiltonian operator satisfies the eigenvalue equation

$$\begin{aligned} \hat{H}|n_{\boldsymbol{Q}s}> &= \sum_{\boldsymbol{Q}s} \hbar\omega_{\boldsymbol{Q}s}(\hat{N}_{\boldsymbol{Q}s} + \frac{1}{2})|n_{\boldsymbol{Q}s}>, \\ &= \sum_{\boldsymbol{Q}s} \hbar\omega_{\boldsymbol{Q}s}(n_{\boldsymbol{Q}s} + \frac{1}{2}|n_{\boldsymbol{Q}s}>, \end{aligned} \tag{9.81}$$

with $n_{\boldsymbol{Q}s} = 0, 1, 2, \cdots$.

The energy eigenvalue for mode $\boldsymbol{Q}s$ is

$$\begin{aligned} \varepsilon_{\boldsymbol{Q}s} &= \hbar\omega_{\boldsymbol{Q}s}(n_{\boldsymbol{Q}s} + \frac{1}{2}), \\ &= \hbar c Q(n_{\boldsymbol{Q}s} + \frac{1}{2}). \end{aligned} \tag{9.82}$$

This expresses that the electromagnetic energy in mode $\boldsymbol{Q}s$ contains $n_{\boldsymbol{Q}s}$ number of excitations, or quasi-particles, called **photons**, each of energy $\hbar cQ$. The term $\frac{1}{2}\hbar cQ$ represents the zero-point energy. The case $n_{\boldsymbol{Q}s} = 0$ is called the vacuum state of the electromagnetic field. A diagrammatic representation for a photon is sketched in figure 9.2.

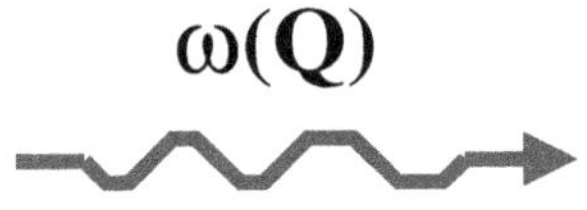

Figure 9.2 A diagrammatic representation for a photon of wavevector $\boldsymbol{Q}$ and frequency $\omega(\boldsymbol{Q})$.

The thermal average of the photon occupation number is given by the Bose-Einstein distribution function

$$\bar{n}_{\boldsymbol{Q}s} = \frac{1}{\exp(\frac{\hbar cQ}{k_B T}) - 1}, \tag{9.83}$$

where k_B is Boltzmann's constant and T is temperature.

9.5 QUANTISATION OF FERMIONIC QUANTUM FIELD

We will keep electrons in mind when discussing fermions in this section. As fermions obey the Pauli exclusion principle, the quantisation scheme discussed in the preceding section for bosonic fields must be appropriately modified to deal with fermionic fields. Jordan and Wigner (1928) showed that the commutation bracket $[\hat{A}, \hat{B}] = \hat{A}\hat{B} - \hat{B}\hat{A}$ for bosonic operators should be replaced with the **anticommutator bracket**

$$\{\hat{C}, \hat{D}\} = \hat{C}\hat{D} + \hat{D}\hat{C}, \tag{9.84}$$

for fermionic operators. In this section, we will present a brief discussion of the $\boldsymbol{N}$-representation for fermions and its application for the quantisation of electron waves in solids.

9.5.1 FERMIONIC CREATION AND ANNIHILATION OPERATORS

The number- (or $\boldsymbol{N}$-) representation operator for fermions in a state characterised with wavevector $\boldsymbol{k}$ is expressed as

$$\hat{N}_{\boldsymbol{k}} = \hat{b}^{\dagger}_{\boldsymbol{k}}\hat{b}_{\boldsymbol{k}}, \tag{9.85}$$

where $\hat{b}_{\boldsymbol{k}}^{\dagger}$ and $\hat{b}_{\boldsymbol{k}}$ are fermionic creation and annihilation operators, respectively. The creation and annihilation operators satisfy the eigenvalue equations

$$\hat{b}_{\boldsymbol{k}}|f_1,f_2,\cdots,f_{\boldsymbol{k}},\cdots> \;=\; \sqrt{f_{\boldsymbol{k}}}|f_1,f_2,\cdots,f_{\boldsymbol{k}}-1,\cdots>, \tag{9.86}$$

$$\hat{b}_{\boldsymbol{k}}^{\dagger}|f_1,f_2,\cdots,f_{\boldsymbol{k}},\cdots> \;=\; \sqrt{1-f_{\boldsymbol{k}}}|f_1,f_2,\cdots,f_{\boldsymbol{k}}+1,\cdots>, \tag{9.87}$$

with $f_{\boldsymbol{k}}$ being the fermion (electron) occupation number in state $\boldsymbol{k}$. Clearly, the eigenvalues of $\hat{b}_{\boldsymbol{k}}$ and $\hat{b}_{\boldsymbol{k}}^{\dagger}$ lie within the range $[0,1]$, which is consistent with Pauli's exclusion principle.

It can be easily verified that the operators $\hat{b}_{\boldsymbol{k}}^{\dagger}$ and $\hat{b}_{\boldsymbol{k}}$ satisfy the following anticommutation relations

$$\begin{aligned}
\{\hat{b}_{\boldsymbol{k}},\hat{b}_{\boldsymbol{l}}\} &= 0,\\
\{\hat{b}_{\boldsymbol{k}}^{\dagger},\hat{b}_{\boldsymbol{l}}^{\dagger}\} &= 0,\\
\{\hat{b}_{\boldsymbol{k}},\hat{b}_{\boldsymbol{l}}^{\dagger}\} &= \hat{I}\delta_{\boldsymbol{k},\boldsymbol{l}}.
\end{aligned} \tag{9.88}$$

The following equations also follow

$$\begin{aligned}
\hat{b}_{\boldsymbol{k}}^{\dagger}\hat{b}_{\boldsymbol{k}}|0_{\boldsymbol{k}}> &= 0; \qquad \hat{b}_{\boldsymbol{k}}\hat{b}_{\boldsymbol{k}}^{\dagger}|0_{\boldsymbol{k}}>=|0_{\boldsymbol{k}}>,\\
\hat{b}_{\boldsymbol{k}}^{\dagger}\hat{b}_{\boldsymbol{k}}|1_{\boldsymbol{k}}> &= |1_{\boldsymbol{k}}>; \quad \hat{b}_{\boldsymbol{k}}\hat{b}_{\boldsymbol{k}}^{\dagger}|1_{\boldsymbol{k}}>=0.
\end{aligned} \tag{9.89}$$

9.5.2 QUANTISATION OF ELECTRON WAVES IN SOLIDS

Let us start with the familiar Schrödinger equation

$$\begin{aligned}
\hat{H}\psi &= \hat{E}\psi,\\
-\frac{\hbar^2}{2m}\nabla^2\psi+\mathscr{V}\psi &= i\hbar\frac{\partial\psi}{\partial t},
\end{aligned} \tag{9.90}$$

where $\mathscr{V}=\mathscr{V}(\boldsymbol{r},t)$ and $\psi=\psi(\boldsymbol{r},t)$. We will first briefly describe how this equation can be derived from the application of the Lagrange equation of motion.

We treat $\psi(\boldsymbol{r},t)$ as a generalised coordinate and express the Lagrangian density as

$$\begin{aligned}
\mathscr{L} &= i\hbar\psi^*\frac{\partial\psi}{\partial t}-\frac{\hbar^2}{2m}\nabla\psi^*\cdot\nabla\psi-\psi^*\mathscr{V}\psi,\\
&= i\hbar\psi^*\frac{\partial\psi}{\partial t}-\frac{\hbar^2}{2m}\sum_j\frac{\partial\psi^*}{\partial r_j}\frac{\partial\psi}{\partial r_j}-\psi^*\mathscr{V}\psi, \quad j=1,2,3.
\end{aligned} \tag{9.91}$$

The Lagrange equation for the field variable ψ is [*cf* equation (8.9)]

$$\frac{\partial}{\partial t}\frac{\partial\mathscr{L}}{\partial\left(\frac{\partial\psi}{\partial t}\right)}+\sum_j\frac{\partial}{\partial r_j}\frac{\partial\mathscr{L}}{\partial\left(\frac{\partial\psi}{\partial r_j}\right)}=\frac{\partial\mathscr{L}}{\partial\psi}. \tag{9.92}$$

It will be left as an exercise to show that the above equation can be expressed as

$$\left(-i\hbar\frac{\partial}{\partial t}+\frac{\hbar^2}{2m}\nabla^2-\mathscr{V}\right)\psi^*=0. \tag{9.93}$$

The Lagrange equation for the field variable ψ^* is

$$\frac{\partial}{\partial t}\frac{\partial\mathscr{L}}{\partial\left(\frac{\partial\psi^*}{\partial t}\right)}+\sum_j\frac{\partial}{\partial r_j}\frac{\partial\mathscr{L}}{\partial\left(\frac{\partial\psi^*}{\partial r_j}\right)}=\frac{\partial\mathscr{L}}{\partial\psi^*}. \tag{9.94}$$

Again, it will be left as an exercise to show that this equation can be expressed as

$$\left(-i\hbar\frac{\partial}{\partial t}+\frac{\hbar^2}{2m}\nabla^2-\mathscr{V}\right)\psi=0. \tag{9.95}$$

Each of equations (9.93) and (9.95) is the Schrödinger equation presented at the start of this subsection.

The Schrödinger equation in (9.90) can also be derived from the application of Hamilton's equation of motion. The Hamiltonian density is

$$\begin{aligned}\mathscr{H} &= \xi\frac{\partial\psi}{\partial t}-\mathscr{L},\\ &= -\frac{i\hbar}{2m}\nabla\xi\cdot\nabla\psi-\frac{i}{\hbar}\xi\mathscr{V}\psi,\end{aligned} \tag{9.96}$$

where

$$\xi=\frac{\partial\mathscr{L}}{\partial\left(\frac{\partial\psi}{\partial t}\right)}=i\hbar\psi^* \tag{9.97}$$

is the momentum variable conjugate to the generalised coordinate ψ. Hamilton's equations of motion [*cf* equations (8.61) and (8.62)] are

$$\begin{aligned}\frac{\partial\psi}{\partial t} &= \frac{\partial\mathscr{H}}{\partial\xi}-\sum_j\frac{\partial}{\partial r_j}\frac{\partial\mathscr{H}}{\partial\left(\frac{\partial\pi}{\partial r_j}\right)},\\ &= -\frac{i}{\hbar}\mathscr{V}\psi+\frac{i\hbar}{2m}\nabla^2\psi,\end{aligned} \tag{9.98}$$

and

$$\begin{aligned}\frac{\partial\xi}{\partial t} &= \frac{\partial\mathscr{H}}{\partial\psi}+\sum_j\frac{\partial}{\partial r_j}\frac{\partial\mathscr{H}}{\partial\left(\frac{\partial\psi}{\partial r_j}\right)},\\ &= \frac{i}{\hbar}\mathscr{V}\xi-\frac{i\hbar}{2m}\nabla^2\xi.\end{aligned} \tag{9.99}$$

Equation (9.98) is the Schrödinger equation and equation (9.99) is its complex conjugate.

In order to develop the quantum field theory for an assembly of fermions (electrons) in potential $V(\boldsymbol{r})$ we treat ψ as an operator. Using equations (9.96) and (9.97) we therefore write down the Hamiltonian operator as

$$\hat{H}=\int d^3r\left[\frac{\hbar^2}{2m}\nabla\psi^\dagger\cdot\nabla\psi+\psi^\dagger\mathscr{V}\psi\right]. \tag{9.100}$$

Using the identity $\nabla\cdot(\psi^\dagger\nabla\psi)=\nabla\psi^\dagger\cdot\nabla\psi+\psi^\dagger\nabla^2\psi$ and noting that $\int d^3r\nabla\cdot(\psi^\dagger\nabla\psi)=0$ (because ψ either vanishes for $r\to\infty$ or obeys periodic boundary condition), we can express the Hamiltonian operator as

$$\hat{H}=\int d^3r\left[-\frac{\hbar^2}{2m}\psi^\dagger\nabla^2\psi+\psi^\dagger\mathscr{V}\psi\right]. \tag{9.101}$$

Using the language of $\boldsymbol{N}$-representation, we express

$$\psi(\boldsymbol{r}) = \sum_{\boldsymbol{k}}\psi_{\boldsymbol{k}}(\boldsymbol{r})\hat{b}_{\boldsymbol{k}}, \tag{9.102}$$

$$\psi^\dagger(\boldsymbol{r}) = \sum_{\boldsymbol{k}}\psi^*_{\boldsymbol{k}}(\boldsymbol{r})\hat{b}^\dagger_{\boldsymbol{k}}, \tag{9.103}$$

where $\{\psi_{\boldsymbol{k}}\}$ form an appropriate set of orthonormal single-particle wave functions and $\hat{b}^{\dagger}_{\boldsymbol{k}}$ and $\hat{b}_k$ are the fermionic creation and annihilation operators described in the preceding section. In the second-quantised notation, therefore, the quantum Hamiltonian operator can be expressed as

$$\begin{aligned}\hat{H} &= \sum_{\boldsymbol{k}}\sum_{\boldsymbol{k}'}\int \mathrm{d}^3r\Big[\psi^*_{\boldsymbol{k}'}(\boldsymbol{r})\{-\frac{\hbar^2}{2m}\nabla^2+\mathscr{V}(\boldsymbol{r})\}\psi_{\boldsymbol{k}}(\boldsymbol{r})\Big]\hat{b}^{\dagger}_{\boldsymbol{k}'}\hat{b}_{\boldsymbol{k}},\\ &= \sum_{\boldsymbol{k}}\sum_{\boldsymbol{k}'}E(\boldsymbol{k})\int \mathrm{d}^3r\psi^*_{\boldsymbol{k}'}(\boldsymbol{r})\psi_{\boldsymbol{k}}(\boldsymbol{r})\hat{b}^{\dagger}_{\boldsymbol{k}'}\hat{b}_{\boldsymbol{k}},\\ &= \sum_{\boldsymbol{k}}E(\boldsymbol{k})\hat{b}^{\dagger}_{\boldsymbol{k}}\hat{b}_{\boldsymbol{k}},\\ &= \sum_{\boldsymbol{k}}E(\boldsymbol{k})\hat{N}_{\boldsymbol{k}},\end{aligned}\tag{9.104}$$

where we have assumed that each $\psi_{\boldsymbol{k}}(\boldsymbol{r})$ satisfies the single-particle Schrödinger equation with eigenvalue $E(\boldsymbol{k})$. Note that we have used the orthonormality condition for $\psi_{\boldsymbol{k}}$ in obtaining the third line in the above equation.

Writing the states of the quantised field as the ket $|f_1,f_2,\cdots,f_{\boldsymbol{k}},\cdots>$, the total energy E of the system can be obtained as

$$\begin{aligned}E &= <f_1,f_2,\cdots,f_{\boldsymbol{k}},\cdots|\hat{H}|f_1,f_2,\cdots,f_{\boldsymbol{k}},\cdots>,\\ &= \sum_{\boldsymbol{k}}E(\boldsymbol{k})<f_1,f_2,\cdots,f_{\boldsymbol{k}},\cdots|\hat{N}_{\boldsymbol{k}}|f_1,f_2,\cdots,f_{\boldsymbol{k}},\cdots>,\\ &= \sum_{\boldsymbol{k}}E(\boldsymbol{k})f_{\boldsymbol{k}}.\end{aligned}\tag{9.105}$$

This results says that in the second quantisation notation the many-body fermionic (electronic) field can be described as an assembly of independent fermions (electrons) of energies $\{E(\boldsymbol{k})\}$ with occupation number $f_{\boldsymbol{k}}$ in the single-particle mode $\boldsymbol{k}$. The diagrammatical representation of an electron is a straight arrow labelled with $E(\boldsymbol{k})$, as shown in figure 9.3.

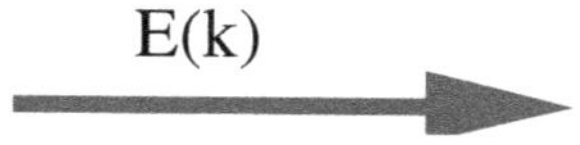

Figure 9.3 A diagrammatic representation for an electron of wavevector $\boldsymbol{k}$ and energy $E(\boldsymbol{k})$.

The thermal average of fermionic occupation number is given by the Fermi-Dirac distribution function

$$\bar{f}_{\boldsymbol{k}} = \frac{1}{\exp[(E(\boldsymbol{k})-E_{\mathrm{F}})/k_{\mathrm{B}}T]+1},\tag{9.106}$$

where E_{F} is the Fermi energy (highest occupied energy) for which the energy occupation factor is one-half (*i.e.* $\bar{f}_{\boldsymbol{k}} = 1/2$ for $E(\boldsymbol{k}) = E_{\mathrm{F}}$). This means that in a solid all electronic states below the Fermi energy are fully occupied and states above the Fermi energy are fully unoccupied.

FURTHER READING

1. Berestetskii V B, Lifshitz L M and Pitaevskii L P 1999 *Quantum Electrodynamics* (Butterworth Heinemann: Oxford)
2. Bransden B H and Joachain C J 2000 *Quantum Mechanics* 2nd edn (Prentice Hall: New York)

3. Dirac P A M 1945 *Rev Mod Phys* **17** 195
4. Griffths D J 1995 *Introduction to Quantum Mechanics* (Prentice Hall: New Jersey)
5. Jordan P and Wigner E 1928 *Z. Physik* **47** 631
6. Lancaster T and Blundell S 2014 *Quantum Field Theory for the Gifted Amateur* (OUP: Oxford)
7. Mandl F 1959 *Introduction to Quantum Field Theory* (Interscience: New York)
8. Peskin M E and Schroeder D V 1995 *An Introduction to Quantum Field Theory* (CRC: Reading, MA)
9. Pines D 1963 *Elementary Excitations in Solids* (Benjamin: New York)
10. Sakurai J J 1985 *Modern Quantum Mechanics* (Addison Wesley: New York)
11. Schiff L I 1968 *Quantum Mechanics* 3rd edn (McGraw Hill: New York)
12. Sebens C T 2015 *Phil Sci* **82** 266
13. Srivastava G P 2022 *The Physics of Phonons* 2nd edn (CRC: Oxon)
14. Ziman J M 1960 *Electrons and Phonons* (Clarendon: Oxford)
15. Ziman J M 1969 *Elements of Advanced Quantum Theory* (Cambridge University Press: Cambridge)

PROBLEMS

1. Use equations (9.8) and (9.9) to derive equation (9.11).
2. Prove the result $[\hat{a},\hat{a}^+]|n>=|n>$, where $\hat{a}$ and $\hat{a}^+$ are bosonic annihilation and creation operators, respectively.
3. Obtain the expectation value of the operator $(\hat{a}-\hat{a}^+)^4$ in the ground state $|0>$, where $\hat{a}$ and $\hat{a}^+$ are bosonic annihilation and creation operators, respectively.
4. Prove the commutation relation

$$[\boldsymbol{X}(\boldsymbol{qb}),\hat{\boldsymbol{P}}(\boldsymbol{q}'\boldsymbol{b}')] = \hat{\boldsymbol{I}}i\hbar\delta_{\boldsymbol{q},\boldsymbol{q}'}\delta_{\boldsymbol{bb}'}$$

   as stated in equation (9.32).
5. Show all necessary steps for obtaining equation (9.35).
6. Using equations (9.35)-(9.38) and other relevant equations, derive each of the three lines in equation (9.44).
7. The Hamiltonian of a monatomic linear chain vibrating along the chain is

$$H = \frac{1}{2m}\sum_j p_j^2 + \frac{1}{2}\Lambda\sum_j (u_j - u_{j+1})^2,$$

   where m is atomic mass, Λ is spring constant and u_j is displacement from equilibrium of jth atom. Following discussions in sections 4.5.1, 8.3.1 and 9.4.2, show that the energy of quantisation is $\sum_q(\bar{n}_q+\frac{1}{2})\hbar\omega(q)$, where q is phonon wavenumber. Show that this turns into the classical result at high temperatures ($k_{\mathrm{B}}T >> \hbar\omega(q)$).
8. Prove the results in equations (9.52)-(9.54).
9. Show step-by-step derivation of equations (9.93) and (9.95) using the Lagrangian density in equation (9.91) and the Lagrange equation in equation (9.92).

10 Classical and Quantum Perturbation Theories

10.1 INTRODUCTION

Analytical solutions of dynamical problems for realistic classical or quantum systems, apart from a few simple or ideal systems, are not possible and intensive numerical or judiciously chosen approximate methods become necessary, as we have seen from the discussion in chapters 6 and 7 of non-linear and chaotic systems. While computationally intensive numerical techniques can always be relied upon, it is useful, and less expensive, to use perturbation theory to obtain approximate solutions of a problem. However, care must be taken to ensure that perturbation theory is applied with appropriate physical considerations. It is important to realise at the outset that the word perturbation should refer to a small change to an exactly solvable situation. For example, for a valid perturbative treatment of an anharmonic oscillator the harmonic potential term (or the linear force term) must be dominant.

In this chapter we will describe how classical and quantum perturbation theories can be applied to a few simple problems. It is hoped that this will help the reader learn and practice other techniques in perturbation theory by consulting advanced level books, some of which are listed at the end of this chapter.

10.2 CLASSICAL PERTURBATION THEORY

We will consider application of classical time-dependent perturbation theory to two systems: (1) a small linear force (or quadratic potential) applied to an otherwise force-free motion of a particle and (2) a small cubic force applied to an undamped harmonic oscillator. We apply the perturbative treatment to system (1) by using two different methods: (a) Hamilton's equations of motion (*cf* section 3.2.3) and (b) Hamilton-Jacobi equations (*cf* section 3.3). Method (a) is called the **Hamiltonian perturbation theory**. As the Hamilton-Jacobi equations are derived by making a canonical transformation of system's Hamiltonian, method (b) is called the **Hamilton-Jacobi perturbation theory**, or the **canonical perturbation theory**. We will treat system (2) by applying the Newtonian force equations (*cf* section 2.2) and the **Poincaré-Lindstedt perturbation method**.

10.2.1 HAMILTONIAN PERTURBATION THEORY

Consider a force-free motion of a particle of mass m along x-axis. The Hamiltonian of the unperturbed system is

$$H_0 = \frac{p_0^2}{2m}, \tag{10.1}$$

where p_0 is particle's momentum canonically conjugate to x. Hamilton's equations of motion are

$$\frac{dx_0}{dt} = \frac{dH_0}{dp} = \frac{p_0}{m} \tag{10.2}$$

and

$$\frac{dp_0}{dt} = -\frac{dH_0}{dx_0} = 0. \tag{10.3}$$

DOI: 10.1201/9781003383314-10

Equation (10.3) suggests that the particle's momentum p_0 is constant, as expected. Hence, the solution of equation (10.2) is

$$x_0(t) = \frac{p_0}{m}t, \quad \text{when } x_0(t=0) = 0. \tag{10.4}$$

Now let the particle be subjected to a *small* linear force $F = -kx$, where $k = m\omega^2$ is a force constant. The potential energy corresponding to this force is

$$\Delta\mathscr{V} = \frac{1}{2}kx^2 = \frac{1}{2}m\omega^2 x^2. \tag{10.5}$$

In the presence of the applied force the particle's Hamiltonian has become

$$H = H_0 + \Delta H = H_0 + \Delta V = \frac{p^2}{2m} + \frac{1}{2}m\omega^2 x^2, \tag{10.6}$$

and Hamilton's equations of motion read

$$\frac{\mathrm{d}x}{\mathrm{d}t} = \frac{p}{m}, \tag{10.7}$$

and

$$\frac{\mathrm{d}p}{\mathrm{d}t} = -m\omega^2 x. \tag{10.8}$$

As ΔH is assumed small, we can expand $x(t)$ and $p(t)$ in convergent series

$$x(t) = x_0(t) + \varepsilon x_1(t) + \varepsilon^2 x_2(t) + \cdots, \tag{10.9}$$

and

$$p(t) = p_0(t) + \varepsilon p_1(t) + \varepsilon^2 p_2(t) + \cdots, \tag{10.10}$$

where ε is a small parameter. The contributions x_1, x_2, p_1, p_2, *etc.* can be determined by substituting equations (10.9) and (10.10) into equations (10.7) and (10.8) and equating terms of similar powers of ε on both sides of the equations. Considering the ε^1 terms, we have

$$\frac{\mathrm{d}p_1}{\mathrm{d}t} = -m\omega^2 x_0 = -\omega^2 p_0 t, \tag{10.11}$$

$$\frac{\mathrm{d}x_1}{\mathrm{d}t} = \frac{p_1}{m}. \tag{10.12}$$

Integrating equations (10.11) and (10.12) and using the initial conditions $x_1 = 0$ and $p_1 = 0$ when $t = 0$, we obtain

$$p_1 = -\frac{1}{2}\omega^2 p_0 t^2, \tag{10.13}$$

$$x_1 = -\frac{p_1}{6}\frac{\omega^2 p_0}{m}t^3. \tag{10.14}$$

Using equations (10.2), (10.3), (10.13) and (10.14) the approximate solutions to equations (10.7) and (10.8) are

$$p(t) \approx p_0[1 - \frac{1}{2}(\omega t)^2], \tag{10.15}$$

and

$$x(t) \approx \frac{p_0}{m\omega}[\omega t - \frac{1}{6}(\omega t)^3]. \tag{10.16}$$

These results can be improved upon by including contributions arising from terms $\varepsilon^j, j > 1$. It will be left as an exercise for the reader to show that the final results are

$$p(t) = p_0 \cos(\omega t), \tag{10.17}$$

and

$$x(t) = \frac{p_0}{m\omega} \sin(\omega t). \tag{10.18}$$

These are the well-known results for the one-dimensional simple harmonic oscillator obtained in section 3.3.8.1 [*cf* equations (3.175) and (3.176) with phase factor $\psi = 0$ and energy $E = p_0^2/2m$].

It is remarkable that the application of the perturbation theory to this problem has yielded exact results. This is because the perturbed part of the Hamiltonian makes the full Hamiltonian of a one-dimensional simple harmonic oscillator and the series solutions for $x(t)$ and $p(t)$ converge. In general, physical considerations must be applied for obtaining useful results from perturbation theory.

10.2.2 HAMILTON-JACOBI PERTURBATION THEORY

We note from section 3.3.6 that a canonical transformation from a phase space (q,p) with unperturbed Hamiltonian $H_0(q,p)$ to another phase space $(Q = \alpha_0, P = \beta_0)$ with Hamiltonian $K_0(\alpha_0, \beta_0, t) = 0$ satisfies the Hamilton-Jacobi equation

$$K_0 = H_0\left(q, \frac{\partial S}{\partial q}, t\right) + \frac{\partial S}{\partial t} = 0, \tag{10.19}$$

where α_0 and β_0 are constants of motion, and $S = S(q, \beta_0, t)$ is a special generating function known as Hamilton's principal function. We also note that for a conservative system

$$S(q, \beta_0, t) = W(q, \beta_0, t = 0) - Et = \int p\mathrm{d}q - Et, \tag{10.20}$$

where E is system's total energy.

When the system is perturbed by adding a small contribution to its Hamiltonian ΔH, the Hamilton-Jacobi equation becomes

$$K(\alpha, \beta, t) = H_0 + \Delta H + \frac{\partial S}{\partial t} = \Delta H(\alpha, \beta, t), \tag{10.21}$$

where we have used the same generating function since the canonical property of a phase space transformation is independent of the form of a system's Hamiltonian. The transformed phase space variables α and β may no longer be constants of motion, but satisfy the Hamilton's equations of motion

$$\frac{\mathrm{d}\alpha}{\mathrm{d}t} = -\frac{\partial K}{\partial \beta} = -\frac{\partial \Delta H}{\partial \beta}, \tag{10.22}$$

$$\frac{\mathrm{d}\beta}{\mathrm{d}t} = \frac{\partial K}{\partial \alpha} = \frac{\partial \Delta H}{\partial \alpha}. \tag{10.23}$$

Considering a small parameter ε, let us expand α and β as

$$\alpha = \alpha_0 + \varepsilon\alpha_1 + \varepsilon^2\alpha_2 + \cdots, \tag{10.24}$$

$$\beta = \beta_0 + \varepsilon\beta_1 + \varepsilon^2\beta_2 + \cdots, \tag{10.25}$$

with the initial conditions $\alpha_j(t = 0) = 0$ and $\beta_j(t = 0) = 0$ for $j > 0$. Writing the Hamiltonian as $H = H_0 + \varepsilon\Delta H$, we note that ε appears on the right-hand side of both equations (10.22) and (10.23).

This means that when seeking solutions of these equations for the jth order (*i.e.* for the ε^j terms) on the left-hand side, the right-hand side should be evaluated for the $(j-1)$th order (*i.e.* for the $\varepsilon^{(j-1)}$ terms). That is, we should express

$$\frac{d\alpha_j}{dt} = -\left.\frac{\partial \Delta H}{\partial \beta}\right|_{\substack{\alpha=\alpha_{j-1}\\ \beta=\beta_{j-1}}}, \tag{10.26}$$

$$\frac{d\beta_j}{dt} = \left.\frac{\partial \Delta H}{\partial \alpha}\right|_{\substack{\alpha=\alpha_{j-1}\\ \beta=\beta_{j-1}}}. \tag{10.27}$$

The essence of the Hamilton-Jacobi perturbation theory, or the canonical perturbation theory, is to determine α_j and β_j for $j > 0$ and use these to obtain the equation of motion for the perturbed system. We will apply this to the system discussed in the preceding section, *viz* a particle of mass m moving along x-axis whose unperturbed Hamiltonian $H_0 = p_0^2/2m$ is perturbed by a small amount $\Delta H = \Delta \mathscr{V} = \frac{1}{2}m\omega^2x^2$.

Hamilton's principal function S can be obtained by solving equation (10.20)

$$\begin{aligned} S(q,\beta_0,t) &= \int p\mathrm{d}q - Et, \\ &= p_0x_0 - Et, \\ &= \alpha_0 x_0 - \frac{\alpha_0^2}{2m}t. \end{aligned} \tag{10.28}$$

From this, we obtain

$$\beta_0 = \frac{\partial S}{\partial \alpha_0} = x_0 - \frac{\alpha_0}{m}t, \tag{10.29}$$

which gives

$$x_0 = \beta_0 + \frac{\alpha_0}{m}t. \tag{10.30}$$

Also,

$$p_0 = \frac{\partial S}{\partial x_0} = \alpha_0. \tag{10.31}$$

For simplicity we set $x_0 = 0$ so that $\beta_0 = 0$ and equations (10.30) and (10.31) produce

$$x_0 = \frac{p_0}{m}t. \tag{10.32}$$

The perturbed Hamiltonian can be expressed as

$$\Delta H(\alpha,\beta,t) = \frac{1}{2}m\omega^2x^2 = \frac{1}{2}m\omega^2\left(\frac{\alpha}{m}t + \beta\right)^2. \tag{10.33}$$

Using this, for the first-order equation of motion (*viz* up to the ε^1 term) we obtain

$$\frac{d\alpha_1}{dt} = -\left.\frac{\partial \Delta H}{\partial \beta}\right|_{\substack{\alpha=\alpha_0\\ \beta=\beta_0}} = -\omega^2\alpha_0 t, \tag{10.34}$$

$$\frac{d\beta_1}{dt} = \left.\frac{\partial \Delta H}{\partial \alpha}\right|_{\substack{\alpha=\alpha_0\\ \beta=\beta_0}} = \frac{\omega^2}{m}\alpha_0 t^2. \tag{10.35}$$

from which

$$\alpha_1 = \alpha_0 - \frac{1}{2}\omega^2\alpha_0 t^2, \tag{10.36}$$

$$\beta_1 = \frac{1}{3}\frac{\omega^2}{m}\alpha_0 t^3. \tag{10.37}$$

Note that in obtaining equation (10.37) we have used the initial condition $\beta_0 = 0$.

Using equations (10.30), (10.31), (10.36) and (10.37), the approximate solution to the perturbed system becomes, at the $j = 1$ stage,

$$\begin{aligned} x(t) &\approx x_0 + \beta_1 + \frac{\alpha_1}{m} t, \\ &= \frac{\alpha_0}{m} t + \frac{\omega^2}{3m} \alpha_0 t^3 - \frac{\omega^2}{2m} \alpha_0 t^3, \\ &= \frac{p_0}{m\omega} \left[\omega t - \frac{1}{3!} (\omega t)^3 \right], \end{aligned} \tag{10.38}$$

and

$$\begin{aligned} p(t) &= \alpha_1 = \alpha_0 \left(1 - \frac{(\omega t)^2}{2} \right), \\ &= p_0 \left[1 - \frac{(\omega t)^2}{2} \right]. \end{aligned} \tag{10.39}$$

The results in equations (10.38) and (10.39) are the same as in equations (10.33) and (10.35) which were obtained from the application of the Hamiltonian perturbation method.

The process described above can be continued to obtain approximate solutions at stages ε^j for $j > 1$ and finally recover the well-known results given in equations (10.17) and (10.18).

10.2.3 POINCARE-LINDSTEDT PERTURBATION METHOD

In section 6.6 we described the one-dimensional anharmonic oscillator, known as the Duffing oscillator. Numerical solutions of such a system was presented in section 6.10. Here we will discuss how an analytic solution to such a system can be obtained by applying perturbation theory.

We start by reminding ourselves that the solution of the harmonic equation

$$\frac{1}{\omega_0^2} \frac{\mathrm{d}^2 x_0}{\mathrm{d}t^2} + x_0 = 0, \tag{10.40}$$

with the initial conditions

$$x_0 = 1 \quad \text{and} \quad \frac{\mathrm{d}x_0}{\mathrm{d}t} = 0 \quad \text{when} \quad t = 0, \tag{10.41}$$

is (see Appendix G or section 3.3.8.1)

$$x_0 = \cos(\omega_0 t), \tag{10.42}$$

where ω_0 is the frequency of the harmonic oscillator.

Now let us write the anharmonic equation (6.21) in the form

$$\frac{1}{\omega_0^2} \frac{\mathrm{d}^2 x}{\mathrm{d}t^2} + x + \varepsilon x^3 = 0, \tag{10.43}$$

where ε is a measure of the strength of a *small* cubic force. Following the discussion in the preceding sections, an approximate solution of this equation can be attempted by expressing $x(t)$ as a convergent series in ε of the form

$$x(t) = x_0(t) + \varepsilon x_1(t) + \varepsilon^2 x_2(t) + \cdots, \tag{10.44}$$

with the initial conditions

$$x_0(t=0) = 1, \quad x_j(t=0) = 0 \text{ for } j > 0, \quad \text{and} \quad \frac{\mathrm{d}x_j}{\mathrm{d}t}(t=0) = 0 \text{ for } j \geq 0. \tag{10.45}$$

However, a straightforward application of the procedure discussed in the preceding sections does not result in a convergent solution for equations of type in (10.43). This is because the solution in the order ε^j contains a *secular term* of the form $(\varepsilon t)^j$, which indicates divergence (José and Saletan, 1998). A convergent and periodic series solution to the problem can be obtained by applying the Poincaré-Lindstedt perturbation method (Poincaré, 1992; Tabor, 1989; José and Saletan, 1998; Gregory, 2006). This is described below.

Introduce a dimensionless time variable $z = \omega(\varepsilon)t$, where $\omega(\varepsilon) = 2\pi/\tau(\varepsilon)$ is the angular frequency of the required solution of equation (10.43). Using the trial solution

$$x = e^{iz} \tag{10.46}$$

we express equation (10.43) in the form

$$\frac{\omega^2}{\omega_0^2}\frac{dx^2}{dz^2} + x + \varepsilon x^3 = 0, \tag{10.47}$$

where $\omega \equiv \omega(\varepsilon)$ and the initial conditions are

$$x = 1 \quad \text{and} \quad \frac{dx}{dz} = 0 \quad \text{when } z = 0. \tag{10.48}$$

Assuming $x = x(z, \varepsilon)$ and $\omega = \omega(\varepsilon)$, we now express x as well as ω in perturbation series as follows

$$x = x_0 + \varepsilon x_1 + \varepsilon^2 x_2 + \cdots, \tag{10.49}$$

$$\frac{\omega}{\omega_0} = 1 + \varepsilon\eta_1 + \varepsilon^2\eta_2 + \cdots, \tag{10.50}$$

where $\{\eta_j\}$ are dimensionless variables.

Substituting equations (10.49) and (10.50) into (10.47), we get

$$\begin{aligned}&(1+\varepsilon\eta_1+\varepsilon^2\eta_2+\cdots)^2\left(\frac{d^2x_0}{dz^2}+\varepsilon\frac{d^2x_1}{dz^2}+\varepsilon^2\frac{d^2x_2}{dz^2}+\cdots\right)\\&+(x_0+\varepsilon x_1+\varepsilon^2x_2+\cdots)+\varepsilon(x_0+\varepsilon x_1+\varepsilon^2x_2+\cdots)^3=0.\end{aligned} \tag{10.51}$$

The initial conditions in equation (10.48) can be expressed as

$$(x_0 + \varepsilon x_1 + \varepsilon^2 x_2 + \cdots) = 1 \quad \text{when } z = 0 \tag{10.52}$$

and

$$\left(\frac{dx_0}{dz} + \varepsilon\frac{dx_1}{dz} + \varepsilon^2\frac{dx_2}{dz} + \cdots\right) = 1 \quad \text{when } z = 0 \tag{10.53}$$

Equating terms of equal power on the two sides of equation (10.51), with $x_0 = 1$ and $\frac{dx_0}{dz} = 0$ when $z = 0$, we obtain

$$\varepsilon^0 \text{ terms}: \quad \frac{d^2x_0}{dz^2} + x_0 = 0, \tag{10.54}$$

$$\varepsilon^1 \text{ terms}: \quad x_0^3 + 2\eta_1\frac{d^2x_0}{dz^2} + x_1 + \frac{d^2x_1}{dz^2} = 0. \tag{10.55}$$

The solution of equation (10.54) is

$$x_0 = \cos z. \tag{10.56}$$

Substitution of equation (10.56) into (10.55) gives

$$\frac{d^2x_1}{dz^2} + x_1 = 2\eta_1 \cos z - \cos^3 z,$$

$$\begin{aligned} &= \frac{1}{4}(8\eta_1 - 3)\cos z - \frac{1}{4}\cos(3z), \\ &= f_1(z) + f_2(z), \end{aligned} \tag{10.57}$$

where we have used the identity $4\cos^3 z = 3\cos z + \cos(3z)$. The complementary solution for this differential equation can be taken in the form

$$x_1^c = B_1 \cos z + B_2 \sin z, \tag{10.58}$$

where B_1 and B_2 are arbitrary constants to be determined from the initial conditions. The particular solutions corresponding to the inhomogeneous terms $f_1(z)$ and $f_2(z)$ can be obtained by following the discussion in Appendix (G) [*cf* equations (G.15) and (G.18)] . The result is

$$x_1^p = \frac{1}{4}(8\eta_1 - 3)\Big[\frac{1}{2}z\sin z\Big] - \frac{1}{4}\Big[-\frac{1}{8}\cos(3z)\Big]. \tag{10.59}$$

The full solution of equation (10.57) is, therefore,

$$x_1 = B_1 \cos z + B_2 \sin z + (\eta_1 - \frac{3}{8})z\sin z + \frac{1}{32}\cos(3z). \tag{10.60}$$

All but the third term on the right-hand side of this equation are periodic functions. In order to ensure periodicity of motion, we should set the third term to zero. This requires setting

$$\eta_1 = \frac{3}{8}. \tag{10.61}$$

With this, the solution x_1 reads

$$x_1 = B_1 \cos z + B_2 \sin z + \frac{1}{32}\cos(3z). \tag{10.62}$$

Using the initial conditions $x_1 = 0$ and $\frac{dx_1}{dz} = 0$ when $z = 0$, we obtain $B_1 = -\frac{1}{32}$ and $B_2 = 0$, and finally write

$$x_1 = \frac{1}{32}[\cos(3z) - \cos z]. \tag{10.63}$$

With the results in equations (10.56), (10.61) and (10.63), the perturbation expansions in equations (10.49) and (10.50) read

$$x = \cos z + \frac{\varepsilon}{32}[\cos(3z) - \cos z] + O(\varepsilon^2), \tag{10.64}$$

$$\omega = \left(1 + \frac{3}{8}\varepsilon\right)\omega_0 + O(\varepsilon^2), \tag{10.65}$$

and

$$z = \omega t = \big(1 + \frac{3}{8}\varepsilon + O(\varepsilon^2)\big)\omega_0 t. \tag{10.66}$$

If desired, solutions up to higher order in ε can be derived, but it is sufficient to look at the results up to the first order and recap on the essential features of an anharmonic oscillator. Using the results in equations (10.64)–(10.66), we write the solution of the cubic anharmonic oscillator to first order in ε as

$$x = \cos\big\{(1 + \frac{3}{8}\varepsilon)\omega_0 t\big\} + \frac{\varepsilon}{32}\big[\cos(3\omega_0 t) - \cos(\omega_0 t)\big], \tag{10.67}$$

and the time period of the motion as

$$\tau = \frac{2\pi}{\omega} = \frac{2\pi}{\omega_0}\Big[1 + \frac{3}{8}\varepsilon\Big]^{-1} \approx \frac{2(1 - \frac{3}{8}\varepsilon)\pi}{\omega_0}. \tag{10.68}$$

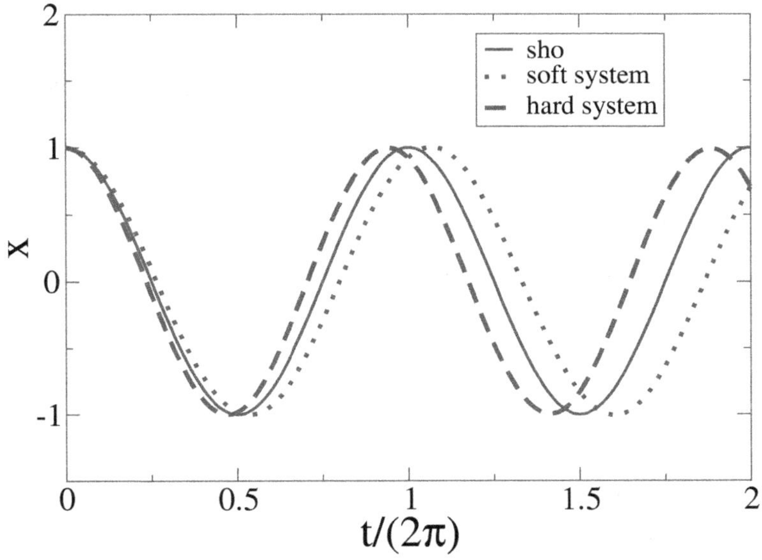

Figure 10.1 Time variations of one-dimensional soft and hard anharmonic oscillators using the perturbative solution in equation (10.67). Plots were made with parameter choices: $\omega_0 = 1$ and $\varepsilon = -\frac{1}{6}$ $(\frac{1}{6})$ for soft (hard) anharmonicity. For comparison, results for the harmonic oscillator, *viz* with $\varepsilon = 0$, are also shown.

Note that the time period of the anharmonic oscillator is different from that of the corresponding harmonic oscillator. Also note that the anharmonic oscillator oscillates with a combination of frequencies ω_0 and $3\omega_0$. Changes similar to these in time period and frequency are characteristics of non-linear oscillators.

It is interesting to compare the analytical results to first order in ε against the numerical results presented in figure 6.16. In order to do that we set $\omega_0 = 1$ and $\varepsilon = -c_3$ so that equation (10.47) becomes equation (6.21). Figure 10.1 shows the variation of the displacement x with time t using the expression in equation (10.67) for the choices $\omega_0 = 1$ and $\varepsilon = \pm\frac{1}{6}$. It can be seen that these results for both $\varepsilon = -\frac{1}{6}$ (soft system) and $\varepsilon = \frac{1}{6}$ (hard system)[1] are almost identical to the numerical results presented in figure 6.16.

The percentage change in the time period of the oscillator due to anharmonicity can be easily expressed from equation (10.68) as

$$\frac{\tau - 2\pi/\omega_0}{2\pi/\omega_0} \times 100 = -\frac{3\varepsilon}{8} \times 100. \tag{10.69}$$

For the choice $|\varepsilon = \frac{1}{6}|$, there is increase (decrease) in the time period by 6.25% for the soft (hard) system, in close agreement with the numerical results obtained in section 6.10. This clearly indicates that the first-order perturbation theory provides a very good description of the motion of the cubic anharmonic oscillator.

[1] See section 6.10 for a description of soft and hard anharmonicities.

10.3 QUANTUM PERTURBATION THEORY

It would be useful to compliment the discussion on the classical perturbation theory with a brief discussion of the quantum perturbation theory. We will start by describing the well-known non-degenerate time-independent Rayleigh-Schrödinger perturbation approach and derive the so-called $(2n+1)$-order energy theorem. We will then derive the non-degenerate time-independent Brillouin-Wigner perturbation theory. This will be followed by a description of time-dependent first-order perturbation theory. Towards the end we will describe two applications of the time-dependent perturbation theory.

10.3.1 RAYLEIGH-SCHRODINGER TIME-INDEPENDENT PERTURBATION THEORY

Consider the time-independent Schrödinger equation for an unperturbed and non-degenerate system[2]

$$\hat{H}_0\Psi_0 = E_0\Psi_0, \tag{10.70}$$

where $\hat{H}_0$, E_0 and Ψ_0 represent, respectively, the Hamiltonian operator, eigenvalue and eigenstate. When the system is 'perturbed' by adding a small additional Hamiltonian operator $\hat{H}'$, we express the Schrödinger equation in the form

$$\hat{H}\Psi = (\hat{H}_0 + \lambda\hat{H}')\Psi = E\Psi, \tag{10.71}$$

where E and Ψ are the eigenvalue and eigenstate to be determined, and λ is a small parameter to indicate the order of perturbation (which will be set to 1 at the end of formulation for each order).

Following the basic idea of perturbation theory, we expand Ψ and E as power series in λ:

$$\Psi = \sum_j \lambda^j \Psi^{(j)}, \tag{10.72}$$

$$E = \sum_j \lambda^j E^{(j)}, \tag{10.73}$$

where, obviously, $E^{(0)} \equiv E_0$ and $\Psi^{(0)} \equiv \Psi_0$.

With equations (10.72) and (10.73), equation (10.71) reads

$$\begin{aligned}(\hat{H}_0 + \lambda\hat{H}')(\Psi^{(0)} + \lambda\Psi^{(1)} + \lambda^2\Psi^{(2)} + \cdots) = \\ (E^{(0)} + \lambda E^{(1)} + \lambda^2 E^{(2)} \cdots)(\Psi^{(0)} + \lambda\Psi^{(1)} + \lambda^2\Psi^{(2)} + \cdots).\end{aligned} \tag{10.74}$$

Equating terms of equal powers of λ on both sides of this equation, we obtain

$$\lambda^0: \quad \hat{H}_0\Psi^{(0)} = E^{(0)}\Psi^{(0)}, \tag{10.75}$$

$$\lambda^1: \quad \hat{H}_0\Psi^{(1)} + \hat{H}'\Psi^{(0)} = E^{(0)}\Psi^{(1)} + E^{(1)}\Psi^{(0)}, \tag{10.76}$$

$$\lambda^2: \quad \hat{H}_0\Psi^{(2)} + \hat{H}'\Psi^{(1)} = E^{(0)}\Psi^{(2)} + E^{(1)}\Psi^{(1)} + E^{(2)}\Psi^{(0)}, \tag{10.77}$$

$$\cdots\cdots.$$

The above expressions can be generalised for $\lambda^j (j \geq 1)$ as

$$\hat{H}_0\Psi^{(j)} + \hat{H}'\Psi^{(j-1)} = E^{(0)}\Psi^{(j)} + E^{(1)}\Psi^{(j-1)} + E^{(2)}\Psi^{(j-2)} + \cdots + E^{(j)}\Psi^{(0)}. \tag{10.78}$$

---

[2]The energy levels of a non-degenerate system are distinct and do not overlap with each other.

Premultiplying equations (10.75)–(10.77) by Ψ_0^* and integrating over all space, noting that $\hat{H}_0$ is Hermitian and making use of the orthonormality of wavefunctions, *viz* $<\Psi^{(i)}|\Psi^{(j)}>=\delta_{ij}$, we obtain

$$\begin{aligned} E^{(0)} &= <\Psi^{(0)}|\hat{H}_0|\Psi^{(0)}>\equiv E_0, && (10.79)\\ E^{(1)} &= <\Psi^{(0)}|\hat{H}'|\Psi^{(0)}>, && (10.80)\\ E^{(2)} &= <\Psi^{(0)}|(\hat{H}'-E^{(1)})|\Psi^{(1)}>, && (10.81)\\ E^{(3)} &= <\Psi^{(0)}|(\hat{H}'-E^{(1)})|\Psi^{(2)}> -E^{(2)}<\Psi^{(0)}|\Psi^{(1)}>. && (10.82)\end{aligned}$$

It will be left as an exercise to show that $E^{(3)}$ can be expressed as

$$E^{(3)}=<\Psi^{(1)}|(\hat{H}'-E^{(1)})|\Psi^{(1)}> -2E^{(2)}<\Psi^{(0)}|\Psi^{(1)}>. \tag{10.83}$$

In general (see, Dalgarno and Stewart, 1956)

$$E^{(n)}=<\Psi^{(0)}|(\hat{H}'-E^{(1)})|\Psi^{(n-1)}> -\sum_{r=1}^{n-2}E^{(r+1)}<\Psi^{(0)}|\Psi^{(n-r-1)}>, \quad n>1. \tag{10.84}$$

Thus, the eigenvalue E of the perturbed system can be obtained from the following series

$$\begin{aligned} E &= E_0+E^{(1)}+E^{(2)}+\cdots,\\ &= E_0+<\Psi^{(0)}|\hat{H}'|\Psi^{(0)}>+<\Psi^{(0)}|(\hat{H}'-E^{(1)})|\Psi^{(1)}>+\cdots && (10.85)\end{aligned}$$

These results state that the energy $E^{(2n+1)}$ can be obtained from the knowledge of the wavefunctions $\Psi^{(0)},\psi^{(1)},\cdots,\Psi^{(n)}$. For this reason the approach described above is known as the **(2*n* + 1)-order perturbation theory**.

In order to proceed, let us consider $\Psi^{(0)}$ to form a complete orthonormal set $\{|k>\}$ and express $\Psi^{(j)}$ in its nth state as

$$\Psi_n^{(j)}=\sum_k a_{nk}^{(j)}|k>. \tag{10.86}$$

Using equations (10.85) and (10.76) we write

$$<l|\hat{H}_0-E_n^{(0)}|\sum_k a_{nk}^{(1)}k>+<l|\hat{H}'-E_n^{(1)}|n>=0. \tag{10.87}$$

This immediately reduces to

$$a_{nl}^{(1)}(E_l^{(0)}-E_n^{(0)})+<l|\hat{H}'|n>-E_n^{(1)}\delta_{nl}=0. \tag{10.88}$$

From this, for $l=n$,

$$E_n^{(1)}=<n|\hat{H}'|n>, \tag{10.89}$$

and for $l\neq n$

$$a_{nl}^{(1)}=\frac{<l|\hat{H}'|n>}{E_n^{(0)}-E_l^{(0)}}. \tag{10.90}$$

The first-order term in equation (10.72), therefore, is

$$\Psi_n^{(1)}=\sum_{l\neq n}\frac{<l|\hat{H}'|n>}{E_n^{(0)}-E_l^{(0)}}|l>. \tag{10.91}$$

The above procedure can be extended to obtain higher order terms in equations (10.72) and (10.73). The results for the second-order terms are (Bransden and Joachain, 2000)

$$E_n^{(2)} = \sum_{k\neq n} \frac{(< k|\hat{H}'|n >)^2}{E_n^{(0)} - E_k^{(0)}}, \tag{10.92}$$

and

$$\begin{aligned}\Psi_n^{(2)} &= \sum_{l\neq n}\Big[\sum_{k\neq n} \frac{< l|\hat{H}'|k >< k|\hat{H}'|n >}{(E_n^{(0)} - E_l^{(0)})(E_n^{(0)} - E_k^{(0)})} - \frac{< n|\hat{H}'|n >< l|\hat{H}'|n >}{(E_n^{(0)} - E_l^{(0)})^2}\Big]|l > \\ &\quad -\frac{1}{2}\sum_{k\neq n} \frac{(< k|\hat{H}'|n >)^2}{(E_n^{(0)} - E_k^{(0)})^2}|n > . \end{aligned} \tag{10.93}$$

The eigenvalue and wavefunction for the nth state are, therefore,

$$E_n = E_n^{(0)} + < n|\hat{H}'|n > + \sum_{k\neq n} \frac{(< k|\hat{H}'|n >)^2}{E_n^{(0)} - E_k^{(0)}} + \cdots, \tag{10.94}$$

and

$$\begin{aligned}\Psi_n &= |n > + \sum_{l\neq n} \frac{< l|\hat{H}'|n >}{E_n^{(0)} - E_l^{(0)}}|n > \\ &\quad + \sum_{l\neq n}\Big[\sum_{k\neq n} \frac{< l|\hat{H}'|k >< k|\hat{H}'|n >}{(E_n^{(0)} - E_l^{(0)})(E_n^{(0)} - E_k^{(0)})} - \frac{< n|\hat{H}'|n >< l|\hat{H}'|n >}{(E_n^{(0)} - E_l^{(0)})^2}\Big]|l > \\ &\quad -\frac{1}{2}\sum_{k\neq n} \frac{(< k|\hat{H}'|n >)^2}{(E_n^{(0)} - E_k^{(0)})^2}|n > + \cdots . \end{aligned} \tag{10.95}$$

Example:

We will present a simple example of the application of the Rayleigh-Schrödinger perturbation method. For this, we consider a one-dimensional anharmonic oscillator with its Hamiltonian operator as

$$\begin{aligned}\hat{H} &= -\frac{\hbar^2}{2m}\frac{\mathrm{d}^2}{\mathrm{d}x^2} + \frac{1}{2}m\omega^2 x^2 + \gamma x^4, \\ &= \hat{H}_0 + \gamma x^4, \\ &= \hat{H}_0 + \hat{H}'. \end{aligned} \tag{10.96}$$

The anharmonicity factor γ will be considered small so that the application of the perturbation method is valid. The wavefunction of the harmonic oscillator in its ground state is (see, Rae (2008))

$$\Psi_0 = \Big(\frac{m\omega}{\pi\hbar}\Big)^{1/4} \exp\big(-\frac{m\omega}{2\hbar}x^2\big). \tag{10.97}$$

The first-order change ΔE in the ground-state energy of the oscillator can be worked out by using the first term in equation (10.94):

$$\begin{aligned}\Delta E^{(0)} &= < n|\hat{H}'|n >, \\ &= < \Psi_0|\hat{H}'\Psi_0 >, \\ &= \gamma\Big(\frac{m\omega}{\pi\hbar}\Big)^{1/2}\int_{-\infty}^{\infty} x^4 \exp\big(-\frac{m\omega}{\hbar}x^2\big)\mathrm{d}x, \\ &= \frac{3\hbar^2}{4m^2\omega^2}\gamma. \end{aligned} \tag{10.98}$$

The ground-state (or, zero-point) energy of the anharmonic oscillator is, therefore, approximately

$$\begin{aligned} E^{(0)}_{\text{anharmonic}} &= E^{(0)}_{\text{harmonic}} + \Delta E^{(0)}, \\ &= \frac{1}{2}\hbar\omega + \frac{3\hbar^2}{4m^2\omega^2}\gamma. \end{aligned} \tag{10.99}$$

From this result it can be seen that the anharmonic correction to the zero-point energy decreases as the frequency of oscillation increases.

10.3.2 BRILLOUIN-WIGNER TIME-INDEPENDENT PERTURBATION THEORY

A modified version of the Rayleigh-Schrödinger perturbation theory is Brillouin-Wigner perturbation theory. We will describe the Brillouin-Wigner perturbation theory for a non-degenerate quantum system.

Let us write the Schrödinger equation for a perturbed non-degenerate system as

$$\hat{H}|\Psi> = E|\Psi>, \tag{10.100}$$

with

$$\hat{H} = \hat{H}_0 + \hat{H}', \tag{10.101}$$

and

$$|\Psi> = |\Psi_0> + |\Phi>, \tag{10.102}$$

where $\hat{H}_0$ and $|\Psi_0>$ are the Hamiltonian operator and wavefunction of the corresponding unperturbed system, $\hat{H}'$ is the perturbed part of the Hamiltonian operator and $|\Psi>$ is the wavefunction of the perturbed system. Let us assume that the solutions E_0 and $|\Psi_0>$ of the Schrödinger equation

$$\hat{H}_0|\Psi_0> = E_0|\Psi_0>, \tag{10.103}$$

for the unperturbed system are known. Also, we can consider the change in the wavefunction due to the perturbation, $|\Phi>$, to be orthogonal to the unperturbed wavefunction $|\Psi_0>$, *i.e.*

$$<\Psi_0|\Phi> = 0, \tag{10.104}$$

which also means that

$$<\Psi_0|\Psi> = 1. \tag{10.105}$$

Defining a *projection operator* $\hat{P}$ and a *complementary operator* $\hat{Q}$

$$\hat{P} = |\Psi_0><\Psi_0|, \tag{10.106}$$

$$\hat{Q} = \hat{I} - \hat{P}, \tag{10.107}$$

we can write

$$\hat{Q}|\Psi> = |\Phi>, \tag{10.108}$$

$$<\Psi_0|\hat{P}|\Psi> = <\Psi_0|\Psi>. \tag{10.109}$$

It is easy to show that $\hat{P}$, and hence $\hat{Q}$, commutes with $\hat{H}_0$.

With the above considerations, we can express equation (10.100) as

$$\begin{aligned} (\hat{H}_0 + \hat{H}')|\Psi> &= E|\Psi>, \\ \hat{H}'|\Psi> &= (E - \hat{H}_0)|\Psi>, \\ \hat{Q}|\Psi> \equiv |\Phi> &= (E - \hat{H}_0)^{-1}\hat{Q}\hat{H}'|\Psi> \\ \hat{Q}|\Psi> &= \hat{R}\hat{H}'|\Psi>, \end{aligned} \tag{10.110}$$

where

$$\hat{R} = (E - \hat{H}_0)^{-1}\hat{Q}\hat{H}' \tag{10.111}$$

is the *resolvent operator.*

The wavefunction of the perturbed system can now be expressed as

$$|\Psi >= |\Psi_0 > +\hat{R}\hat{H}'|\Psi >, \tag{10.112}$$

which can be solved iteratively, as follows

$$\begin{aligned} |\Psi > &= |\Psi_0 > +\hat{R}\hat{H}'(|\Psi_0 > +\hat{R}\hat{H}'|\Psi >), \\ &= |\Psi_0 > +\hat{R}\hat{H}'|\Psi_0 > +(\hat{R}\hat{H}')^2(\Psi >, \\ &= |\Psi_0 > +\hat{R}\hat{H}'|\Psi_0 > +(\hat{R}\hat{H}')^2|\Psi_0 > +(\hat{R}\hat{H}')^3|\Psi_0 > +\cdots . \end{aligned} \tag{10.113}$$

Using the Schrödinger equation in (10.100) we write

$$< \Psi_0|\hat{H}_0|\Psi > + < \Psi_0|\hat{H}'|\Psi >= E < \Psi_0|\Psi >, \tag{10.114}$$

which, as $\hat{H}_0$ is Hermitian, can be expressed as

$$E = E_0+ < \Psi_0|\hat{H}'|\Psi > . \tag{10.115}$$

This, using equation (10.113), can be solved iteratively in the form

$$E = E_0+ < \Psi_0|\hat{H}'|\Psi_0 > + < \Psi_0|\hat{H}'\hat{R}\hat{H}'|\Psi_0 > +\cdots . \tag{10.116}$$

The results in equations (10.113) and (10.115) can be expressed for individual states by considering $|\Psi_0 >$ to form a complete orthonormal set $\{|k >\}$ and expressing $|\Psi >$ in its nth state as a linear combination of the set $\{|k >\}$. [This is what we had done in equation (10.86) when discussing the Rayleigh-Schrödinger perturbation theory]. The resolvent operator can then be expressed for the nth state as

$$\hat{R} = \sum_{m\neq 0} \frac{|m >< m|}{E_n - E_m^{(0)}} . \tag{10.117}$$

The energy and wavefunction expressions for the perturbed system can be expressed as

$$E_n = E_n^{(0)} + < n|\hat{H}'|n > + \sum_{k\neq n} \frac{< n|\hat{H}'|k >< k|\hat{H}'|n >}{E_n - E_k^{(0)}} + \cdots , \tag{10.118}$$

and

$$|\Psi_n >= |n > + \sum_{l\neq n} \frac{< l|\hat{H}'|n >}{E_n - E_l^{(0)}}|l > + \sum_{l\neq n}\sum_{k\neq n} \frac{< l|\hat{H}'|k >< k|\hat{H}'|n >}{(E_n - E_l^{(0)})(E_n - E_k^{(0)})}|l > + \cdots . \tag{10.119}$$

We make the following observations. Firstly, when E_n is replaced by $E_n^{(0)}$ in the right-hand side of the Brillouin-Wigner perturbation expansion, it turns into a Rayleigh-Schrödinger-like perturbation series. Secondly, higher-order terms are easier to derive and are simpler in form for the Brillouin-Wigner series than for the Rayleigh-Schrödinger series. However there is a disadvantage in applying the Brillouin-Wigner perturbation method. This is because the 'unknown' E_n appears on the right-hand side of the higher-order terms. In order to apply this method, either a trial value or only the lowest-order result for E_n may be used in the right-hand side.

10.3.3 TIME-DEPENDENT PERTURBATION THEORY

In the preceding two sub-sections we described two variants of time-independent quantum perturbation theory. In this sub-section we will briefly describe the time-dependent quantum perturbation theory and obtain Fermi's golden rule formula for transition rate from an energy eigenstate of a system to a continuum of energy eigenstates. We will then describe how the Fermi golden rule formula can be applied by considering two specific examples in condensed matter: (1) phonon scattering rate due to random isotopic impurities and (2) phonon scattering rate in solids due to cubic anharmonicity in crystal potential.

10.3.3.1 Fermi's golden rule formula

Let us consider the time-dependent Schrödinger equation for a perturbed quantum system

$$\begin{aligned} i\hbar\frac{\partial\Psi}{\partial t} &= \hat{H}\Psi, \\ &= (\hat{H}_0+\lambda\hat{H}'(t))\Psi, \end{aligned} \tag{10.120}$$

where $\hat{H}_0$ is the time-independent unperturbed part, $\hat{H}'(t)$ is the time-dependent perturbed part and λ is a parameter to account for the order of perturbation. Let us assume that the unperturbed Hamiltonian satisfies the time-dependent Schrödinger equation

$$i\hbar\frac{\partial\Psi_0}{\partial t} = \hat{H}_0\Psi_0, \tag{10.121}$$

with the general solution

$$\Psi_0 = \sum_k A_k^{(0)} \mathrm{e}^{iE_k^{(0)}t/\hbar}|k>, \tag{10.122}$$

where $|k>$ is a member of a complete and orthonormal set of eigenfunctions corresponding to the energy eigenvalue $E_k^{(0}$.

Let us express

$$\Psi = \sum_k A_k(t)\mathrm{e}^{iE_k^{(0)}t/\hbar}|k>, \tag{10.123}$$

with

$$\sum_k |A_k(t)|^2 = 1, \tag{10.124}$$

and

$$<\Psi|\Psi>= 1. \tag{10.125}$$

Then, $|A_k(t)|^2 = |<k|\Psi>|^2$ is the probability of finding the system in the state $|k>$ at time t with the probability amplitude $A_k(t)$. For the kth state, the Schrödinger equation in (10.120) can be expressed as

$$\frac{\mathrm{d}A_k(t)}{\mathrm{d}t} = \frac{1}{i\hbar}\lambda\sum_m <k|\hat{H}'(t)|m> \mathrm{e}^{i(E_k^{(0)}-E_m^{(0)})t/\hbar}A_m(t). \tag{10.126}$$

Expressing $A_k(t)$ in powers of λ as

$$A_k(t) = A_k^{(0)} + \lambda A_k^{(1)}(t) + \lambda^2 A_k^{(2)}(t) + \cdots, \tag{10.127}$$

we can see that the equation of motion of the first-order term $A_k^{(1)}(t)$ is

$$\frac{\mathrm{d}A_k^{(1)}(t)}{\mathrm{d}t} = \frac{1}{i\hbar} <k|\hat{H}'(t)|m> \mathrm{e}^{i(E_k^{(0)}-E_m^{(0)})t/\hbar}. \tag{10.128}$$

For $k \neq m$, the solution of the above equation is

$$A_k^{(1)}(t) = \frac{1}{i\hbar}\int_0^t < k|\hat{H}'(t')|m > \mathrm{e}^{i(E_k^{(0)}-E_m^{(0)})t'/\hbar}\mathrm{d}t', \tag{10.129}$$

with $A_k^{(1)}(t=0) = A_m^{(1)}(t=0) = 0$. Thus the probability that the system initially in kth state at $t=0$ can transit to mth state ($m \neq k$) at time t is

$$T_{km}^{(1)} = \frac{1}{\hbar^2}\left|\int_0^t < m|\hat{H}'(t')|k > \mathrm{e}^{i(E_k^{(0)}-E_m^{(0)})t'/\hbar}\mathrm{d}t'\right|^2. \tag{10.130}$$

In many physical systems we are required to consider the probability of transition from an initial state $|i>$ to a final state $|f>$ which lies in a band of energies. The number of energy states in the interval E_f to $E_f + \mathrm{d}E_f$ is given by $\mathrm{d}N = g(E_f)\mathrm{d}E_f$, where $g(E_f)$ is the density of final states. With this consideration, the transition probability is given by **Fermi's golden rule formula** (Fermi, 1950)

$$P_i^f = \frac{2\pi}{\hbar}|< f|\hat{H}'|i >|^2 g(E_f), \tag{10.131}$$

$$= \frac{2\pi}{\hbar}|< f|\hat{H}'|i >|^2 \delta(E_f - E_i). \tag{10.132}$$

10.3.3.2 Example 1: Mass-defect scattering of phonons

Let us consider a harmonic crystal containing substitutional isotopic atoms. The Hamiltonian function for this system is

$$H = \sum_{\boldsymbol{l}b}\frac{\boldsymbol{p}(\boldsymbol{l}b)\cdot\boldsymbol{p}(\boldsymbol{l}b)}{2m(\boldsymbol{l}b)} + \mathscr{V}_2, \tag{10.133}$$

where $\boldsymbol{l}$ represents the location of a unit cell, $\boldsymbol{b}$ locates the position of the bth atom in the lth unit cell, $m(\boldsymbol{l}b)$ is mass of $\{\boldsymbol{l}b\}$th atom and $\mathscr{V}_2$ is the harmonic crystal potential. Let us define the average mass of the bth atom as

$$\bar{m}(b) = \frac{1}{N_0}\sum_{\boldsymbol{l}} m(\boldsymbol{l}b) = \sum_i f_i(b)m_i(b), \tag{10.134}$$

where N_0 is the number of unit cells in the crystal and $f_i(b)$ is the fraction of ith isotope of the bth atom of mass $m_i(b)$. We can then separate the Hamiltonian function into an unperturbed part H_0 and a perturbed part due to mass difference $H' = H_{\mathrm{md}}$ as

$$H_0 = \sum_{\boldsymbol{l}b}\frac{\boldsymbol{p}(\boldsymbol{l}b)\cdot\boldsymbol{p}(\boldsymbol{l}b)}{2\bar{m}(b)} + \mathscr{V}_2, \tag{10.135}$$

$$H_{\mathrm{md}} = \frac{1}{2}\sum_{\boldsymbol{l}b}\Delta m(\boldsymbol{l}b)\frac{\boldsymbol{p}(\boldsymbol{l}b)\cdot\boldsymbol{p}(\boldsymbol{l}b)}{[\bar{m}(b)]^2}, \tag{10.136}$$

where

$$\Delta m(\boldsymbol{l}b) = m(\boldsymbol{l}b) - \bar{m}(b). \tag{10.137}$$

Following the procedure developed in section 9.4.2 we can express the momentum operator in second quantised notation as

$$\hat{\boldsymbol{P}}(\boldsymbol{l}b) = \sum_{\boldsymbol{q}s}\sqrt{\frac{\bar{m}(\boldsymbol{l}b)\hbar\omega(\boldsymbol{q}s)}{2N_0}}\,\boldsymbol{e}^*(\boldsymbol{b}|\boldsymbol{q}s)(\hat{a}_{\boldsymbol{q}s} + \hat{a}^{\dagger}_{-\boldsymbol{q}s})\mathrm{e}^{i\boldsymbol{q}\cdot\boldsymbol{l}}, \tag{10.138}$$

where $e(b|qs)$ is the orthonormal polarisation vector. With this, the perturbed Hamiltonian operator can be expressed in second quantisation notation as

$$\begin{aligned}
\hat{H}_{\rm md} &= \frac{1}{2}\sum_{lb}\sum_{qsq's'}\frac{\Delta m(lb)}{[\bar{m}(b)]^2}\frac{\bar{m}(b)\hbar}{2N_0}\sqrt{\omega(qs)\omega(q's')}e^*_{qs}\cdot e_{q's'} \\
&\quad (\hat{a}_{qs}+\hat{a}^\dagger_{-qs})(\hat{a}_{q's'}+\hat{a}^\dagger_{-q's'})e^{i(q-q').l}, \\
&= \frac{\hbar}{4}\sum_{b}\sum_{qsq's'}\sqrt{\omega(qs)\omega(q's')}e^*_{qs}\cdot e_{q's'}\mathscr{M}_{qq'} \\
&\quad (\hat{a}_{qs}+\hat{a}^\dagger_{-qs})(\hat{a}_{q's'}+\hat{a}^\dagger_{-q's'}), \qquad (10.139)\\
&= \sum_{qsq's'}\hat{H}_{\rm md}(qs,q's'), \qquad (10.140)
\end{aligned}$$

where

$$\mathscr{M}_{qq'} = \frac{1}{N_0}\sum_{l}\frac{\Delta m(lb)}{\bar{m}(b)}\exp[\mathrm{i}(q-q')\cdot l]. \qquad (10.141)$$

The above expression for $\hat{H}_{\rm md}$ can be simplified by noting that in the expansion

$$(\hat{a}_{qs}+\hat{a}^\dagger_{-qs})(\hat{a}_{q's'}+a^\dagger_{-q's'}) = \hat{a}_{qs}\hat{a}^\dagger_{-q's'}+\hat{a}^\dagger_{-qs}\hat{a}_{q's'}+\hat{a}_{qs}a_{q's'}+\hat{a}^\dagger_{-qs}a^\dagger_{-q's'}$$

the last two terms with two annihilators or two creators do not conserve energy and only the first two terms are physically meaningful. Thus

$$\begin{aligned}
\hat{H}_{\rm md} &= \frac{\hbar}{4}\sum_{b}\sum_{qsq's'}\sqrt{\omega(qs)\omega(q's')}e^*_{qs}\cdot e_{q's'}\mathscr{M}_{qq'}(\hat{a}_{qs}\hat{a}^\dagger_{-q's'}+\hat{a}^\dagger_{-qs}\hat{a}_{q's'}), \\
&= \frac{\hbar}{2}\sum_{b}\sum_{qsq's'}\sqrt{\omega(qs)\omega(q's')}e^*_{qs}\cdot e_{q's'}\mathscr{M}_{qq'}(\hat{a}_{qs}\hat{a}^\dagger_{q's'}+\hat{a}^\dagger_{qs}\hat{a}_{q's'}), \qquad (10.142)
\end{aligned}$$

where we have noted that the sum over $-q$ duplicates the sum over q.

As discussed in chapter 5, phonon frequencies in a crystals are spread continuously in bands across the Brillouin zone. We therefore apply Fermi's golden rule formula to write the transition probability for a phonon in an initial state $|i>$ to scatter into a final state $|f>$ due to the mass-defect perturbation. We express the initial and final states as

$$|i>=|n_{qs},n_{q's'}>, \quad |f>=|n_{qs}-1,n_{q's'}+1>, \qquad (10.143)$$

where n_{qs} is the number of phonons in state qs. The probability for the transition $qs\rightarrow q's'$ is

$$\begin{aligned}
P^{q's'}_{qs}(\mathrm{md}) &= \frac{\pi}{2}\Big| < n_{qs}-1,n_{q's'}+1|\sum_{b}\sqrt{\omega(qs)\omega(q's')}e^*_{qs}\cdot e_{q's'} \\
&\quad \hat{a}_{qs}\hat{a}^\dagger_{q's'}\mathscr{M}_{qq'}|n_{qs},n_{q's'}>\Big|^2\delta(\omega(q's')-\omega(qs)) \\
&= \frac{\pi}{2}n_{qs}(n_{q's'}+1)\omega(qs)\omega(q's') \\
&\quad \sum_{b}\left(e^*_{qs}(b)\cdot e_{q's'}(b)\right)^2|\mathscr{M}_{qq'}|^2\delta(\omega(qs)-\omega(q's')). \qquad (10.144)
\end{aligned}$$

If we assume that the isotopes are randomly distributed, then

$$|\mathscr{M}_{qq'}|^2 = \frac{1}{N_0^2}\sum_{ll'}\frac{\Delta m(lb)\Delta m(l'b)}{(\bar{m}(b))^2}\exp[i(q-q')\cdot(l-l')]$$

$$
\begin{aligned}
&= \frac{1}{N_0^2}\sum_{l}\left(\frac{\Delta m(lb)}{\bar{m}(b)}\right)^2 + \frac{1}{N_0^2}\sum_{ll'\neq l}\frac{\Delta m(lb)\Delta m(l'b)}{(\bar{m}(b))^2}\exp[i(\boldsymbol{q}-\boldsymbol{q}')\cdot(\boldsymbol{l}-\boldsymbol{l}')] \\
&= \frac{1}{N_0^2}\sum_{l}\left(\frac{\Delta m(lb)}{\bar{m}(b)}\right)^2 \\
&= \frac{1}{N_0}\sum_{i} f_i(b)\left(\frac{\Delta m_i(b)}{\bar{m}(b)}\right)^2 \\
&= \frac{1}{N_0}\Gamma_{\mathrm{md}}(b).
\end{aligned}
\tag{10.145}
$$

The total probability of transition from $\boldsymbol{qs}$ into all allowed energy range for $\boldsymbol{q}'s'$ is

$$
\begin{aligned}
P_{\boldsymbol{qs}}(\mathrm{md}) = \sum_{\boldsymbol{q}'s'} P_{\boldsymbol{qs}}^{\boldsymbol{q}'s'}(\mathrm{md}) &= \frac{\pi}{2N_0}\sum_{b}\Gamma_{\mathrm{md}}(b)\sum_{\boldsymbol{q}'s'} n_{\boldsymbol{qs}}(n_{\boldsymbol{q}'s'}+1)\omega(\boldsymbol{qs})\omega(\boldsymbol{q}'s') \\
&\quad \left(\boldsymbol{e}_{\boldsymbol{qs}}^{*}(b)\cdot\boldsymbol{e}_{\boldsymbol{q}'s'}(b)\right)^2 \delta\left(\omega(\boldsymbol{qs})-\omega(\boldsymbol{q}'s')\right) \\
&\propto \omega_{\boldsymbol{qs}}^2 g(\omega_{\boldsymbol{qs}}).
\end{aligned}
\tag{10.146}
$$

A quantity of physical interest is scattering rate $\tau_{\boldsymbol{qs}}^{-1}$ for a phonon in state $\boldsymbol{qs}$. This is defined as (see, *e.g.* chapter 6 in Srivastava, 2022)

$$
\tau_{\boldsymbol{qs}}^{-1} = \frac{P_{\boldsymbol{qs}}}{\bar{n}_{\boldsymbol{qs}}(\bar{n}_{\boldsymbol{qs}}+1)},
\tag{10.147}
$$

where $\bar{n}_{\boldsymbol{qs}}$ is the Bose-Einstein distribution function describing the thermal average of the phonon occupation number at temperature T.

Numerical evaluation of phonon scattering rate due to mass defects can be carried out by performing the sum over the wavevector q' in equation (10.146). In general, this would require employing a reliable technique for Brillouin zone summation (see, *e.g.* Srivastava, 2022). We will not do that, but simply mention that the scattering rate is proportional to the product of the phonon density of states and the frequency square. If we disregard crystal structure and treat the solid as an elastic continuum, then we only have phonon acoustic branches with the long-wavelength dispersion relation $\omega_{\boldsymbol{qs}} = c_s q$ which produces the density of states $g(\omega_{\boldsymbol{qs}}) \propto \omega_{\boldsymbol{qs}}^2$ (Kittel, 1996; Srivastava, 2022) and the mass-defect scattering rate becomes proportional to $\omega_{\boldsymbol{qs}}^4$, which is the well-known Rayleigh scattering formula.

10.3.3.3 Example 2: Anharmonic scattering of phonons

Phonons in an infinitely long and ideally perfect harmonic crystal have infinitely long lifetimes. This is not so in a realistic anharmonic crystal. As long as anharmonicity is a weak effect, we can account for the finite phonon lifetime resulting from multi-phonon interactions. Theoretically, such interactions can be analysed by using quantum perturbation techniques. Expressing crystal potential energy as $\mathscr{V} = \mathscr{V}_2 + \mathscr{V}_3 + \mathscr{V}_4 + \cdots$, we will expect the effect on phonon states from $\mathscr{V}_n$ to be smaller compared to that from $\mathscr{V}_{n-1}$. In other words, we will expect the largest perturbation to phonon states arising from the cubic anharmonicity $\mathscr{V}_3$. In this section we will apply Fermi's golden rule formula to examine the probability of first-order change in a phonon state due to $\mathscr{V}_3$. We can consider two cases. In one scenario (class 1) a phonon combines with another phonon, both get annihilated and a third phonon of higher energy is created. In another scenario (class 2) an energetic phonon decays into two lower energy phonons. A proper theoretical discussion of these three-phonon processes is quite involved and lies much outside the scope of the present book. Interested reader is directed to the books by Ziman (1960) and Srivastava (2022) for details. Here we will present a very simplified treatment.

Let us treat a solid as an elastic continuum. In equations (8.53) and (8.56) we expressed the cubic anharmonic potential $\mathscr{V}_3$ as a function of second- and third-order elastic constants. A simplified form of $\mathscr{V}_3$ in phase space, using the second quantisation notation, can be taken as [see, Srivastava (2022)]

$$\begin{aligned}\mathscr{V}_3 &= R \sum_{\boldsymbol{q}s,\boldsymbol{q}'s',\boldsymbol{q}''s''} \sqrt{\omega(\boldsymbol{q}s)\omega(\boldsymbol{q}'s')\omega(\boldsymbol{q}''s'')}\,\delta_{\boldsymbol{q}+\boldsymbol{q}'+\boldsymbol{q}'',\boldsymbol{0}} \\ &\quad (a^{\dagger}_{\boldsymbol{q}s} - a_{-\boldsymbol{q}s})(a^{\dagger}_{\boldsymbol{q}'s'} - a_{-\boldsymbol{q}'s'})(a^{\dagger}_{\boldsymbol{q}''s''} - a_{-\boldsymbol{q}''s''}),\end{aligned} \tag{10.148}$$

where R is some constant that measures the strength of cubic anharmonicity and the Kronecker symbol $\delta_{\boldsymbol{q}+\boldsymbol{q}'+\boldsymbol{q}'',\boldsymbol{0}}$ indicates that all phonons taking part in three-phonon processes lie within the Debye sphere (which represents spherical Brillouin zone for the continuum medium). Within this constraint, three-phonon interactions are known as normal (N) processes.

As stated in chapter 9, in an elastic continuum there are three acoustic phonon branches (longitudinal acoustic, or LA, and doubly-degenerate transverse acoustic, or TA) governed by the dispersion relation $\omega(\boldsymbol{q}s) = c_s q$, with c_s being the phonon speed for polarisation s. For the present discussion, we will restrict ourselves to only the LA branch. The momentum and energy conservations relations for the class 1 and class 2 three-phonon interactions are:

$$\text{class 1 process}: \quad \boldsymbol{q}+\boldsymbol{q}' \to \boldsymbol{q}'', \quad \omega_{\boldsymbol{q}}(\mathsf{LA}) + \omega_{\boldsymbol{q}'}(\mathsf{LA}) = \omega_{\boldsymbol{q}''}(\mathsf{LA}), \tag{10.149}$$

and

$$\text{class 2 process}: \quad \boldsymbol{q} \to \boldsymbol{q}'+\boldsymbol{q}'', \quad \omega_{\boldsymbol{q}}(\mathsf{LA}) = \omega_{\boldsymbol{q}'}(\mathsf{LA}) + \omega_{\boldsymbol{q}''}(\mathsf{LA}). \tag{10.150}$$

With an initial state as

$$|i> = |n_{\boldsymbol{q}}, n_{\boldsymbol{q}'}, n_{\boldsymbol{q}''}> \tag{10.151}$$

the final states we consider are

$$\text{class 1}: \quad |f> = |n_{\boldsymbol{q}}-1, n_{\boldsymbol{q}'}-1, n_{\boldsymbol{q}''}+1> \tag{10.152}$$

and

$$\text{class 2}: \quad |f> = |n_{\boldsymbol{q}}-1, n_{\boldsymbol{q}'}+1, n_{\boldsymbol{q}''}+1>. \tag{10.153}$$

The probabilities that the phonon state $\boldsymbol{q}$ is involved in three-phonon class 1 and class 2 processes are then (we will not use the polarisation symbol label as all states are being considered as longitudinal acoustic)

$$\begin{aligned}\text{class 1}: & \\ P^{\boldsymbol{q}''}_{\boldsymbol{q},\boldsymbol{q}'} &= \frac{2\pi}{\hbar^2}\left| < n_{\boldsymbol{q}}-1, n_{\boldsymbol{q}'}-1, n_{\boldsymbol{q}''}+1 | \mathscr{V}_3 | n_{\boldsymbol{q}}, n_{\boldsymbol{q}'}, n_{\boldsymbol{q}''} > \right|^2 \\ &\quad \delta(\omega_{\boldsymbol{q}} + \omega_{\boldsymbol{q}'} - \omega_{\boldsymbol{q}''}), \\ &= \frac{2\pi}{\hbar^2} R^2 \omega_{\boldsymbol{q}}\omega_{\boldsymbol{q}'}\omega_{\boldsymbol{q}''} \\ &\quad \left| < n_{\boldsymbol{q}}-1, n_{\boldsymbol{q}'}-1, n_{\boldsymbol{q}''}+1 | a_{\boldsymbol{q}} a_{\boldsymbol{q}'} a^{\dagger}_{\boldsymbol{q}''} | n_{\boldsymbol{q}}, n_{\boldsymbol{q}'}, n_{\boldsymbol{q}''} > \right|^2 \\ &\quad \delta_{\boldsymbol{q}+\boldsymbol{q}',\boldsymbol{q}''}\,\delta(\omega_{\boldsymbol{q}} + \omega_{\boldsymbol{q}'} - \omega_{\boldsymbol{q}''}), \\ &= \frac{2\pi}{\hbar^2} R^2 \omega_{\boldsymbol{q}}\omega_{\boldsymbol{q}'}\omega_{\boldsymbol{q}''} n_{\boldsymbol{q}} n_{\boldsymbol{q}'} (n_{\boldsymbol{q}''}+1) \\ &\quad \delta_{\boldsymbol{q}+\boldsymbol{q}',\boldsymbol{q}''}\,\delta(\omega_{\boldsymbol{q}} + \omega_{\boldsymbol{q}'} - \omega_{\boldsymbol{q}''}),\end{aligned} \tag{10.154}$$

$$\begin{aligned}\text{class 2}: & \\ P^{\boldsymbol{q}',\boldsymbol{q}''}_{\boldsymbol{q}} &= \frac{2\pi}{\hbar^2}\left| < n_{\boldsymbol{q}}-1, n_{\boldsymbol{q}'}+1, n_{\boldsymbol{q}''}+1 | \mathscr{V}_3 | n_{\boldsymbol{q}}, n_{\boldsymbol{q}'}, n_{\boldsymbol{q}''} > \right|^2 \\ &\quad \delta(\omega_{\boldsymbol{q}} - \omega_{\boldsymbol{q}'} - \omega_{\boldsymbol{q}''}) \\ &= \frac{2\pi}{\hbar^2} R^2 \omega_{\boldsymbol{q}}\omega_{\boldsymbol{q}'}\omega_{\boldsymbol{q}''} \\ &\quad \left| < n_{\boldsymbol{q}}-1, n_{\boldsymbol{q}'}+1, n_{\boldsymbol{q}''}+1 | a_{\boldsymbol{q}} a^{\dagger}_{\boldsymbol{q}'} a^{\dagger}_{\boldsymbol{q}''} | n_{\boldsymbol{q}}, n_{\boldsymbol{q}'}, n_{\boldsymbol{q}''} > \right|^2\end{aligned}$$

$$\begin{aligned} & \delta_{\boldsymbol{q}-\boldsymbol{q}',\boldsymbol{q}''}\delta\left(\omega_{\boldsymbol{q}}-\omega_{\boldsymbol{q}'}-\omega_{\boldsymbol{q}''}\right), \\ = & \frac{2\pi}{\hbar^2}R^2\omega_{\boldsymbol{q}}\omega_{\boldsymbol{q}'}\omega_{\boldsymbol{q}''}n_{\boldsymbol{q}}(n_{\boldsymbol{q}'}+1)(n_{\boldsymbol{q}''}+1) \\ & \delta_{\boldsymbol{q}-\boldsymbol{q}',\boldsymbol{q}''}\delta\left(\omega_{\boldsymbol{q}}-\omega_{\boldsymbol{q}'}-\omega_{\boldsymbol{q}''}\right). \end{aligned} \tag{10.155}$$

Using the above results, the total probability of the state $\boldsymbol{q}$ getting involved in three-phonon processes can be expressed as

$$\begin{aligned} P_{\boldsymbol{q}} = & \frac{2\pi}{\hbar^2}R^2 n_{\boldsymbol{q}}\omega_{\boldsymbol{q}}\sum_{\boldsymbol{q}'\boldsymbol{q}''}\omega_{\boldsymbol{q}'}\omega_{\boldsymbol{q}''} \\ & \Big[n_{\boldsymbol{q}'}(n_{\boldsymbol{q}''}+1)\delta_{\boldsymbol{q}+\boldsymbol{q}',\boldsymbol{q}''}\delta\left(\omega_{\boldsymbol{q}}+\omega_{\boldsymbol{q}'}-\omega_{\boldsymbol{q}''}\right) \\ & +\frac{1}{2}(n_{\boldsymbol{q}'}+1)(n_{\boldsymbol{q}''}+1)\delta_{\boldsymbol{q}-\boldsymbol{q}',\boldsymbol{q}''}\delta\left(\omega_{\boldsymbol{q}}-\omega_{\boldsymbol{q}'}-\omega_{\boldsymbol{q}''}\right)\Big]. \end{aligned} \tag{10.156}$$

Note that we have introduced the factor $\frac{1}{2}$ in the second term on the right-hand side of the above equation to avoid double counting in the summation. To proceed, we can get rid of one of the summations in view of the Kronecker delta symbols $\delta_{\boldsymbol{q}\pm\boldsymbol{q}',\boldsymbol{q}''}$ and perform the other summation either numerically or analytically (if manageable) by making use of the Dirac delta functions $\delta\left(\omega_{\boldsymbol{q}}\pm\omega_{\boldsymbol{q}'}-\omega_{\boldsymbol{q}''}\right)$. As we have restricted ourselves to the consideration of only the LA phonon branch and the linear dispersion relation $\omega \propto q$, we will adopt an analytical approach to carry out the summations.

Using the dispersion relations $\omega_{\boldsymbol{q}} = qc_{\mathrm{LA}}$, $\omega_{\boldsymbol{q}'} = q'c_{\mathrm{LA}}$ and $\omega_{\boldsymbol{q}''} = q''c_{\mathrm{LA}}$, we express the momentum condition (imposed by the Kronecker symbol) and energy conservation condition (imposed by the Dirac delta function) for class 1 and class 2 processes in the form

$$|\boldsymbol{q}\pm\boldsymbol{q}'| = |\boldsymbol{q}''|, \tag{10.157}$$

$$q\pm q' = q''. \tag{10.158}$$

The sum over $\boldsymbol{q}''$ can be removed using the momentum conservation conditions $\boldsymbol{q}'' = \boldsymbol{q}\pm\boldsymbol{q}'$, reducing equation (10.156) to

$$\begin{aligned} P_{\boldsymbol{q}} \propto & \; n_q c_{\mathrm{LA}}^2 \sum_{\boldsymbol{q}'}\Big[qq'|q+q'|n_{q'}(n_{q''}+1)\delta\left(q+q'-q''\right) \\ & +\frac{1}{2}qq'|q-q'|(n_{q'}+1)(n_{q''}+1)\delta\left(q-q'-q''\right)\Big]. \end{aligned} \tag{10.159}$$

For carrying out the sum over $\boldsymbol{q}'$ we use spherical polar coordinates (q',θ',ϕ') and assume that $\boldsymbol{q}$ points along the z-direction and $\boldsymbol{q}'$ lies in a $\phi' =$ constant plane at the polar angle θ'. Then using $\sum_{\boldsymbol{q}'} \to \int \mathrm{d}\boldsymbol{q}' = \int \mathrm{d}\phi' \int \mathrm{d}\theta' \sin\theta' \int \mathrm{d}q' q'^2$, we write

$$\begin{aligned} P_{\boldsymbol{q}} \propto & \; n_q c_{\mathrm{LA}}^2 \Big[\int \mathrm{d}q' qq'^3(q+q')n_{q'}(n_{q''}+1)\int_0^{\pi}\mathrm{d}\theta'\sin\theta'\delta\left(q+q'-q''\right) \\ & +\frac{1}{2}\int \mathrm{d}q' qq'^3(q-q')n_{q'}(n_{q''}+1)\int_0^{\pi}\mathrm{d}\theta'\sin\theta'\delta\left(q-q'-q''\right)\Big], \\ \propto & \; n_q c_{\mathrm{LA}}^2 \Big[\int \mathrm{d}x' xx'^3(x+x')n_{q'}(n_{(x+x')}+1)I_1 \\ & +\frac{1}{2}\int \mathrm{d}x' xx'^3(x-x')n_{q'}(n_{(x-x')}+1)I_2\Big]. \end{aligned} \tag{10.160}$$

where $\boldsymbol{x} = \boldsymbol{q}/q_D$, $\boldsymbol{x}' = \boldsymbol{q}'/q_D$ and q_D is the Debye radius.

The integrals I_1 and I_2 can be carried out as follows. Let us define

$$\Delta_+ = x'' - x - x' = |\boldsymbol{x}+\boldsymbol{x}'| - x - x'$$

and

$$\Delta_- = x'' - x + x' = |\boldsymbol{x} - \boldsymbol{x}'| - x + x'$$

so that

$$-xx' \sin\theta' \mathrm{d}\theta' = |\boldsymbol{x} + \boldsymbol{x}'| \mathrm{d}\Delta_+$$

and

$$xx' \sin\theta' \mathrm{d}\theta' = |\boldsymbol{x} - \boldsymbol{x}'| \mathrm{d}\Delta_- .$$

The results for the integrals are, therefore,

$$I_1 = -\int_0^{-2x'} \mathrm{d}\Delta_+ |\boldsymbol{x} + \boldsymbol{x}'| \delta(\Delta_+) = x + x' \tag{10.161}$$

$$I_2 = \int_0^{2x'} \mathrm{d}\Delta_- |\boldsymbol{x} - \boldsymbol{x}'| \delta(\Delta_-) = x - x'. \tag{10.162}$$

The limits for the integration on the variable x' can be determined from the joint consideration of momentum and energy conservation conditions in equations (10.157) and (10.158) subject to $x'' \leq 1$. With these considerations

$$\text{for the first integral}: \quad 0 \leq x + x' \leq 1, \tag{10.163}$$

$$\text{for the second integral}: \quad 0 \leq x - x' \leq 1. \tag{10.164}$$

Putting things together, we can now express the total probability for a LA phonon to undergo three-phonon interactions of the types $\mathsf{LA} \pm \mathsf{LA} \rightarrow \mathsf{LA}$ as

$$P_{\boldsymbol{q}} \propto n_q c_{\mathsf{LA}}^2 \Big[\int_0^{(1-x)} \mathrm{d}x' x'^2 (x+x')^2 n_{q'} (n_{(x+x')} + 1) + \frac{1}{2} \int_0^x \mathrm{d}x' x'^2 (x-x')^2 n_{q'} (n_{(x-x')} + 1) \Big]. \tag{10.165}$$

Note that in the above discussion we have carried out the sums over $\boldsymbol{q}'$ and $\boldsymbol{q}''$ in equation (10.156) by restricting all three participating phonon wavevectors ($\boldsymbol{q}$, $\boldsymbol{q}'$ and $\boldsymbol{q}''$) to be coplanar. A general three-dimensional geometry has been adopted in a study by Srivastava (1980).

As mentioned in the previous sub-section, a quantity of physical interest is scattering rate $\tau_{\boldsymbol{q}s}^{-1}$ for a phonon in state $\boldsymbol{q}s$, defined as $\tau_{\boldsymbol{q}s}^{-1} = \frac{P_{\boldsymbol{q}s}}{\bar{n}_{\boldsymbol{q}s}(\bar{n}_{\boldsymbol{q}s}+1)}$. Using equation (10.165), and remembering that we have used $\omega = cq = cq_D x$, it can be shown that the scattering rate is a composite function $\mathscr{F}(\omega, T)$ of the phonon frequency and the crystal temperature, *i.e.* $\tau_{\boldsymbol{q}}^{-1} \propto \mathscr{F}(\omega, T)$. A detailed discussion of the low- and high-temperature approximations of the function $\mathscr{F}(\omega, T)$ has been presented in Srivastava (2022). Here we consider the high-temperature approximation $\bar{n}_{\boldsymbol{q}} \approx k_B T/\hbar\omega$. Substituting this in equation (10.165), we can show that the composite function $\mathscr{F}(\omega, T)$ takes a simpler form $\mathscr{F}(\omega, T) = Tf(\omega)$, where f is a function of frequency ω. This would lead to $\tau_{\boldsymbol{q}}^{-1} \propto Tf(\omega)$, *i.e.* at high temperatures the cubic anharmonic scattering rate of a phonon varies as some function of its frequency and increases linearly with crystal temperature.

This is a good place to point out that quantum calculations, non-perturbative as well as perturbative, involve wavevector summations such as those seen in equations (10.146) and (10.159). In general, such summations cannot be carried out analytically without invoking a great many simplifying assumptions (as we have done here). Such summations in condensed matter are known as Brillouin zone summations. Several techniques have been discussed in the literature for carrying out such summations. One method that is routinely employed in the study of the physics of electrons and phonons is the so-called special wavevector scheme. Monkhorst and Pack (1976) and Fehlner and Vosko (1977) have described schemes for generating special wavevectors within the Brillouin zone for crystals of cubic as well as hexagonal symmetries.

FURTHER READING

1. Bransden B H and Joachain C J 2000 *Quantum Mechanics* 2nd edn (Prentice Hall: New York)
2. Dalgarno A and Stewart A L 1956 *Proc. Roy. Soc. London, Ser. A* **238** 269
3. Fermi E 1950 *Nuclear Physics* (University of Chicago Press: Chicago)
4. Goldstein H, Poole C and Safko J 2002 *Classical Mechanics* 3rd edn (Addison-Wesley: New York)
5. Gregory R D 2006 *Classical Mechanics* (Cambridge University Press: Cambridge)
6. José J V and Saletan E J 1998 *Classical Dynamics – A Contemporary Approach* (Cambridge University Press: Cambridge)
7. Kittel C 1996 *Introduction to Solid State Physics* 7th edn (Wiley: New York)
8. Liboff R L 1998 *Introductory Quantum Mechanics* 3rd edn (Addison Wesley: New York)

PROBLEMS

1. Following the discussion in section 10.2.2, show that the solution of the perturbed one-dimensional problem up to the second-order (*i.e.* up to ε^2) is given by the equations

$$x(t) = \frac{p_0}{m\omega}\left[\omega t - \frac{(\omega t)^3}{3!} + \frac{(\omega t)^5}{5!}\right]$$

and

$$p(t) = p_0\left[1 - \frac{(\omega t)^2}{2!} + \frac{(\omega t)^4}{4!}\right].$$

2. Consider the non-linear oscillator equation

$$\frac{1}{\omega_0^2}\frac{\mathrm{d}^2 x}{\mathrm{d}t^2} + x + \varepsilon x^5 = 0,$$

where ω_0 is the frequency of the corresponding harmonic oscillator and ε is a small parameter. Use the Poincarè-Lindstedt perturbation method to obtain the approximate solution to the first order in ε.

3. Prove that the expression for $E^{(3)}$ in equation (10.82) can be written as the expression given in equation (10.83).
4. An electric field of strength $\mathscr{E}$ is applied along the z-axis to the hydrogen atom in its ground state. This generates a perturbation to the ground-state Hamiltonian by $H' = e\mathscr{E}z$. Applying the Rayleigh-Schrödinger perturbation method, show that there is no first-order change to the energy of this state.
5. Follow the approach discussed in section 10.3.3.3 to show that the probability of an acoustic transverse phonon of wavevector $\boldsymbol{q}$ to take part in the three-phonon normal (N) process $(\boldsymbol{q}, \mathsf{TA}) + (\boldsymbol{q}', \mathsf{LA}) \to (\boldsymbol{q}'', \mathsf{LA})$ can be expressed as

$$P_{\boldsymbol{q}} \propto n_q c_{\mathsf{TA}} c_{\mathsf{LA}} \int_{(1-\alpha)x/2}^{(1-\alpha x)} \mathrm{d}x' x'^2 (\alpha x + x')^2 n_{q'} (n_{(\alpha q + q')} + 1),$$

where $\alpha = c_{\mathsf{TA}}/c_{\mathsf{LA}}$, $x = q/q_D$, $x' = q'/q_D$ and q_D is the Debye radius.

A Mathematical Preliminaries

A.1 FRAMES OF REFERENCE

A.1.1 ORGANISATION OF NUMERICAL VALUES OF QUANTITIES IN A COORDINATE SYSTEM

In order to deal with a mathematical or numerical description of any event it is essential to set up a frame of reference, or a coordinate system. Let us choose a k-dimensional coordinate system. A collection of k^n entries of numerical values is a **tensor** of order (degree, or rank) n to represent a physical quantity. A tensor is given different names, depending on its order n, as listed below.

(i) $n = 0$: Tensor of the zero rank is a **scalar**, as $k^0 = 1$ is just a single number.

(ii) $n = 1$: Tensor of the first rank is a **vector**, as k^1 is a collection of k numbers. With $k = 2$, we can consider the two-dimensional Cartesian coordinate system and express a two-component vector in x-y plane as

$$\boldsymbol{x} + \boldsymbol{y} = x\hat{\boldsymbol{x}} + y\hat{\boldsymbol{y}} = (x, y). \tag{A.1}$$

With $k = 3$, we can consider the (x, y, z) three-dimensional Cartesian coordinate system and express a three-component vector as

$$\boldsymbol{r} = \boldsymbol{x} + \boldsymbol{y} + \boldsymbol{z} = x\hat{\boldsymbol{x}} + y\hat{\boldsymbol{y}} + z\hat{\boldsymbol{z}} = (x, y, z). \tag{A.2}$$

Note that we have used three different notations to express these vectors. Here, $\hat{\boldsymbol{x}}$, $\hat{\boldsymbol{y}}$ and $\hat{\boldsymbol{z}}$ represent unit vectors along the x-, y- and z-axes, respectively.

(iii) $n = 2$: Tensor of the second rank is a square **matrix**, representing a collection of k^2 numbers. With $k = 2$, we can express a 2×2 matrix $\mathscr{B}$ as

$$\mathscr{B} = \begin{bmatrix} b_{11} & b_{12} \\ b_{21} & b_{22} \end{bmatrix}, \tag{A.3}$$

where b_{ij} is the element of the matrix in it's ith row and jth column.

(iv) $n = 3$: Tensor of the third rank represents a collection of k^3 numbers in the k-dimensional space. A simple example is the alternating tensor, or isotropic tensor, of rank 3. In three-dimensional space ($k = 3$), the 27 elements of this tensor are given by the Levi-Civita symbol ε_{ijk}, defined as

$$\begin{aligned} \varepsilon_{123} = \varepsilon_{231} = \varepsilon_{312} &= 1, \\ \varepsilon_{132} = \varepsilon_{213} = \varepsilon_{321} &= -1, \\ \varepsilon_{ijk} &= 0 \quad \text{for other permutations of } i, j, k. \end{aligned} \tag{A.4}$$

A.1.2 DIFFERENT COORDINATE SYSTEMS

Description of physical systems is possible by using a variety of coordinate systems. We routinely use the (orthogonal) Cartesian coordinates (x, y, z). Other types of orthogonal coordinates are called curvilinear systems. Depending on the shape of the physical system under consideration, its description may be more conveniently carried out by using a particular curvilinear coordinate system. Three most frequently used curvilinear coordinate systems are the plane polar, cylindrical polar and spherical polar. A note of caution is in order. As opposed to the Cartesian coordinates, a curvilinear coordinate may not have the units of length.

DOI: 10.1201/9781003383314-A

A.1.2.1 Plane polar coordinates

As mentioned earlier, the Cartesian coordinates of a point in the x-y plane are (x,y). In the plane polar coordinate system, the coordinates of a point are expressed as (ρ,ϕ), where

$$x=\rho\cos\phi,\quad y=\rho\sin\phi, \tag{A.5}$$

as illustrated in figure A.1. From these equations, the plane polar variables ρ and ϕ are expressed in terms of the Cartesian coordinates as

$$\rho=(x^2+y^2)^{1/2},\quad \phi=\arctan\left(\frac{y}{x}\right). \tag{A.6}$$

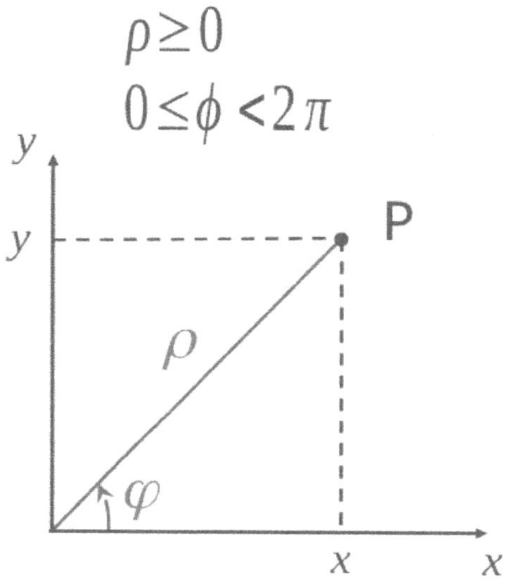

Figure A.1 Plane polar coordinates.

The position vector of the point P in figure A.1 can be written as

$$\boldsymbol{x}+\boldsymbol{y},\quad \text{or}\quad x\hat{\boldsymbol{x}}+y\hat{\boldsymbol{y}}\quad :\quad \text{Cartesian coordinates}, \tag{A.7}$$

$$\boldsymbol{\rho}+\boldsymbol{\phi},\quad \text{or}\quad \rho\hat{\boldsymbol{\rho}}+\phi\hat{\boldsymbol{\phi}}\quad :\quad \text{plane polar coordinates}. \tag{A.8}$$

Note that in the polar coordinates $\boldsymbol{\phi}$ is an angle vector, $\hat{\boldsymbol{\rho}}$ is a unit vector in the direction of increasing $\boldsymbol{\rho}$ and $\hat{\boldsymbol{\phi}}$ is a unit vector in the direction of increasing ϕ. The variables ρ and ϕ are restricted as $\rho\geq 0$ and $0\leq\phi<2\pi$.

A.1.2.2 Cylindrical polar coordinates

In the cylindrical polar coordinate system, the coordinates of a point are expressed as (ρ,ϕ,z), where

$$x=\rho\cos\phi,\quad y=\rho\sin\phi,\quad z=z, \tag{A.9}$$

as illustrated in figure A.2. From these equations, the cylindrical polar variables ρ, ϕ and z are expressed in terms of the Cartesian coordinates as

$$\rho=(x^2+y^2)^{1/2},\quad \phi=\arctan\left(\frac{y}{x}\right),\quad z=z. \tag{A.10}$$

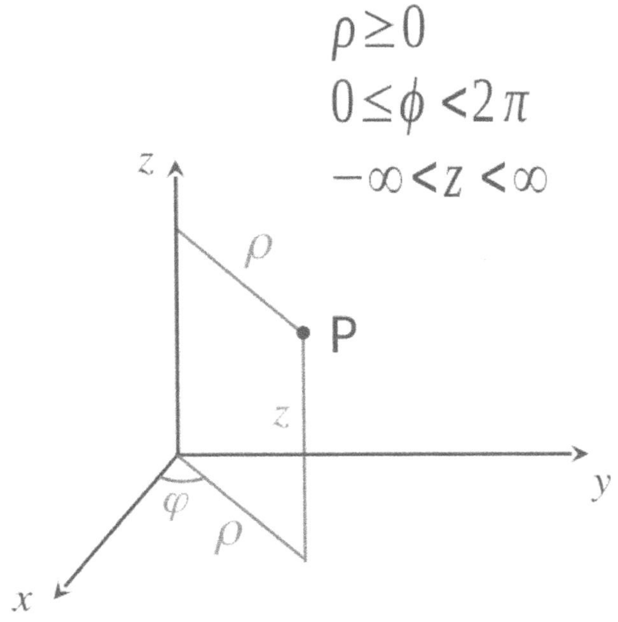

Figure A.2 Cylindrical coordinates.

The position vector of the point P in figure A.2 can be written as

$$\boldsymbol{x}+\boldsymbol{y}+\boldsymbol{z}, \quad \text{or} \quad x\hat{\boldsymbol{x}}+y\hat{\boldsymbol{y}}+z\hat{\boldsymbol{z}} \quad : \quad \text{Cartesian coordinates}\,, \tag{A.11}$$

$$\boldsymbol{\rho}+\boldsymbol{\phi}+\boldsymbol{z}, \quad \text{or} \quad \rho\hat{\boldsymbol{\rho}}+\phi\hat{\boldsymbol{\phi}}+z\hat{\boldsymbol{z}} \quad : \quad \text{cylindrical polar coordinates}\,. \tag{A.12}$$

The variables ρ, ϕ and z are restricted as $\rho \geq 0$, $0 \leq \phi < 2\pi$ and $-\infty < z < \infty$.

A.1.2.3 Spherical polar coordinates

In the spherical polar coordinate system, the coordinates of a point are expressed as (r, θ, ϕ), where

$$x = r\sin\theta\cos\phi, \quad y = r\sin\theta\sin\phi, \quad z = r\cos\theta, \tag{A.13}$$

as illustrated in figure A.3. From these equations, the spherical polar variables r, θ and ϕ are expressed in terms of the Cartesian coordinates as

$$r = (x^2+y^2+z^2)^{1/2}, \quad \phi = \arctan\left(\frac{y}{x}\right), \quad \theta = \arccos\left(\frac{z}{r}\right). \tag{A.14}$$

The position vector of the point P in figure A.3 can be written as

$$\boldsymbol{x}+\boldsymbol{y}+\boldsymbol{z}, \quad \text{or} \quad x\hat{\boldsymbol{x}}+y\hat{\boldsymbol{y}}+z\hat{\boldsymbol{z}} \quad : \quad \text{Cartesian coordinates}\,, \tag{A.15}$$

$$\boldsymbol{r}+\boldsymbol{\theta}+\boldsymbol{\phi}, \quad \text{or} \quad r\hat{\boldsymbol{r}}+\theta\hat{\boldsymbol{\theta}}+\phi\hat{\boldsymbol{\phi}} \quad : \quad \text{spherical polar coordinates}\,. \tag{A.16}$$

The variables r, θ and ϕ are restricted as $r \geq 0$, $0 \leq \theta \leq \pi$ and $0 \leq \phi < 2\pi$.

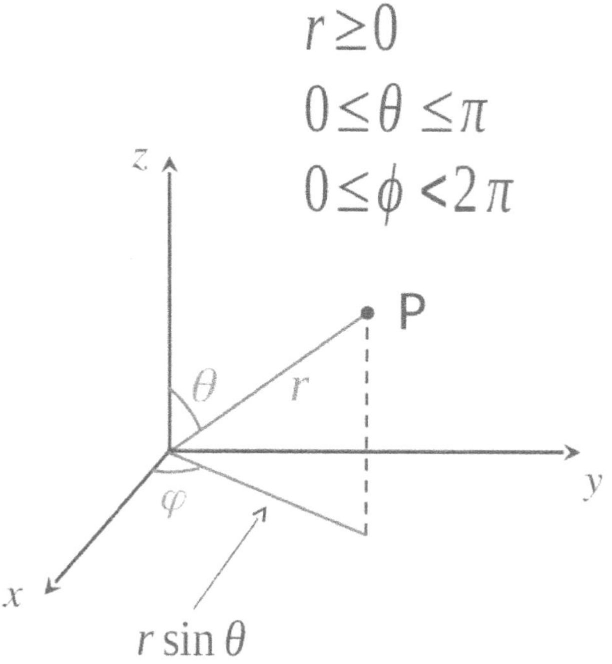

Figure A.3 Spherical polar coordinates.

A.2 FULL AND PARTIAL DIFFERENTIATIONS

Consider a function $f(x)$ of a single variable x. A small change $\mathrm{d}x$ in x makes a small change $\mathrm{d}f$ in the function, which is expressed as

$$\mathrm{d}f = \frac{\mathrm{d}f}{\mathrm{d}x}\mathrm{d}x, \tag{A.17}$$

with $\frac{\mathrm{d}f}{\mathrm{d}x}$ called the (full) differential of f with respect to x. Now consider a multivariable function, such as $f(\boldsymbol{r})$ with $\boldsymbol{r} = x\hat{\boldsymbol{x}} + y\hat{\boldsymbol{y}} + z\hat{\boldsymbol{z}}$, where x, y and z are the three independent variables in the Cartesian system. The change $\mathrm{d}f$ due to a small change $\mathrm{d}\boldsymbol{r} = \mathrm{d}x\hat{\boldsymbol{x}} + \mathrm{d}y\hat{\boldsymbol{y}} + \mathrm{d}z\hat{\boldsymbol{z}}$ can be expressed as

$$\mathrm{d}f = \boldsymbol{\nabla} f \cdot \mathrm{d}\boldsymbol{r} = f_x\mathrm{d}x + f_y\mathrm{d}y + f_z\mathrm{d}z, \tag{A.18}$$

where

$$\boldsymbol{\nabla} = \hat{\boldsymbol{x}}\frac{\partial}{\partial x} + \hat{\boldsymbol{y}}\frac{\partial}{\partial y} + \hat{\boldsymbol{z}}\frac{\partial}{\partial z}, \tag{A.19}$$

is the Del, or Nabla, operator. In this case the full change in f is contributed by the small changes in the independent variables x, y and z. The terms $f_x = \frac{\partial f}{\partial x}|_{y,z}$, $f_y = \frac{\partial f}{\partial y}|_{z,x}$ and $f_z = \frac{\partial f}{\partial z}|_{x,y}$ are the partial differentials of f with respect to x, y and z, respectively. The subscript 'y,z' in $\frac{\partial f}{\partial x}|_{y,z}$ indicates that while differentiating f with respect to x, the variables y and z are treated as constants. Sometimes it becomes convenient to write $\frac{\partial f}{\partial x}|_{y,z}$ simply as $\frac{\partial f}{\partial x}$, or f_x. Similar considerations are implied for other partial differentials. We will adopt the assumed implicitness in this book.

The expressions for the Del operator in the cylindrical and spherical polar systems are as follows:

$$\text{Cylindrical polar}: \boldsymbol{\nabla} = \hat{\boldsymbol{\rho}}\frac{\partial}{\partial \rho} + \hat{\boldsymbol{\theta}}\frac{1}{\rho}\frac{\partial}{\partial \phi} + \hat{\boldsymbol{\phi}}\frac{\partial}{\partial \phi}, \tag{A.20}$$

$$\text{Spherical polar}: \boldsymbol{\nabla} = \hat{\boldsymbol{r}}\frac{\partial}{\partial r} + \hat{\boldsymbol{\theta}}\frac{1}{r}\frac{\partial}{\partial \theta} + \hat{\boldsymbol{\phi}}\frac{1}{r\sin\theta}\frac{\partial}{\partial \phi}. \tag{A.21}$$

The operator $\nabla^2 = \boldsymbol{\nabla}\cdot\boldsymbol{\nabla}$ is called the Laplacian. The expressions for the Laplacian in the Cartesian, cylindrical and spherical polar systems as follows:

$$\text{Cartesian}: \nabla^2 = \frac{\partial^2}{\partial x^2} + \frac{\partial^2}{\partial y^2} + \frac{\partial^2}{\partial z^2}, \tag{A.22}$$

$$\text{Cylindrical polar}: \nabla^2 = \frac{1}{\rho}\frac{\partial}{\partial \rho}\left(\rho\frac{\partial}{\partial \rho}\right) + \frac{1}{\rho^2}\frac{\partial^2}{\partial \phi^2} + \frac{\partial^2}{\partial z^2}, \tag{A.23}$$

$$\text{Spherical polar}: \nabla^2 = \frac{1}{r^2}\frac{\partial}{\partial r}\left(r^2\frac{\partial}{\partial r}\right) + \frac{1}{r^2\sin\theta}\frac{\partial}{\partial \theta}\left(\sin\theta\frac{\partial}{\partial \theta}\right) + \frac{1}{r^2\sin^2\theta}\frac{\partial^2}{\partial \phi^2}. \tag{A.24}$$

A.3 ALGEBRAIC METHOD FOR CHANGE OF VARIABLES FROM CARTESIAN TO CURVILINEAR COORDINATES

Let us derive expressions for elementary area and elementary volume by developing an algebraic method for change of variables from the Cartesian coordinate system to the three types of curvilinear coordinate systems discussed in the previous section.

A.3.1 ELEMENTARY AREA IN PLANE POLAR COORDINATE SYSTEM

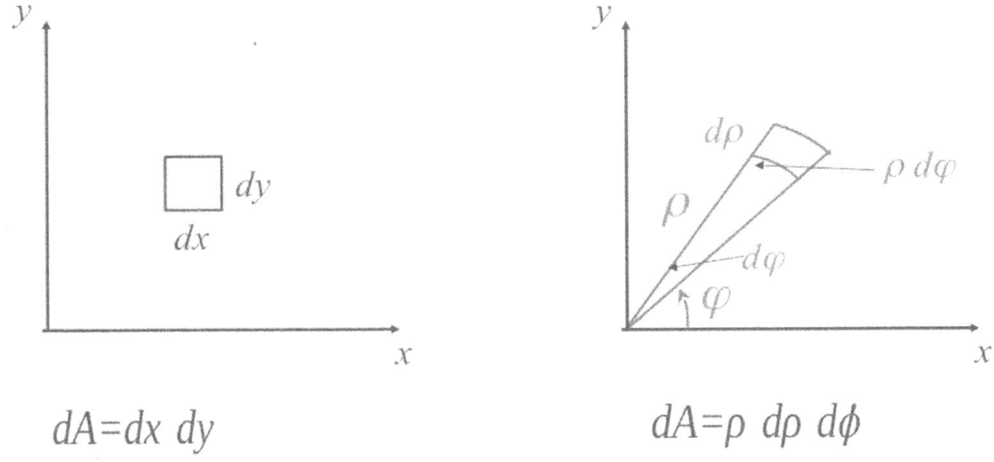

Figure A.4 An elementary area in the Cartesian system (left panel) and the plane polar system (right panel).

Consider an elementary area $dA = dxdy$ in the Cartesian system. Using equation (A.5), we express

$$dx = \frac{\partial x}{\partial \rho}d\rho - \frac{\partial x}{\partial \phi}d\phi = \cos\phi d\phi - (-\rho\sin\phi)d\phi,$$

$$dy = \frac{\partial y}{\partial \rho}d\rho + \frac{\partial y}{\partial \phi}d\phi = \sin\phi d\phi + \rho\cos\phi d\phi, \tag{A.25}$$

where it is acknowledged that as the angle ϕ increases, the distance x decreases. The expression for the elementary area dA in the plane polar coordinate system is thus

$$\begin{aligned} dA = dxdy &= [\cos\phi d\phi - (-\rho\sin\phi)][\sin\phi d\phi + \rho\cos\phi d\phi] \\ &\simeq \rho d\rho d\phi. \end{aligned} \tag{A.26}$$

The elementary area in both the Cartesian system and the polar system is shown in figure A.4.

We can express the result in equation (A.26) as

$$dxdy = |J| d\rho d\phi, \tag{A.27}$$

where $|J|$ is determinant of the Jacobian matrix

$$J = \begin{bmatrix} \frac{\partial x}{\partial \rho} & \frac{\partial x}{\partial \phi} \\ \frac{\partial y}{\partial \rho} & \frac{\partial y}{\partial \phi} \end{bmatrix}. \tag{A.28}$$

A.3.2 ELEMENTARY AREA AND VOLUME IN CYLINDRICAL POLAR COORDINATE SYSTEM

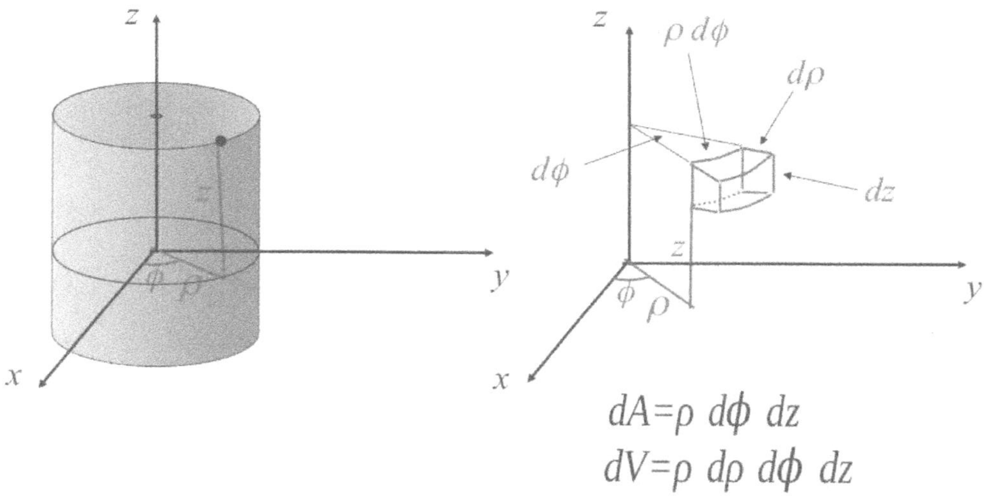

Figure A.5 Elementary area and elementary volume in the cylindrical polar coordinate system.

We can generalise the statements made in equations (A.27) and (A.28). If x and y are given as functions of two new (and independent) variables u and v, then the elemental area $dxdy$ can be expressed as

$$dA = dxdy = |J| du dv, \tag{A.29}$$

where

$$J = \begin{bmatrix} \frac{\partial x}{\partial u} & \frac{\partial x}{\partial v} \\ \frac{\partial y}{\partial u} & \frac{\partial y}{\partial v} \end{bmatrix}. \tag{A.30}$$

Similarly, if x, y and z are given as functions of three new (and independent) variables u, v and w, then the elemental volume $\mathrm{d}V = \mathrm{d}x\mathrm{d}y\mathrm{d}z$ can be expressed as

$$\mathrm{d}V = \mathrm{d}x\mathrm{d}y\mathrm{d}z = |J|\mathrm{d}u\mathrm{d}v\mathrm{d}w, \tag{A.31}$$

where

$$J = \begin{bmatrix} \frac{\partial x}{\partial u} & \frac{\partial x}{\partial v} & \frac{\partial x}{\partial w} \\ \frac{\partial y}{\partial u} & \frac{\partial y}{\partial v} & \frac{\partial y}{\partial w} \end{bmatrix}. \tag{A.32}$$

Using equation (A.9), we can obtain the following expressions for elementary area $\mathrm{d}A$ and elementary volume $\mathrm{d}V$ in the cylindrical polar coordinate system.

$$\mathrm{d}A = \mathrm{d}x\mathrm{d}z = |J|\mathrm{d}\rho\mathrm{d}\phi = \rho\mathrm{d}\phi\mathrm{d}z, \tag{A.33}$$

$$\mathrm{d}V = \mathrm{d}x\mathrm{d}y\mathrm{d}z = |J|\mathrm{d}\rho\mathrm{d}\phi\mathrm{d}z = \rho\mathrm{d}\rho\mathrm{d}\phi\mathrm{d}z. \tag{A.34}$$

Figure A.5 shows a graphical illustration of $\mathrm{d}A$ and $\mathrm{d}V$.

A.3.3 ELEMENTARY AREA AND VOLUME IN SPHERICAL POLAR COORDINATE SYSTEM

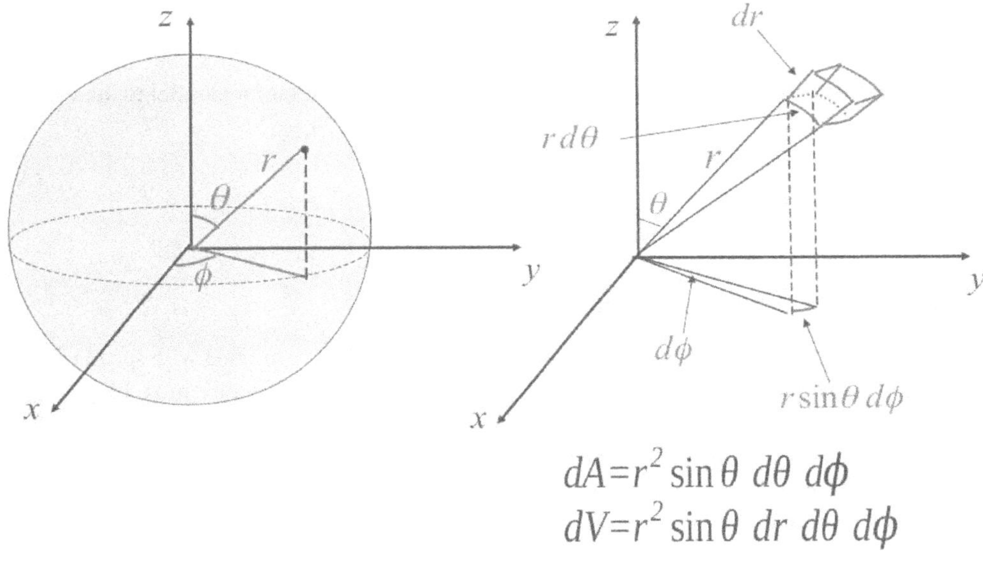

Figure A.6 Illustration of elementary area and elementary volume in the spherical polar coordinate system.

Using the spherical polar coordinates (r, θ, ϕ) defined in equation (A.13), we can obtain

$$J = \begin{bmatrix} \frac{\partial x}{\partial r} & \frac{\partial x}{\partial \theta} & \frac{\partial x}{\partial \phi} \\ \frac{\partial y}{\partial r} & \frac{\partial y}{\partial \theta} & \frac{\partial y}{\partial \phi} \\ \frac{\partial z}{\partial r} & \frac{\partial z}{\partial \theta} & \frac{\partial z}{\partial \phi} \end{bmatrix} = \begin{bmatrix} \sin\theta\cos\phi & r\cos\theta\cos\phi & -r\sin\theta\sin\phi \\ \sin\theta\sin\phi & r\cos\theta\sin\phi & r\sin\theta\cos\phi \\ \cos\theta & -r\sin\theta & 0 \end{bmatrix}. \tag{A.35}$$

The expressions for elementary area $\mathrm{d}A$ and elementary volume $\mathrm{d}V$ are, thus

$$\mathrm{d}A = \mathrm{d}x\mathrm{d}y = |J|\mathrm{d}\theta\mathrm{d}\phi = r^2\sin\theta\mathrm{d}\theta\mathrm{d}\phi, \tag{A.36}$$

$$\mathrm{d}V = \mathrm{d}x\mathrm{d}y\mathrm{d}z = |J|\mathrm{d}r\mathrm{d}\theta\mathrm{d}\phi = r^2\sin\theta\mathrm{d}r\mathrm{d}\theta\mathrm{d}\phi. \tag{A.37}$$

These are illustrated in figure A.6.

A.4 MOMENT OF INERTIA

The geometrical quantity

$$\mathrm{I} = MR^2 \tag{A.38}$$

is called the moment of inertia of a point mass M that is rotating at a distance R from an axis of rotation. Table A.1 lists expressions for the moment of inertia of a few rigid bodies[1].

Table A.1
Moment of inertia of bodies about the axis through their centre of mass.

Object and its geometrical parameters	Moment of inertia
Slender rod of length L	$\frac{1}{12}ML^2$
Rectangular plate of side lengths a and b	$\frac{1}{12}M(a^2+b^2)$
Hollow cylinder of radii R_1 and R_2	$\frac{1}{2}M(R_1^2+R_2^2)$
Solid cylinder of radius R	$\frac{1}{2}MR^2$
Thin walled hollow cylinder of radius R	MR^2
Solid sphere of radius R	$\frac{2}{5}MR^2$
Thin walled hollow sphere of radius R	$\frac{2}{3}MR^2$

The moment of inertia of an object of mass M about an axis of rotation parallel to, and distant d from, the axis through its centre of mass is

$$\mathrm{I}_{\text{parallel axis}} = \mathrm{I}_{\text{cm}} + Md^2, \tag{A.39}$$

where I_{cm} is the moment of inertia about the centre of mass.

A.5 LINEAR AND ANGULAR KINETIC ENERGIES

The linear, or translational, kinetic energy T_{trans} of an object of mass m moving with speed v is $\frac{1}{2}mv^2$. The rotational kinetic energy T_{rot} of an object, rotating with angular velocity ω is $\frac{1}{2}\mathrm{I}_{\text{cm}}\omega^2$, where I_{cm} is the moment of inertia about it centre of mass. The total kinetic energy T of a rigid object is the sum of its linear (*i.e.* of its centre of mass) and rotational (*i.e.* about the axis through its centre of mass) contributions:

$$T = T_{\text{trans}} + T_{\text{rot}} = \frac{1}{2}mv^2 + \frac{1}{2}\mathrm{I}_{\text{cm}}\omega^2. \tag{A.40}$$

[1] For the purpose of the discussion of classical dynamics in this book we have treated the moment of inertia as a scalar. It should be noted, however, that in general it is a tensor quantity.

B Conservative Property of Central Forces

In this appendix we will prove that: (1) work done by a central force in moving a particle from a point $\boldsymbol{r}_1$ to another point $\boldsymbol{r}_2$ is path independent and (2) curl of a central force is zero.

B.1 PROOF THAT CENTRAL FORCES ARE CONSERVATIVE

In order to show that central forces are conservative, we will use the spherical coordinates[1] (r, θ, ϕ) and show that the work done by the central force $\boldsymbol{F}(\boldsymbol{r}) = F(r)\frac{\boldsymbol{r}}{r}$ depends only on the radial coordinate r and not on the angular coordinates θ and ϕ. Using equation (A.16), we write

$$\boldsymbol{r} = r\hat{\boldsymbol{r}} + \theta\hat{\boldsymbol{\theta}} + \phi\hat{\boldsymbol{\phi}}, \tag{B.1}$$

where r, θ and ϕ are the coordinates in the spherical polar coordinate system. The work done by the force $\boldsymbol{F}(\boldsymbol{r})$ is

$$\begin{aligned} \text{Work done} &= \int_{\boldsymbol{r}_1}^{\boldsymbol{r}_2} \boldsymbol{F}(\boldsymbol{r}) \cdot \mathrm{d}\boldsymbol{r}, \\ &= \int_{\boldsymbol{r}_1}^{\boldsymbol{r}_2} \boldsymbol{F}(\boldsymbol{r}) \cdot (\hat{\boldsymbol{r}}\mathrm{d}r + \hat{\boldsymbol{\theta}}\mathrm{d}\theta + \hat{\boldsymbol{\phi}}\mathrm{d}\phi), \\ &= \int_{\boldsymbol{r}_1}^{\boldsymbol{r}_2} F(r)\hat{\boldsymbol{r}} \cdot (\hat{\boldsymbol{r}}\mathrm{d}r + \hat{\boldsymbol{\theta}}\mathrm{d}\theta + \hat{\boldsymbol{\phi}}\mathrm{d}\phi), \\ &= \int_{\boldsymbol{r}_1}^{\boldsymbol{r}_2} F(r)\mathrm{d}r. \end{aligned} \tag{B.2}$$

As the integral depends only on the radial coordinate r and not on the angular coordinates θ and ϕ, it has been proven that the work done is path independent. In other words, we have proved that a central force is conservative.

B.2 CURL OF A CENTRAL FORCE

Consider a central force in the form $\boldsymbol{F}(\boldsymbol{r}) = F(r)\frac{\boldsymbol{r}}{r}$. From equation (A.21), the expression for the Del operator in the spherical polar coordinate system is

$$\boldsymbol{\nabla} = \hat{\boldsymbol{r}}\frac{\partial}{\partial r} + \hat{\boldsymbol{\theta}}\frac{1}{r}\frac{\partial}{\partial \theta} + \hat{\boldsymbol{\phi}}\frac{1}{r\sin\theta}\frac{\partial}{\partial \phi}. \tag{B.3}$$

The curl of $\boldsymbol{F}(\boldsymbol{r})$ is

$$\begin{aligned} \boldsymbol{\nabla} \times F(r)\frac{\boldsymbol{r}}{r} &= \begin{vmatrix} \hat{\boldsymbol{r}} & \hat{\boldsymbol{\theta}} & \hat{\boldsymbol{\phi}} \\ \frac{\partial}{\partial r} & \frac{1}{r}\frac{\partial}{\partial \theta} & \frac{1}{r\sin\theta}\frac{\partial}{\partial \phi} \\ F(r) & 0 & 0 \end{vmatrix}, \\ &= \mathbf{0}. \end{aligned} \tag{B.4}$$

This proves that the curl of a central force, and hence of a conservative force, is zero.

---

[1] The Cartesian coordinate system can also be employed, but it is much simpler to use the spherical polar coordinates for this problem.

DOI: 10.1201/9781003383314-B

C Euler-Lagrange Variational Principle

C.1 CALCULUS OF VARIATIONS

We start by defining a few terms relevant to calculus of variations before describing the Euler-Lagrange variational principle.

C.1.1 FUNCTIONAL

A *functional* is a function of variables and expressions. For example, $f\left(y(x), \frac{\mathrm{d}y}{\mathrm{d}x}, x\right)$ is a functional, with $y = y(x)$ being a function of x.

C.1.2 FUNCTIONAL DERIVATIVE

The functional derivative $\frac{\delta f}{\delta y(x)}$ of $f\left(y(x)\right)$ with respect to $y(x)$ is defined using the equation

$$\delta f(y) = f(y+\delta y) - f(y) = \int \frac{\delta f}{\delta y(x)} \delta y(x) \mathrm{d}x, \tag{C.1}$$

where δ refers to a small change.

C.1.3 EXTREMAL VALUE

Consider a functional $J(\alpha)$ of parameter α. If

$$\left(\frac{\mathrm{d}J}{\mathrm{d}\alpha}\right)\bigg|_{\tilde{\alpha}} = 0, \tag{C.2}$$

then J is *stationary* (or, has an *extremum*) at $\tilde{\alpha}$, and $J(\tilde{\alpha})$ is the *extremal value* of J.

At the extremum, J can be a minimum, a maximum or a saddle point:

$$\left(\frac{\mathrm{d}^2 J}{\mathrm{d}\alpha^2}\right)\bigg|_{\tilde{\alpha}} \begin{cases} > 0 & \text{minimum,} \\ < 0 & \text{maximum,} \\ = 0 & \text{saddle point.} \end{cases} \tag{C.3}$$

A variation in the functional J, due to a small change $\delta\alpha$, is

$$\delta J = \left(\frac{\mathrm{d}J}{\mathrm{d}\alpha}\right)\bigg|_{\tilde{\alpha}} \delta\alpha. \tag{C.4}$$

For an arbitrary $\delta\alpha$, the condition for an extremum of J is thus

$$\delta J = 0. \tag{C.5}$$

DOI: 10.1201/9781003383314-C

C.2 EULER-LAGRANGE VARIATIONAL PRINCIPLE

Consider a functional J in the following integral form

$$J(y) = \int_{x_1}^{x_2} f(y,y',x)\mathrm{d}x, \tag{C.6}$$

where $y' = \frac{\mathrm{d}y}{\mathrm{d}x}$ and the functional $f(x,y,y')$ is defined using a function

$$y = y(x) \tag{C.7}$$

between two points $x = x_1$ and $x = x_2$ with fixed values $y(x_1) = Y_1$ and $y(x_2) = Y_2$. Here we have assumed that f is twice continuously differentiable. We wish to find the optimum function $y(x)$ which results in the integral $J(y)$ taking an extremum value.

Let us consider a *δ-variation* $\delta y(x)$ in the function $y(x)$, subject to

$$\delta y(x_1) = \delta y(x_2) = 0. \tag{C.8}$$

To first order, the variation in J is[1]

$$\begin{aligned}\delta J &= \delta \int_{x_1}^{x_2} f(y,y',x)\mathrm{d}x \\ &= \int_{x_1}^{x_2} \delta f(y,y',x)\mathrm{d}x \\ &= \int_{x_1}^{x_2} (\frac{\partial f}{\partial y}\delta y + \frac{\partial f}{\partial y'}\delta y')\mathrm{d}x.\end{aligned} \tag{C.9}$$

The second term in equation (C.9) can be integrated by parts:

$$\begin{aligned}\int_{x_1}^{x_2} \frac{\partial f}{\partial y'}\delta y'\mathrm{d}x &= \frac{\partial f}{\partial y'}\delta y\Big|_{x_1}^{x_2} - \int_{x_1}^{x_2} \frac{\mathrm{d}}{\mathrm{d}x}(\frac{\partial f}{\partial y'})\delta y\mathrm{d}x, \\ &= -\int_{x_1}^{x_2} \frac{\mathrm{d}}{\mathrm{d}x}(\frac{\partial f}{\partial y'})\delta y(x)\mathrm{d}x \quad \text{(for a small } \delta \text{ variation)}.\end{aligned} \tag{C.10}$$

Using equations (C.9) and (C.10), we can now express

$$\delta J(y) = \int_{x_1}^{x_2} \left[\frac{\partial f}{\partial y} - \frac{\mathrm{d}}{\mathrm{d}x}(\frac{\partial f}{\partial y'})\right]\delta y(x)\mathrm{d}x. \tag{C.11}$$

Therefore, for an arbitrary $\delta y(x)$, $\delta J = 0$ only if

$$\frac{\partial f}{\partial y} - \frac{\mathrm{d}}{\mathrm{d}x}(\frac{\partial f}{\partial y'}) = 0. \tag{C.12}$$

This is the Euler-Lagrange partial differential equation that must be satisfied by the functional $f(y,y',x)$, and whose solution will result in the optimum function $y = y(x)$. Equation (C.12) can be generalised for several (n) variables $\{y_i, y_i'\}$:

$$\frac{\partial f}{\partial y_i} - \frac{\mathrm{d}}{\mathrm{d}x}(\frac{\partial f}{\partial y_i'}) = 0, \quad i = 1,2,\ldots,n. \tag{C.13}$$

This method of the calculus of variations is known as the Euler-Lagrange variational principle.

Note that the Euler-Lagrange variational principle provides the *necessary condition* for the integral J to be stationary. It does not tell whether the integral J is at its minimum or maximum, or is at a saddle point, for the optimum function $y = y(x)$ obtained from solving the Euler-Lagrange equation (C.12). Additional physical requirement(s) must be invoked to decide whether J is a lower bound or an upper bound. In some situations, it is possible to obtain both a lower bound and an upper bound, thus deriving complementary variational principles [see, Arthurs (1970)].

[1]Note: $\int \delta f \mathrm{d}x = \delta \int f \mathrm{d}x - [\int \frac{\mathrm{d}\delta}{\mathrm{d}x}\mathrm{d}x]\Big|_{x_1}^{x_2} \int f \mathrm{d}x = \delta \int f \mathrm{d}x - \delta\Big|_{x_1}^{x_2} \int f \mathrm{d}x = \delta \int f \mathrm{d}x + 0.$

D Introduction to Matrix Eigensolutions

D.1 SPECIAL TYPE OF LINEAR INHOMOGENEOUS EQUATIONS

In this appendix we will consider a system of inhomogeneous linear equations of the form

$$DA = \lambda A, \tag{D.1}$$

where D is a square matrix, $\boldsymbol{A}$ is a vector and λ is a constant (or, a scale factor). Considering D as a $n \times n$ matrix and $\boldsymbol{A}$ as a $n \times 1$ matrix (*i.e.* a column vector), we can express this equation in detail, as follows

$$\begin{pmatrix} d_{11} & d_{12} & \cdots & d_{1n} \\ d_{21} & d_{22} & \cdots & d_{2n} \\ \cdots & \cdots & \cdots & \cdots \\ d_{n1} & d_{n2} & \cdots & d_{nn} \end{pmatrix} \begin{pmatrix} a_1 \\ a_2 \\ \cdot\cdot \\ a_n \end{pmatrix} = \lambda \begin{pmatrix} a_1 \\ a_2 \\ \cdot\cdot \\ a_n \end{pmatrix}. \tag{D.2}$$

or

$$\begin{aligned} d_{11}a_1 + d_{12}a_2 + \cdots + d_{1n}a_n &= \lambda a_1, \\ d_{21}a_1 + d_{12}a_2 + \cdots + d_{1n}a_n &= \lambda a_2, \\ &\vdots \\ d_{21}a_1 + d_{12}a_2 + \cdots + d_{1n}a_n &= \lambda a_n, \end{aligned} \tag{D.3}$$

where d_{ij} and a_i are the elements of D and $\boldsymbol{A}$, respectively.

Equation (D.1) is called a matrix eigenvalue equation. For non-trivial solutions (*i.e.* for $\boldsymbol{A} \neq \boldsymbol{0}$), λ represents the *eigenvalues*, *characteristic values* or *latent roots* of the matrix D, and $\boldsymbol{A}$ represents the *eigenvectors* or *characteristic vectors* of the matrix D. Eigenvalues are obtained by solving the *characteristic* (or, *secular*) equation

$$|D - \lambda I| = 0, \tag{D.4}$$

where I is a $n \times n$ unit matrix. Eigenvectors are solutions of $\boldsymbol{A}$ belonging to eigenvalues λ.

It is important to note that for a $n \times n$ matrix D, there will be n number of eigenvalues which we may label as $\lambda_1, \lambda_2, , \lambda_3, ..., \lambda_n$. and there will be n number of eigenvectors which we may label as $\boldsymbol{A}_1, \boldsymbol{A}_2, \boldsymbol{A}_3, ..., \boldsymbol{A}_n$. An eigenvector $\boldsymbol{A}_i$ will correspond to the eigenvalue λ_i. Finally, the size of each eigenvector (*i.e.* the number of components of each of $\boldsymbol{A}_i$) will be n. We may, therefore, express the ith eigenvector $\boldsymbol{A}_i$ as $\{A_{ij}\}$, with component indices $j = 1, 2, 3, ..., n$. Eigenvectors of distinct eigenvalues are linearly independent.

We will provide a proof for the linear independence of eigenvectors. We will also present an example of evaluation of eigenvalues and eigenvectors.

D.1.1 LINEAR INDEPENDENCE OF EIGENVECTORS

Let us consider D as a 3×3 matrix whose eigenvalues are $\lambda_1, \lambda_2, \lambda_3$ and eigenvectors are $\boldsymbol{A}_1, \boldsymbol{A}_2, \boldsymbol{A}_3$.

To prove that $\boldsymbol{A}_1$, $\boldsymbol{A}_2$ and $\boldsymbol{A}_3$ are linearly independent we should prove that the relationship

$$c_1\boldsymbol{A}_1 + c_2\boldsymbol{A}_2 + c_3\boldsymbol{A}_3 = \boldsymbol{0}, \tag{D.5}$$

DOI: 10.1201/9781003383314-D

to hold, where c_1, c_2, c_3 are a set of constants, requires $c_1 = c_2 = c_3 = 0$. To prove this we will assume

$$c_1\boldsymbol{A}_1 + c_2\boldsymbol{A}_2 + c_3\boldsymbol{A}_3 = \boldsymbol{0}$$

and then show that $c_1 = c_2 = c_3 = 0$.

Let us operate with the matrix $(D - \lambda_1 I)$ on both sides of equation (D.5)

$$(D - \lambda_1 I)(c_1\boldsymbol{A}_1 + c_2\boldsymbol{A}_2 + c_3\boldsymbol{A}_3) = \boldsymbol{0}. \tag{D.6}$$

Using the eigenvalue equations $D\boldsymbol{A}_1 = \lambda_1\boldsymbol{A}_1$, $D\boldsymbol{A}_2 = \lambda_2\boldsymbol{A}_2$, and $D\boldsymbol{A}_3 = \lambda_3\boldsymbol{A}_3$, equation (D.6) can be expressed as

$$c_2(\lambda_2 - \lambda_1)\boldsymbol{A}_2 + c_3(\lambda_3 - \lambda_1)\boldsymbol{A}_3 = \boldsymbol{0}. \tag{D.7}$$

Now we operate with the matrix $(D - \lambda_2 I)$ on both sides of equation (D.7)

$$(D - \lambda_2 I)\Big(c_2(\lambda_2 - \lambda_1)\boldsymbol{A}_2 + c_3(\lambda_3 - \lambda_1)\boldsymbol{A}_3\Big) = \boldsymbol{0}. \tag{D.8}$$

and express it as

$$c_3(\lambda_3 - \lambda_1)(\lambda_3 - \lambda_2)\boldsymbol{A}_3 = \boldsymbol{0}. \tag{D.9}$$

As $\boldsymbol{A}_3$ is a nonzero vector and $\lambda_1, \lambda_2, \lambda_3$ are distinct eigenvalues, we have

$$(\lambda_3 - \lambda_1)(\lambda_3 - \lambda_2) \neq 0$$

and thus from equation (D.9) we obtain

$$c_3 = 0.$$

With this equation (D.7) reads

$$c_2(\lambda_2 - \lambda_1)\boldsymbol{A}_2 = \boldsymbol{0}.$$

As $\boldsymbol{A}_2$ is a nonzero vector and $\lambda_2 \neq \lambda_1$, this gives

$$c_2 = 0.$$

With $c_2 = c_3 = 0$ and assuming that $\boldsymbol{A}_1 \neq 0$, from equation (D.5) we obtain

$$c_1 = 0.$$

Thus, we have proven that equation (D.5) holds true provided that $c_1 = c_2 = c_3 = 0$. In other words, we have proven that eigenvectors of distinct eigenvalues of a square matrix are linearly independent.

D.1.2 EXAMPLE OF EVALUATION OF EIGENVALUES AND EIGENVECTORS

Let us consider equation (D.1) with the following example for the matrix

$$D = \begin{pmatrix} 2 & 2 & -2 \\ 1 & 3 & 1 \\ 1 & 2 & 2 \end{pmatrix}. \tag{D.10}$$

We will determine the eigenvalues of the matrix D and an eigenvector corresponding to each eigenvalue.

As the size of the matrix D is 3×3, there will be three eigenvalues which we will label λ_1, λ_2 and λ_3. And for each of these eigenvalues there will be three eigenvectors. We will label the three eigenvectors as $A_{i,1}$, $A_{i,2}$ and $A_{i,3}$ corresponding to the eigenvalues λ_i, $i = 1, 2, 3$.

It is easy to show that the characteristic equation $|D - \lambda I| = 0$ can be expressed as

$$(2 - \lambda)(\lambda - 1)(\lambda - 4) = 0. \tag{D.11}$$

The three solutions of this equation are

$$\lambda_1 = 2, \quad \lambda_2 = 1, \quad \text{and} \quad \lambda_3 = 4. \tag{D.12}$$

For $\lambda_1 = 2$, we write the equation $(D - \lambda_1 I)\boldsymbol{A}_1 = \mathbf{0}$ as

$$\begin{pmatrix} 0 & 2 & -2 \\ 1 & 1 & 1 \\ 1 & 2 & 0 \end{pmatrix} \begin{bmatrix} A_{11} \\ A_{12} \\ A_{13} \end{bmatrix} = \begin{bmatrix} 0 \\ 0 \\ 0 \end{bmatrix}. \tag{D.13}$$

Equating the vectors on the two sides of the equation, we obtain

$$A_{11} = -2A_{12} \quad \text{and} \quad A_{12} == A_{13}. \tag{D.14}$$

From these, we can express the simplest form of the vector $\boldsymbol{A}_1$ as

$$\begin{bmatrix} A_{11} \\ A_{12} \\ A_{13} \end{bmatrix} = \begin{bmatrix} -2 \\ 1 \\ 1 \end{bmatrix}. \tag{D.15}$$

For $\lambda_2 = 1$, we write the equation $(D - \lambda_2 I)\boldsymbol{A}_2 = \mathbf{0}$ as

$$\begin{pmatrix} 1 & 2 & -2 \\ 1 & 2 & 1 \\ 1 & 2 & 1 \end{pmatrix} \begin{bmatrix} A_{21} \\ A_{22} \\ A_{23} \end{bmatrix} = \begin{bmatrix} 0 \\ 0 \\ 0 \end{bmatrix}. \tag{D.16}$$

From this we obtain

$$A_{21} = 0 \quad \text{and} \quad A_{22} == A_{23}. \tag{D.17}$$

The eigenvector $\boldsymbol{A}_2$ is thus

$$\begin{bmatrix} A_{21} \\ A_{22} \\ A_{23} \end{bmatrix} = \begin{bmatrix} -2 \\ 1 \\ 0 \end{bmatrix}. \tag{D.18}$$

For $\lambda_3 = 4$, we write the equation $(D - \lambda_3 I)\boldsymbol{A}_3 = \mathbf{0}$ as

$$\begin{pmatrix} -2 & 2 & -2 \\ 1 & -1 & 1 \\ 1 & 2 & -2 \end{pmatrix} \begin{bmatrix} A_{31} \\ A_{32} \\ A_{33} \end{bmatrix} = \begin{bmatrix} 0 \\ 0 \\ 0 \end{bmatrix}. \tag{D.19}$$

From this

$$A_{31} = 0 \quad \text{and} \quad A_{32} == A_{33}. \tag{D.20}$$

The eigenvector $\boldsymbol{A}_3$ is thus

$$\begin{bmatrix} A_{31} \\ A_{32} \\ A_{33} \end{bmatrix} = \begin{bmatrix} 0 \\ 1 \\ 1 \end{bmatrix}. \tag{D.21}$$

E Reciprocal Lattice and Brillouin Zone

In section 4.5 it was mentioned that the dynamical behaviour of an infinitely periodic three-dimensional system is discussed by examining the relationship between the frequency ω and the wavevector $\boldsymbol{q}$ of travelling waves. Results from scattering based experimental techniques in condensed matter are also usually interpreted using the momentum space language. Similar to the concept of real space lattice, we develop the concept of a lattice in momentum space, called **reciprocal lattice**. The two spaces are conjugate to each other, with a mapping achieved by the mathematical relation

$$\exp(i\boldsymbol{G}.\boldsymbol{T}) = 1, \tag{E.1}$$

where $\boldsymbol{T}$ is a translation vector in direct space and $\boldsymbol{G}$ is a translation vector in reciprocal space. The reciprocal lattice corresponding to a given real space lattice can be constructed both geometrically and mathematically, in conformity with the relation in equation (E.1).

Let us view a real space lattice as a network of interpenetrating planes. Starting from a lattice point, imagine drawing normals to all possible planes. If a plane is at a distance d, then represent this plane by a new point at a distance $1/d$ along its normal. A collection of an infinite number of new points thus assembled produces the reciprocal lattice. This is the graphical, or geometrical, approach for constructing the reciprocal lattice corresponding to a direct lattice.

For the mathematical construction of the reciprocal lattice corresponding to a direct lattice, we collect lattice points at the tips of all possible vectors

$$\boldsymbol{G} = m_1\boldsymbol{b}_1 + m_2\boldsymbol{b}_2 + m_3\boldsymbol{b}_3, \tag{E.2}$$

where $\{m_j\}$ are any integers and $\{\boldsymbol{b}_j\}$ are the primitive translation vectors of the reciprocal lattice. Due to translational symmetry of the reciprocal lattice, the relationship between the frequency ω and the wavevector $\boldsymbol{q}$ of travelling waves need only be examined for values of $\boldsymbol{q}$ lying within the smallest region spanned by $\boldsymbol{b}_1$, $\boldsymbol{b}_2$ and $\boldsymbol{b}_3$. If $\{\boldsymbol{a}_j\}$ are the primitive translation vectors of the direct lattice, then the primitive translation vectors of the reciprocal lattice are defined as

$$\begin{aligned} \boldsymbol{b}_1 &= \frac{2\pi}{\Omega}(\boldsymbol{a}_2 \times \boldsymbol{a}_3), \\ \boldsymbol{b}_2 &= \frac{2\pi}{\Omega}(\boldsymbol{a}_3 \times \boldsymbol{a}_1), \\ \boldsymbol{b}_3 &= \frac{2\pi}{\Omega}(\boldsymbol{a}_1 \times \boldsymbol{a}_2), \end{aligned} \tag{E.3}$$

where

$$\Omega = |\boldsymbol{a}_1 \cdot (\boldsymbol{a}_2 \times \boldsymbol{a}_3)| = |\boldsymbol{a}_2 \cdot (\boldsymbol{a}_3 \times \boldsymbol{a}_1)| = |\boldsymbol{a}_3 \cdot (\boldsymbol{a}_1 \times \boldsymbol{a}_2)| \neq 0 \tag{E.4}$$

is the volume of the primitive unit cell of the direct lattice. It can be shown that the vectors $\{\boldsymbol{b}_i\}$ satisfy the relation $\boldsymbol{a}_i \cdot \boldsymbol{b}_j = 2\pi\delta_{ij}$, where δ_{ij} is the Kronecker delta symbol ($\delta_{ij} = 1$ when $i = j$ and $\delta_{ij} = 0$ when $i \neq j$).

A general translation vector of the direct lattice can be written as

$$\boldsymbol{T} = n_1\boldsymbol{a}_1 + n_2\boldsymbol{a}_2 + n_3\boldsymbol{a}_3, \tag{E.5}$$

with $\{n_j\}$ as any integers. Using equations (E.2)-(E.5), it is easily noted that the mathematical relation in equation (E.1) is verified, *viz* $\exp(i\boldsymbol{G} \cdot \boldsymbol{T}) = \exp(2\pi Ni) = 1$, where N is an integer.

DOI: 10.1201/9781003383314-E

The primitive unit cell of a direct lattice is the parallelopiped defined by its primitive translation vectors $\boldsymbol{a}_1$, $\boldsymbol{a}_2$ and $\boldsymbol{a}_3$. However, it is customary to construct the primitive unit cell of a reciprocal lattice in the form of Wigner-Seitz cell and refer to that as the central, or first, **Brillouin zone**. By default, the central Brillouin zone is simply called the Brillouin zone. This is the smallest volume enclosed by the bisector planes to the reciprocal lattice vectors.

The discussion above is for three-dimensional lattices. We will make some modifications when dealing with two- and one-dimensional lattices.

FCC lattice:
The primitive translation vectors of the three-dimensional FCC lattice, shown in figure 5.9, may be taken as

$$\begin{aligned} \boldsymbol{a}_1 &= \frac{a}{2}(0,1,1), \\ \boldsymbol{a}_2 &= \frac{a}{2}(1,0,1), \\ \boldsymbol{a}_3 &= \frac{a}{2}(1,1,0), \end{aligned} \tag{E.6}$$

where a is the cubic lattice constant. Using equation (E.3), the primitive translation vectors of the reciprocal lattice are

$$\begin{aligned} \boldsymbol{b}_1 &= \frac{2\pi}{a}(-1,1,1), \\ \boldsymbol{b}_2 &= \frac{2\pi}{a}(1,-1,1), \\ \boldsymbol{b}_3 &= \frac{2\pi}{a}(1,1,-1). \end{aligned} \tag{E.7}$$

Using equations (E.2) and (E.7), a general reciprocal translation vector can be expressed as

$$\boldsymbol{G} = \frac{2\pi}{a}(-m_1+m_2+m_3, m_1-m_2+m_3, m_1+m_2-m_3). \tag{E.8}$$

Assigning positive, zero and negative integer values to $\{m_j\}$, we obtain $(\frac{2\pi}{a})(\pm1,\pm1,\pm1)$ as the eight shortest and $\pm(\frac{2\pi}{a})(2,0,0)$, $\pm(\frac{2\pi}{a})(0,2,0)$, $\pm(\frac{2\pi}{a})(0,0,2)$ as the six next-shortest $\boldsymbol{G}$ vectors. Plane perpendicularly bisecting these fourteen vectors generates the Brillouin zone in the shape of a truncated octahedron as shown in figure 5.10.

Square lattice:
Figure 5.2(b) shows a square lattice. Being a two-dimensional lattice, a square lattice has only two primitive translation vectors $\boldsymbol{a}_1$ and $\boldsymbol{a}_2$. To proceed with the scheme for obtaining reciprocal lattice vectors, we would need to artificially construct the third primitive translation vector $\boldsymbol{a}_3$. This can be done by invoking an artificial periodicity normal to the lattice plane such that the third primitive translation vector is of a very large magnitude and points in a direction normal to the lattice plane.

For a square lattice in the x-y plane, we take

$$\begin{aligned} \boldsymbol{a}_1 &= a(1,0,0), \\ \boldsymbol{a}_2 &= a(0,1,0), \\ \boldsymbol{a}_3 &= c(0,0,1), \end{aligned} \tag{E.9}$$

where a is the square side length and $c \to \infty$ is an artificially chosen very large separation between two neighbouring lattice planes along the z axis.

Using equations (E.9) and (E.3), the primitive translation vectors of the corresponding reciprocal lattice are

$$\boldsymbol{b}_1 = \frac{2\pi}{a}(1,0,0),$$

$$\begin{aligned} \boldsymbol{b}_2 &= \frac{2\pi}{a}(0,1,0), \\ \boldsymbol{b}_3 &= \frac{2\pi}{c}(0,0,1). \end{aligned} \tag{E.10}$$

The in-plane vectors in equations (E.9) and (E.10) can also be written as

$$\begin{aligned} \boldsymbol{a}_1 &= a(1,0), \\ \boldsymbol{a}_2 &= a(0,1), \end{aligned} \tag{E.11}$$

and

$$\begin{aligned} \boldsymbol{b}_1 &= \frac{2\pi}{a}(1,0), \\ \boldsymbol{b}_2 &= \frac{2\pi}{a}(0,1), \end{aligned} \tag{E.12}$$

which is what has been adopted in section 5.2.1. Figure 5.2(c) shows the square Brillouin zone and its principal symmetry directions.

It is worth mentioning that the consideration of infinitely large $\boldsymbol{a}_3$ means that we are describing the physics of a single two-dimensional lattice, as there is essentially no interaction between neighbouring two-dimensional lattices. The corresponding reciprocal lattice is also essentially two-dimensional, as the length of $\boldsymbol{b}_3$ is infinitesimally small.

Hexagonal lattice:
Figure 5.4(b) shows a hexagonal lattice in the x-y plane, with one choice for its primitive translation vectors $\boldsymbol{a}_1$ and $\boldsymbol{a}_2$. As we did for the square lattice, we invoke an artificial periodicity normal to the lattice plane and set up a third primitive translation vector $\boldsymbol{a}_3$ of a very large magnitude and points in a direction normal to the lattice plane. Thus we express

$$\begin{aligned} \boldsymbol{a}_1 &= a(1,0,0), \\ \boldsymbol{a}_2 &= a(\frac{1}{2},\frac{\sqrt{3}}{2},0), \\ \boldsymbol{a}_3 &= c(0,0,1), \end{aligned} \tag{E.13}$$

where a is the distance between nearest lattice points in the plane and $c \to \infty$ is an artificially chosen very large separation between two neighbouring lattice planes along the z axis.

Using equations (E.9) and (E.3), the primitive translation vectors of the corresponding reciprocal lattice are

$$\begin{aligned} \boldsymbol{b}_1 &= \frac{2\pi}{a}(1,-\frac{1}{\sqrt{3}},0), \\ \boldsymbol{b}_2 &= \frac{2\pi}{a}(0,\frac{2}{\sqrt{3}},0), \\ \boldsymbol{b}_3 &= \frac{2\pi}{c}(0,0,1). \end{aligned} \tag{E.14}$$

As $c \to \infty$, the magnitude of $\boldsymbol{b}_3$ is infinitesimally small and we can proceed to express $\boldsymbol{b}_1$ and $\boldsymbol{b}_2$ as the two primitive translation vectors of the reciprocal lattice in the form $\boldsymbol{b}_1 = \frac{2\pi}{a}(1,-\frac{1}{\sqrt{3}})$ and $\boldsymbol{b}_2 = \frac{2\pi}{a}(0,\frac{2}{\sqrt{3}})$ as stated in section 5.2.2. Figure 5.4(c) shows the hexagonal Brillouin zone and its principal symmetry directions.

One-dimensional lattice:
Figure 4.5(a) shows the one-dimensional lattice with one choice of its primitive unit cell. In section 4.5 we expressed its primitive translation vector of magnitude a along the line of the lattice points.

Here we extend the scheme described above for two-dimensional lattices and artificially construct the second and third primitive translation vectors along directions perpendicular to the line of the lattice points and of infinitely large magnitudes. In other words, we choose three primitive translation vectors as

$$\begin{aligned} \boldsymbol{a}_1 &= a(1,0,0), \\ \boldsymbol{a}_2 &= b(0,1,0), \\ \boldsymbol{a}_3 &= c(0,0,1), \end{aligned} \tag{E.15}$$

where $b \to \infty$ and $c \to \infty$.

Using equations (E.9) and (E.3), the primitive translation vectors of the corresponding reciprocal lattice are

$$\begin{aligned} \boldsymbol{b}_1 &= \frac{2\pi}{a}(1,0,0), \\ \boldsymbol{b}_2 &= \frac{2\pi}{b}(0,1,0), \\ \boldsymbol{b}_3 &= \frac{2\pi}{c}(0,0,1). \end{aligned} \tag{E.16}$$

As $b \to \infty$ and $c \to \infty$, the vectors $\boldsymbol{b}_2$ and $\boldsymbol{b}_3$ become too small to be of physical importance and we are left with only $\boldsymbol{b}_1$ as the only meaningful primitive translation vector. As we have a single vector and a single direction to describe the system, we can safely drop the suffix 1 and express $\boldsymbol{b}_1$ as simply $\boldsymbol{b} = \frac{2\pi}{a}\hat{\boldsymbol{x}}$, where $\hat{\boldsymbol{x}}$ is a unit vector along the line of the lattice points. This is what has been used in section 4.5.

The perpendicular bisector planes to the lattice line are the $\{y\text{-}z\}$ planes at the tips of the vectors $\pm\frac{1}{2}\boldsymbol{b}$. Thus the Brillouin zone is the region between $-\frac{1}{2}\boldsymbol{b}$ and $\frac{1}{2}\boldsymbol{b}$, as shown in figure 5.4(c).

F Numerical Solution of Differential Equations

For obtaining numerical solution a differential equation is basically converted to a *finite difference equation*. We will explain this by considering first a first-order differential equation and then a second-order differential equation.

F.1 FIRST-ORDER DIFFERENTIAL EQUATIONS

Let us consider a first-order differential equation

$$\frac{\mathrm{d}y}{\mathrm{d}x} = f(x,y), \tag{F.1}$$

where x is a single independent variable and y is given an initial condition $y = y_0$ for $x = x_0$. The general idea behind the finite difference equation method is to compute the value $y = y_n$ at $x = x_n = x_0 + nh$, where n is *iteration number* and h is sufficiently small. To see how this can be done, let us expand $y(x)$ using Taylor's theorem

$$y(x+h) = y(x) + h\frac{\mathrm{d}y}{\mathrm{d}x}|_{x_0} + \frac{1}{2}h^2\frac{\mathrm{d}^2y}{\mathrm{d}x^2}|_{x_0} + \cdots. \tag{F.2}$$

F.1.1 FIRST-ORDER METHOD

Euler algorithm:
Euler's algorithm only considers the first two terms on the right-hand side of equation (F.2), meaning that it is a first-order method. In this scheme, then

$$\begin{aligned} y(x_0+h) &\approx y(x_0) + h\frac{\mathrm{d}y}{\mathrm{d}x}|_{x_0}, \\ &= y(x_0) + hf(x_0,y_0). \end{aligned} \tag{F.3}$$

With this, we set up a system of difference equations relating the increment in y to the increment in x at each step

$$\begin{aligned} x_{n+1} &= x_n + h, \\ y_{n+1} &= y_n + hf(x_n,y_n). \end{aligned} \tag{F.4}$$

The iteration process is stopped when convergence is achieved, *i.e.* after sufficiently large n for which $|y_{n+1} - y_n|$ becomes acceptably negligible.

F.1.2 HIGHER-ORDER METHODS

As noted earlier, the Euler method of integrating a first-order differential equation is first-order, as it contains h^p with $p = 1$ as the largest order in the small increment h. Higher-order integration methods will contain terms up to h^p, with $p > 1$.

DOI: 10.1201/9781003383314-F

Second-order Runge-Kutta algorithms:
An example is the second-order *Runge-Kutta* method. Let us consider equation (F.2) up to the second order

$$y(x+h) = y(x) + h\frac{\mathrm{d}y}{\mathrm{d}x}|_{x_0} + \frac{1}{2}h^2\frac{\mathrm{d}^2y}{\mathrm{d}x^2}|_{x_0} + \mathcal{O}(h^3). \quad \text{(F.5)}$$

Expressing

$$\frac{\mathrm{d}^2y}{\mathrm{d}x^2} = \frac{\mathrm{d}f(y,x)}{\mathrm{d}x} = \frac{\partial f}{\partial x} + \frac{\partial f}{\partial y}\frac{\mathrm{d}y}{\mathrm{d}x} = \frac{\partial f}{\partial x} + f\frac{\partial f}{\partial y},$$

and following the discussion in the previous sub-section, we can express equation (F.5) as the following difference scheme

$$\begin{aligned} y_{n+1} &= y_n + hf(x_n,y_n) + \frac{h^2}{2}g(x_n,y_n), \\ \text{where} \quad g(x,y) &= \frac{\partial f(x,y)}{\partial x} + f(x,y)\frac{\partial f(x,y)}{\partial y}. \end{aligned} \quad \text{(F.6)}$$

Let us now think of expressing the solution y_{n+1} as follows

$$\begin{aligned} y_{n+1} &= y_n + ak_1 + bk_2, \\ k_1 &= hf(x_n,y_n), \\ k_2 &= hf(x_n+\alpha h, y_n+\beta k_1), \end{aligned} \quad \text{(F.7)}$$

where a,b,α and β are four constants. The choice $a = 1$ and $k_2 = 0$ leads to Euler's first-order scheme. In order for equation (F.7) to retain terms up to h^2, we can expand the term k_2 as

$$\begin{aligned} k_2 &= hf(x_n+\alpha h, y_n+\beta k_1), \\ &\approx h\Big(f(x_n,y_n) + \alpha h\frac{\partial f}{\partial x}(x_n,y_n) + \beta k_1\frac{\partial f}{\partial x}(x_n,y_n)\Big), \end{aligned} \quad \text{(F.8)}$$

and substitute this in equation (F.7) to get

$$y_{n+1} = y_n + (a+b)hf(x_n,y_n) + bh^2(\alpha\frac{\partial f}{\partial x} + \beta f\frac{\partial f}{\partial y})(x_n,y_n). \quad \text{(F.9)}$$

Comparison of the terms with identical coefficients in equations (F.6) and (F.9) produces the conditions

$$a+b = 1; \quad \alpha b = \frac{1}{2}; \quad \beta b = \frac{1}{2}. \quad \text{(F.10)}$$

There are many choices of a,b,α and β that satisfy the conditions in equation (F.10). The choice $a = b = \frac{1}{2}$ and $\alpha = \beta = 1$ produces the well-known second-order Runge-Kutta iterative scheme

$$\begin{aligned} y_{n+1} &= y_n + \frac{1}{2}(k_1+k_2), \\ k_1 &= hf(x_n,y_n), \\ k_2 &= hf(x_n+h, y_n+k_1) \quad \text{(RK2 – version 1)}. \end{aligned} \quad \text{(F.11)}$$

Another choice is $a = 0, b = 1$ and $\alpha = \beta = \frac{1}{2}$. This results in the following form of the second-order Runge-Kutta iterative scheme

$$\begin{aligned} y_{n+1} &= y_n + k_2, \\ k_1 &= hf(x_n,y_n), \\ k_2 &= hf(x_n+\frac{1}{2}h, y_n+\frac{1}{2}k_1) \quad \text{(RK2 – version 2)}. \end{aligned} \quad \text{(F.12)}$$

This version (RK2) amounts to using a modified Euler method. To understand this, we note that in the Euler method the equation $y_{n+1} = y_n + hf(x_n,y_n)$ extrapolates the solution y_n to the next iteration by evaluating the slope $f(x_n, y_n)$ at the point (x_n, y_n). In the RK2 method the extrapolation is achieved by evaluating the slope at the mid-point $(x_n + \frac{1}{2}h, y_n + \frac{1}{2}k_1)$. where $k_1 = hf(x_n, y_n)$.

The procedure described above can be extended to derive even higher-order Runge-Kutta schemes. Without going into details, we will simply list the commonly used third-order (RK3) and fourth-order (RK4) Runge-Kutta algorithms. For details the reader may wish to follow text books on numerical analysis, amongst others, by Ralston and Rabinowitz (1965), Lambert (1973), Dodes (1978), Gould and Tobochnik (1988), Atkinson *et al.* (1989), Press *et al.* (1992) and Flowers (1996).

Third-order Runge-Kutta algorithms:

$$\begin{aligned} y_{n+1} &= y_n + \frac{1}{6}k_1 + \frac{2}{3}k_2 + \frac{1}{6}k_3, \\ k_1 &= hf(x_n, y_n), \\ k_2 &= hf(x_n + \frac{1}{2}h, y_n + \frac{1}{2}k_1), \\ k_3 &= hf(x_n + h, y_n - k_1 + 2k_2) \quad \text{(RK3 – version 1)}. \end{aligned} \tag{F.13}$$

or

$$\begin{aligned} y_{n+1} &= y_n + \frac{1}{4}k_1 + \frac{3}{4}k_3, \\ k_1 &= hf(x_n, y_n), \\ k_2 &= hf(x_n + \frac{1}{3}h, y_n + \frac{1}{3}k_1), \\ k_3 &= hf(x_n + \frac{2}{3}h, y_n + \frac{2}{3}k_2) \quad \text{(RK3 – version 2)}. \end{aligned} \tag{F.14}$$

Fourth-order Runge-Kutta algorithms:

$$\begin{aligned} y_{n+1} &= y_n + \frac{1}{6}k_1 + \frac{1}{3}k_2 + \frac{1}{3}k_3 + \frac{1}{6}k_4, \\ k_1 &= hf(x_n, y_n), \\ k_2 &= hf(x_n + \frac{1}{2}h, y_n + \frac{1}{2}k_1), \\ k_3 &= hf(x_n + \frac{1}{2}h, y_n + \frac{1}{2}k_2), \\ k_4 &= hf(x_n + h, y_n + k_3) \quad \text{(RK4 – version 1)}. \end{aligned} \tag{F.15}$$

or

$$\begin{aligned} y_{n+1} &= y_n + \frac{1}{8}k_1 + \frac{3}{8}k_2 + \frac{3}{8}k_3 + \frac{1}{8}k_4, \\ k_1 &= hf(x_n, y_n), \\ k_2 &= hf(x_n + \frac{1}{3}h, y_n + \frac{1}{3}k_1), \\ k_3 &= hf(x_n + \frac{2}{3}h, y_n - \frac{1}{3}k_1 + k_2), \\ k_4 &= hf(x_n + h, y_n + k_1 - k_2 + k_3) \quad \text{(RK4 – version 2)}. \end{aligned} \tag{F.16}$$

F.2 SECOND-ORDER DIFFERENTIAL EQUATIONS

We will consider two types of second-order differential equations.

F.2.1 DIFFERENTIAL EQUATIONS OF TYPE $\frac{d^2Y}{dX^2} = G(X, Y)$

Let us consider a second-order differential equation of the type

$$\frac{d^2y}{dx^2} = g(x, y). \tag{F.17}$$

In order to solve this numerically, we express it as the following two coupled first-order differential equations

$$\frac{dy}{dx} = v, \tag{F.18}$$

$$\frac{dv}{dx} = g(x,y). \tag{F.19}$$

These equations can be numerically solved by using a double set of one of the algorithms discussed above. We will list four algorithms without providing any derivation. Details can be found in text books on numerical analysis, as mentioned earlier.

Euler algorithm:

The Euler algorithm discussed in the previous section can be applied to each of the two equations (F.18) and (F.19). Accordingly, with steps $x_n = x_0 + nh$, where n is iteration number and h is sufficiently small,

$$v_{n+1} = v_n + hg(x_n, y_n), \tag{F.20}$$

$$y_{n+1} = y_n + hv_n. \tag{F.21}$$

Note that at any iteration during computation the variables y and v can be updated independently, *i.e.* either of these can be updated before the other.

Euler-Cromer algorithm:

Cromer (1981) noted that the Euler algorithm for solving second-order differential equations can be made *more stable* if the current, rather than the previous, estimate of v is used in equation (F.21). The procedure, known as the Euler-Cromer algorithm, is thus

$$v_{n+1} = v_n + hg(x_n, y_n), \tag{F.22}$$

$$y_{n+1} = y_n + hv_{n+1}. \tag{F.23}$$

Note that, in contrast to the Euler algorithm, at any iteration during computation the Euler-Cromer algorithm requires the variable v to be updated before the variable y is updated.

Verlet algorithm: The Verlet algorithm expresses the solutions as (Gould and Tobochnik, 1988)

$$y_{n+1} = y_n + hv_n + \frac{h^2}{2} g(x_n, y_n), \tag{F.24}$$

$$v_{n+1} = v_n + \frac{h}{2}\left(g(x_{n+1}, y_{n+1}) + g(x_n, y_n)\right). \tag{F.25}$$

Second-order Runge-Kutta algorithm:

The second-order Runge-Kutta algorithm can be expressed as (Gould and Tobochnik, 1988)

$$y_{n+1} = y_n + k_{2,y}, \tag{F.26}$$

$$v_{n+1} = v_n + k_{2,v}, \tag{F.27}$$

$$k_{1,y} = hv_n, \tag{F.28}$$

$$k_{1,v} = hg(x_n, y_n), \tag{F.29}$$

$$k_{2,y} = h(v_n + \frac{1}{2} k_{1,v}), \tag{F.30}$$

$$k_{2,v} = hg(y_n + \frac{1}{2} k_{1,y}). \tag{F.31}$$

Third-order Runge-Kutta algorithm:

The third-order Runge-Kutta algorithm can be expressed as (Abramowitz and Stegun, 1972)

$$y_{n+1} = y_n + h\big(v_n + \frac{1}{6}(k_1 + 2k_2)\big), \tag{F.32}$$

$$v_{n+1} = v_n + \frac{1}{6}(k_1 + 4k_2 + k3), \tag{F.33}$$

$$k_1 = hf(x_n, y_n), \tag{F.34}$$

$$k_2 = hf\big(x_n + \frac{h}{2}, y_n + \frac{h}{2}, v_n + \frac{h}{8}k_1\big), \tag{F.35}$$

$$k_3 = hf\big(x_n + h, y_n + hv_n + \frac{h}{2}k_2\big). \tag{F.36}$$

Expressions for higher-order Runge-Kutta algorithms can be found in text books.

F.2.2 DIFFERENTIAL EQUATIONS OF TYPE $\frac{d^2Y}{dX^2} = G(X, Y, \frac{dY}{dX})$

Now, let us now consider a general second-order differential equation of the type

$$\frac{d^2y}{dx^2} = g(x, y, \frac{dy}{dx}). \tag{F.37}$$

We express this as the following two coupled first-order differential equations

$$\frac{dy}{dx} = v, \tag{F.38}$$

$$\frac{dv}{dx} = g(x, y, v). \tag{F.39}$$

We will list three numerical algorithms to solve these equations. Details can be found in text books on numerical analysis as mentioned earlier.

Euler algorithm:

With steps $x_n = x_0 + nh$, where n is iteration number and h is sufficiently small, the Euler algorithm is

$$v_{n+1} = v_n + hg(x_n, y_n, v_n), \tag{F.40}$$

$$y_{n+1} = y_n + hv_n. \tag{F.41}$$

Euler-Cromer algorithm:

The Euler-Cromer algorithm is

$$v_{n+1} = v_n + hg(x_n, y_n, v_n), \tag{F.42}$$

$$y_{n+1} = y_n + hv_{n+1}. \tag{F.43}$$

Fourth-order Runge-Kutta algorithm:

The fourth-order Runge-Kutta algorithm can be expressed as (Abramowitz and Stegun, 1972)

$$y_{n+1} = y_n + h\big[v_n + \frac{1}{6}(k_1 + k_2 + k_3)\big], \tag{F.44}$$

$$v_{n+1} = v_n + \frac{1}{6}(k_1 + 2k_2 + 2k_3 + k_4), \tag{F.45}$$

$$k_1 = hf(x_n, y_n, v_n), \tag{F.46}$$

$$k_2 = hf\big(x_n+\frac{h}{2}, y_n+\frac{h}{2}v_n+\frac{h}{8}k_1, v_n+\frac{k_1}{2}\big), \tag{F.47}$$

$$k_3 = hf\big(x_n+\frac{h}{2}, y_n+\frac{h}{2}v_n+\frac{h}{8}k_1, v_n+\frac{k_2}{2}\big), \tag{F.48}$$

$$k_4 = hf\big(x_n+h, y_n+hv_n+\frac{h}{2}k_3, v_n+k_3\big). \tag{F.49}$$

FURTHER READING

1. Abramowitz M and Stegun I A (Eds) 1972 *Handbook of Mathematical Functions – with Formulas, Graphs, and Mathematical Tables* (Dover: New York)
2. Cromer A 1981 *Am J Phys* **49** 455
3. Dodes I A 1978 *Numerical Analysis for Computer Science* (North-Holland: New York)
4. Flowers B H 1996 *An Introduction to Numerical Methods in C++* (Clarendon Press: Oxford)
5. Gould H and Tobochnik J 1988 *An Introduction to Computer Simulation Methods – Application to Physical Systems Part 1* (Addison-Wesley: New York)
6. Press W H, Teukolsky S A, Vetterling W T and Flannery B P 1992 *Numerical Recipes in Fortran 77 – The Art of Scientific Computing* 2nd edn (Cambridge University Press: Cambridge)
7. Ralston A and Rabinowitz P 1978 *A First Course in Numerical Analysis* 2nd edn (McGraw-Hill: London)

G Analytic Solution of Second Order Linear Inhomogeneous Differential Equations

Analytic solution of a general inhomogeneous ordinary differential equation may not always be possible to obtain. In Appendix F we discussed numerical solution of a general second-order ordinary differential equation of type $\frac{d^2y}{dx^2} = g(x, y, \frac{dy}{dx})$ [see, equation (F.37)]. In this appendix we will discuss analytic solution of inhomogeneous second-order linear differential equations of the form

$$a_0 \frac{d^2y}{dx^2} + a_1 \frac{dy}{dx} + a_3 y = f(x), \tag{G.1}$$

where a_1, a_2 and a_3 are constants, and $f(x) \neq 0$. The general solution of equation (G.1) is

$$y(x) = y_c(x) + y_p(x), \tag{G.2}$$

where $y_c(x)$ is the solution (called the complementary function) of the homogeneous part of the differential equation (*viz* with $f(x) = 0$) and $y_p(x)$ is the particular solution (called the particular integral) of the inhomogeneous problem.

For solving the homogeneous second-order linear differential equation

$$a_0 \frac{d^2y}{dx^2} + a_1 \frac{dy}{dx} + a_2 y = 0, \tag{G.3}$$

we substitute

$$y = e^{mx}, \tag{G.4}$$

and obtain the auxiliary equation

$$a_0 m^2 + a_1 m + a_2 = 0. \tag{G.5}$$

As this is a quadratic equation in m, there are two solutions of equation (G.3). Assuming that the two roots are independent, we can write the general solution (the complementary function) of equation (G.3) as

$$y_c(x) = A_1 e^{m_1 x} + A_2 e^{m_2 x}, \tag{G.6}$$

where A_1 and A_2 are arbitrary constants that may be determined if two boundary (or initial) conditions are provided.

The particular solution y_p can be any function that satisfies the inhomogeneous differential equation, provided it is independent of y_c. There are two principal methods of determining a particular solution: the method of undetermined coefficients and the method of variation of parameters. Here we will discuss the method of undetermined coefficients. In this method $y_p(x)$ is considered as a trial function which is dependent on the form of the function $f(x)$ and contains a finite number of arbitrary constants. The trial function is substituted into the differential equation (G.3) and the constants are chosen so that the equation is satisfied. The following three schemes are useful.

DOI: 10.1201/9781003383314-G

(i) If $f(x) = ae^{bx}$, where a and b are constants, and if the auxiliary equation (G.5) has $m = b$ as a root of multiplicity k, then one can take

$$y_p(x) = Ax^k e^{bx}, \tag{G.7}$$

where A is a constant to be determined. Obviously, if $m = b$ is not a root, then $k = 0$.

(ii) If $f(x) = a\sin(bx)$ or $a\cos(bx)$, where a and b are constants, and $(m^2 + b^2)$ occurs k times as a factor of the auxiliary equation, then one can take

$$y_p(x) = x^k[A\sin(bx) + B\cos(bx)], \tag{G.8}$$

where A and B are constants. Again, $k = 0$ if $(m^2 + b^2)$ is not a factor.

(iii) If $f(x) = ax^s$, where a and b are constants, and if $m = 0$ occurs k times as a root of the auxiliary equation, then one can take

$$y_p(x) = x^k(Ax^s + Bx^{s-1} + \cdots + Px + Q), \tag{G.9}$$

where A, B, $\cdots$, P, Q are constants to be determined. If $m = 0$ is not a root, then $k = 0$.

Example

We will consider obtaining analytic solution for the following differential equation, which appears in section 10.2.3 of chapter 10.

$$\frac{d^2y}{dx^2} + y = a\cos(bx), \tag{G.10}$$

with the initial conditions $y = 1$ when $x = 0$ and $\frac{dy}{dx} = 0$ when $x = 0$.

In this case the auxiliary equation is

$$m^2 + 1 = 0, \tag{G.11}$$

so that $m = \pm i$ and the complementary function becomes

$$\begin{aligned} y_c(x) &= A_1 e^{ix} + A_2 e^{-ix}, \\ &= B_1 \cos x + B_2 \sin x \end{aligned} \tag{G.12}$$

The initial conditions $y_c = 1$ and $\frac{dy_c}{dx} = 0$ when $x = 0$ make $B_a = 1$ and $B_2 = 0$. Therefore

$$y_c(x) = \cos x. \tag{G.13}$$

In order to determine the particular solution, we note that $m^2 + b^2 = -1 + b^2 = 0$ coincides with $m^2 + 1$ when $b = 1$, otherwise not. Therefore, we can use the technique (ii) with two separate considerations.

Case 1: $b = 1$.

In this case $k = 1$ and we can write

$$y_p(x) = x[A\sin x + B\cos x]. \tag{G.14}$$

Substitution of equation (G.14) in (G.10) gives $B = 0$ and $A = a/2$, so that the particular solution is

$$y_p(x) = \frac{a}{2} x \sin x. \tag{G.15}$$

Thus the full solution of the differential equation $\frac{d^2y}{dx^2} + y = a\cos x$ is

$$y(x) = \cos x + \frac{a}{2} x \sin x. \tag{G.16}$$

Case 2: $b \neq 1$.
In this case $k = 0$ and we can write

$$y_p(x) = x[A\sin(bx) + B\cos(bx)]. \tag{G.17}$$

Substitution of equation (G.17) in (G.10) gives $A = 0$ and $B = \frac{a}{1-b^2}$, so that the particular solution is

$$y_p(x) = \frac{a}{1-b^2}\cos(bx). \tag{G.18}$$

Thus the full solution of the differential equation $\frac{d^2y}{dx^2} + y = a\cos(bx)$, with $b \neq 1$, is

$$y(x) = \cos x + \frac{a}{1-b^2}\cos(bx). \tag{G.19}$$

Note that the solutions in equations (G.15) and (G.19) are consistent with the well-known result that the solution of the harmonic equation of motion $\frac{d^2y}{dx^2} + y = 0$, with initial conditions $y = 1$ and $\frac{dy}{dx} = 0$ when $x = 0$, is $y = \cos x$.

FURTHER READING

1. Lea S M 2004 *Mathematics for Physicists* (Thompson Learning: London)
2. Riley K F and Hobson M P 2011 *Foundation Mathematics for the Physical Sciences* (Cambridge University Press: Cambridge)
3. Stephenson G 1969 *Mathematical Methods for Science Students* (Longmans: London)

Answers to Selected Problems

Chapter 2

Question 6: The expression for the acceleration of the reel is

$$\ddot{x} = \frac{Pb(b-a)}{m(k^2+b^2)} > 0.$$

This is in the direction of the pull.

Question 8: Lagrange's equations of motion are:

$$m\ddot{r} = mr(\dot{\theta}^2 + \sin^2\theta\dot{\phi}^2) - \frac{\partial V}{\partial r},$$

$$2mr\dot{r}\dot{\theta} + mr^2\ddot{\theta} = mr^2\sin\theta\cos\theta\dot{\phi}^2 - \frac{\partial V}{\partial \theta},$$

and

$$2mr\dot{r}\sin^2\theta\dot{\phi} + 2mr^2\sin\theta\cos\theta\dot{\theta}\dot{\phi} + mr^2\sin^2\theta\ddot{\phi} = -\frac{\partial V}{\partial \phi}.$$

Question 10: Stable point is $x = 0$. The frequency of oscillations is $\omega = \sqrt{\frac{aF_0}{m}}$.

Question 12: (a) Generalised coordinates are: x and α.
(d) For small angle of oscillation $\ddot{\alpha} = -\frac{2g}{2a}\alpha$. The period of oscillations is $T = 2\pi\sqrt{\frac{2a}{3g}}$.

Chapter 3

Question 2: Hamilton's equations of motion are:

$$\begin{aligned} \dot{x} &= \frac{\partial H}{\partial p} = \frac{p}{m_0[1+(\frac{p}{m_0c})^2]^{1/2}}, \\ \dot{p} &= -\frac{\partial H}{\partial x} = -\frac{\mathrm{d}V}{\mathrm{d}x}. \end{aligned}$$

Question 4: The Hamiltonian is

$$H = p_\theta\dot{\theta} - L = \frac{3}{2}\frac{p_\theta^2}{Ml^2} - \frac{Mgl}{2}\cos\theta.$$

Hamilton's equations of motion are:

$$\dot{x} = \frac{\partial H}{\partial p} = \frac{p/m_0}{\sqrt{1+(p/m_0c)^2}}$$

and

$$\dot{p} = -\frac{\partial H}{\partial x} = -\frac{\partial V}{\partial x}.$$

Time period is

$$T = 2\pi\sqrt{\frac{2l}{3g}} = 1.16\text{s}.$$

Question 5: (a) Generalised momentum components are

$$p_x = \frac{\partial L}{\partial \dot{x}} = m(a\dot{x} + b\dot{y})$$

and

$$p_y = \frac{\partial L}{\partial \dot{y}} = m(b\dot{x} + c\dot{y}).$$

(b) Hamilton's equations of motion are

$$\ddot{x} = -\frac{k}{m}x \text{ and } \ddot{y} = -\frac{k}{m}y.$$

These can be written together as

$$\ddot{\boldsymbol{r}}_{||} = -\frac{k}{m}\boldsymbol{r}_{||},$$

which expresses simple harmonic motion in 2D.

Question 6: Generalised potential is

$$V = -\frac{b}{r}(1 + \frac{\dot{r}^2}{c^2}).$$

The Hamiltonian is

$$H = -\frac{b}{r} + \frac{p_r^2}{2(m + \frac{2b}{rc^2})}.$$

Hamilton's equations of motion are

$$\begin{aligned}
\dot{r} &= \frac{\partial H}{\partial p_r} = \frac{p_r}{m + \frac{2b}{rc^2}} \\
\dot{\theta} &= \frac{\partial H}{\partial p_\theta} = \frac{p_\theta}{mr^2} \\
\dot{p}_r &= -\frac{\partial H}{\partial r} = -\frac{b}{r^2} - \frac{bp_r^2}{r^2c^2(m + \frac{2b}{rc^2})^2} + \frac{p_\theta^2}{mr^3} \\
\dot{p}_\theta &= -\frac{\partial H}{\partial \theta} = 0.
\end{aligned}$$

Question 7: The Hamiltonian is

$$H = \frac{p_\varphi^2}{2MR^2} + \frac{p_z^2}{2M} + \frac{k}{2}(R^2 + z^2).$$

Hamilton's equations of motion are

$$\dot{\varphi} = \frac{\partial H}{\partial p_\varphi} = \frac{\varphi}{MR^2},$$

$$\begin{aligned}
\dot{z} &= \frac{\partial H}{\partial p_z} = \frac{p_z}{M}, \\
\dot{p}_\varphi &= -\frac{\partial H}{\partial \varphi} = 0, \\
\dot{p}_z &= -\frac{\partial H}{\partial z} = -kz.
\end{aligned}$$

From these equations we obtain

$$\ddot{z} = -\frac{k}{M}z,$$

which represents simple harmonic motion.

Question 8: Hamilton's equations of motion are:

$$\dot{\theta} = \frac{\partial H}{\partial p_\theta} = \frac{5p_\theta}{7m(R-a)^2}$$

and

$$\dot{p}_\vartheta = -\frac{\partial H}{\partial \theta} = -mg(R-a)\sin\theta.$$

The time period of oscillations is

$$\begin{aligned}
T = \frac{2\pi}{\omega} &= 2\pi\sqrt{\frac{7}{5}\frac{R-a}{g}} \\
&\simeq 2\pi\sqrt{\frac{7}{5}\frac{R}{g}} \quad \text{for } R >> a.
\end{aligned}$$

Question 9: Hamilton's equations of motion are

$$\begin{aligned}
\dot{\theta}_1 &= \frac{\partial H}{\partial p_{\theta_1}} = (p_{\theta_1} - p_{\theta_2})/ml^2, \\
\dot{\theta}_2 &= \frac{\partial H}{\partial p_{\theta_2}} = (2p_{\theta_2} - p_{\theta_1})/ml^2, \\
\dot{p}_{\theta_1} &= -\frac{\partial H}{\partial \theta_1} = -2mgl\theta_1, \\
\dot{p}_{\theta_2} &= -\frac{\partial H}{\partial \theta_2} = -mgl\theta_2.
\end{aligned}$$

The angular accelerations of the two bobs can be expressed as

$$\begin{aligned}
\ddot{\theta}_1 &= (\dot{p}_{\theta_1} - \dot{p}_{\theta_2})/ml^2 = \frac{g}{l}(-2\theta_2 + \theta_2), \\
\ddot{\theta}_2 &= (2\dot{p}_{\theta_2} - \dot{p}_{\theta_1})/ml^2 = \frac{g}{l}(-2\theta_2 + 2\theta_1).
\end{aligned} \tag{G.15}$$

Question 17: $f(P) = \sqrt{2m\omega P}$.
Generating function of type 1: $F_1 = \frac{1}{2}m\omega q^2 \cot Q$.

Question 18: $\alpha = 1/2$ and $\beta = 2$. Generating function of type 4, $F_4 = F_4(p,P)$, is $F_4 = \frac{1}{2}P^2\cot(2p) + \text{constant}$.

Question 19: Hamilton's principal function S is

$$S = A + c_1 x + c_2 y - \frac{1}{3m^2 g}\left(2mE - c_1^2 - c_2^2 - 2m^2 gz\right)^{3/2} - Et,$$

where A is a constant.

Question 20: Hamilton's principal function $S = W - Et$, where W is Hamilton's characteristic function and E is system's total energy. $W = W_\rho + W_\phi$. $W_\phi = C\phi = p_\phi \phi$, where C is a constant.

$$W_\rho = \sqrt{2m}\int \mathrm{d}\rho \sqrt{E - V - \frac{C^2}{2m\rho^2}}.$$

As $p_\phi = \frac{\partial W_\phi}{\partial \phi} = C =$ constant, it immediately follows that ϕ is a constant. Also, as H does not contain ϕ, the angular momentum is conserved.

Chapter 4

Question 2: Because both atoms oscillate in phase with frequency ω_1 without any change in the length of the interatomic spring.

Chapter 5

Question 1: The speed ratio is $\frac{v_{\mathrm{LA}}}{v_{\mathrm{TA}}} = \sqrt{\frac{\Lambda_1+\Lambda_2}{\Lambda_2}}$. As the ratio $\frac{\Lambda_2}{\Lambda_1}$ increases from 0.5 to 1.0, the speed ratio $\frac{v_{\mathrm{LA}}}{v_{\mathrm{TA}}}$ decreases from 1.732 to 1.414.

Question 3: As $\boldsymbol{q} \parallel [1,1]$, neither $\boldsymbol{\varepsilon}_1$ nor $\boldsymbol{\varepsilon}_2$ is longitudinal or transverse. Therefore the titles 'transverse' and 'longitudinal' are purely formal along this direction.

Question 5: The two solutions are:
$\omega_1^2 = \frac{3\Lambda}{2\mu}\left[1 - \sqrt{1 - \frac{4\mu^2}{m_1 m_2}}\right]$ and $\omega_2^2 = \frac{3\Lambda}{2\mu}\left[1 + \sqrt{1 - \frac{4\mu^2}{m_1 m_2}}\right]$.
Given $\Delta f = \frac{\omega_2 - \omega_1}{2\pi} = 3 \times 10^{12}\ \mathrm{s}^{-1}$, we estimate that $\Lambda \approx 139$ N/m.

Chapter 6

Question 2: $\theta_0 = 0.29\pi$

Question 3: $p_\theta = ml^2\dot{\theta} = \sqrt{2ml^2}\left[E - mgl(1-\cos\theta)\right]^{1/2}$.

Question 4: $p_\theta = m\sqrt{2gl^3}\left[1+\cos\theta\right]^{1/2}$.

Question 6: $\omega' = 29.58$ rad/s, or $\nu' = \omega'/(2\pi) = 4.7\ \mathrm{s}^{-1}$.

Question 10:

$$\begin{aligned} E(t) &= \frac{1}{2}mA^2 e^{-2\beta t}\left[(\beta^2+\omega_0^2)\cos^2(\omega' t - \phi) + \omega'^2 \sin^2(\omega' t - \phi)\right. \\ &\quad \left. + 2\beta\omega' \cos(\omega' t - \phi)\sin(\omega' t - \phi)\right]. \end{aligned}$$

Question 12: Both maps look alike, with zero at $x = 0$ and $x = 1$, and maximum at $x = 1/2$. The difference is that the maximum values of the sine and logistic maps are 1 and $1/4$, respectively.

Question 13: For chaotic behaviour $\lambda > 0$, which requires $r > 1/2$.

Chapter 7

Question 11: Refer to the paper by Cheffer *et al.* . Figure 18 in that paper provides a good overview of the differences between normal, flutter and ventricular ECG rhythms.

Question 12: Analytical solution is $V = V_R + I_c R + (V_0 - V_R - I_c R)\exp(-t/\tau)$.

Question 14: Equation (7.75) reduces to a one-dimensional form $C\frac{dV}{dt} = -\frac{(V-V_{\rm rest})(V_{\rm th}-V)}{R(V_{\rm th}-V_{\rm rest})} - W + I_{\rm inj}$ which is a quadratic integrate-and-fire model.

Question 18: There is a single equilibrium point for set 1 (Hopf). There are 3 equilibrium points for $I_{\rm inj} = 0$ and $I_{\rm inj} < 41\ \mu$A for set 2 (SNLC). For $I_{\rm inj} = 41\ \mu$A there is a saddle-equilibrium point and an equilibrium point. For $I_{\rm inj} > 41\ \mu$A there is a single equilibrium point.

Question 20:

$$b = b - 2\eta[(w-1)x^2 + b],$$
$$w = w - 2\eta[(w-1)x^2 + b]x^2.$$

Chapter 8

Question 3: Equation of motion is

$$\rho\frac{\partial^2 u}{\partial t^2} = -Au(x) - Bu^3(x) + Y\frac{\partial^2 u}{\partial x^2}.$$

Question 4: The Hamiltonian density is

$$\mathscr{H} = \frac{p^2}{2} + \frac{1}{2}\left(\frac{\partial\phi}{\partial x}\right)^2 + \frac{1}{2}m^2\phi^2.$$

Hamilton's equations of motion are

$$\begin{aligned}\frac{\partial p}{\partial t} &= -m^2\phi + \frac{\partial^2\phi}{\partial x^2},\\ \frac{\partial^2\phi}{\partial t^2} &= \frac{\partial^2\phi}{\partial x^2} - m^2\phi,\end{aligned}$$

Chapter 9

Question 3: The expectation value is $<0|(\hat{a} - \hat{a}^+)^4|0> = 3$.

Chapter 10

Question 2: The solution, up to order ε^1, is

$$x = \cos z - \left[\frac{1}{24}\cos z_0 - \frac{5}{128}\cos(3z_0) - \frac{1}{384}\cos(5z_0)\right]\varepsilon,$$

with $z = \omega_0(1 + \frac{5}{16}\varepsilon)t$ and $z_0 = \omega_0 t$.

References

Abramowitz M and Stegun I A (Eds) 1972 *Handbook of Mathematical Functions – with Formulas, Graphs, and Mathematical Tables* (Dover: New York)

Aggarwal C C 2023 *Neural Networks and Deep Learning* 2nd edn (Springer: Switzerland)

Ashcroft N W and Mermin N D 1976 *Solid State Physics* (Sounders College: Philadelphia)

Atkinson L V, Harley P J and Hudson J D 1989 *Numerical Methods with Fortran 77 – A Practical Introduction* (Addison Wesley: Wokingham, England)

Azizi T and Mugabi R 2020 *Applied Mathematics* **11** 203

Arthurs A M 1970 *Complementary Variational Principles* (Oxford: Clarendon)

Badrulhisham F, Pogatzki-Zahn E, Segelcke D, Spisak T and Vollert J 2024 *Brain, Behaviour, and Immunity* **115** 470

Barger V and Olsson M G 1973 *Classical Mechanics: A Modern Perspective* (McGraw Hill: New York)

Berestetskii V B, Lifshitz L M and Pitaevskii L P 1999 *Quantum Electrodynamics* (Butterworth Heinemann: Oxford)

Born M and Huang K 1954 *Dynamical Theory of Crystal Lattices* (oxford: Clarendon)

Boussinesq J 1903 *Théorie Analytique de la Chaleur* (Gauthier-Villars: Paris)

Bransden B H and Joachain C J 2000 *Quantum Mechanics* 2nd edn (Prentice Hall: New York)

Broyden C G 1965 *Math Comput* **19** 577

Brugger K 1964 *Phys Rev* **133** A1611

Cheffer A, Savi M A, Pereira T L and de Paula A S 2021 *Appl Math Modelling* **96** 152

Chua L O 1984 *IEEE Transactions on Circuits and Systems* **CAS-31** 69

Cromer A 1981 *Am J Phys* **49** 455

Dalgarno A and Stewart A L 1956 *Proc Roy Soc London, A* **328** 269

de Launay J 1956 *Solid State Physics* (New York: Academic) **2** 220

Dirac P A M 1945 *Rev Mod Phys* **17** 195

Dodes I A 1978 *Numerical Analysis for Computer Science* (North-Holland: New York)

Drabble J R 1966 *Semiconductors and Semimetals* vol 2, eds R K Willardson and A C Beer (Academic: New York)

Duffing G 1918 *Erzwungene Schwingungen bei Veränderlicher Eigenfrequenz und ihre Technische Bedeutung* (Vieweg: Braunschweig)

Ermentrout G B and Terman D H 2010 *Mathematical Foundations of Neuroscience* Vol 35 (Springer Science & Business Media: Berlin)

Eyert V 1996 *J Comput Phys* **124** 271

Fang T and Dowell E H 1987 *Int J Non-Linear Mechanics* **22** 401

Fehlner W R and Vosko S H 1977 *Can J Phys* **55** 2041

Feigenbaum M J 1978 *J Stat Phys* **19** 25

Fermi E 1950 *Nuclear Physics* (University of Chicago Press: Chicago)

Flowers B H 1996 *An Introduction to Numerical Methods in C++* (Clarendon Press: Oxford)

FitzHugh R 1960 *Biophys Journ* **1** 445

Gois S R F S M and Savi M A 2009 *Chaos, Solitons and Fractals* **41** 2553

Goldstein H, Poole C and Safko J 2002 *Classical Mechanics* 3rd edn (Addison-Wesley)

Gould H and Tobochnik J 1988 *An Introduction to Computer Simulation Methods – Application to Physical Systems Part 1* (Addison-Wesley: New York)

Gregory R D 2006 *Classical Mechanics* (Cambridge University Press: Cambridge)

Griffths D J 1995 *Introduction to Quantum Mechanics* (Prentice Hall: New York)

Grudziński K and Żebrowski J J 2004 *Physica* A **336** 153
Hamilton R A H and Parrott J E 1969 *Phys Rev* **178** 1284
Hansel D and Mato G 2001 *Phys Rev Lett* **86** 4175
Hauer T 2022 *Humanit Soc Sci Commun* **9** 272
Hebb D O 1949 *The Organization of Behavior: A Neuropsychological Theory* (Wiley: New York)
Hénon M 1976 *Commun Math Phys* **50** 69
Herman F J 1959 *J Phys Chem Solids* **8** 405
Herrmann L, Jokeit M, Weeger O and Kollmannsberger S 2025 *Deep Learning in Computational Mechanics – An Introductory Course* 2nd edn (Springer: Switzerland)
Hertz J, Krogh A and Palmer R G 2018 *Introduction to the Theory of Neural Computation* (CRC: Boca Raton)
Hindmarsh J L and Rose R M 1982 *Nature* **296** 162
——1984 *Proc Roy Soc London B* **221** 87
Hodgkin A L and Huxley A F 1952 *J Physiol London* **117** 500
Homes P 1979 *Phil Trans Roy Soc London: A. Math & Phys Sciences* **292** 419
Hori J and Asahi T 1964 *Prog Theo Phys* **31** 49
Hori M, Wijerathne L, Riaz R and Ichimura T 2018 *Journal of JSCE* **6** 1
Horner J *etal* 2020 *The Astronomical Journal* **159** 10
Izhikevich E M 2003 *IEEE Trans Neural Networks* **14** 1569
——2004 *IEEE Trans Neural Networks* **15** 1063
——2007 *Dynamical Systems in Neuroscience* (MIT Press, Cambridge, Massachusetts)
Jones B K and Trefan G 2001 *Am J Phys* **69** 464
Jordan P and Wigner E 1928 *Z. Physik* **47** 631
José J V and Saletan E J 1998 *Classical Dynamics – A Contemporary Approach* (Cambridge University Press: Cambridge)
Kennedy M P 1993 *IEEE Trans Circuits and Systems – I: Fundamental Theory and Applications* **40** 657
Kesavaswamy K and Krishnamurthy N 1978 *Am J Phys* **46** 815
Kittel C 1996 {*Introduction to Solid State Physics*} 7th edn (Wiley: New York)
Knight B W 1972 *J Gen Physio* **59** 734
Kuate G C G and Fotsin H B (2022) Phys Scr **97** 045205
Lambert J D 1973 *Methods in Ordinary Differential Equations* (Wiley: New York)
Lancaster T and Blundell S 2014 *Quantum Field Theory for the Gifted Amateur* (OUP: Oxford)
Landau L D and Lifshitz E M 1976 *Mechanics* 3rd edn (Pergamon: Oxford)
Landau L D and Lifshitz E M 1986 *Theory of Elasticity*, 3rd edn (Butterworth and Heinemann: Oxford)
Lapicque L 1907 *J Physiol Paris* **9** 620
Latham P E et al. 2000 *J Neurophysiol* **83** 808
Li T-Y and Yorke J A 1975 *Am Math Mon* **82** 985
Liapounoff M A 1949 *Problême Général du Mouvement* (Princeton University Press: Princeton)
Lorentz E N 1963 *J Atmos Sci* **42** 433
——1975 *WMO, GARP Publication Series* bf 16 132
Mandal A K 2020 *Am J Engg Res* **9** 87
Mandl F 1959 *Introduction to Quantum Field Theory* (Interscience: New York)
May R M and Oster G F 1980 *Phys Lett* **78A** 1
Maradudin A A et al. 1971 *Solid State Physics* Supplement 3 (New York: Academic)
McSharry P E et al. 2003 *IEEE Trans on Biomed Engg* **50** 289
Monkhorst H J and Pack J D 1976 *Phys Rev* B **13** 5189
Morris C and Lecar H 1981 Biophys J **35** 193
Oberbeck O 1879 *Annu Rev Phys Chem* **7** 271

Olsen L F and Deng H 1985 *Quart Rev Biophys* **18** 165
Palmer T N 1991 *The New Scientist Guide to Chaos* ed Nina Hall (Penguin Books: London)
——1993 *Weather* **48** 314
——1999 *J Clim* **12** 575
——2000 *Rep Prog Phys* **63** 71
Parker T S and Chua L O 1987 *Proc IEEE* **75** 1081
Parrott J E 1969 *Solid State Theory - Methods and Applications* (Ed. P T Landsberg) (Wiley - Interscience: London)
Peskin M E and Schroeder D V 1995 *An Introduction to Quantum Field Theory* (CRC: Reading, MA)
Pines D 1963 *Elementary Excitations in Solids* (Benjamin: New York)
Poincaré H 1992 it New methods in Celestial Mechanics, Vol 2, Ed D L Goroff, AIP. [Original: *Les Mèthodes Nouvelles de la Mècanique Celeste* (1892-9)
Press W H, Teukolsky S A, Vetterling W T, Flannery B P 1992 *Numerical Recipies in Fortran 77* (cambridge University Press)
Qian Y, Siau K L and Nah F F 2024 *Societal Impacts* **3** 100040
Rae A I M 2008 *Quantum Mechanics* 5th edn (Taylor and Francis: Florida)
Ralston A and Rabinowitz P 1978 *A First Course in Numerical Analysis* 2nd edn (McGraw-Hill: London)
Rayleigh L 1885 *Proc London Math Soc* **17** 4
Rosenblatt E 1958 *Psychol Rev* **65** 386
Sahoo R and Mishra R 2012 *J Expt and Theor Physics* **114** 805
Saltzmann B 1962 *J Atmos Sci* **19** 329
Schiff L I 1968 *Quantum Mechanics* 3rd edn (McGraw Hill: New York)
Sebens C T 2015 *Phil Sci* **62** 266
Smith H M J 1948 *Phil Trans Roy Soc Lond* **241** 105
Srivastava D P 2024 (Private communication)
Srivastava G P 1980 *J Phys Chem Solids* **41** 357
——1984 *J Phys A: Math Gen* **17** L317 and **17** 2737
——1999 *Theoretical Modelling of Semiconductor Surfaces* (World Scientific: Singapore)
——2022 *The Physics of Phonons*, 2nd edn (CRC: Oxon)
Stein R B 1967 *Proc R Soc Lond* **167** 64
Sterratt D, Graham B, Gillies A and Willshaw D 2023 *Principles of Computational Modelling in Neuroscience* 2nd edn (Cambridge University Press: Cambridge)
Stillinger F H and Weber T A 1985 Phys Rev B **31** 5262
Strogatz S H 1994 *Nonlinear Dynamics and Chaos* 2nd edn (Westview: Cambridge, MA)
Surianarayanan C, Lawrence J J, Chelliah P R, Prakash E and Hewage C 2023 *Sensors* **23** 3062
Tabor M 1989 *Chaos and Integrability in Nonlinear Dynamics* (Wiles: New York)
Tamaševičiūte *et al.* 2008 Nonlinear Analysis: Modelling and Control **13** 241
Thornton S T and Marion J B 2004 *Classical dynamics of Particles and Systems* 5th edn (Thomson: Belmont CA)
Tsumoto K *et al.* 2006 Neurocomputing **69** 293
van der Pol B 1926 *Phil Mag* **2** 978
Wallis R F 1957 *Phys Rev* **105** 540
——1964 *Surf Sci* **2** 146
——1974 *Prog Surf Sci* **4** 233
Wiggins S 2003 *Introduction to Applied Nonlinear Dynamical Systems and Chaos* 2nd edn (Springer: New York)
Wild J 1980 *Am J Phys* **48** 297
Widrow B and Hoff Jr M E 1960 *IRE WESCON Conv Rec* **4** 96

Yang G R and Wang X-J 2020 *Neuron* **107** 1048
Zhong G Q and Ayrom F 1985 *Int J Circuit Theory Appl* **13** 93
Ziman J M 1960 *Electrons and Phonons* (Clarendon: Oxford)
——1969 *Elements of Advanced Quantum Theory* (Cambridge University Press: Cambridge)

Index

For Product Safety Concerns and Information please contact our EU representative GPSR@taylorandfrancis.com
Taylor & Francis Verlag GmbH, Kaufingerstraße 24, 80331 München, Germany

www.ingramcontent.com/pod-product-compliance
Lightning Source LLC
Chambersburg PA
CBHW080926220326
41598CB00034B/5692
* 9 7 8 1 0 3 2 4 6 5 2 7 2 *